STATA BASE REFERENCE MANUAL
VOLUME 2
I–P
RELEASE 10

A Stata Press Publication
StataCorp LP
College Station, Texas

Stata Press, 4905 Lakeway Drive, College Station, Texas 77845

Version 10
Typeset in TeX
Printed in the United States of America

10 9 8 7 6 5 4 3 2 1

ISBN-10: 1-59718-027-0 (volumes 1–3)
ISBN-10: 1-59718-024-6 (volume 1)
ISBN-10: 1-59718-025-4 (volume 2)
ISBN-10: 1-59718-026-2 (volume 3)
ISBN-13: 978-1-59718-027-6 (volumes 1–3)
ISBN-13: 978-1-59718-024-5 (volume 1)
ISBN-13: 978-1-59718-025-2 (volume 2)
ISBN-13: 978-1-59718-026-9 (volume 3)

The suggested citation for this software is

StataCorp. 2007. *Stata Statistical Software: Release 10*. College Station, TX: StataCorp LP.

Title

inequality — Inequality measures

Remarks

Stata does not have commands for inequality measures, except roctab has an option to report Gini and Pietra indices; see [R] **roc**. Stata users, however, have developed an excellent suite of commands, many of which have been published in the *Stata Journal* (SJ) and in the *Stata Technical Bulletin* (STB).

Issue	insert	author(s)	command	description
STB-48	gr35	N. J. Cox	psm, qsm, pdagum, qdagum	Diagnostic plots for assessing Singh–Maddala and Dagum distributions fitted by MLE
STB-23	sg31	R. Goldstein	rspread	Measures of diversity: Absolute and relative
STB-48	sg104	S. P. Jenkins	sumdist, xfrac, ineqdeco, geivars, ineqfac, povdeco	Analysis of income distributions
STB-48	sg106	S. P. Jenkins	smfit, dagumfit	Fitting Singh–Maddala and Dagum distributions by maximum likelihood
STB-48	sg107	S. P. Jenkins, P. Van Kerm	glcurve	Generalized Lorenz curves and related graphs
STB-49	sg107.1	S. P. Jenkins, P. Van Kerm	glcurve	update of sg107
SJ-1-1	gr0001	S. P. Jenkins, P. Van Kerm	glcurve7	update for Stata 7 of sg107.1
SJ-4-4	gr0001_1	S. P. Van Kerm, P. Jenkins	glcurve	update for Stata 8 of gr0001; *install this version*
STB-48	sg108	P. Van Kerm	poverty	Computing poverty indices
STB-51	sg115	D. Jolliffe, B. Krushelnytskyy	ineqerr	Bootstrap standard errors for indices of inequality
STB-51	sg117	D. Jolliffe, A. Semykina	sepov	Robust standard errors for the Foster–Greer–Thorbecke class of poverty indices
SJ-6-1	st0100	A. López-Feldman	descogini	Decomposing inequality and obtaining marginal effects
SJ-6-4	snp15_7	R. Newson	somersd	Gini coefficient is a special case of Somers' *D*
STB-23	sg30	E. Whitehouse	lorenz, inequal, atkinson, relsgini	Measures of inequality in Stata

More commands may be available; enter Stata and type search inequality measure, all.

Max Otto Lorenz (1876–1959) was born in Iowa and studied at the Universities of Iowa and Wisconsin. He proposed what is now known as the Lorenz curve in 1905. Lorenz worked for the Interstate Commerce Commission between 1911 and 1944, mainly with transportation data. His hobbies included calendar reform and Interlingua, a proposed international language.

To download and install the Jenkins and Van Kerm `glcurve` command from the Internet, for instance, you could

1. Select **Help > SJ and User-written Programs**.
2. Click on *Stata Journal*.
3. Click on *sj4-4*.
4. Click on *gr0001_1*.
5. Click on *click here to install*.

or you could instead do the following:

1. Navigate to the appropriate SJ issue:
   a. Type `net from http://www.stata-journal.com/software`
      Type `net cd sj4-4`

   or

   b. Type `net from http://www.stata-journal.com/software/sj4-4`
2. Type `net describe gr0001_1`
3. Type `net install gr0001_1`

To download and install the Jenkins `sumdist` command from the Internet, for instance, you could

1. Select **Help > SJ and User-written Programs**.
2. Click on *STB*.
3. Click on *stb48*.
4. Click on *sg104*.
5. Click on *click here to install*.

or you could instead do the following:

1. Navigate to the appropriate STB issue:
   a. Type `net from http://www.stata.com`
      Type `net cd stb`
      Type `net cd stb48`

   or

   b. Type `net from http://www.stata.com/stb/stb48`
2. Type `net describe sg104`
3. Type `net install sg104`

References

Cox, N. J. 1999. gr35: Diagnostic plots for assessing Singh–Maddala and Dagum distributions fitted by MLE. *Stata Technical Bulletin* 48: 2–4. Reprinted in *Stata Technical Bulletin Reprints*, vol. 8, pp. 72–74.

Goldstein, R. 1995. sg31: Measures of diversity: Absolute and relative. *Stata Technical Bulletin* 23: 23–26. Reprinted in *Stata Technical Bulletin Reprints*, vol. 4, pp. 150–154.

Jenkins, S. P. 1999a. sg104: Analysis of income distributions. *Stata Technical Bulletin* 48: 4–18. Reprinted in *Stata Technical Bulletin Reprints*, vol. 8, pp. 243–260.

——. 1999b. sg106: Fitting Singh–Maddala and Dagum distributions by maximum likelihood. *Stata Technical Bulletin* 48: 19–25. Reprinted in *Stata Technical Bulletin Reprints*, vol. 8, pp. 261–268.

Jenkins, S. P., and P. Van Kerm. 1999a. sg107: Generalized Lorenz curves and related graphs. *Stata Technical Bulletin* 48: 25–29. Reprinted in *Stata Technical Bulletin Reprints*, vol. 8, pp. 269–274.

——. 1999b. sg107.1: Generalized Lorenz curves and related graphs. *Stata Technical Bulletin* 49: 23. Reprinted in *Stata Technical Bulletin Reprints*, vol. 9, p. 171.

——. 2001. gr0001: Generalized Lorenz curves and related graphs: Update for Stata 7. *Stata Journal* 1: 107–112.

Jolliffe, D., and B. Krushelnytskyy. 1999. sg115: Bootstrap standard errors for indices of inequality. *Stata Technical Bulletin* 51: 28–32. Reprinted in *Stata Technical Bulletin Reprints*, vol. 9, pp. 191–196.

Jolliffe, D., and A. Semykina. 1999. sg117: Robust standard errors for the Foster–Greer–Thorbecke class of poverty indices. *Stata Technical Bulletin* 51: 34–36. Reprinted in *Stata Technical Bulletin Reprints*, vol. 9, pp. 200–203.

Kleiber, C., and S. Kotz. 2003. *Statistical Size Distributions in Economics and Actuarial Sciences*. Hoboken, NJ: Wiley.

López-Feldman, A. 2006. Decomposing inequality and obtaining marginal effects. *Stata Journal* 6: 106–111.

Lorenz, M. O. 1905. Methods of measuring the concentration of wealth. *Publications, American Statistical Association* 9: 209–219.

Newson, R. 2006. Confidence intervals for rank statistics: Percentile slopes, differences, and ratios. *Stata Journal* 6: 497–520.

Van Kerm, P. 1999. sg108: Computing poverty indices. *Stata Technical Bulletin* 48: 29–33. Reprinted in *Stata Technical Bulletin Reprints*, vol. 8, pp. 274–278.

Whitehouse, E. 1995. sg30: Measures of inequality in Stata. *Stata Technical Bulletin* 23: 20–23. Reprinted in *Stata Technical Bulletin Reprints*, vol. 4, pp. 146–150.

Title

intreg — Interval regression

Syntax

`intreg` *depvar*$_1$ *depvar*$_2$ [*indepvars*] [*if*] [*in*] [*weight*] [, *options*]

options	description
Model	
`noconstant`	suppress constant term
`het(`*varlist* [, `noconstant`]`)`	independent variables to model the variance; use `noconstant` to suppress constant term
`offset(`*varname*`)`	include *varname* in model with coefficient constrained to 1
`constraints(`*constraints*`)`	apply specified linear constraints
`collinear`	keep collinear variables
SE/Robust	
`vce(`*vcetype*`)`	*vcetype* may be `oim`, `robust`, `cluster` *clustvar*, `opg`, `bootstrap`, or `jackknife`
Reporting	
`level(`*#*`)`	set confidence level; default is `level(95)`
Max options	
maximize_options	control the maximization process; seldom used

depvar$_1$, *depvar*$_2$, *indepvars*, and *varlist* may contain time-series operators; see [U] **11.4.3 Time-series varlists**.

`bootstrap`, `by`, `jackknife`, `nestreg`, `rolling`, `statsby`, `stepwise`, `svy`, and `xi` are allowed; see [U] **11.1.10 Prefix commands**.

Weights are not allowed with the `bootstrap` prefix.

`aweight`s are not allowed with the `jackknife` prefix.

`vce()` and weights are not allowed with the `svy` prefix.

`aweight`s, `fweight`s, `iweight`s, and `pweight`s are allowed; see [U] **11.1.6 weight**.

See [U] **20 Estimation and postestimation commands** for more capabilities of estimation commands.

Description

`intreg` fits a model of $y = [\,depvar_1, depvar_2\,]$ on *indepvars*, where y for each observation is point data, interval data, left-censored data, or right-censored data.

depvar$_1$ and *depvar*$_2$ should have the following form:

type of data		*depvar*$_1$	*depvar*$_2$
point data	$a = [\,a, a\,]$	a	a
interval data	$[\,a, b\,]$	a	b
left-censored data	$(\,-\infty, b\,]$	.	b
right-censored data	$[\,a, +\infty\,)$	a	.

Options

Model

noconstant; see [R] **estimation options**.

het(*varlist* [, noconstant]) specifies that *varlist* be included in the specification of the conditional variance. This *varlist* enters the variance specification collectively as multiplicative heteroskedasticity.

offset(*varname*), constraints(*constraints*), collinear; see [R] **estimation options**.

SE/Robust

vce(*vcetype*) specifies the type of standard error reported, which includes types that are derived from asymptotic theory, that are robust to some kinds of misspecification, that allow for intragroup correlation, and that use bootstrap or jackknife methods; see [R] ***vce_option***.

Reporting

level(*#*); see [R] **estimation options**.

Max options

maximize_options: difficult, technique(*algorithm_spec*), iterate(*#*), [no]log, trace, gradient, showstep, hessian, shownrtolerance, tolerance(*#*), ltolerance(*#*), gtolerance(*#*), nrtolerance(*#*), nonrtolerance, from(*init_specs*); see [R] **maximize**. These options are seldom used.

Setting the optimization type to technique(bhhh) resets the default *vcetype* to vce(opg).

Remarks

intreg is a generalization of the models fitted by cnreg and tobit. If you know that the value for the jth individual is somewhere in the interval $[\,y_{1j}, y_{2j}\,]$, then the likelihood contribution from this individual is simply $\Pr(y_{1j} \leq Y_j \leq y_{2j})$. For censored data, their likelihoods contain terms of the form $\Pr(Y_j \leq y_j)$ for left-censored data and $\Pr(Y_j \geq y_j)$ for right-censored data, where y_j is the observed censoring value and Y_j denotes the random variable representing the dependent variable in the model.

Hence, intreg can fit models for data where each observation represents interval data, left-censored data, right-censored data, or point data. Regardless of the type of observation, the data should be stored in the dataset as interval data; that is, two dependent variables, *depvar*$_1$ and *depvar*$_2$, are used to hold the endpoints of the interval. If the data are left-censored, the lower endpoint is $-\infty$ and is represented by a missing value '.' or an extended missing value '.a, .b, ..., .z' in *depvar*$_1$. If the data are right-censored, the upper endpoint is $+\infty$ and is represented by a missing value '.' (or an extended missing value) in *depvar*$_2$. Point data are represented by the two endpoints being equal.

(*Continued on next page*)

type of data		$depvar_1$	$depvar_2$
point data	$a = [a, a]$	a	a
interval data	$[a, b]$	a	b
left-censored data	$(-\infty, b]$	.	b
right-censored data	$[a, +\infty)$	a	.

Truly missing values of the dependent variable must be represented by missing values in both $depvar_1$ and $depvar_2$.

Interval data arise naturally in many contexts, such as wage data. Often you only know that, for example, a person's salary is between $30,000 and $40,000. Below we give an example for wage data and show how to set up $depvar_1$ and $depvar_2$.

▷ Example 1

We have a dataset that contains the yearly wages of working women. Women were asked via a questionnaire to indicate a category for their yearly income from employment. The categories were less than 5,000, 5,001–10,000, ..., 25,001–30,000, 30,001–40,000, 40,001–50,000, and more than 50,000. The wage categories are stored in the variable **wagecat**.

```
. use http://www.stata-press.com/data/r10/womenwage
(Wages of women)

. tab wagecat

       Wage |
   category |
   ($1000s) |      Freq.     Percent        Cum.
------------+-----------------------------------
          5 |         14        2.87        2.87
         10 |         83       17.01       19.88
         15 |        158       32.38       52.25
         20 |        107       21.93       74.18
         25 |         57       11.68       85.86
         30 |         30        6.15       92.01
         40 |         19        3.89       95.90
         50 |         14        2.87       98.77
         51 |          6        1.23      100.00
------------+-----------------------------------
      Total |        488      100.00
```

A value of 5 for **wagecat** represents the category less than 5,000, a value of 10 represents 5,001–10,000, ..., and a value of 51 represents greater than 50,000.

To use **intreg**, we must create two variables, **wage1** and **wage2**, containing the lower and upper endpoints of the wage categories. Here's one way to do it. We first create a little dataset containing the nine wage categories, lag the wage categories into **wage1**, and match-merge this dataset with nine observations back into the main one.

```
. by wagecat: keep if _n==1
(479 observations deleted)

. generate wage1 = wagecat[_n-1]
(1 missing value generated)

. keep wagecat wage1
```

```
. save lagwage
file lagwage.dta saved
. use http://www.stata-press.com/data/r10/womenwage
(Wages of women)
. merge wagecat using lagwage
variable wagecat does not uniquely identify observations in the master data
. tab _merge
```

_merge	Freq.	Percent	Cum.
3	488	100.00	100.00
Total	488	100.00	

```
. drop _merge
```

The variable _merge created by merge indicates that all the observations in the merge were matched (you should always check this; see [D] **merge** for more information).

Now we create the upper endpoint and list the new variables:

```
. generate wage2 = wagecat
. replace wage2 = . if wagecat == 51
(6 real changes made, 6 to missing)
. sort age, stable
. list wage1 wage2 in 1/10
```

	wage1	wage2
1.	.	5
2.	5	10
3.	5	10
4.	10	15
5.	.	5
6.	.	5
7.	.	5
8.	5	10
9.	5	10
10.	5	10

We can now run intreg:

```
. intreg wage1 wage2 age age2 nev_mar rural school tenure
Fitting constant-only model:
Iteration 0:   log likelihood = -967.24956
Iteration 1:   log likelihood =  -967.1368
Iteration 2:   log likelihood =  -967.1368
Fitting full model:
Iteration 0:   log likelihood = -856.65324
Iteration 1:   log likelihood = -856.33294
Iteration 2:   log likelihood = -856.33293
```

(Continued on next page)

```
Interval regression                               Number of obs   =        488
                                                  LR chi2(6)      =     221.61
Log likelihood = -856.33293                       Prob > chi2     =     0.0000

------------------------------------------------------------------------------
             |      Coef.   Std. Err.      z    P>|z|     [95% Conf. Interval]
-------------+----------------------------------------------------------------
         age |   .7914438   .4433604     1.79   0.074    -.0775265    1.660414
        age2 |  -.0132624   .0073028    -1.82   0.069    -.0275757    .0010509
     nev_mar |  -.2075022   .8119581    -0.26   0.798    -1.798911    1.383906
       rural |  -3.043044   .7757324    -3.92   0.000    -4.563452   -1.522637
      school |   1.334721   .1357873     9.83   0.000     1.068583    1.600859
      tenure |   .8000664   .1045077     7.66   0.000     .5952351    1.004898
       _cons |  -12.70238   6.367117    -1.99   0.046     -25.1817   -.2230583
-------------+----------------------------------------------------------------
    /lnsigma |   1.987823   .0346543    57.36   0.000     1.919902    2.055744
-------------+----------------------------------------------------------------
       sigma |   7.299626   .2529634                       6.82029     7.81265
------------------------------------------------------------------------------

  Observation summary:         14  left-censored observations
                                0     uncensored observations
                                6 right-censored observations
                              468       interval observations
```

We could also model these data by using an ordered probit model using oprobit:

```
. oprobit wagecat age age2 nev_mar rural school tenure
Iteration 0:   log likelihood =  -881.1491
Iteration 1:   log likelihood = -764.31729
Iteration 2:   log likelihood = -763.31191
Iteration 3:   log likelihood = -763.31049
Ordered probit regression                         Number of obs   =        488
                                                  LR chi2(6)      =     235.68
                                                  Prob > chi2     =     0.0000
Log likelihood = -763.31049                       Pseudo R2       =     0.1337

------------------------------------------------------------------------------
     wagecat |      Coef.   Std. Err.      z    P>|z|     [95% Conf. Interval]
-------------+----------------------------------------------------------------
         age |   .1674519   .0620333     2.70   0.007     .0458689     .289035
        age2 |  -.0027983   .0010214    -2.74   0.006    -.0048001   -.0007964
     nev_mar |  -.0046417   .1126736    -0.04   0.967     -.225478    .2161946
       rural |  -.5270036   .1100448    -4.79   0.000    -.7426875   -.3113197
      school |   .2010587   .0201189     9.99   0.000     .1616263    .2404911
      tenure |   .0989916   .0147887     6.69   0.000     .0700063     .127977
-------------+----------------------------------------------------------------
       /cut1 |   2.650637   .8957242                        .89505    4.406225
       /cut2 |   3.941018   .8979164                      2.181134    5.700902
       /cut3 |   5.085205   .9056579                      3.310149    6.860262
       /cut4 |   5.875534    .912093                      4.087864    7.663203
       /cut5 |   6.468723   .9181166                      4.669247    8.268198
       /cut6 |   6.922726   .9215452                      5.116531    8.728922
       /cut7 |    7.34471   .9237624                      5.534169    9.155251
       /cut8 |   7.963441   .9338878                      6.133054    9.793827
------------------------------------------------------------------------------
```

We can directly compare the log likelihoods for the intreg and oprobit models since both likelihoods are discrete. If we had point data in our intreg estimation, the likelihood would be a mixture of discrete and continuous terms, and we could not compare it directly with the oprobit likelihood.

Here the `oprobit` log likelihood is significantly larger (i.e., less negative), so it fits better than the `intreg` model. The `intreg` model assumes normality, but the distribution of wages is skewed and definitely nonnormal. Normality is more closely approximated if we model the log of wages.

```
. generate logwage1 = log(wage1)
(14 missing values generated)

. generate logwage2 = log(wage2)
(6 missing values generated)

. intreg logwage1 logwage2 age age2 nev_mar rural school tenure
Fitting constant-only model:

Iteration 0:   log likelihood = -889.23647
Iteration 1:   log likelihood = -889.06346
Iteration 2:   log likelihood = -889.06346

Fitting full model:

Iteration 0:   log likelihood = -773.81968
Iteration 1:   log likelihood = -773.36566
Iteration 2:   log likelihood = -773.36563

Interval regression                               Number of obs   =        488
                                                  LR chi2(6)      =     231.40
Log likelihood = -773.36563                       Prob > chi2     =     0.0000

------------------------------------------------------------------------------
             |      Coef.   Std. Err.      z    P>|z|     [95% Conf. Interval]
-------------+----------------------------------------------------------------
         age |   .0645589   .0249954     2.58   0.010     .0155689    .1135489
        age2 |  -.0010812   .0004115    -2.63   0.009    -.0018878   -.0002746
     nev_mar |  -.0058151   .0454867    -0.13   0.898    -.0949674    .0833371
       rural |  -.2098361   .0439454    -4.77   0.000    -.2959675   -.1237047
      school |   .0804832   .0076783    10.48   0.000     .0654341    .0955323
      tenure |   .0397144   .0058001     6.85   0.000     .0283464    .0510825
       _cons |   .7084023   .3593193     1.97   0.049     .0041495    1.412655
-------------+----------------------------------------------------------------
    /lnsigma |   -.906989   .0356265   -25.46   0.000    -.9768157   -.8371623
-------------+----------------------------------------------------------------
       sigma |   .4037381   .0143838                      .3765081    .4329373
------------------------------------------------------------------------------

  Observation summary:        14  left-censored observations
                               0     uncensored observations
                               6 right-censored observations
                             468       interval observations
```

The log likelihood of this `intreg` model is close to the `oprobit` log likelihood, and the z statistics for both models are similar. ◁

❑ Technical Note

`intreg` has two parameterizations for the log-likelihood function: the transformed parameterization $(\beta/\sigma,\ 1/\sigma)$ and the untransformed parameterization $(\beta,\ \ln(\sigma))$. By default, the log likelihood for `intreg` is parameterized in the transformed parameter space. This parameterization tends to be more convergent, but it requires that any starting values and constraints have the same parameterization, and it prevents the estimation with multiplicative heteroskedasticity. Therefore, when the option `het()` is specified, `intreg` switches to the untransformed log likelihood for the fit of the conditional-variance model. Similarly, specifying `from()` or `constraints()` causes the optimization in the untransformed parameter space to allow constraints on (and starting values for) the coefficients on the covariates without reference to σ.

The estimation results are all saved in the $(\beta,\ \ln(\sigma))$ metric. ❑

Saved Results

intreg saves the following in e():

Scalars

e(N)	# of observations	e(ll_0)	log likelihood, constant-only model
e(N_unc)	# of uncensored observations	e(p)	p-value for model χ^2 test
e(N_lc)	# of left-censored observations	e(sigma)	sigma
e(N_rc)	# of right-censored observations	e(se_sigma)	standard error of sigma
e(N_int)	# of interval observations	e(chi2)	χ^2
e(k)	# of parameters	e(df_m)	model degrees of freedom
e(k_eq)	# of equations	e(rank)	rank of e(V)
e(k_eq_model)	# of equations in model Wald test	e(rank0)	rank of e(V) for constant-only model
e(k_aux)	# of auxiliary parameters		
e(k_dv)	# of dependent variables	e(ic)	# of iterations
e(N_clust)	# of clusters	e(rc)	return code
e(ll)	log likelihood	e(converged)	1 if converged, 0 otherwise

Macros

e(cmd)	intreg	e(diparm#)	display transformed parameter #
e(cmdline)	command as typed	e(opt)	type of optimization
e(depvar)	names of dependent variables	e(ml_method)	type of ml method
e(wtype)	weight type	e(ml_score)	program used to implement scores
e(wexp)	weight expression	e(user)	name of likelihood-evaluator program
e(title)	title in estimation output	e(technique)	maximization technique
e(clustvar)	name of cluster variable	e(crittype)	optimization criterion
e(offset)	offset	e(properties)	b V
e(chi2type)	Wald or LR; type of model χ^2 test	e(footnote)	program used to implement the footnote display
e(vce)	*vcetype* specified in vce()		
e(vcetype)	title used to label Std. Err.	e(predict)	program used to implement predict

Matrices

e(b)	coefficient vector	e(V)	variance–covariance matrix of the estimators
e(ilog)	iteration log (up to 20 iterations)		
e(gradient)	gradient vector		

Functions

e(sample)	marks estimation sample

Methods and Formulas

intreg is implemented as an ado-file using the ml commands (see [R] **ml**); its robust variance computation is performed by _robust (see [P] **_robust**).

The likelihood for intreg subsumes that of the tobit and cnreg models.

Let $\mathbf{y} = \mathbf{X}\boldsymbol{\beta} + \boldsymbol{\epsilon}$ be the model. $\mathbf{y}$ represents continuous outcomes—either observed or not observed. Our model assumes $\boldsymbol{\epsilon} \sim N(\mathbf{0}, \sigma^2\mathbf{I})$.

For observations $j \in \mathcal{C}$, we observe y_j, i.e., point data. Observations $j \in \mathcal{L}$ are left-censored; we know only that the unobserved y_j is less than or equal to $y_{\mathcal{L}j}$, a censoring value that we do know. Similarly, observations $j \in \mathcal{R}$ are right-censored; we know only that the unobserved y_j is greater than or equal to $y_{\mathcal{R}j}$. Observations $j \in \mathcal{I}$ are intervals; we know only that the unobserved y_j is in the interval $[\, y_{1j},\, y_{2j}\,]$.

The log likelihood is

$$
\begin{aligned}
\ln L = &-\frac{1}{2}\sum_{j\in\mathcal{C}} w_j \left\{ \left(\frac{y_j - \mathbf{x}\boldsymbol{\beta}}{\sigma} \right)^2 + \log 2\pi\sigma^2 \right\} \\
&+ \sum_{j\in\mathcal{L}} w_j \log \Phi\left(\frac{y_{\mathcal{L}j} - \mathbf{x}\boldsymbol{\beta}}{\sigma} \right) \\
&+ \sum_{j\in\mathcal{R}} w_j \log \left\{ 1 - \Phi\left(\frac{y_{\mathcal{R}j} - \mathbf{x}\boldsymbol{\beta}}{\sigma} \right) \right\} \\
&+ \sum_{j\in\mathcal{I}} w_j \log \left\{ \Phi\left(\frac{y_{2j} - \mathbf{x}\boldsymbol{\beta}}{\sigma} \right) - \Phi\left(\frac{y_{1j} - \mathbf{x}\boldsymbol{\beta}}{\sigma} \right) \right\}
\end{aligned}
$$

where $\Phi()$ is the standard cumulative normal and w_j is the weight for the jth observation. If no weights are specified, $w_j = 1$. If `aweights` are specified, $w_j = 1$, and σ is replaced by $\sigma/\sqrt{a_j}$ in the above, where a_j are the `aweights` normalized to sum to N.

Maximization is as described in [R] **maximize**; the estimate reported as `_sigma` is $\widehat{\sigma}$. See [U] **20.15 Obtaining robust variance estimates** and [P] **_robust** for descriptions of the computation performed when `vce(robust)` is specified as an option to `intreg`.

See Amemiya (1973) for a generalization of the tobit model to variable, but known, cutoffs.

References

Amemiya, T. 1973. Regression analysis when the dependent variable is truncated normal. *Econometrica* 41: 997–1016.

Conroy, R. M. 2005. Stings in the tails: Detecting and dealing with censored data. *Stata Journal* 5: 395–404.

Goldberger, A. S. 1983. Abnormal selection bias. In *Studies in Econometrics, Time Series, and Multivariate Statistics*, ed. S. Karlin, T. Amemiya, and L. A. Goodman, 67–84. New York: Academic Press.

Hurd, M. 1979. Estimation in truncated samples when there is heteroscedasticity. *Journal of Econometrics* 11: 247–258.

Johnston, J., and J. DiNardo. 1997. *Econometric Methods*. 4th ed. New York: McGraw–Hill.

Kendall, M. G., and A. Stuart. 1973. *The Advanced Theory of Statistics: Inference and Relationship, Vol. 2*. New York: Hafner.

Kmenta, J. 1997. *Elements of Econometrics*. 2nd ed. Ann Arbor: University of Michigan Press.

Long, J. S. 1997. *Regression Models for Categorical and Limited Dependent Variables*. Thousand Oaks, CA: Sage.

Maddala, G. S. 1992. *Introduction to Econometrics*. 2nd ed. New York: Macmillan.

Stewart, M. B. 1983. On least-squares estimation when the dependent variable is grouped. *Review of Economic Studies* 50: 737–753.

Also See

[R] **intreg postestimation** — Postestimation tools for intreg

[R] **cnreg** — Censored-normal regression

[R] **tobit** — Tobit regression

[R] **regress** — Linear regression

[SVY] **svy estimation** — Estimation commands for survey data

[XT] **xtintreg** — Random-effects interval-data regression models

[XT] **xttobit** — Random-effects tobit models

[U] **20 Estimation and postestimation commands**

Title

intreg postestimation — Postestimation tools for intreg

Description

The following postestimation commands are available for `intreg`:

command	description
`adjust`[1]	adjusted predictions of $\mathbf{x}\beta$
`estat`	AIC, BIC, VCE, and estimation sample summary
`estat` (svy)	postestimation statistics for survey data
`estimates`	cataloging estimation results
`lincom`	point estimates, standard errors, testing, and inference for linear combinations of coefficients
`lrtest`[2]	likelihood-ratio test
`mfx`	marginal effects or elasticities
`nlcom`	point estimates, standard errors, testing, and inference for nonlinear combinations of coefficients
`predict`	predictions, residuals, influence statistics, and other diagnostic measures
`predictnl`	point estimates, standard errors, testing, and inference for generalized predictions
`suest`	seemingly unrelated estimation
`test`	Wald tests for simple and composite linear hypotheses
`testnl`	Wald tests of nonlinear hypotheses

[1] `adjust` is not appropriate with time-series operators.
[2] `lrtest` is not appropriate with `svy` estimation results.

See the corresponding entries in the *Stata Base Reference Manual* for details, but see [SVY] **estat** for details about `estat` (svy).

Syntax for predict

`predict` [*type*] *newvar* [*if*] [*in*] [, *statistic* `nooffset`]

`predict` [*type*] { *stub** | $newvar_{\text{reg}}$ $newvar_{\text{lnsigma}}$ } [*if*] [*in*] , `scores`

statistic	description
Main	
`xb`	linear prediction; the default
`stdp`	standard error of the prediction
`stdf`	standard error of the forecast
`pr(`a`,`b`)`	$\Pr(a < y_j < b)$
`e(`a`,`b`)`	$E(y_j \mid a < y_j < b)$
`ystar(`a`,`b`)`	$E(y_j^*)$, $y_j^* = \max\{a, \min(y_j, b)\}$

These statistics are available both in and out of sample; type `predict ... if e(sample) ...` if wanted only for the estimation sample.

`stdf` is not allowed with `svy` postestimation results.

where a and b may be numbers or variables; a missing ($a \geq .$) means $-\infty$, and b missing ($b \geq .$) means $+\infty$; see [U] **12.2.1 Missing values**.

Options for predict

Main

`xb`, the default, calculates the linear prediction.

`stdp` calculates the standard error of the prediction, which can be thought of as the standard error of the predicted expected value or mean for the observation's covariate pattern. The standard error of the prediction is also referred to as the standard error of the fitted value.

`stdf` calculates the standard error of the forecast, which is the standard error of the point prediction for 1 observation. It is commonly referred to as the standard error of the future or forecast value. By construction, the standard errors produced by `stdf` are always larger than those produced by `stdp`; see [R] **regress** *Methods and Formulas*.

`pr(`a`,`b`)` calculates $\Pr(a < \mathbf{x}_j\mathbf{b} + u_j < b)$, the probability that $y_j|\mathbf{x}_j$ would be observed in the interval (a, b).

a and b may be specified as numbers or variable names; *lb* and *ub* are variable names;
`pr(20,30)` calculates $\Pr(20 < \mathbf{x}_j\mathbf{b} + u_j < 30)$;
`pr(`*lb*`,`*ub*`)` calculates $\Pr(lb < \mathbf{x}_j\mathbf{b} + u_j < ub)$; and
`pr(20,`*ub*`)` calculates $\Pr(20 < \mathbf{x}_j\mathbf{b} + u_j < ub)$.

a missing ($a \geq .$) means $-\infty$; `pr(.,30)` calculates $\Pr(-\infty < \mathbf{x}_j\mathbf{b} + u_j < 30)$;
`pr(`*lb*`,30)` calculates $\Pr(-\infty < \mathbf{x}_j\mathbf{b} + u_j < 30)$ in observations for which $lb \geq .$
and calculates $\Pr(lb < \mathbf{x}_j\mathbf{b} + u_j < 30)$ elsewhere.

b missing ($b \geq .$) means $+\infty$; `pr(20,.)` calculates $\Pr(+\infty > \mathbf{x}_j\mathbf{b} + u_j > 20)$;
`pr(20,`*ub*`)` calculates $\Pr(+\infty > \mathbf{x}_j\mathbf{b} + u_j > 20)$ in observations for which $ub \geq .$
and calculates $\Pr(20 < \mathbf{x}_j\mathbf{b} + u_j < ub)$ elsewhere.

`e(`a`,`b`)` calculates $E(\mathbf{x}_j\mathbf{b} + u_j \mid a < \mathbf{x}_j\mathbf{b} + u_j < b)$, the expected value of $y_j|\mathbf{x}_j$ conditional on $y_j|\mathbf{x}_j$ being in the interval (a, b), meaning that $y_j|\mathbf{x}_j$ is censored.
a and b are specified as they are for `pr()`.

`ystar(`a`,`b`)` calculates $E(y_j^*)$, where $y_j^* = a$ if $\mathbf{x}_j\mathbf{b} + u_j \leq a$, $y_j^* = b$ if $\mathbf{x}_j\mathbf{b} + u_j \geq b$, and $y_j^* = \mathbf{x}_j\mathbf{b} + u_j$ otherwise, meaning that y_j^* is truncated. a and b are specified as they are for `pr()`.

`nooffset` is relevant only if you specified `offset(`*varname*`)`. It modifies the calculations made by `predict` so that they ignore the offset variable; the linear prediction is treated as $\mathbf{x}_j\mathbf{b}$ rather than as $\mathbf{x}_j\mathbf{b} + \text{offset}_j$.

`scores` calculates equation-level score variables.

The first new variable will contain $\partial \ln L/\partial(\mathbf{x}_j\boldsymbol{\beta})$.

The second new variable will contain $\partial \ln L/\partial \ln\sigma$.

Methods and Formulas

All postestimation commands listed above are implemented as ado-files.

Also See

[R] **intreg** — Interval regression

[U] **20 Estimation and postestimation commands**

Title

ivprobit — Probit model with endogenous regressors

Syntax

Maximum likelihood estimator

`ivprobit` *depvar* [*varlist*$_1$] (*varlist*$_2$ = *varlist*$_{\text{iv}}$) [*if*] [*in*] [*weight*] [, *mle_options*]

Two-step estimator

`ivprobit` *depvar* [*varlist*$_1$] (*varlist*$_2$ = *varlist*$_{\text{iv}}$) [*if*] [*in*] [*weight*], `twostep` [*tse_options*]

mle_options	description
Model	
`mle`	use conditional maximum-likelihood estimator; the default
`asis`	retain perfect predictor variables
`constraints(`*constraints*`)`	apply specified linear constraints
SE/Robust	
`vce(`*vcetype*`)`	*vcetype* may be `oim`, `robust`, `cluster` *clustvar*, `opg`, `bootstrap`, or `jackknife`
Reporting	
`level(`*#*`)`	set confidence level; default is `level(95)`
`first`	report first-stage regression
Max options	
maximize_options	control the maximization process

(*Continued on next page*)

tse_options	description
Model	
*twostep	use Newey's two-step estimator; the default is mle
asis	retain perfect predictor variables
SE	
vce(*vcetype*)	*vcetype* may be standard, bootstrap, or jackknife
Reporting	
level(#)	set confidence level; default is level(95)
first	report first-stage regression

*twostep is required.

depvar, $varlist_1$, $varlist_2$, and $varlist_{iv}$ may contain time-series operators; see [U] **11.4.3 Time-series varlists**.

bootstrap, by, jackknife, rolling, statsby, svy, and xi are allowed; see [U] **11.1.10 Prefix commands**.

Weights are not allowed with the bootstrap prefix.

vce(), first, twostep, and weights are not allowed with the svy prefix.

fweights, iweights, and pweights are allowed with the maximum likelihood estimator. fweights are allowed with Newey's two-step estimator. See [U] **11.1.6 weight**.

See [U] **20 Estimation and postestimation commands** for more capabilities of estimation commands.

Description

ivprobit fits probit models where one or more of the regressors are endogenously determined. By default, ivprobit uses maximum likelihood estimation. Alternatively, Newey's (1987) minimum chi-squared estimator can be invoked with the twostep option. See [R] **ivtobit** for tobit estimation with endogenous regressors and [R] **probit** for probit estimation when the model contains no endogenous regressors.

Options for ML estimator

Model

mle requests that the conditional maximum-likelihood estimator be used. This is the default.

asis requests that all specified variables and observations be retained in the maximization process. This option is typically not used and may introduce numerical instability. Normally, ivprobit drops any endogenous or exogenous variables that perfectly predict success or failure in the dependent variable. The associated observations are also dropped. For more information, see the discussion of model identification in [R] **probit**.

constraints(*constraints*); see [R] **estimation options**.

SE/Robust

vce(*vcetype*) specifies the type of standard error reported, which includes types that are derived from asymptotic theory, that are robust to some kinds of misspecification, that allow for intragroup correlation, and that use bootstrap or jackknife methods; see [R] ***vce_option***.

Reporting

level(#); see [R] **estimation options**.

first requests that the parameters for the reduced-form equations showing the relationships between the endogenous variables and instruments be displayed. For the two-step estimator, first shows the first-stage regressions. For the maximum likelihood estimator, these parameters are estimated jointly with the parameters of the probit equation. The default is not to show these parameter estimates.

Max options

maximize_options: difficult, technique(*algorithm_spec*), iterate(#), [no]log, trace, gradient, showstep, hessian, shownrtolerance, tolerance(#), ltolerance(#), gtolerance(#), nrtolerance(#), nonrtolerance, from(*init_specs*); see [R] **maximize**. This model's likelihood function can be difficult to maximize, especially with multiple endogenous variables. The difficult and technique(bfgs) options may be helpful in achieving convergence.

Setting the optimization type to technique(bhhh) resets the default *vcetype* to vce(opg).

Options for two-step estimator

Model

twostep is required and requests that Newey's (1987) efficient two-step estimator be used to obtain the coefficient estimates.

asis requests that all specified variables and observations be retained in the maximization process. This option is typically not used and may introduce numerical instability. Normally, ivprobit drops any endogenous or exogenous variables that perfectly predict success or failure in the dependent variable. The associated observations are also dropped. For more information, see the discussion of model identification in [R] **probit**.

SE

vce(*vcetype*) specifies the type of standard error reported, which includes types that are derived from asymptotic theory and that use bootstrap or jackknife methods; see [R] ***vce_option***.

Reporting

level(#); see [R] **estimation options**.

first requests that the parameters for the reduced-form equations showing the relationships between the endogenous variables and instruments be displayed. For the two-step estimator, first shows the first-stage regressions. For the maximum likelihood estimator, these parameters are estimated jointly with the parameters of the probit equation. The default is not to show these parameter estimates.

Remarks

Remarks are presented under the following headings:

Model setup
Model identification

Model setup

`ivprobit` fits models with dichotomous dependent variables and endogenous regressors. You can use it to fit a probit model when you suspect that one or more of the regressors is correlated with the error term. `ivprobit` is to probit modeling what `ivregress` is to linear regression analysis; see [R] **ivregress** for more information.

Formally, the model is

$$y_{1i}^* = \boldsymbol{y}_{2i}\boldsymbol{\beta} + \boldsymbol{x}_{1i}\boldsymbol{\gamma} + u_i$$
$$\boldsymbol{y}_{2i} = \boldsymbol{x}_{1i}\boldsymbol{\Pi}_1 + \boldsymbol{x}_{2i}\boldsymbol{\Pi}_2 + \boldsymbol{v}_i$$

where $i = 1, \dots, N$, $\boldsymbol{y}_{2i}$ is a $1 \times p$ vector of endogenous variables, $\boldsymbol{x}_{1i}$ is a $1 \times k_1$ vector of exogenous variables, $\boldsymbol{x}_{2i}$ is a $1 \times k_2$ vector of additional instruments, and the equation for $\boldsymbol{y}_{2i}$ is written in reduced form. By assumption $(u_i, \boldsymbol{v}_i) \sim \mathrm{N}(\mathbf{0}, \boldsymbol{\Sigma})$ where σ_{11} is normalized to one to identify the model. $\boldsymbol{\beta}$ and $\boldsymbol{\gamma}$ are vectors of structural parameters, and $\boldsymbol{\Pi}_1$ and $\boldsymbol{\Pi}_2$ are matrices of reduced-form parameters. We do not observe y_{1i}^*; instead, we observe

$$y_{1i} = \begin{cases} 0 & y_{1i}^* < 0 \\ 1 & y_{1i}^* \geq 0 \end{cases}$$

The order condition for identification of the structural parameters requires that $k_2 \geq p$. Presumably, $\boldsymbol{\Sigma}$ is not block diagonal between u_i and $\boldsymbol{v}_i$; otherwise, $\boldsymbol{y}_{2i}$ would not be endogenous.

❑ Technical Note

This model is derived under the assumption that $(u_i, \boldsymbol{v}_i)$ is independently and identically distributed multivariate normal for all i. The `vce(cluster` *clustvar*`)` option can be used to control for a lack of independence. As with most probit models, if u_i is heteroskedastic, point estimates will be inconsistent.
❑

▷ Example 1

We have hypothetical data on 500 two-parent households, and we wish to model whether the woman is employed. We have a variable `fem_work` that is equal to one if she has a job and zero otherwise. Her decision to work is a function of the number of children at home (`kids`), number of years of schooling completed (`fem_educ`), and other household income measured in thousands of dollars (`other_inc`). We suspect that unobservable shocks affecting the woman's decision to hold a job also affect the household's other income. Therefore, we treat `other_inc` as endogenous. As an instrument, we use the number of years of schooling completed by the man (`male_educ`).

The syntax for specifying the exogenous, endogenous, and instrumental variables is identical to that used in `ivregress`; see [R] **ivregress** for details.

```
. use http://www.stata-press.com/data/r10/laborsup
. ivprobit fem_work fem_educ kids (other_inc = male_educ)
Fitting exogenous probit model
Iteration 0:   log likelihood = -344.63508
Iteration 1:   log likelihood = -260.38957
Iteration 2:   log likelihood = -255.37712
Iteration 3:   log likelihood = -255.31445
Iteration 4:   log likelihood = -255.31444
Fitting full model
Iteration 0:   log likelihood = -2371.4753
Iteration 1:   log likelihood = -2369.3178
Iteration 2:   log likelihood = -2368.2198
Iteration 3:   log likelihood = -2368.2062
Iteration 4:   log likelihood = -2368.2062
```

```
Probit model with endogenous regressors           Number of obs   =        500
                                                  Wald chi2(3)    =     163.88
Log likelihood = -2368.2062                       Prob > chi2     =     0.0000

------------------------------------------------------------------------------
    fem_work |      Coef.   Std. Err.      z    P>|z|     [95% Conf. Interval]
-------------+----------------------------------------------------------------
   other_inc |  -.0542756   .0060854    -8.92   0.000    -.0662027   -.0423485
    fem_educ |    .211111   .0268648     7.86   0.000     .1584569    .2637651
        kids |  -.1820929   .0478267    -3.81   0.000    -.2758316   -.0883543
       _cons |   .3672083   .4480724     0.82   0.412    -.5109975    1.245414
-------------+----------------------------------------------------------------
     /athrho |   .3907858   .1509443     2.59   0.010     .0949403    .6866313
    /lnsigma |   2.813383   .0316228    88.97   0.000     2.751404    2.875363
-------------+----------------------------------------------------------------
         rho |   .3720374   .1300519                      .0946561    .5958135
       sigma |   16.66621   .5270318                      15.66461    17.73186
------------------------------------------------------------------------------
Instrumented:  other_inc
Instruments:   fem_educ kids male_educ
------------------------------------------------------------------------------
Wald test of exogeneity (/athrho = 0): chi2(1) =     6.70 Prob > chi2 = 0.0096
```

Since we did not specify `mle` or `twostep`, `ivprobit` used the maximum likelihood estimator by default. At the top of the output we see the iteration log. `ivprobit` fits a probit model ignoring endogeneity to obtain starting values for the endogenous model. The header of the output contains the sample size as well as a Wald statistic and p-value for the test of the hypothesis that all the slope coefficients are jointly zero. Below the table of coefficients, Stata reminds us that the endogenous variable is `other_inc` and that `fem_educ`, `kids`, and `male_educ` were used as instruments.

At the bottom of the output is a Wald test of the exogeneity of the instrumented variables. If the test statistic is not significant, there is not sufficient information in the sample to reject the null that there is no endogeneity. Then a regular probit regression may be appropriate; the point estimates from `ivprobit` are consistent, though those from `probit` are likely to have smaller standard errors.

◁

Various two-step estimators have also been proposed for the endogenous probit model, and Newey's (1987) minimum chi-squared estimator is available with the `twostep` option.

▷ Example 2

Refitting our labor-supply model with the two-step estimator yields

```
. ivprobit fem_work fem_educ kids (other_inc = male_educ), twostep
Two-step probit with endogenous regressors        Number of obs   =        500
                                                  Wald chi2(3)    =      93.97
                                                  Prob > chi2     =     0.0000

------------------------------------------------------------------------------
    fem_work |      Coef.   Std. Err.      z    P>|z|     [95% Conf. Interval]
-------------+----------------------------------------------------------------
   other_inc |   -.058473   .0093364    -6.26   0.000    -.0767719    -.040174
    fem_educ |    .227437   .0281628     8.08   0.000     .1722389     .282635
        kids |  -.1961748   .0496323    -3.95   0.000    -.2934522   -.0988973
       _cons |   .3956061   .4982649     0.79   0.427    -.5809751    1.372187
------------------------------------------------------------------------------
Instrumented:  other_inc
Instruments:   fem_educ kids male_educ
------------------------------------------------------------------------------
Wald test of exogeneity:      chi2(1) =      6.50          Prob > chi2 = 0.0108
```

All the coefficients have the same signs as their counterparts in the maximum likelihood model. The Wald test at the bottom of the output confirms our earlier finding of endogeneity.

◁

❑ Technical Note

In a standard probit model, the error term is assumed to have a variance of one. In the probit model with endogenous regressors, we assume that $(u_i, \boldsymbol{v}_i)$ is multivariate normal with covariance matrix

$$\mathrm{Var}(u_i, \boldsymbol{v}_i) = \boldsymbol{\Sigma} = \begin{bmatrix} 1 & \boldsymbol{\Sigma}_{21}' \\ \boldsymbol{\Sigma}_{21} & \boldsymbol{\Sigma}_{22} \end{bmatrix}$$

With the properties of the multivariate normal distribution, $\mathrm{Var}(u_i|\boldsymbol{v}_i) = 1 - \boldsymbol{\Sigma}_{21}'\boldsymbol{\Sigma}_{22}^{-1}\boldsymbol{\Sigma}_{21}$. As a result, Newey's estimator and other two-step probit estimators do not yield estimates of $\boldsymbol{\beta}$ and $\boldsymbol{\gamma}$ but rather $\boldsymbol{\beta}/\sigma$ and $\boldsymbol{\gamma}/\sigma$, where σ is the square root of $\mathrm{Var}(u_i|\boldsymbol{v}_i)$. Hence, we cannot directly compare the estimates obtained from Newey's estimator with those obtained via maximum likelihood or with those obtained from `probit`. See Wooldridge (2002, 472–477) for a discussion of Rivers and Vuong's (1988) two-step estimator. The issues raised pertaining to the interpretation of the coefficients of that estimator are identical to those that arise with Newey's estimator. Wooldridge also discusses ways to obtain marginal effects from two-step estimators.

❑

Despite the coefficients not being directly comparable to their maximum likelihood counterparts, the two-step estimator is nevertheless useful. The maximum likelihood estimator may have difficulty converging, especially with multiple endogenous variables. The two-step estimator, consisting of nothing more complicated than a probit regression, will almost certainly converge. Moreover, although the coefficients from the two models are not directly comparable, the two-step estimates can still be used to test for statistically significant relationships.

Model identification

As in the linear simultaneous-equation model, the order condition for identification requires that the number of excluded exogenous variables (i.e., the additional instruments) be at least as great as the number of included endogenous variables. `ivprobit` checks this for you and issues an error message if the order condition is not met.

Like `probit`, `logit`, and `logistic`, `ivprobit` checks the exogenous and endogenous variables to see if any of them predict the outcome variable perfectly. It will then drop offending variables and observations and fit the model on the remaining data. Instruments that are perfect predictors do not affect estimation, so they are not checked. See *Model identification* in [R] **probit** for more information.

`ivprobit` will also occasionally display messages such as

```
Note: 4 failures and 0 successes completely determined.
```

For an explanation of this message, see [R] **logit**.

Saved Results

`ivprobit` saves the following in `e()`:

Scalars

`e(N)`	number of observations	`e(ll)`	log likelihood
`e(k)`	number of parameters	`e(N_clust)`	number of clusters
`e(k_eq)`	number of equations	`e(endog_ct)`	number of endogenous regressors
`e(k_eq_model)`	number of equations in model Wald test	`e(p)`	model Wald p-value
		`e(p_exog)`	exogeneity test Wald p-value
`e(k_aux)`	number of auxiliary parameters	`e(chi2)`	model Wald χ^2
`e(k_dv)`	number of dependent variables	`e(chi2_exog)`	Wald χ^2 test of exogeneity
`e(N_cds)`	number of completely determined successes	`e(rank)`	rank of `e(V)`
		`e(ic)`	number of iterations
`e(N_cdf)`	number of completely determined failures	`e(rc)`	return code
		`e(converged)`	`1` if converged, `0` otherwise
`e(df_m)`	model degrees of freedom		

Macros

`e(cmd)`	`ivprobit`	`e(diparm#)`	display transformed parameter #
`e(cmdline)`	command as typed	`e(opt)`	type of optimization
`e(depvar)`	name of dependent variable	`e(method)`	`ml` or `twostep`
`e(instd)`	instrumented variables	`e(ml_method)`	type of `ml` method
`e(insts)`	instruments	`e(user)`	name of likelihood-evaluator program
`e(wtype)`	weight type	`e(technique)`	maximization technique
`e(wexp)`	weight expression	`e(crittype)`	optimization criterion
`e(title)`	title in estimation output	`e(properties)`	`b V`
`e(clustvar)`	name of cluster variable	`e(estat_cmd)`	program used to implement `estat`
`e(chi2type)`	`Wald`; type of model χ^2 test	`e(footnote)`	program used to implement the footnote display
`e(vce)`	*vcetype* specified in `vce()`		
`e(vcetype)`	title used to label Std. Err.	`e(predict)`	program used to implement `predict`

Matrices

`e(b)`	coefficient vector	`e(V)`	variance–covariance matrix of the estimators
`e(ilog)`	iteration log (up to 20 iterations)	`e(gradient)`	gradient vector
`e(Sigma)`	$\widehat{\boldsymbol{\Sigma}}$		

Functions

`e(sample)`	marks estimation sample

Methods and Formulas

`ivprobit` is implemented as an ado-file.

Fitting limited dependent variable models with endogenous regressors has received considerable attention in the econometrics literature. Building on the results of Amemiya (1978, 1979), Newey (1987) developed an efficient method of estimation that encompasses both Rivers and Vuong's (1988) simultaneous-equations probit model and Smith and Blundell's (1986) simultaneous-equations tobit model. With modern computers, maximum likelihood estimation is feasible as well. For compactness, we write the model as

$$y_{1i}^* = \boldsymbol{z}_i\boldsymbol{\delta} + u_i \tag{1a}$$

$$\boldsymbol{y}_{2i} = \boldsymbol{x}_i\boldsymbol{\Pi} + \boldsymbol{v}_i \tag{1b}$$

where $\boldsymbol{z}_i = (\boldsymbol{y}_{2i}, \boldsymbol{x}_{1i})$, $\boldsymbol{x}_i = (\boldsymbol{x}_{1i}, \boldsymbol{x}_{2i})$, $\boldsymbol{\delta} = (\boldsymbol{\beta}', \boldsymbol{\gamma}')'$, and $\boldsymbol{\Pi} = (\boldsymbol{\Pi}_1', \boldsymbol{\Pi}_2')'$.

Deriving the likelihood function is straightforward since we can write the joint density $f\left(y_{1i}, \boldsymbol{y}_{2i}|\boldsymbol{x}_i\right)$ as $f\left(y_{1i}|\boldsymbol{y}_{2i}, \boldsymbol{x}_i\right) f\left(\boldsymbol{y}_{2i}|\boldsymbol{x}_i\right)$. When there is endogenous regressor, the log likelihood for observation i is

$$\ln L_i = w_i\left[y_{1i}\ln\Phi\left(m_i\right) + \left(1-y_{1i}\right)\ln\left\{1-\Phi\left(m_i\right)\right\} + \ln\phi\left(\frac{y_{2i}-\boldsymbol{x}_i\boldsymbol{\Pi}}{\sigma}\right) - \ln\sigma\right]$$

where

$$m_i = \frac{\boldsymbol{z}_i\boldsymbol{\delta} + \rho\left(y_{2i}-\boldsymbol{x}_i\boldsymbol{\Pi}\right)/\sigma}{\left(1-\rho^2\right)^{\frac{1}{2}}}$$

$\Phi(\cdot)$ and $\phi(\cdot)$ are the standard normal distribution and density functions, respectively; σ is the standard deviation of v_i; ρ is the correlation coefficient between u_i and v_i; and w_i is the weight for observation i or one if no weights were specified. Instead of estimating σ and ρ, we estimate $\ln\sigma$ and $\operatorname{atanh}\rho$, where

$$\operatorname{atanh}\rho = \frac{1}{2}\ln\left(\frac{1+\rho}{1-\rho}\right)$$

For multiple endogenous regressors, let

$$\operatorname{Var}(u_i, \boldsymbol{v}_i) = \boldsymbol{\Sigma} = \begin{bmatrix} 1 & \boldsymbol{\Sigma}_{21}' \\ \boldsymbol{\Sigma}_{21} & \boldsymbol{\Sigma}_{22} \end{bmatrix}$$

As in any probit model, we have imposed the normalization $\operatorname{Var}(u_i) = 1$ to identify the model. The log likelihood for observation i is

$$\ln L_i = w_i\left[y_{1i}\ln\Phi\left(m_i\right) + \left(1-y_{1i}\right)\ln\left\{1-\Phi\left(m_i\right)\right\} + \ln f(\boldsymbol{y}_{2i}|\boldsymbol{x}_i)\right]$$

where

$$\ln f(\boldsymbol{y}_{2i}|\boldsymbol{x}_i) = -\frac{p}{2}\ln 2\pi - \frac{1}{2}\ln|\boldsymbol{\Sigma}_{22}| - \frac{1}{2}\left(\boldsymbol{y}_{2i}-\boldsymbol{x}_i\boldsymbol{\Pi}\right)\boldsymbol{\Sigma}_{22}^{-1}\left(\boldsymbol{y}_{2i}-\boldsymbol{x}_i\boldsymbol{\Pi}\right)'$$

and

$$m_i = \left(1-\boldsymbol{\Sigma}_{21}'\boldsymbol{\Sigma}_{22}^{-1}\boldsymbol{\Sigma}_{21}\right)^{-\frac{1}{2}}\left\{\boldsymbol{z}_i\boldsymbol{\delta} + \left(\boldsymbol{y}_{2i}-\boldsymbol{x}_i\boldsymbol{\Pi}\right)\boldsymbol{\Sigma}_{22}^{-1}\boldsymbol{\Sigma}_{21}\right\}$$

Instead of maximizing the log-likelihood function with respect to $\boldsymbol{\Sigma}$, we maximize with respect to the Cholesky decomposition $\boldsymbol{S}$ of $\boldsymbol{\Sigma}$; that is, there exists a lower triangular matrix $\boldsymbol{S}$ such that $\boldsymbol{S}\boldsymbol{S}' = \boldsymbol{\Sigma}$. This maximization ensures that $\boldsymbol{\Sigma}$ is positive definite, as a covariance matrix must be. Let

$$\boldsymbol{S} = \begin{bmatrix} 1 & 0 & 0 & \dots & 0 \\ s_{21} & s_{22} & 0 & \dots & 0 \\ s_{31} & s_{32} & s_{33} & \dots & 0 \\ \vdots & \vdots & \vdots & \ddots & \vdots \\ s_{p+1,1} & s_{p+1,2} & s_{p+1,3} & \dots & s_{p+1,p+1} \end{bmatrix}$$

The two-step estimates are obtained using Newey's (1987) minimum chi-squared estimator. The reduced-form equation for y_{1i}^* is

$$
\begin{aligned}
y_{1i}^* &= (\boldsymbol{x}_i\boldsymbol{\Pi} + \boldsymbol{v}_i)\boldsymbol{\beta} + \boldsymbol{x}_{1i}\boldsymbol{\gamma} + u_i \\
&= \boldsymbol{x}_i\boldsymbol{\alpha} + \boldsymbol{v}_i\boldsymbol{\beta} + u_i \\
&= \boldsymbol{x}_i\boldsymbol{\alpha} + \nu_i
\end{aligned}
$$

where $\nu_i = \boldsymbol{v}_i\boldsymbol{\beta} + u_i$. Since u_i and $\boldsymbol{v}_i$ are jointly normal, ν_i is also normal. Note that

$$
\boldsymbol{\alpha} = \begin{bmatrix} \boldsymbol{\Pi}_1 \\ \boldsymbol{\Pi}_2 \end{bmatrix} \boldsymbol{\beta} + \begin{bmatrix} \boldsymbol{I} \\ \boldsymbol{0} \end{bmatrix} \boldsymbol{\gamma} = D(\boldsymbol{\Pi})\boldsymbol{\delta}
$$

where $D(\boldsymbol{\Pi}) = (\boldsymbol{\Pi}, \boldsymbol{I}_1)$ and $\boldsymbol{I}_1$ is defined such that $\boldsymbol{x}_i\boldsymbol{I}_1 = \boldsymbol{x}_{1i}$. Letting $\widehat{\boldsymbol{z}}_i = (\boldsymbol{x}_i\widehat{\boldsymbol{\Pi}}, \boldsymbol{x}_{1i})$, $\widehat{\boldsymbol{z}}_i\boldsymbol{\delta} = \boldsymbol{x}_i D(\widehat{\boldsymbol{\Pi}})\boldsymbol{\delta}$, where $D(\widehat{\boldsymbol{\Pi}}) = (\widehat{\boldsymbol{\Pi}}, \boldsymbol{I}_1)$. Thus one estimator of $\boldsymbol{\alpha}$ is $D(\widehat{\boldsymbol{\Pi}})\boldsymbol{\delta}$; denote this estimator $\widehat{\boldsymbol{D}}\boldsymbol{\delta}$.

$\boldsymbol{\alpha}$ could also be estimated directly as the solution to

$$
\max_{\boldsymbol{\alpha},\boldsymbol{\lambda}} \sum_{i=1}^{N} l(y_{1i}, \boldsymbol{x}_i\boldsymbol{\alpha} + \widehat{\boldsymbol{v}}_i\boldsymbol{\lambda}) \tag{2}
$$

where $l(\cdot)$ is the log likelihood for probit. Denote this estimator $\widetilde{\boldsymbol{\alpha}}$. The inclusion of the $\widehat{\boldsymbol{v}}_i\boldsymbol{\lambda}$ term follows because the multivariate normality of $(u_i, \boldsymbol{v}_i)$ implies that, conditional on $\boldsymbol{y}_{2i}$, the expected value of u_i is nonzero. Since $\boldsymbol{v}_i$ is unobservable, the least-squares residuals from fitting equation $(1b)$ are used.

Amemiya (1978) shows that the estimator of $\boldsymbol{\delta}$ defined by

$$
\max_{\boldsymbol{\delta}} \; (\widetilde{\boldsymbol{\alpha}} - \widehat{\boldsymbol{D}}\boldsymbol{\delta})'\widehat{\boldsymbol{\Omega}}^{-1}(\widetilde{\boldsymbol{\alpha}} - \widehat{\boldsymbol{D}}\boldsymbol{\delta})
$$

where $\widehat{\boldsymbol{\Omega}}$ is a consistent estimator of the covariance of $\sqrt{N}(\widetilde{\boldsymbol{\alpha}} - \widehat{\boldsymbol{D}}\boldsymbol{\delta})$, is asymptotically efficient relative to all other estimators that minimize the distance between $\widetilde{\boldsymbol{\alpha}}$ and $D(\widehat{\boldsymbol{\Pi}})\boldsymbol{\delta}$. Thus an efficient estimator of $\boldsymbol{\delta}$ is

$$
\widehat{\boldsymbol{\delta}} = (\widehat{\boldsymbol{D}}'\widehat{\boldsymbol{\Omega}}^{-1}\widehat{\boldsymbol{D}})^{-1}\widehat{\boldsymbol{D}}'\widehat{\boldsymbol{\Omega}}^{-1}\widetilde{\boldsymbol{\alpha}} \tag{3}
$$

and

$$
\text{Var}(\widehat{\boldsymbol{\delta}}) = (\widehat{\boldsymbol{D}}'\widehat{\boldsymbol{\Omega}}^{-1}\widehat{\boldsymbol{D}})^{-1} \tag{4}
$$

To implement this estimator, we need $\widehat{\boldsymbol{\Omega}}^{-1}$.

Consider the two-step maximum likelihood estimator that results from first fitting equation $(1b)$ by OLS and computing the residuals $\widehat{\boldsymbol{v}}_i = \boldsymbol{y}_{2i} - \boldsymbol{x}_i\widehat{\boldsymbol{\Pi}}$. The estimator is then obtained by solving

$$
\max_{\boldsymbol{\delta},\boldsymbol{\lambda}} \sum_{i=1}^{N} l(y_{1i}, \boldsymbol{z}_i\boldsymbol{\delta} + \widehat{\boldsymbol{v}}_i\boldsymbol{\lambda})
$$

This is the two-step instrumental variables (2SIV) estimator proposed by Rivers and Vuong (1988), and its role will become apparent shortly.

From Proposition 5 of Newey (1987), $\sqrt{N}(\widetilde{\boldsymbol{\alpha}} - \widehat{\boldsymbol{D}}\boldsymbol{\delta}) \xrightarrow{d} \text{N}(\boldsymbol{0}, \boldsymbol{\Omega})$, where

$$
\boldsymbol{\Omega} = \boldsymbol{J}_{\alpha\alpha}^{-1} + (\boldsymbol{\lambda} - \boldsymbol{\beta})'\boldsymbol{\Sigma}_{22}(\boldsymbol{\lambda} - \boldsymbol{\beta})\boldsymbol{Q}^{-1}
$$

and $\boldsymbol{\Sigma}_{22} = E\{\boldsymbol{v}_i'\boldsymbol{v}_i\}$. $\boldsymbol{J}_{\alpha\alpha}^{-1}$ is simply the covariance matrix of $\widetilde{\boldsymbol{\alpha}}$, ignoring that $\widehat{\boldsymbol{\Pi}}$ is an estimated parameter matrix. Moreover, Newey shows that the covariance matrix from an OLS regression of $\boldsymbol{y}_{2i}(\widehat{\boldsymbol{\lambda}} - \widehat{\boldsymbol{\beta}})$ on $\boldsymbol{x}_i$ is a consistent estimator of the second term. $\widehat{\boldsymbol{\lambda}}$ can be obtained from solving (2), and the 2SIV estimator yields a consistent estimate, $\widehat{\boldsymbol{\beta}}$.

Mechanically, estimation proceeds in several steps.

1. Each of the endogenous right-hand-side variables is regressed on all the exogenous variables, and the fitted values and residuals are calculated. The matrix $\widehat{D} = D(\widehat{\boldsymbol{\Pi}})$ is assembled from the estimated coefficients.
2. `probit` is used to solve (2) and obtain $\widetilde{\boldsymbol{\alpha}}$ and $\widehat{\boldsymbol{\lambda}}$. The portion of the covariance matrix corresponding to $\boldsymbol{\alpha}$, $\boldsymbol{J}_{\alpha\alpha}^{-1}$, is also saved.
3. The 2SIV estimator is evaluated, and the parameters $\widehat{\boldsymbol{\beta}}$ corresponding to $\boldsymbol{y}_{2i}$ are collected.
4. $\boldsymbol{y}_{2i}(\widehat{\boldsymbol{\lambda}} - \widehat{\boldsymbol{\beta}})$ is regressed on $\boldsymbol{x}_i$. The covariance matrix of the parameters from this regression is added to $\boldsymbol{J}_{\alpha\alpha}^{-1}$, yielding $\widehat{\boldsymbol{\Omega}}$.
5. Evaluating equations (3) and (4) yields the estimates $\widehat{\boldsymbol{\delta}}$ and $\text{Var}(\widehat{\boldsymbol{\delta}})$.
6. A Wald test of the null hypothesis $H_0: \boldsymbol{\lambda} = \boldsymbol{0}$, using the 2SIV estimates, serves as our test of exogeneity.

The two-step estimates are not directly comparable to those obtained from the maximum likelihood estimator or from `probit`. The argument is the same for Newey's efficient estimator as for Rivers and Vuong's (1988) 2SIV estimator, so we consider the simpler 2SIV estimator. From the properties of the normal distribution,

$$E(u_i|\boldsymbol{v}_i) = \boldsymbol{v}_i\boldsymbol{\Sigma}_{22}^{-1}\boldsymbol{\Sigma}_{21} \qquad \text{and} \qquad \text{Var}(u_i|\boldsymbol{v}_i) = 1 - \boldsymbol{\Sigma}_{21}'\boldsymbol{\Sigma}_{22}^{-1}\boldsymbol{\Sigma}_{21}$$

We write u_i as $u_i = \boldsymbol{v}_i\boldsymbol{\Sigma}_{22}^{-1}\boldsymbol{\Sigma}_{21} + e_i = \boldsymbol{v}_i\boldsymbol{\lambda} + e_i$ where $e_i \sim \text{N}(0, 1-\rho^2)$, $\rho^2 = \boldsymbol{\Sigma}_{21}'\boldsymbol{\Sigma}_{22}^{-1}\boldsymbol{\Sigma}_{21}$, and e_i is independent of $\boldsymbol{v}_i$. In the second stage of 2SIV, we use a probit regression to estimate the parameters of

$$y_{1i} = \boldsymbol{z}_i\boldsymbol{\delta} + \boldsymbol{v}_i\boldsymbol{\lambda} + e_i$$

Because $\boldsymbol{v}_i$ is unobservable, we use the sample residuals from the first-stage regressions.

$$\Pr(y_{1i} = 1|\boldsymbol{z}_i, \boldsymbol{v}_i) = \Pr(\boldsymbol{z}_i\boldsymbol{\delta} + \boldsymbol{v}_i\boldsymbol{\lambda} + e_i > 0|\boldsymbol{z}_i, \boldsymbol{v}_i) = \Phi\left\{(1-\rho^2)^{-\frac{1}{2}}(\boldsymbol{z}_i\boldsymbol{\delta} + \boldsymbol{v}_i\boldsymbol{\lambda})\right\}$$

Hence, as mentioned previously, 2SIV and Newey's estimator do not estimate $\boldsymbol{\delta}$ and $\boldsymbol{\lambda}$ but rather

$$\boldsymbol{\delta}_\rho = \frac{1}{(1-\rho^2)^{\frac{1}{2}}}\boldsymbol{\delta} \qquad \text{and} \qquad \boldsymbol{\lambda}_\rho = \frac{1}{(1-\rho^2)^{\frac{1}{2}}}\boldsymbol{\lambda}$$

Acknowledgments

The two-step estimator is based on the `probitiv` command written by Jonah Gelbach and the `ivprob` command written by Joe Harkness.

References

Amemiya, T. 1978. The estimation of a simultaneous equation generalized probit model. *Econometrica* 46: 1193–1205.

——. 1979. The estimation of a simultaneous equation tobit model. *International Economic Review* 20: 169–181.

Miranda, A., and S. Rabe-Hesketh. 2006. Maximum likelihood estimation of endogenous switching and sample selection models for binary, ordinal, and count variables. *Stata Journal* 6: 285–308.

Newey, W. K. 1987. Efficient estimation of limited dependent variable models with endogenous explanatory variables. *Journal of Econometrics* 36: 231–250.

Rivers, D., and Q. H. Vuong. 1988. Limited information estimators and exogeneity tests for simultaneous probit models. *Journal of Econometrics* 39: 347–366.

Smith, R., and R. Blundell. 1986. An exogeneity test for the simultaneous equation tobit model with an application to labor supply. *Econometrica* 54: 679–685.

Wooldridge, J. M. 2002. *Econometric Analysis of Cross Section and Panel Data*. Cambridge, MA: MIT Press.

Also See

[R] **ivprobit postestimation** — Postestimation tools for ivprobit

[R] **ivregress** — Single-equation instrumental-variables regression

[R] **ivtobit** — Tobit model with endogenous regressors

[R] **probit** — Probit regression

[XT] **xtprobit** — Random-effects and population-averaged probit models

[U] **20 Estimation and postestimation commands**

Title

ivprobit postestimation — Postestimation tools for ivprobit

Description

The following postestimation commands are of special interest after `ivprobit`:

command	description
`estat clas`	`estat classification` reports various summary statistics, including the classification table
`lroc`	graphs the ROC curve and calculates the area under the curve
`lsens`	graphs sensitivity and specificity versus probability cutoff

For information about these commands, see [R] **logistic postestimation**.
These commands are not appropriate after the two-step estimator or the `svy` prefix.

The following postestimation commands are also available:

command	description
`adjust`[1]	adjusted predictions of $\mathbf{x}\beta$ and probabilities
`estat`[2]	AIC, BIC, VCE, and estimation sample summary
`estat` (svy)	postestimation statistics for survey data
`estimates`	cataloging estimation results
`hausman`	Hausman's specification test
`lincom`	point estimates, standard errors, testing, and inference for linear combinations of coefficients
`lrtest`[3]	likelihood-ratio test; not available with two-step estimator
`mfx`	marginal effects or elasticities
`nlcom`	point estimates, standard errors, testing, and inference for nonlinear combinations of coefficients
`predict`	predictions, residuals, influence statistics, and other diagnostic measures
`predictnl`	point estimates, standard errors, testing, and inference for generalized predictions
`suest`[2]	seemingly unrelated estimation
`test`	Wald tests for simple and composite linear hypotheses
`testnl`	Wald tests of nonlinear hypotheses

[1] `adjust` is not appropriate with time-series operators.
[2] `estat ic` and `suest` are not appropriate after `ivprobit, twostep`.
[3] `lrtest` is not appropriate with `svy` estimation results.

See the corresponding entries in the *Stata Base Reference Manual* for details, but see [SVY] **estat** for details about `estat` (svy).

Syntax for predict

After ML or twostep

predict [*type*] *newvar* [*if*] [*in*] [, *statistic* rules asif]

After ML

predict [*type*] { *stub** | *newvarlist* } [*if*] [*in*] , scores

statistic	description
Main	
xb	linear prediction; the default
stdp	standard error of the linear prediction
pr	probability of a positive outcome; not available with two-step estimator

These statistics are available both in and out of sample; type `predict ... if e(sample) ...` if wanted only for the estimation sample.

Options for predict

Main

`xb`, the default, calculates the linear prediction.

`stdp` calculates the standard error of the linear prediction.

`pr` calculates the probability of a positive outcome. `pr` is not available with the two-step estimator.

`rules` requests that Stata use any rules that were used to identify the model when making the prediction. By default, Stata calculates missing for excluded observations. `rules` is not available with the two-step estimator.

`asif` requests that Stata ignore the rules and the exclusion criteria and calculate predictions for all observations possible using the estimated parameters from the model. `asif` is not available with the two-step estimator.

`scores`, not available with `twostep`, calculates equation-level score variables.

For models with one endogenous regressor, four new variables are created.

The first new variable will contain $\partial \ln L/\partial(\boldsymbol{z}_i\boldsymbol{\delta})$.

The second new variable will contain $\partial \ln L/\partial(\boldsymbol{x}_i\boldsymbol{\Pi})$.

The third new variable will contain $\partial \ln L/\partial$ atanh ρ.

The fourth new variable will contain $\partial \ln L/\partial \ln\sigma$.

For models with p endogenous regressors, $p + \{(p+1)(p+2)\}/2$ new variables are created.

The first new variable will contain $\partial \ln L/\partial(\boldsymbol{z}_i\boldsymbol{\delta})$.

The second through $(p+1)$-th new variables will contain $\partial \ln L/\partial(\boldsymbol{x}_i\boldsymbol{\Pi}_k)$, $k = 1, \ldots, p$, where $\boldsymbol{\Pi}_k$ is the kth column of $\boldsymbol{\Pi}$.

The remaining score variables will contain the partial derivatives of $\ln L$ with respect to s_{21}, s_{31}, ..., $s_{p+1,1}$, s_{22}, ..., $s_{p+1,2}$, ..., $s_{p+1,p+1}$, where $s_{m,n}$ denotes the (m,n) element of the Cholesky decomposition of the error covariance matrix.

Remarks

Remarks are presented under the following headings:

Marginal effects
Obtaining predicted values

Marginal effects

▷ Example 1

We can obtain marginal effects by using the `mfx` command after `ivprobit`. We will calculate marginal effects by using the labor-supply model of example 1 in [R] **ivprobit**.

```
. mfx compute, predict(p) eqlist(fem_work) force
Marginal effects after ivprobit
      y  = Probability of positive outcome (predict, p)
         =  .44363395
```

variable	dy/dx	Std. Err.	z	P>|z|	[95% C.I.	]	X
other_~c	-.0214364	.00242	-8.87	0.000	-.026176	-.016697	49.6023
fem_educ	.0833791	.01057	7.89	0.000	.062664	.104094	12.046
kids	-.0719183	.01888	-3.81	0.000	-.108927	-.03491	1.976

Here we see that a $1,000 increase in `other_inc` leads to a decrease of 0.0214 in the probability that the woman has a job, when all variables are held equal to their sample means. If we had not used the `eqlist()` option, `mfx` would have included a row in the table for `male_educ`, for which the marginal effect is identically zero since that variable is an instrument. We used the `force` option so that `mfx` would calculate standard errors. Otherwise, `mfx` would refuse to do so because `other_inc` is an endogenous variable.

◁

Obtaining predicted values

After fitting your model with `ivprobit`, you can obtain the linear prediction and its standard error for both the estimation sample and other samples by using the `predict` command; see [U] **20 Estimation and postestimation commands** and [R] **predict**. If you had used the maximum likelihood estimator, you could also obtain the probability of a positive outcome.

`predict`'s `p` option calculates the probability of a positive outcome, remembering any rules used to identify the model, and calculates missing for excluded observations. `predict`'s `rules` option uses the rules in predicting probabilities, whereas `predict`'s `asif` option ignores both the rules and the exclusion criteria and calculates probabilities for all possible observations using the estimated parameters from the model. See *Obtaining predicted values* in [R] **probit postestimation** for an example.

Methods and Formulas

All postestimation commands listed above are implemented as ado-files.

The linear prediction is calculated as $z_i\widehat{\boldsymbol{\delta}}$, where $\widehat{\boldsymbol{\delta}}$ is the estimated value of $\boldsymbol{\delta}$, and z_i and $\boldsymbol{\delta}$ are defined in ($1a$) of [R] **ivprobit**. The probability of a positive outcome is $\Phi(z_i\widehat{\boldsymbol{\delta}})$, where $\Phi(\cdot)$ is the standard normal distribution function.

Also See

[R] **ivprobit** — Probit model with endogenous regressors

[U] **20 Estimation and postestimation commands**

Title

ivregress — Single-equation instrumental-variables regression

Syntax

`ivregress` *estimator depvar* [*varlist*$_1$] (*varlist*$_2$ = *varlist*$_{iv}$) [*if*] [*in*] [*weight*] [, *options*]

estimator	description
2sls	two-stage least squares (2SLS)
liml	limited-information maximum likelihood (LIML)
gmm	generalized method of moments (GMM)

options	description
Model	
noconstant	suppress constant term
hascons	has user-supplied constant
GMM[1]	
wmatrix(*wmtype*)	*wmtype* may be robust, cluster *clustvar*, hac *kernel*, or unadjusted
center	center moments in weight matrix computation
igmm	use iterative instead of two-step GMM estimator
eps(#)[2]	specify # for parameter convergence criterion; default is eps(1e-6)
weps(#)[2]	specify # for weight matrix convergence criterion; default is weps(1e-6)
optimization_options[2]	control the optimization process; seldom used
SE/Robust	
vce(*vcetype*)	*vcetype* may be unadjusted, robust, cluster *clustvar*, bootstrap, jackknife, or hac *kernel*
Reporting	
level(#)	set confidence level; default is level(95)
first	report first-stage regression
small	make degrees-of-freedom adjustments and report small-sample statistics
noheader	display only the coefficient table
depname(*depname*)	substitute dependent variable name
eform(*string*)	report exponentiated coefficients and use *string* to label them

[1]These options may be specified only when gmm is specified.

[2]These options may be specified only when igmm is specified.

depvar, $varlist_1$, $varlist_2$, and $varlist_{iv}$ may contain time-series operators; see [U] **11.4.3 Time-series varlists**.

bootstrap, by, jackknife, rolling, statsby, svy, and xi are allowed; see [U] **11.1.10 Prefix commands**.

Weights are not allowed with the bootstrap prefix.

aweights are not allowed with the jackknife prefix.

hascons, vce(), noheader, depname(), and weights are not allowed with the svy prefix.

aweights, fweights, iweights, and pweights are allowed; see [U] **11.1.6 weight**.

See [U] **20 Estimation and postestimation commands** for more capabilities of estimation commands.

Description

ivregress fits a linear regression of *depvar* on $varlist_1$ and $varlist_2$, using $varlist_{iv}$ (along with $varlist_1$) as instruments for $varlist_2$. ivregress supports estimation via two-stage least squares (2SLS), limited-information maximum likelihood (LIML), and generalized method of moments (GMM).

In the language of instrumental variables, $varlist_1$ and $varlist_{iv}$ are the exogenous variables, and $varlist_2$ are the endogenous variables.

Options

Model

noconstant; see [R] **estimation options**.

hascons indicates that a user-defined constant or its equivalent is specified among the independent variables.

GMM

wmatrix(*wmtype*) specifies the type of weighting matrix to be used in conjunction with the GMM estimator.

Specifying wmatrix(robust) requests a weighting matrix that is optimal when the error term is heteroskedastic. wmatrix(robust) is the default.

Specifying wmatrix(cluster *clustvar*) requests a weighting matrix that accounts for arbitrary correlation among observations within clusters identified by *clustvar*.

Specifying wmatrix(hac *kernel #*) requests a heteroskedasticity- and autocorrelation-consistent (HAC) weighting matrix using the specified kernel (see below) with # lags. The bandwidth of a kernel is equal to $\# + 1$.

Specifying wmatrix(hac *kernel* opt) requests a HAC weighting matrix using the specified kernel, and the lag order is selected using Newey and West's (1994) optimal lag-selection algorithm.

Specifying wmatrix(hac *kernel*) requests a HAC weighting matrix using the specified kernel and $N - 2$ lags, where N is the sample size.

There are three kernels available for HAC weighting matrices, and you may request each one by using the name used by statisticians or the name perhaps more familiar to economists:

bartlett or nwest requests the Bartlett (Newey–West) kernel;

parzen or gallant requests the Parzen (Gallant) kernel; and

quadraticspectral or andrews requests the quadratic spectral (Andrews) kernel.

Specifying wmatrix(unadjusted) requests a weighting matrix that is suitable when the errors are homoskedastic. The two-step GMM estimator with this weighting matrix is equivalent to the 2SLS estimator.

center requests that the sample moments be centered (demeaned) when computing GMM weight matrices. By default, centering is not done.

igmm requests that the iterative GMM estimator be used instead of the default two-step GMM estimator. Convergence is declared when the relative change in the parameter vector from one iteration to the next is less than eps() or the relative change in the weight matrix is less than weps().

eps(*#*) specifies the convergence criterion for successive parameter estimates when the iterative GMM estimator is used. The default is eps(1e-6). Convergence is declared when the relative difference between successive parameter estimates is less than eps() and the relative difference between successive estimates of the weighting matrix is less than weps().

weps(*#*) specifies the convergence criterion for successive estimates of the weighting matrix when the iterative GMM estimator is used. The default is weps(1e-6). Convergence is declared when the relative difference between successive parameter estimates is less than eps() and the relative difference between successive estimates of the weighting matrix is less than weps().

optimization_options: iterate(*#*), [no]log. iterate() specifies the maximum number of iterations to perform in conjunction with the iterative GMM estimator. The default is 16,000 or the number set using set maxiter. log/nolog specifies whether to show the iteration log. These options are seldom used.

SE/Robust

vce(*vcetype*) specifies the type of standard error reported, which includes types that are robust to some kinds of misspecification, that allow for intragroup correlation, and that use bootstrap or jackknife methods; see [R] ***vce_option***.

vce(unadjusted), the default for 2sls and liml, specifies that an unadjusted (nonrobust) VCE matrix be used. The default for gmm is based on the *wmtype* specified in the wmatrix() option; see wmatrix(*wmtype*) above. If wmatrix() is specified with gmm but vce() is not, then *vcetype* is set equal to *wmtype*. To override this behavior and obtain an unadjusted (nonrobust) VCE matrix, specify vce(unadjusted).

ivregress also allows the following:

vce(hac *kernel* [*#*|opt]) specifies that a HAC covariance matrix be used. The syntax used with vce(hac *kernel* ...) is identical to that used with wmatrix(hac *kernel* ...); see wmatrix(*wmtype*) above.

Reporting

level(*#*); see [R] **estimation options**.

first requests that the first-stage regression results be displayed.

small requests that the degrees-of-freedom adjustment $N/(N-k)$ be made to the variance–covariance matrix of parameters and that small-sample F and t statistics be reported, where N is the sample size and k is the number of parameters estimated. By default, no degrees-of-freedom adjustment is made, and Wald and z statistics are reported. Even with this option, no degrees-of-freedom adjustment is made to the weighting matrix when the GMM estimator is used.

noheader suppresses the display of the summary statistics at the top of the output, displaying only the coefficient table.

`depname(`*depname*`)` is used only in programs and ado-files that use `ivregress` to fit models other than instrumental-variables regression. `depname()` may be specified only at estimation time. *depname* is recorded as the identity of the dependent variable, even though the estimates are calculated using *depvar*. This method affects the labeling of the output—not the results calculated—but could affect later calculations made by `predict`, where the residual would be calculated as deviations from *depname* rather than *depvar*. `depname()` is most typically used when *depvar* is a temporary variable (see [P] **macro**) used as a proxy for *depname*.

`eform(`*string*`)` is used only in programs and ado-files that use `ivregress` to fit models other than instrumental-variables regression. `eform()` specifies that the coefficient table be displayed in "exponentiated form", as defined in [R] **maximize**, and that *string* be used to label the exponentiated coefficients in the table.

Remarks

`ivregress` performs instrumental-variables regression and weighted instrumental-variables regression. For a general discussion of instrumental variables, see Baum (2006), Cameron and Trivedi (2005), Davidson and MacKinnon (1993, 2004), Greene (2003), and Wooldridge (2002a,b). See Hall (2005) for a lucid presentation of GMM estimation. Some of the earliest work on simultaneous systems can be found in Cowles Commission monographs—Koopmans and Marschak (1950) and Koopmans and Hood (1953)—with the first development of 2SLS appearing in Theil (1953) and Basmann (1957). However, Stock and Watson (2003, 334–337) present an example of the method of instrumental variables that was first published in 1928 by Philip Wright.

The syntax for `ivregress` assumes that you want to fit one equation from a system of equations or an equation for which you do not want to specify the functional form for the remaining equations of the system. To fit a full system of equations, using either 2SLS equation-by-equation or three-stage least squares, see [R] **reg3**. An advantage of `ivregress` is that you can fit one equation of a multiple-equation system without specifying the functional form of the remaining equations.

Formally, the model fitted by `ivregress` is

$$y_i = \mathbf{y}_i \beta_1 + \mathbf{x}_{1i} \beta_2 + u_i \tag{1}$$

$$\mathbf{y}_i = \mathbf{x}_{1i} \mathbf{\Pi}_1 + \mathbf{x}_{2i} \mathbf{\Pi}_2 + \mathbf{v}_i \tag{2}$$

Here y_i is the dependent variable for the ith observation, $\mathbf{y}_i$ represents the endogenous regressors ($varlist_2$ in the syntax diagram), $\mathbf{x}_{1i}$ represents the included exogenous regressors ($varlist_1$ in the syntax diagram), and $\mathbf{x}_{2i}$ represents the excluded exogenous regressors ($varlist_{\text{iv}}$ in the syntax diagram). $\mathbf{x}_{1i}$ and $\mathbf{x}_{2i}$ are collectively called the instruments. u_i and $\mathbf{v}_i$ are zero-mean error terms, and the correlations between u_i and the elements of $\mathbf{v}_i$ are presumably nonzero.

The rest of the discussion is presented under the following headings:

2SLS and LIML estimators
GMM estimator

2SLS and LIML estimators

The most common instrumental-variables estimator is 2SLS.

▷ Example 1: 2SLS estimator

We have state data from the 1980 census on the median dollar value of owner-occupied housing (**hsngval**) and the median monthly gross rent (**rent**). We want to model **rent** as a function of **hsngval** and the percentage of the population living in urban areas (**pcturban**):

$$\texttt{rent}_i = \beta_0 + \beta_1 \texttt{hsngval}_i + \beta_2 \texttt{pcturban}_i + u_i$$

where i indexes states and u_i is an error term.

Because random shocks that affect rental rates in a state probably also affect housing values, we treat **hsngval** as endogenous. We believe that the correlation between **hsngval** and u is not equal to zero. On the other hand, we have no reason to believe that the correlation between **pcturban** and u is nonzero, so we assume that **pcturban** is exogenous.

Since we are treating **hsngval** as an endogenous regressor, we must have one or more additional variables available that are correlated with **hsngval** but uncorrelated with u. Moreover, these excluded exogenous variables must not affect **rent** directly, because if they do then they should be included in the regression equation we specified above. In our dataset, we have a variable for family income (**faminc**) and dummies for region of the country (**reg2** through **reg4**) that we believe are correlated with **hsngval** but not the error term. Together, **pcturban**, **faminc**, **reg2**, **reg3**, and **reg4** constitute our set of instruments.

To fit the equation in Stata, we specify the dependent variable and the list of included exogenous variables. In parentheses, we specify the endogenous regressors, an equals sign, and the excluded exogenous variables. Only the additional exogenous variables must be specified to the right of the equals sign; the exogenous variables that appear in the regression equation are automatically included as instruments.

Here we fit our model with the 2SLS estimator:

```
. use http://www.stata-press.com/data/r10/hsng2
(1980 Census housing data)

. ivregress 2sls rent pcturban (hsngval = faminc reg2-reg4)

Instrumental variables (2SLS) regression               Number of obs =      50
                                                       Wald chi2(2)  =   90.76
                                                       Prob > chi2   =  0.0000
                                                       R-squared     =  0.5989
                                                       Root MSE      =  22.166

------------------------------------------------------------------------------
        rent |      Coef.   Std. Err.      z    P>|z|     [95% Conf. Interval]
-------------+----------------------------------------------------------------
     hsngval |   .0022398   .0003284     6.82   0.000     .0015961    .0028836
    pcturban |    .081516   .2987652     0.27   0.785     -.504053     .667085
       _cons |   120.7065   15.22839     7.93   0.000     90.85942    150.5536
------------------------------------------------------------------------------
Instrumented:  hsngval
Instruments:   pcturban faminc reg2 reg3 reg4
```

As we would expect, states with higher housing values have higher rental rates. The proportion of a state's population that is urban does not have a significant effect on rents.

◁

❑ Technical Note

In a simultaneous equations framework, we could write the model we just fitted as

$$\texttt{hsngval}_i = \pi_0 + \pi_1\texttt{faminc}_i + \pi_2\texttt{reg2}_i + \pi_3\texttt{reg3}_i + \pi_4\texttt{reg4}_i + v_i$$
$$\texttt{rent}_i = \beta_0 + \beta_1\texttt{hsngval}_i + \beta_2\texttt{pcturban}_i + u_i$$

which here happens to be recursive (triangular), because `hsngval` appears in the equation for `rent` but `rent` does not appear in the equation for `hsngval`. In general, however, systems of simultaneous equations are not recursive. Because this system is recursive, we could fit the two equations individually via OLS if we were willing to assume that u and v were independent. For a more detailed discussion of triangular systems, see Kmenta (1997, 719–720).

Historically, instrumental-variables estimation and systems of simultaneous equations were taught concurrently, and older textbooks describe instrumental-variables estimation solely in the context of simultaneous equations. However, in recent decades the treatment of endogeneity and instrumental-variables estimation has taken on a much broader scope, while interest in the specification of complete systems of simultaneous equations has waned. Most recent textbooks, such as Cameron and Trivedi (2005), Davidson and MacKinnon (1993, 2004), and Wooldridge (2002a,b), treat instrumental-variables estimation as an integral part of the modern economists' toolkit and introduce it long before shorter discussions on simultaneous equations.

❑

In addition to the 2SLS member of the κ-class estimators, `ivregress` implements the LIML estimator. Both theoretical and Monte Carlo exercises indicate that the LIML estimator may yield less bias and confidence intervals with better coverage rates than the 2SLS estimator. See Poi (2006) and Stock, Wright, and Yogo (2002) (and the papers cited therein) for Monte Carlo evidence.

▷ Example 2: LIML estimator

Here we refit our model with the LIML estimator:

```
. ivregress liml rent pcturban (hsngval = faminc reg2-reg4)
Instrumental variables (LIML) regression               Number of obs =        50
                                                       Wald chi2(2)  =     75.71
                                                       Prob > chi2   =    0.0000
                                                       R-squared     =    0.4901
                                                       Root MSE      =    24.992

------------------------------------------------------------------------------
        rent |      Coef.   Std. Err.      z    P>|z|     [95% Conf. Interval]
-------------+----------------------------------------------------------------
     hsngval |   .0026686   .0004173     6.39   0.000     .0018507    .0034865
    pcturban |  -.1827391   .3571132    -0.51   0.609    -.8826681    .5171899
       _cons |   117.6087   17.22625     6.83   0.000     83.84587    151.3715
------------------------------------------------------------------------------
Instrumented:  hsngval
Instruments:   pcturban faminc reg2 reg3 reg4
```

These results are qualitatively similar to the 2SLS results, although the coefficient on `hsngval` is about 19% higher.

◁

GMM estimator

Since the celebrated paper of Hansen (1982), the GMM has been a popular method of estimation in economics and finance, and it lends itself well to instrumental-variables estimation. The basic principle is that we have some *moment* or *orthogonality* conditions of the form

$$E(\mathbf{z}_i u_i) = \mathbf{0} \tag{3}$$

From (1), we have $u_i = y_i - \mathbf{y}_i\boldsymbol{\beta}_1 - \mathbf{x}_{1i}\boldsymbol{\beta}_2$. What are the elements of the instrument vector $\mathbf{z}_i$? By assumption, $\mathbf{x}_{1i}$ is uncorrelated with u_i, as are the excluded exogenous variables $\mathbf{x}_{2i}$, and so we use $\mathbf{z}_i = [\mathbf{x}_{1i}\ \mathbf{x}_{2i}]$. The moment conditions are simply the mathematical representation of the assumption that the instruments are exogenous—that is, the instruments are orthogonal to (uncorrelated with) u_i.

If the number of elements in $\mathbf{z}_i$ is just equal to the number of unknown parameters, then we can apply the analogy principle to (3) and solve

$$\frac{1}{N}\sum_i \mathbf{z}_i u_i = \frac{1}{N}\sum_i \mathbf{z}_i\left(y_i - \mathbf{y}_i\boldsymbol{\beta}_1 - \mathbf{x}_{1i}\boldsymbol{\beta}_2\right) = \mathbf{0} \tag{4}$$

This equation is known as the method-of-moments estimator. Here where the number of instruments equals the number of parameters, the method-of-moments estimator coincides with the 2SLS estimator, which also coincides with what has historically been called the indirect least-squares estimator (Judge et al. 1985, 595).

The "generalized" in GMM addresses the case in which the number of instruments (columns of $\mathbf{z}_i$) exceeds the number of parameters to be estimated. Here there is no unique solution to the population moment conditions defined in (3), so we cannot use (4). Instead, we define the objective function

$$Q(\boldsymbol{\beta}_1, \boldsymbol{\beta}_2) = \left(\frac{1}{N}\sum_i \mathbf{z}_i u_i\right)' \mathbf{W} \left(\frac{1}{N}\sum_i \mathbf{z}_i u_i\right) \tag{5}$$

where $\mathbf{W}$ is a positive-definite matrix with the same number of rows and columns as the number of columns of $\mathbf{z}_i$. $\mathbf{W}$ is known as the weighting matrix, and we specify its structure with the `wmatrix()` option. The GMM estimator of $(\boldsymbol{\beta}_1, \boldsymbol{\beta}_2)$ minimizes $Q(\boldsymbol{\beta}_1, \boldsymbol{\beta}_2)$; that is, the GMM estimator chooses $\boldsymbol{\beta}_1$ and $\boldsymbol{\beta}_2$ to make the moment conditions as close to zero as possible for a given $\mathbf{W}$.

A well-known result is that if we define the matrix $\mathbf{S}_0$ to be the covariance of $\mathbf{z}_i u_i$ and set $\mathbf{W} = \mathbf{S}_0^{-1}$, then we obtain the optimal two-step GMM estimator, where by optimal estimator we mean the one that results in the smallest variance given the moment conditions defined in (3).

Suppose that the errors u_i are heteroskedastic but independent among observations. Then

$$\mathbf{S}_0 = E(\mathbf{z}_i u_i u_i \mathbf{z}_i') = E(u_i^2 \mathbf{z}_i \mathbf{z}_i')$$

and the sample analogue is

$$\widehat{\mathbf{S}} = \frac{1}{N}\sum_i \widehat{u}_i^2 \mathbf{z}_i \mathbf{z}_i' \tag{6}$$

To implement this estimator, we need estimates of the sample residuals $\widehat{u}_i$. `ivregress gmm` obtains the residuals by estimating $\boldsymbol{\beta}_1$ and $\boldsymbol{\beta}_2$ by 2SLS and then evaluates (6) and sets $\mathbf{W} = \widehat{\mathbf{S}}^{-1}$. Equation (6) is the same as the center term of the "sandwich" robust covariance matrix available from most Stata estimation commands through the `robust` option.

▷ Example 3: GMM estimator

Here we refit our model of rents by using the GMM estimator, allowing for heteroskedasticity in u_i:

```
. ivregress gmm rent pcturban (hsngval = faminc reg2-reg4), wmatrix(robust)
Instrumental variables (GMM) regression               Number of obs =       50
                                                      Wald chi2(2)  =   112.09
                                                      Prob > chi2   =   0.0000
                                                      R-squared     =   0.6616
GMM weight matrix: Robust                             Root MSE      =   20.358

------------------------------------------------------------------------------
             |               Robust
        rent |      Coef.   Std. Err.      z    P>|z|     [95% Conf. Interval]
-------------+----------------------------------------------------------------
     hsngval |   .0014643   .0004473     3.27   0.001     .0005877     .002341
    pcturban |   .7615482   .2895105     2.63   0.009     .1941181    1.328978
       _cons |   112.1227   10.80234    10.38   0.000     90.95052    133.2949
------------------------------------------------------------------------------
Instrumented:  hsngval
Instruments:   pcturban faminc reg2 reg3 reg4
```

Since we requested that a heteroskedasticity-consistent weighting matrix be used during estimation but did not specify the `vce()` option, `ivregress` reported standard errors that are robust to heteroskedasticity. Had we specified `vce(unadjusted)`, we would have obtained standard errors that would be correct only if the weighting matrix $\mathbf{W}$ does in fact converge to $\mathbf{S}_0^{-1}$.

◁

❑ Technical Note

Many software packages that implement GMM estimation use the same heteroskedasticity-consistent weighting matrix we used in the previous example to obtain the optimal two-step estimates but do not use a heteroskedasticity-consistent VCE, even though they may label the standard errors as being "robust". To replicate results obtained from other packages, you may have to use the `vce(unadjusted)` option. See *Methods and Formulas* below for a discussion of robust covariance matrix estimation in the GMM framework.

❑

By changing our definition of $\mathbf{S}_0$, we can obtain GMM estimators suitable for use with other types of data that violate the assumption that the errors are independently and identically distributed. For example, you may have a dataset that consists of multiple observations for each person in a sample. The observations that correspond to the same person are likely to be correlated, and the estimation technique should account for that lack of independence. Say that in your dataset people are identified by the variable `personid` and you type

```
. ivregress gmm ..., wmatrix(cluster personid)
```

Here `ivregress` estimates $\mathbf{S}_0$ as

$$\widehat{\mathbf{S}} = \frac{1}{N}\sum_{c\in C}\mathbf{q}_c\mathbf{q}_c'$$

where C denotes the set of clusters and

$$\mathbf{q}_c = \sum_{i\in c_j}\widehat{u}_i\mathbf{z}_i$$

where c_j denotes the jth cluster. This weighting matrix accounts for the within-person correlation among observations, so the GMM estimator that uses this version of $\mathbf{S}_0$ will be more efficient than the estimator that ignores this correlation.

▷ Example 4: GMM estimator with clustering

We have data from the National Longitudinal Survey on young womens' wages as reported in a series of interviews from 1968 through 1988, and we want to fit a model of wages as a function of each woman's age and age squared, job tenure, birth year, and level of education. We believe that random shocks that affect a woman's wage also affect her job tenure, so we treat tenure as endogenous. As additional instruments, we use her union status, number of weeks worked in the past year, and a dummy indicating whether she lives in a metropolitan area. Because we have several observations for each woman (corresponding to interviews done over several years), we want to control for clustering on each person.

```
. use http://www.stata-press.com/data/r10/nlswork
(National Longitudinal Survey.  Young Women 14-26 years of age in 1968)

. generate agesq = age^2
(24 missing values generated)

. ivregress gmm ln_wage age agesq birth_yr grade (tenure = union wks_work msp),
> wmatrix(cluster idcode)

Instrumental variables (GMM) regression              Number of obs =    18625
                                                     Wald chi2(5)  =  1807.17
                                                     Prob > chi2   =   0.0000
                                                     R-squared     =        .
GMM weight matrix: Cluster (idcode)                  Root MSE      =   .46951

                             (Std. Err. adjusted for 4110 clusters in idcode)
------------------------------------------------------------------------------
             |               Robust
     ln_wage |      Coef.   Std. Err.      z    P>|z|     [95% Conf. Interval]
-------------+----------------------------------------------------------------
      tenure |    .099221   .0037764    26.27   0.000     .0918194    .1066227
         age |   .0171146   .0066895     2.56   0.011     .0040034    .0302259
       agesq |  -.0005191    .000111    -4.68   0.000    -.0007366   -.0003016
    birth_yr |  -.0085994   .0021932    -3.92   0.000     -.012898   -.0043008
       grade |    .071574   .0029938    23.91   0.000     .0657062    .0774417
       _cons |   .8575071   .1616274     5.31   0.000     .5407231    1.174291
------------------------------------------------------------------------------
Instrumented:  tenure
Instruments:   age agesq birth_yr grade union wks_work msp
```

Both job tenure and years of schooling have significant positive effects on wages.

◁

Time-series data are often plagued by serial correlation. In these cases we can construct a weighting matrix to account for the fact that the error in period t is probably correlated with the errors in periods $t-1$, $t-2$, etc. A HAC weighting matrix can be used to account for both serial correlation and potential heteroskedasticity.

To request a HAC weighting matrix, you specify the option `wmatrix(hac` *kernel* [#|`opt`]`)`. *kernel* specifies which of three kernels to use: `bartlett`, `parzen`, or `quadraticspectral`. *kernel* determines the amount of weight given to lagged values when computing the HAC matrix, and # denotes the maximum number of lags to use. Many texts refer to the bandwidth of the kernel instead of the number of lags; the bandwidth is equal to the number of lags plus one. If neither `opt` nor # is specified, then $N-2$ lags are used, where N is the sample size.

If you specify `wmatrix(hac` *kernel* `opt)`, then `ivregress` uses Newey and West's (1994) algorithm for automatically selecting the number of lags to use. Although the authors' Monte Carlo simulations do show that the procedure may result in size distortions of hypothesis tests, the procedure is still useful when little other information is available to help choose the number of lags.

For more on GMM estimation, see Baum (2006); Baum, Schaffer, and Stillman (2003); Cameron and Trivedi (2005); Davidson and MacKinnon (1993, 2004); Hayashi (2000); or Wooldridge (2002b). See Newey and West (1987) for an introduction to HAC covariance matrix estimation.

Saved Results

`ivregress` saves the following in `e()`:

Scalars			
`e(N)`	number of observations	`e(r2)`	R^2
`e(mss)`	model sum of squares	`e(r2_a)`	adjusted R^2
`e(df_m)`	model degrees of freedom	`e(F)`	F statistic
`e(rss)`	residual sum of squares	`e(rmse)`	root mean squared error
`e(df_r)`	residual degrees of freedom	`e(N_clust)`	number of clusters
`e(chi2)`	χ^2 statistic	`e(iterations)`	number of GMM iterations (`0` if not applicable)
`e(J)`	value of GMM objective function		
`e(vcelagopt)`	lags used in HAC VCE matrix (if Newey–West algorithm used)	`e(wlagopt)`	lags used in HAC weight matrix (if Newey–West algorithm used)
`e(kappa)`	κ used in LIML estimator		
Macros			
`e(cmd)`	`ivregress`	`e(estimator)`	estimator used
`e(cmdline)`	command as typed	`e(vce)`	*vcetype* specified in `vce()`
`e(depvar)`	name of dependent variable	`e(vcetype)`	title used to label Std. Err.
`e(instd)`	instrumented variable	`e(small)`	`small` if small-sample statistics
`e(insts)`	instruments	`e(footnote)`	`ivreg_footnote`
`e(wtype)`	weight type	`e(properties)`	`b V`
`e(wexp)`	weight expression	`e(predict)`	program used to implement `predict`
`e(title)`	title in estimation output		
`e(wmatrix)`	*wmtype* specified in `wmatrix()`	`e(estat_cmd)`	program used to implement `estat`
`e(clustvar)`	name of cluster variable		
Matrices			
`e(b)`	coefficient vector	`e(V)`	variance–covariance matrix of the estimators
`e(W)`	weight matrix used to compute GMM estimates		
`e(S)`	moment covariance matrix used to compute GMM variance–covariance matrix		
Functions			
`e(sample)`	marks estimation sample		

Methods and Formulas

`ivregress` is implemented as an ado-file.

Items printed in lowercase and not boldfaced (e.g., x) are scalars. Items printed in lowercase and boldfaced (e.g., $\mathbf{x}$) are vectors. Items printed in uppercase and boldfaced (e.g., $\mathbf{X}$) are matrices.

The model is

$$\mathbf{y} = \mathbf{Y}\boldsymbol{\beta}_1 + \mathbf{X}_1\boldsymbol{\beta}_2 + \mathbf{u} = \mathbf{X}\boldsymbol{\beta} + \mathbf{u}$$

$$\mathbf{Y} = \mathbf{X}_1\boldsymbol{\Pi}_1 + \mathbf{X}_2\boldsymbol{\Pi}_2 + \mathbf{v} = \mathbf{Z}\boldsymbol{\Pi} + \mathbf{V}$$

where $\mathbf{y}$ is an $N \times 1$ vector of the left-hand-side variable; N is the sample size; $\mathbf{Y}$ is an $N \times p$ matrix of p endogenous regressors; $\mathbf{X}_1$ is an $N \times k_1$ matrix of k_1 included exogenous regressors; $\mathbf{X}_2$ is an $N \times k_2$ matrix of k_2 excluded exogenous variables, $\mathbf{X} = [\mathbf{Y}\ \mathbf{X}_1]$, $\mathbf{Z} = [\mathbf{X}_1\ \mathbf{X}_2]$; $\mathbf{u}$ is an $N \times 1$ vector of errors; $\mathbf{V}$ is an $N \times p$ matrix of errors; $\boldsymbol{\beta} = [\boldsymbol{\beta}_1\ \boldsymbol{\beta}_2]$ is a $k = (p + k_1) \times 1$ vector of parameters; and $\boldsymbol{\Pi}$ is a $(k_1 + k_2) \times p$ vector of parameters. If a constant term is included in the model, then one column of $\mathbf{X}_1$ contains all ones.

Let $\mathbf{v}$ be a column vector of weights specified by the user. If no weights are specified, $\mathbf{v} = \mathbf{1}$. Let $\mathbf{w}$ be a column vector of normalized weights. If no weights are specified or if the user specified `fweights` or `iweights`, $\mathbf{w} = \mathbf{v}$; otherwise, $\mathbf{w} = \left\{\mathbf{v}/(\mathbf{1}'\mathbf{v})\right\}(\mathbf{1}'\mathbf{1})$. Let $\mathbf{D}$ denote the $N \times N$ matrix with $\mathbf{w}$ on the main diagonal and zeros elsewhere. If no weights are specified, $\mathbf{D}$ is the identity matrix.

The weighted number of observations n is defined as $\mathbf{1}'\mathbf{w}$. For `iweights`, this is truncated to an integer. The *sum of the weights* is $\mathbf{1}'\mathbf{v}$. Define $c = 1$ if there is a constant in the regression and zero otherwise.

The order condition for identification requires that $k_2 \geq p$: the number of excluded exogenous variables must be at least as great as the number of endogenous regressors.

In the following formulas, if weights are specified, $\mathbf{X}_1'\mathbf{X}_1$, $\mathbf{X}'\mathbf{X}$, $\mathbf{X}'\mathbf{y}$, $\mathbf{y}'\mathbf{y}$, $\mathbf{Z}'\mathbf{Z}$, $\mathbf{Z}'\mathbf{X}$, and $\mathbf{Z}'\mathbf{y}$ are replaced with $\mathbf{X}_1'\mathbf{D}\mathbf{X}_1$, $\mathbf{X}'\mathbf{D}\mathbf{X}$, $\mathbf{X}'\mathbf{D}\mathbf{y}$, $\mathbf{y}'\mathbf{D}\mathbf{y}$, $\mathbf{Z}'\mathbf{D}\mathbf{Z}$, $\mathbf{Z}'\mathbf{D}\mathbf{X}$, and $\mathbf{Z}'\mathbf{D}\mathbf{y}$, respectively. We suppress the $\mathbf{D}$ below to simplify the notation.

2SLS and LIML estimators

Define the κ-class estimator of $\boldsymbol{\beta}$ as

$$\mathbf{b} = \left\{\mathbf{X}'(\mathbf{I} - \kappa\mathbf{M}_\mathbf{Z})^{-1}\mathbf{X}\right\}^{-1}\mathbf{X}'(\mathbf{I} - \kappa\mathbf{M}_\mathbf{Z})^{-1}\mathbf{y}$$

where $\mathbf{M}_\mathbf{Z} = \mathbf{I} - \mathbf{Z}(\mathbf{Z}'\mathbf{Z})^{-1}\mathbf{Z}'$. The 2SLS estimator results from setting $\kappa = 1$. The LIML estimator results from selecting κ to be the minimum eigenvalue of $(\mathbf{Y}'\mathbf{M}_\mathbf{Z}\mathbf{Y})^{-1/2}\mathbf{Y}'\mathbf{M}_{\mathbf{X}_1}\mathbf{Y}(\mathbf{Y}'\mathbf{M}_\mathbf{Z}\mathbf{Y})^{-1/2}$, where $\mathbf{M}_{\mathbf{X}_1} = \mathbf{I} - \mathbf{X}_1(\mathbf{X}_1'\mathbf{X}_1)^{-1}\mathbf{X}_1'$.

The total sum of squares, TSS, equals $\mathbf{y}'\mathbf{y}$ if there is no intercept and $\mathbf{y}'\mathbf{y} - \left\{(\mathbf{1}'\mathbf{y})^2/n\right\}$ otherwise. The degrees of freedom are $n-c$. The error sum of squares, ESS, is defined as $\mathbf{y}'\mathbf{y} - 2\mathbf{b}\mathbf{X}'\mathbf{y} + \mathbf{b}'\mathbf{X}'\mathbf{X}\mathbf{b}$. The model sum of squares, MSS, equals TSS − ESS. The degrees of freedom are $k - c$.

The mean squared error, s^2, is defined as $\text{ESS}/(n-k)$ if `small` is specified and ESS/n otherwise. The root mean squared error is s, its square root.

If $c = 1$ and `small` is not specified, a Wald statistic W of the joint significance of the $k - 1$ parameters of $\boldsymbol{\beta}$ except the constant term is calculated; $W \sim \chi^2(k-1)$. If $c = 1$ and `small` is specified, then an F statistic is calculated as $F = W/(k-1)$; $F \sim F(k-1, n-k)$.

The R-squared is defined as $R^2 = 1 - \text{ESS}/\text{TSS}$.

The adjusted R-squared is $R_\text{a}^2 = 1 - (1 - R^2)(n-c)/(n-k)$.

If robust is not specified, then $\text{Var}(\mathbf{b}) = s^2\{\mathbf{X}'(\mathbf{I} - \kappa\mathbf{M_Z})^{-1}\mathbf{X}\}^{-1}$. For a discussion of robust variance estimates in regression and regression with instrumental variables, see *Methods and Formulas* in [R] **regress**. If small is not specified, then $k = 0$ in the formulas given there.

GMM estimator

We obtain an initial consistent estimate of $\boldsymbol{\beta}$ by using the 2SLS estimator; see above. Using this estimate of $\boldsymbol{\beta}$, we compute the weighting matrix $\mathbf{W}$ and calculate the GMM estimator

$$\mathbf{b}_{\text{GMM}} = \{\mathbf{X}'\mathbf{Z}\mathbf{W}\mathbf{Z}'\mathbf{X}\}^{-1}\mathbf{X}'\mathbf{Z}\mathbf{W}\mathbf{Z}'\mathbf{y}$$

The variance of $\mathbf{b}_{\text{GMM}}$ is

$$\text{Var}(\mathbf{b}_{\text{GMM}}) = n\{\mathbf{X}'\mathbf{Z}\mathbf{W}\mathbf{Z}'\mathbf{X}\}^{-1}\mathbf{X}'\mathbf{Z}\mathbf{W}\widehat{\mathbf{S}}\mathbf{W}\mathbf{Z}'\mathbf{X}\{\mathbf{X}'\mathbf{Z}\mathbf{W}\mathbf{Z}'\mathbf{X}\}^{-1}$$

$\text{Var}(\mathbf{b}_{\text{GMM}})$ is of the sandwich form $\mathbf{DMD}$; see [P] **_robust**. If the user specifies the small option, ivregress implements a small-sample adjustment by multiplying the VCE by $N/(N-k)$.

If vce(unadjusted) is specified, then we set $\widehat{\mathbf{S}} = \mathbf{W}^{-1}$ and the VCE reduces to the "optimal" GMM variance estimator

$$\text{Var}(\boldsymbol{\beta}_{\text{GMM}}) = n\{\mathbf{X}'\mathbf{Z}\mathbf{W}\mathbf{Z}'\mathbf{X}\}^{-1}$$

However, if $\mathbf{W}^{-1}$ is not a good estimator of $E(\mathbf{z}_i u_i u_i \mathbf{z}_i')$, then the optimal GMM estimator is inefficient, and inference based on the optimal variance estimator could be misleading.

$\mathbf{W}$ is calculated using the residuals from the initial 2SLS estimates, whereas $\mathbf{S}$ is estimated using the residuals based on $\mathbf{b}_{\text{GMM}}$. The wmatrix() option affects the form of $\mathbf{W}$, whereas the vce() option affects the form of $\mathbf{S}$. Except for different residuals being used, the formulas for $\mathbf{W}^{-1}$ and $\mathbf{S}$ are identical, so we focus on estimating $\mathbf{W}^{-1}$.

If wmatrix(unadjusted) is specified, then

$$\mathbf{W}^{-1} = \frac{s^2}{n}\sum_i \mathbf{z}_i\mathbf{z}_i'$$

where $s^2 = \sum_i u_i^2/n$. This weight matrix is appropriate if the errors are homoskedastic.

If wmatrix(robust) is specified, then

$$\mathbf{W}^{-1} = \frac{1}{n}\sum_i u_i^2\mathbf{z}_i\mathbf{z}_i'$$

which is appropriate if the errors are heteroskedastic.

If wmatrix(cluster *clustvar*) is specified, then

$$\mathbf{W}^{-1} = \frac{1}{n}\sum_c \mathbf{q}_c\mathbf{q}_c'$$

where c indexes clusters,

$$\mathbf{q}_c = \sum_{i\in c_j} u_i\mathbf{z}_i$$

and c_j denotes the jth cluster.

If `wmatrix(hac` *kernel* [#]`)` is specified, then

$$\mathbf{W}^{-1} = \frac{1}{n}\sum_i u_i^2 \mathbf{z}_i\mathbf{z}_i' + \frac{1}{n}\sum_{l=1}^{l=n-1}\sum_{i=l+1}^{i=n} K(l,m)u_i u_{i-l}\left(\mathbf{z}_i\mathbf{z}_{i-l}' + \mathbf{z}_{i-l}\mathbf{z}_i'\right)$$

where $m = \#$ if # is specified and $m = n-2$ otherwise. Define $z = l/(m+1)$. If *kernel* is `nwest`, then

$$K(l,m) = \begin{cases} 1-z & 0 \le z \le 1 \\ 0 & \text{otherwise} \end{cases}$$

If *kernel* is `gallant`, then

$$K(l,m) = \begin{cases} 1-6z^2+6z^3 & 0 \le z \le 0.5 \\ 2(1-z)^3 & 0.5 < z \le 1 \\ 0 & \text{otherwise} \end{cases}$$

If *kernel* is `quadraticspectral`, then

$$K(l,m) = \begin{cases} 1 & z = 0 \\ 3\left\{\sin(\theta)/\theta - \cos(\theta)\right\}/\theta^2 & \text{otherwise} \end{cases}$$

where $\theta = 6\pi z/5$.

If `wmatrix(hac` *kernel* `opt)` is specified, then `ivregress` uses Newey and West's (1994) automatic lag-selection algorithm, which proceeds as follows. Define $\mathbf{h}$ to be a $(k_1+k_2)\times 1$ vector containing ones in all rows except for the row corresponding to the constant term (if present); that row contains a zero. Define

$$f_i = (u_i\mathbf{z}_i)\mathbf{h}$$

$$\widehat{\sigma}_j = \frac{1}{n}\sum_{i=j+1}^{n} f_i f_{i-j} \qquad j = 0,\dots,m^*$$

$$\widehat{s}^{(q)} = 2\sum_{j=1}^{m^*}\widehat{\sigma}_j j^q$$

$$\widehat{s}^{(0)} = \widehat{\sigma}_0 + 2\sum_{j=1}^{m^*}\widehat{\sigma}_j$$

$$\widehat{\gamma} = c_\gamma\left\{\left(\frac{\widehat{s}^{(q)}}{\widehat{s}^{(0)}}\right)^2\right\}^{1/2q+1}$$

$$m = \widehat{\gamma}n^{1/(2q+1)}$$

where q, m^*, and c_γ depend on the kernel specified:

Kernel	q	m^*	c_γ
Bartlett	1	$\text{int}\left\{20(T/100)^{2/9}\right\}$	1.4117
Parzen	2	$\text{int}\left\{20(T/100)^{4/25}\right\}$	2.6614
Quadratic spectral	2	$\text{int}\left\{20(T/100)^{2/25}\right\}$	1.3221

where $\text{int}(x)$ denotes the integer obtained by truncating x toward zero. For the Bartlett and Parzen kernels, the optimal lag is $\min\{\text{int}(m), m^*\}$. For the quadratic spectral, the optimal lag is $\min\{m, m^*\}$.

If `center` is specified, when computing weighting matrices `ivregress` replaces the term $u_i z_i$ in the formulas above with $u_i\mathbf{z}_i - \overline{u\mathbf{z}}$, where $\overline{u\mathbf{z}} = \sum_i u_i\mathbf{z}_i/N$.

References

Andrews, R. W. K. 1991. Heteroscedasticity and autocorrelation consistent covariance matrix estimation. *Econometrica* 59: 817–858.

Basmann, R. L. 1957. A generalized classical method of linear estimation of coefficients in a structural equation. *Econometrica* 25: 77–83.

Baum, C. F. 2006. *An Introduction to Modern Econometrics Using Stata*. College Station, TX: Stata Press.

Baum, C. F., M. E. Schaffer, and S. Stillman. 2003. Instrumental variables and GMM: Estimation and testing. *Stata Journal* 3: 1–31.

Cameron, A. C., and P. K. Trivedi. 2005. *Microeconometrics: Methods and Applications*. New York: Cambridge University Press.

Davidson, R., and J. G. MacKinnon. 1993. *Estimation and Inference in Econometrics*. New York: Oxford University Press.

——. 2004. *Econometric Theory and Methods*. New York: Oxford University Press.

Gallant, A. R. 1987. *Nonlinear Statistical Models*. New York: Wiley.

Greene, W. H. 2003. *Econometric Analysis*. 5th ed. Upper Saddle River, NJ: Prentice Hall.

Hall, A. R. 2005. *Generalized Method of Moments*. Oxford: Oxford University Press.

Hansen, L. P. 1982. Large sample properties of generalized method of moments estimators. *Econometrica* 50: 1029–1054.

Hayashi, F. 2000. *Econometrics*. Princeton, NJ: Princeton University Press.

Judge, G. G., W. E. Griffiths, R. C. Hill, H. Lütkepohl, and T.-C. Lee. 1985. *The Theory and Practice of Econometrics*. 2nd ed. New York: Wiley.

Kmenta, J. 1997. *Elements of Econometrics*. 2nd ed. Ann Arbor: University of Michigan Press.

Koopmans, T. C., and W. C. Hood. 1953. *Studies in Econometric Method*. New York: Wiley.

Koopmans, T. C., and J. Marschak. 1950. *Statistical Inference in Dynamic Economic Models*. New York: Wiley.

Newey, W. K., and K. D. West. 1987. A simple positive semi-definite heteroscedasticity and autocorrelation consistent covariance matrix. *Econometrica* 55: 703–708.

——. 1994. Automatic lag selection in covariance matrix estimation. *Review of Economic Studies* 61: 631–653.

Poi, B. P. 2006. Jackknife instrumental variables estimation in Stata. *Stata Journal* 6: 364–376.

Stock, J. H., and M. W. Watson. 2003. *Introduction to Econometrics*. Boston: Addison–Wesley.

Stock, J. H., J. H. Wright, and M. Yogo. 2002. A survey of weak instruments and weak identification in generalized method of moments. *Journal of Business and Economic Statistics* 20: 518–529.

Theil, H. 1953. *Repeated Least Squares Applied to Complete Equation Systems*. Mimeograph from the Central Planning Bureau, The Hague.

Wooldridge, J. M. 2002a. *Introductory Econometrics: A Modern Approach.* 2nd ed. Cincinnati, OH: South-Western.

——. 2002b. *Econometric Analysis of Cross Section and Panel Data.* Cambridge, MA: MIT Press.

Wright, P. G. 1928. *The Tariff on Animal and Vegetable Oils.* New York: Macmillan.

Also See

[R] **ivregress postestimation** — Postestimation tools for ivregress

[R] **ivprobit** — Probit model with endogenous regressors

[R] **ivtobit** — Tobit model with endogenous regressors

[R] **regress** — Linear regression

[XT] **xtivreg** — Instrumental variables and two-stage least squares for panel-data models

[U] **20 Estimation and postestimation commands**

Title

ivregress postestimation — Postestimation tools for ivregress

Description

The following postestimation commands are of special interest after ivregress:

command	description
estat firststage	report "first-stage" regression statistics
estat overid	perform tests of overidentifying restrictions

These commands are not appropriate after the svy prefix.

For information about these commands, see below.

The following postestimation commands are also available:

command	description
adjust[1]	adjusted predictions of $\mathbf{x}\beta$ or $\exp(\mathbf{x}\beta)$
estat	VCE and estimation sample summary
estat (svy)	postestimation statistics for survey data
estimates	cataloging estimation results
hausman	Hausman's specification test
lincom	point estimates, standard errors, testing, and inference for linear combinations of coefficients
mfx	marginal effects or elasticities
nlcom	point estimates, standard errors, testing, and inference for nonlinear combinations of coefficients
predict	predictions, residuals, influence statistics, and other diagnostic measures
predictnl	point estimates, standard errors, testing, and inference for generalized predictions
test	Wald tests for simple and composite linear hypotheses
testnl	Wald tests of nonlinear hypotheses

[1] adjust is not appropriate with time-series operators.

See the corresponding entries in the *Stata Base Reference Manual* for details, but see [SVY] **estat** for details about estat (svy).

Special-interest postestimation commands

estat firststage reports various statistics that measure the relevance of the excluded exogenous variables. By default, whether the equation has one or more than one endogenous regressor determines what statistics are reported.

estat overid performs tests of overidentifying restrictions. If the 2SLS estimator was used, Sargan's (1958) and Basmann's (1960) χ^2 tests are reported, as is Wooldridge's (1995) robust score test; if the LIML estimator was used, Anderson and Rubin's (1950) χ^2 test and Basmann's F test are reported; and if the GMM estimator was used, Hansen's (1982) J statistic χ^2 test is reported. A statistically significant test statistic always indicates that the instruments may not be valid.

Syntax for predict

```
predict [type] newvar [if] [in] [, statistic]

predict [type] {stub* | newvarlist} [if] [in], scores
```

statistic	description
Main	
xb	linear prediction; the default
residuals	residuals
stdp	standard error of the prediction
stdf	standard error of the forecast
pr(*a*,*b*)	$\Pr(a < y_j < b)$
e(*a*,*b*)	$E(y_j \mid a < y_j < b)$
ystar(*a*,*b*)	$E(y_j^*)$, $y_j^* = \max\{a, \min(y_j, b)\}$

These statistics are available both in and out of sample; type `predict ... if e(sample) ...` if wanted only for the estimation sample.

`stdf` is not allowed with `svy` estimation results.

where *a* and *b* may be numbers or variables; *a* missing ($a \geq .$) means $-\infty$, and *b* missing ($b \geq .$) means $+\infty$; see [U] **12.2.1 Missing values**.

Options for predict

Main

`xb`, the default, calculates the linear prediction.

`residuals` calculates the residuals, that is, $y_j - \mathbf{x}_j\mathbf{b}$. These are based on the estimated equation when the observed values of the endogenous variables are used—not the projections of the instruments onto the endogenous variables.

`stdp` calculates the standard error of the prediction, which can be thought of as the standard error of the predicted expected value or mean for the observation's covariate pattern. This is also referred to as the standard error of the fitted value.

`stdf` calculates the standard error of the forecast, which is the standard error of the point prediction for 1 observation. It is commonly referred to as the standard error of the future or forecast value. By construction, the standard errors produced by `stdf` are always larger than those produced by `stdp`; see *Methods and Formulas* in [R] **regress**.

`pr(`*a*`,`*b*`)` calculates $\Pr(a < \mathbf{x}_j\mathbf{b} + u_j < b)$, the probability that $y_j|\mathbf{x}_j$ would be observed in the interval (a, b).

a and *b* may be specified as numbers or variable names; *lb* and *ub* are variable names;
`pr(20,30)` calculates $\Pr(20 < \mathbf{x}_j\mathbf{b} + u_j < 30)$;
`pr(`*lb*`,`*ub*`)` calculates $\Pr(lb < \mathbf{x}_j\mathbf{b} + u_j < ub)$; and
`pr(20,`*ub*`)` calculates $\Pr(20 < \mathbf{x}_j\mathbf{b} + u_j < ub)$.

a missing ($a \geq .$) means $-\infty$; `pr(.,30)` calculates $\Pr(-\infty < \mathbf{x}_j\mathbf{b} + u_j < 30)$;
`pr(`*lb*`,30)` calculates $\Pr(-\infty < \mathbf{x}_j\mathbf{b} + u_j < 30)$ in observations for which $lb \geq .$
and calculates $\Pr(lb < \mathbf{x}_j\mathbf{b} + u_j < 30)$ elsewhere.

b missing ($b \geq .$) means $+\infty$; pr(20,.) calculates $\Pr(+\infty > \mathbf{x}_j\mathbf{b} + u_j > 20)$;
pr(20,*ub*) calculates $\Pr(+\infty > \mathbf{x}_j\mathbf{b} + u_j > 20)$ in observations for which $ub \geq .$
and calculates $\Pr(20 < \mathbf{x}_j\mathbf{b} + u_j < ub)$ elsewhere.

e(*a*,*b*) calculates $E(\mathbf{x}_j\mathbf{b} + u_j \mid a < \mathbf{x}_j\mathbf{b} + u_j < b)$, the expected value of $y_j|\mathbf{x}_j$ conditional on $y_j|\mathbf{x}_j$ being in the interval (a, b), meaning that $y_j|\mathbf{x}_j$ is censored.
a and b are specified as they are for pr().

ystar(*a*,*b*) calculates $E(y_j^*)$, where $y_j^* = a$ if $\mathbf{x}_j\mathbf{b} + u_j \leq a$, $y_j^* = b$ if $\mathbf{x}_j\mathbf{b} + u_j \geq b$, and $y_j^* = \mathbf{x}_j\mathbf{b} + u_j$ otherwise, meaning that y_j^* is truncated. a and b are specified as they are for pr().

scores calculates the scores for the model. A new score variable is created for each endogenous regressor, as well as an equation-level score that applies to all exogenous variables and constant term (if present).

Syntax for estat firststage

```
estat firststage [, all forcenonrobust]
```

Options for estat firststage

all requests that all first-stage goodness-of-fit statistics be reported regardless of whether the model contains one or more endogenous regressors. By default, if the model contains one endogenous regressor, then the first-stage R^2, adjusted R^2, partial R^2, and F statistics are reported, whereas if the model contains multiple endogenous regressors, then Shea's partial R^2 and adjusted partial R^2 are reported instead.

forcenonrobust requests that the minimum eigenvalue statistic and its critical values be reported even though a robust VCE was used at estimation time. The reported critical values assume that the errors are independently and identically distributed normal, so the user must determine whether the critical values are appropriate for a given application.

Syntax for estat overid

```
estat overid [, lags(#) forceweights forcenonrobust]
```

Options for estat overid

lags(*#*) specifies the number of lags to use for prewhitening when computing the heteroskedasticity- and autocorrelation-consistent (HAC) version of the score test of overidentifying restrictions. Specifying lags(0) requests no prewhitening. This option is valid only when the model was fitted via 2SLS and a HAC covariance matrix was requested when the model was fitted. The default is lags(1).

forceweights requests that the tests of overidentifying restrictions be computed even though aweights, pweights, or iweights were used in the previous estimation. By default, these tests are conducted only after unweighted or frequency-weighted estimation. The reported critical values may be inappropriate for weighted data, so the user must determine whether the critical values are appropriate for a given application.

`forcenonrobust` requests that the Sargan and Basmann tests of overidentifying restrictions be performed after 2SLS or LIML estimation even though a robust VCE was used at estimation time. These tests assume that the errors are independently and identically distributed normal, so the user must determine whether the critical values are appropriate for a given application.

Remarks

Remarks are presented under the following headings:

estat firststage
estat overid

estat firststage

For an excluded exogenous variable to be a valid instrument, it must be sufficiently correlated with the included endogenous regressors but uncorrelated with the error term. In recent decades, researchers have paid considerable attention to the issue of instruments that are only weakly correlated with the endogenous regressors. In such cases, the usual 2SLS, GMM, and LIML estimators are biased toward the OLS estimator, and inference based on the standard errors reported by, for example, `ivregress` can be severely misleading. For more information on the theory behind instrumental-variables estimation with weak instruments, see Nelson and Startz (1990), Staiger and Stock (1997), Hahn and Hausman (2003), and the survey article by Stock, Wright, and Yogo (2002).

When the instruments are only weakly correlated with the endogenous regressors, some Monte Carlo evidence suggests that the LIML estimator performs better than the 2SLS and GMM estimators; see, for example, Poi (2006) and Stock, Wright, and Yogo (2002) (and the papers cited therein). On the other hand, the LIML estimator often results in confidence intervals that are somewhat larger than those from the 2SLS estimator.

Moreover, using more instruments is not a solution, because the biases of instrumental-variables estimators increase with the number of instruments. See Hahn and Hausman (2003).

`estat firststage` produces several statistics for judging the explanatory power of the instruments and is most easily explained with examples.

▷ Example 1

In example 1 of [R] **ivregress**, we fitted a model of the average rental rate for housing in a state as a function of the percentage of the population living in urban areas and the average value of houses. We treated `hsngval` as endogenous because unanticipated shocks that affect rental rates probably affect house prices as well. We used family income and region dummies as additional instruments for `hsngval`. Now we explore how highly correlated those instruments are with `hsngval`:

```
. use http://www.stata-press.com/data/r10/hsng2
(1980 Census housing data)
. ivregress 2sls rent pcturban (hsngval = faminc reg2-reg4)
  (output omitted)
. estat firststage
  First-stage regression summary statistics
```

Variable	R-sq.	Adjusted R-sq.	Partial R-sq.	F(4,44)	Prob > F
hsngval	0.6908	0.6557	0.5473	13.2978	0.0000

```
Minimum eigenvalue statistic = 13.2978

Critical Values                      # of endogenous regressors:    1
Ho: Instruments are weak             # of excluded instruments:     4

                                          5%     10%     20%     30%
2SLS relative bias                     16.85   10.27    6.71    5.34

                                         10%     15%     20%     25%
2SLS Size of nominal 5% Wald test      24.58   13.96   10.26    8.31
LIML Size of nominal 5% Wald test       5.44    3.87    3.30    2.98
```

To understand these results, recall that the first-stage regression is

$$\texttt{hsngval}_i = \pi_0 + \pi_1\texttt{pcturban}_i + \pi_2\texttt{faminc} + \pi_3\texttt{reg2} + \pi_4\texttt{reg3} + \pi_5\texttt{reg4} + v_i$$

where v_i is an error term. The column marked "R-sq." is the simple R^2 from fitting the first-stage regression by OLS, and the column marked "Adjusted R-sq." is the adjusted R^2 from that regression. Higher values purportedly indicate stronger instruments, and instrumental-variables estimators exhibit less bias when the instruments are strongly correlated with the endogenous variable.

Looking at just the R^2 and adjusted R^2 can be misleading, however. If `hsngval` were strongly correlated with the included exogenous variable `pcturban` but only weakly correlated with the additional instruments, then these statistics could be large even though a weak-instrument problem is present.

The partial R^2 statistic measures the correlation between `hsngval` and the additional instruments after *partialling out* the effect of `pcturban`. Unlike the R^2 and adjusted R^2 statistics, the partial R^2 statistic will not be inflated because of strong correlation between `hsngval` and `pcturban`. Bound, Jaeger, and Baker (1995) and others have promoted using this statistic.

The column marked "F(4, 44)" is an F statistic for the joint significance of π_2, π_3, π_4, and π_5, the coefficients on the additional instruments. Its p-value is listed in the column marked "Prob > F". If the F statistic is not significant, then the additional instruments have no significant explanatory power for `hsngval` after controlling for the effect of `pcturban`. However, Hall, Rudebusch, and Wilcox (1996) used Monte Carlo simulation to show that simply having an F statistic that is significant at the typical 5% or 10% level is not sufficient. Stock, Wright, and Yogo (2002) suggest that the F statistic should exceed 10 for inference based on the 2SLS estimator to be reliable when there is one endogenous regressor.

`estat firststage` also presents the Cragg and Donald (1993) minimum eigenvalue statistic as a further test of weak instruments. Stock and Yogo (2005) discuss two characterizations of weak instruments: first, weak instruments cause instrumental-variables estimators to be biased; second, hypothesis tests of parameters estimated by instrumental-variables estimators may suffer from severe size distortions. The test statistic in our example is 13.30, which is identical to the F statistic just discussed because our model contains one endogenous regressor.

The null hypothesis of each of Stock and Yogo's tests is that the set of instruments is weak. To perform these tests, we must first choose either the largest relative bias of the 2SLS estimator we are willing to tolerate or the largest rejection rate of a nominal 5% Wald test we are willing to tolerate. If the test statistic exceeds the critical value, we can conclude that our instruments are not weak.

The row marked "2SLS relative bias" contains critical values for the test that the instruments are weak based on the bias of the 2SLS estimator *relative to* the bias of the OLS estimator. For example, from past experience we might know that the OLS estimate of a parameter β may be 50% too high. Saying that we are willing to tolerate a 10% relative bias means that we are willing to tolerate a bias of the 2SLS estimator no greater than 5% (i.e., 10% of 50%). In our rental rate model, if we are willing to tolerate a 10% relative bias, then we can conclude that our instruments are not weak because the test statistic of 13.30 exceeds the critical value of 10.22. However, if we were only willing to tolerate a relative bias of 5%, we would conclude that our instruments are weak because $13.30 < 16.85$.

The rows marked "2SLS Size of nominal 5% Wald test" and "LIML Size of nominal 5% Wald test" contain critical values pertaining to Stock and Yogo's (2005) second characterization of weak instruments. This characterization defines a set of instruments to be weak if a Wald test at the 5% level can have an actual rejection rate of no more than 10%, 15%, 20%, or 25%. Using the current example, suppose that we are willing to accept a rejection rate of at most 10%. Because $13.30 < 24.58$, we cannot reject the null hypothesis of weak instruments. On the other hand, if we use the LIML estimator instead, then we can reject the null hypothesis since $13.30 > 5.44$.

◁

❑ Technical Note

Stock and Yogo (2005) tabulated critical values for 2SLS relative biases of 5%, 10%, 20%, and 30% for models with 1, 2, or 3 endogenous regressors and between 3 and 30 excluded exogenous variables (instruments). They also provide critical values for worst-case rejection rates of 5%, 10%, 20%, and 25% for nominal 5% Wald tests of the endogenous regressors with 1 or 2 endogenous regressors and between 1 and 30 instruments. If the model previously fitted by `ivregress` has more instruments or endogenous regressors than these limits, the critical values are not shown. Stock and Yogo did not consider GMM estimators.

❑

When the model being fitted contains more than one endogenous regressor, the R^2 and F statistics described above can overstate the relevance of the excluded instruments. Suppose that there are two endogenous regressors, Y_1 and Y_2, and that there are two additional instruments, z_1 and z_2. Say that z_1 is highly correlated with both Y_1 and Y_2 but z_2 is not correlated with either Y_1 or Y_2. Then the first-stage regression of Y_1 on z_1 and z_2 (along with the included exogenous variables) will produce large R^2 and F statistics, as will the regression of Y_2 on z_1, z_2, and the included exogenous variables. Nevertheless, the lack of correlation between z_2 and Y_1 and Y_2 is problematic. Here, although the order condition indicates that the model is just identified (the number of excluded instruments equals the number of endogenous regressors), the irrelevance of z_2 implies that the model is in fact not identified. Even if the model is overidentified, including irrelevant instruments can adversely affect the properties of instrumental-variables estimators, since their biases increase as the number of instruments increases.

▷ Example 2

`estat firststage` presents different statistics when the model contains multiple endogenous regressors. For illustration, we refit our model of rental rates, assuming that both `hsngval` and `faminc` are endogenously determined. We use the three region dummies along with `popden`, a measure of population density, as additional instruments.

```
. ivregress 2sls rent pcturban (hsngval faminc = reg2-reg4 popden)
  (output omitted )
. estat firststage
  Shea's partial R-squared

                      Shea's              Shea's
   Variable       Partial R-sq.   Adj. Partial R-sq.

    hsngval          0.3477              0.2897
     faminc          0.1893              0.1173

  Minimum eigenvalue statistic = 2.51666

  Critical Values                        # of endogenous regressors:    2
  Ho: Instruments are weak               # of excluded instruments:     4

                                            5%      10%      20%      30%
  2SLS relative bias                      11.04     7.56     5.57     4.73

                                           10%      15%      20%      25%
  2SLS Size of nominal 5% Wald test      16.87     9.93     7.54     6.28
  LIML Size of nominal 5% Wald test       4.72     3.39     2.99     2.79
```

Consider the endogenous regressor **hsngval**. Part of its variation is attributable to its correlation with the other regressors **pcturban** and **faminc**. The other component of **hsngval**'s variation is peculiar to it and orthogonal to the variation in the other regressors. Similarly, we can think of the instruments as predicting the variation in **hsngval** in two ways, one stemming from the fact that the predicted values of **hsngval** are correlated with the predicted values of the other regressors and one from the variation in the predicted values of **hsngval** that is orthogonal to the variation in the predicted values of the other regressors.

What really matters for instrumental-variables estimation is whether the component of **hsngval** that is orthogonal to the other regressors can be explained by the component of the predicted value of **hsngval** that is orthogonal to the predicted values of the other regressors in the model. Shea's (1997) partial R^2 statistic measures this correlation. Because the bias of instrumental-variables estimators increases as more instruments are used, Shea's adjusted partial R^2 statistic is often used instead, as it makes a degrees-of-freedom adjustment for the number of instruments, analogous to the adjusted R^2 measure used in OLS regression. Although what constitutes a "low" value for Shea's partial R^2 depends on the specifics of the model being fitted and the data used, these results, taken in isolation, do not strike us as being a particular cause for concern.

However, with this specification the minimum eigenvalue statistic is low. We cannot reject the null hypothesis of weak instruments for either of the characterizations we have discussed.

◁

By default, **estat firststage** determines which statistics to present based on the number of endogenous regressors in the model previously fitted. However, you can specify the option **all** to obtain all the statistics.

❑ Technical Note

If the previous estimation was conducted using **aweights**, **pweights**, or **iweights**, then the first-stage regression summary statistics are computed using those weights. However, in these cases the minimum eigenvalue statistic and its critical values are not available.

If the previous estimation included a robust VCE, then the first-stage F statistic is based on a robust VCE as well; for example, if you fitted your model with a HAC VCE using the Bartlett kernel and four lags, then the F statistic reported is based on regression results using a HAC VCE using the Bartlett kernel and four lags. By default, the minimum eigenvalue statistic and its critical values are not displayed. You can use the `forcenonrobust` option to obtain them in these cases; the minimum eigenvalue statistic is computed using the weights, though the critical values reported may not be appropriate.

❑

estat overid

In addition to the requirement that instrumental variables be correlated with the endogenous regressors, the instruments must also be uncorrelated with the structural error term. If the model is overidentified, meaning that the number of additional instruments exceeds the number of endogenous regressors, then we can test whether the instruments are uncorrelated with the error term. If the model is just identified, then we cannot perform a test of overidentifying restrictions.

The estimator you used to fit the model determines which tests of overidentifying restrictions `estat overid` reports. If you used the 2SLS estimator without a robust VCE, `estat overid` reports Sargan's (1958) and Basmann's (1960) χ^2 tests. If you used the 2SLS estimator and requested a robust VCE, Wooldridge's robust score test of overidentifying restrictions is performed instead; without a robust VCE, Wooldridge's test statistic is identical to Sargan's test statistic. If you used the LIML estimator, `estat overid` reports the Anderson–Rubin (1950) likelihood-ratio test and Basmann's (1960) F test. `estat overid` reports Hansen's (1982) J statistic if you used the GMM estimator. Davidson and MacKinnon (1993, 235–236) give a particularly clear explanation of the intuition behind tests of overidentifying restrictions. Also see Judge et al. (1985, 614–616) for a summary of tests of overidentifying restrictions for the 2SLS and LIML estimators.

Tests of overidentifying restrictions actually test two different things simultaneously. One, as we have discussed, is whether the instruments are uncorrelated with the error term. The other is that the equation is misspecified and that one or more of the excluded exogenous variables should in fact be included in the structural equation. Thus, a significant test statistic could represent either an invalid instrument or an incorrectly specified structural equation.

▷ Example 3

Here we refit the model that treated just `hsngval` as endogenous using 2SLS, and then we perform tests of overidentifying restrictions:

```
. ivregress 2sls rent pcturban (hsngval = faminc reg2-reg4)
  (output omitted)
. estat overid
  Tests of overidentifying restrictions:
  Sargan (score) chi2(3) =  11.2877  (p = 0.0103)
  Basmann chi2(3)        =  12.8294  (p = 0.0050)
```

Both test statistics are significant at the 5% test level, which means that either one or more of our instruments is invalid or that our structural model is specified incorrectly.

One possibility is that the error term in our structural model is heteroskedastic. Both Sargan's and Basmann's tests assume that the errors are independently and identically distributed (i.i.d.); if the errors are not i.i.d., then these tests are not valid. Here we refit the model by requesting heteroskedasticity-robust standard errors, and then we use `estat overid` to obtain Wooldridge's score test of overidentifying restrictions, which is robust to heteroskedasticity.

```
. ivregress 2sls rent pcturban (hsngval = faminc reg2-reg4), vce(robust)
  (output omitted )
. estat overid
  Test of overidentifying restrictions:
  Score chi2(3)            =   6.8364  (p = 0.0773)
```

Here we no longer reject the null hypothesis that our instruments are valid at the 5% significance level, though we do reject the null at the 10% level. You can verify that the robust standard error on the coefficient for **hsngval** is more than twice as large as its nonrobust counterpart and that the robust standard error for **pcturban** is nearly 50% larger.

◁

❑ Technical Note

The test statistic for the test of overidentifying restrictions performed after GMM estimation is simply the sample size times the value of the objective function $Q(\beta_1, \beta_2)$ defined in (5) of [R] **ivregress**, evaluated at the GMM parameter estimates. If the weighting matrix $\mathbf{W}$ is optimal, meaning that $\mathbf{W} = \text{Var}(\mathbf{z}_i u_i)$, then $Q(\beta_1, \beta_2) \overset{A}{\sim} \chi^2(q)$, where q is the number of overidentifying restrictions. However, if the estimated $\mathbf{W}$ is not optimal, then the test statistic will not have an asymptotic χ^2 distribution.

Like the Sargan and Basmann tests of overidentifying restrictions for the 2SLS estimator, the Anderson–Rubin and Basmann tests after LIML estimation are predicated on the errors' being i.i.d. If the previous LIML results were reported with robust standard errors, then `estat overid` by default issues an error message and refuses to report the Anderson–Rubin and Basmann test statistics. You can use the `forcenonrobust` option to override this behavior. You can also use `forcenonrobust` to obtain the Sargan and Basmann test statistics after 2SLS estimation with robust standard errors.

❑

By default `estat overid` issues an error message if the previous estimation was conducted using `aweights`, `pweights`, or `iweights`. You can use the `forceweights` option to override this behavior, though the test statistics may no longer have the expected χ^2 distributions.

Saved Results

`estat firststage` saves the following in `r()`:

Scalars

`r(mineig)`	minimum eigenvalue statistic

Matrices

`r(mineigcv)`	critical values for minimum eigenvalue statistic
`r(multiresults)`	Shea's partial R^2 statistics
`r(singleresults)`	first-stage R^2 and F statistics

(Continued on next page)

After 2SLS estimation, `estat overid` saves the following in `r()`:

Scalars

`r(lags)`	lags used in prewhitening
`r(df)`	χ^2 degrees of freedom
`r(score)`	score χ^2 statistic
`r(basmann)`	Basmann χ^2 statistic
`r(sargan)`	Sargan χ^2 statistic

After LIML estimation, `estat overid` saves the following in `r()`:

Scalars

`r(ar)`	Anderson–Rubin χ^2 statistic
`r(ar_df)`	χ^2 degrees of freedom
`r(basmann)`	Basmann F statistic
`r(basmann_df_n)`	F numerator degrees of freedom
`r(basmann_df_d)`	F denominator degrees of freedom

After GMM estimation, `estat overid` saves the following in `r()`:

Scalars

`r(HansenJ)`	Hansen's $J\chi^2$ statistic
`r(J_df)`	χ^2 degrees of freedom

Methods and Formulas

All postestimation commands listed above are implemented as ado-files.

See *Methods and Formulas* of [R] **ivregress** for definitions of the notation used here.

When the structural equation includes one endogenous regressor, `estat firststage` fits the regression

$$\mathbf{y} = \mathbf{X}_1\boldsymbol{\pi}_1 + \mathbf{X}_2\boldsymbol{\pi}_2 + \mathbf{v}$$

via OLS where $\mathbf{y}$ is an $N \times 1$ column vector of the observations on the endogenous regressor, $\mathbf{X}_1$ is an $N \times k_1$ matrix of included exogenous variables, and $\mathbf{X}_2$ is an $N \times k_2$ matrix of excluded exogenous variables. The R^2 and adjusted R^2 from that regression are reported in the output, as well as the F statistic from the Wald test of H_0: $\boldsymbol{\pi}_2 = \mathbf{0}$. To obtain the partial R^2 statistic, `estat firststage` fits the regression

$$\mathbf{M}_{\mathbf{X}_1}\mathbf{y} = \mathbf{M}_{\mathbf{X}_1}\mathbf{X}_2\boldsymbol{\xi} + \mathbf{e}$$

by OLS where $\mathbf{e}$ is an error term, $\boldsymbol{\xi}$ is a $k_2 \times 1$ parameter vector, and $\mathbf{M}_{\mathbf{X}_1} = \mathbf{I} - \mathbf{X}_1(\mathbf{X}_1'\mathbf{X}_1)^{-1}\mathbf{X}_1'$; i.e., the partial R^2 is the R^2 between $\mathbf{y}$ and $\mathbf{X}_2$ after eliminating the effects of $\mathbf{X}_1$. If the model contains multiple endogenous regressors and the `all` option is specified, these statistics are calculated for each endogenous regressor in turn.

To calculate Shea's partial R^2, let $\mathbf{y}_1$ denote the endogenous regressor whose statistic is being calculated and $\mathbf{Y}_0$ denote the other endogenous regressors. Define $\widetilde{\mathbf{y}}_1$ as the residuals obtained from regressing $\mathbf{y}_1$ on $\mathbf{Y}_0$ and $\mathbf{X}_1$. Let $\widehat{\mathbf{y}}_1$ denote the fitted values obtained from regressing $\mathbf{y}_1$ on $\mathbf{X}_1$ and $\mathbf{X}_2$; i.e., $\widehat{\mathbf{y}}_1$ are the fitted values from the first-stage regression for $\mathbf{y}_1$, and define the columns of $\widehat{\mathbf{Y}}_0$ analogously. Finally, let $\widetilde{\widehat{\mathbf{y}}}_1$ denote the residuals from regressing $\widehat{\mathbf{y}}_1$ on $\widehat{\mathbf{Y}}_0$ and $\mathbf{X}_1$. Shea's partial R^2 is the simple R^2 from the regression of $\widetilde{\mathbf{y}}_1$ on $\widetilde{\widehat{\mathbf{y}}}_1$; denote this as R_S^2. Shea's adjusted partial R^2 is equal to $1-(1-R_S^2)(N-1)/(N-k_Z+1)$ if a constant term is included and $1-(1-R_S^2)(N-1)/(N-k_Z)$ if there is no constant term included in the model, where $k_Z = k_1 + k_2$. For one endogenous regressor, one instrument, no exogenous regressors, and a constant term, R_S^2 equals the adjusted R_S^2.

The Stock and Yogo minimum eigenvalue statistic, first proposed by Cragg and Donald (1993) as a test for underidentification, is the minimum eigenvalue of the matrix

$$\mathbf{G} = \frac{1}{k_Z}\widehat{\boldsymbol{\Sigma}}_{\mathbf{VV}}^{-1/2}\mathbf{Y}'\mathbf{M}'_{\mathbf{X}_1}\mathbf{X}_2(\mathbf{X}'_2\mathbf{M}_{\mathbf{X}_1}\mathbf{X}_2)^{-1}\mathbf{X}'_2\mathbf{M}_{\mathbf{X}_1}\mathbf{Y}\widehat{\boldsymbol{\Sigma}}_{\mathbf{VV}}^{-1/2}$$

where

$$\widehat{\boldsymbol{\Sigma}}_{\mathbf{VV}} = \frac{1}{N-k_Z}\mathbf{Y}'\mathbf{M_Z}\mathbf{Y}$$

$\mathbf{M_Z} = \mathbf{I} - \mathbf{Z}(\mathbf{Z}'\mathbf{Z})^{-1}\mathbf{Z}'$, and $\mathbf{Z} = [\mathbf{X}_1\ \mathbf{X}_2]$. Critical values are obtained from the tables in Stock and Yogo (2005).

The Sargan (1958) and Basmann (1960) χ^2 statistics are calculated by running the auxiliary regression

$$\widehat{\mathbf{u}} = \mathbf{Z}\boldsymbol{\delta} + \mathbf{e}$$

where $\widehat{\mathbf{u}}$ are the sample residuals from the model and $\mathbf{e}$ is an error term. Then Sargan's statistic is

$$S = N\left(1 - \frac{\widehat{\mathbf{e}}'\widehat{\mathbf{e}}}{\widehat{\mathbf{u}}'\widehat{\mathbf{u}}}\right)$$

where $\widehat{\mathbf{e}}$ are the residuals from that auxiliary regression. Basmann's statistic is calculated as

$$B = S\frac{N-k_Z}{N-S}$$

Both S and B are distributed $\chi^2(m)$, where m, the number of overidentifying restrictions, is equal to $k_Z - k$, where k is the number of endogenous regressors.

Wooldridge's (1995) score test of overidentifying restrictions is identical to Sargan's (1958) statistic under the assumption of i.i.d. and therefore is not recomputed unless a robust VCE was used at estimation time. If a heteroskedasticity-robust VCE was used, Wooldridge's test proceeds as follows. Let $\widehat{\mathbf{Y}}$ denote the $N \times k$ matrix of fitted values obtained by regressing the endogenous regressors on $\mathbf{X}_1$ and $\mathbf{X}_2$. Let $\mathbf{Q}$ denote an $N \times m$ matrix of excluded exogenous variables; the test statistic to be calculated is invariant to which m of the k_2 excluded exogenous variables is chosen. Define the ith element of $\widehat{\mathbf{k}}_j$, $i = 1, \dots, N$, $j = 1, \dots, m$, as

$$k_{ij} = \widehat{q}_{ij} u_i$$

where $\widehat{q}_{ij}$ is the ith element of $\widehat{\mathbf{q}}_j$, the fitted values from regressing the jth column of $\mathbf{Q}$ on $\widehat{\mathbf{Y}}$ and $\mathbf{X}_1$. Finally, fit the regression

$$\mathbf{1} = \theta_1\widehat{\mathbf{k}}_1 + \cdots + \theta_m\widehat{\mathbf{k}}_m + \mathbf{v}$$

where $\mathbf{1}$ is an $N \times 1$ vector of ones and $\mathbf{v}$ is a regression error term, and calculate the residual sum of squares, RSS. Then the test statistic is $W = N - \text{RSS}$. $W \sim \chi^2(m)$. If a HAC VCE was used at estimation, then the $\widehat{\mathbf{k}}_j$ are prewhitened using a VAR(p) model, where p is specified using the `lags()` option.

The Anderson–Rubin (1950), AR, test of overidentifying restrictions for use after the LIML estimator is calculated as $\text{AR} = N(\kappa - 1)$, where κ is the minimal eigenvalue of a certain matrix defined in the *Methods and Formulas* of [R] **ivregress**. $\text{AR} \sim \chi^2(m)$. (Some texts define this statistic as $N \ln(\kappa)$ since $\ln(x) \approx (x - 1)$ for x near one.) Basmann's F statistic for use after the LIML estimator is calculated as $B_F = (\kappa - 1)(N - k_Z)/m$. $B_F \sim F(m, N - k_Z)$.

Hansen's J statistic is simply the value of the GMM objective function defined in (5) of [R] **ivregress**, evaluated at the estimated parameter values. Under the null hypothesis that the overidentifying restrictions are valid, $J \sim \chi^2(m)$.

References

Anderson, T. W., and H. Rubin. 1950. The asymptotic properties of estimates of the parameters of a single equation in a complete system of stochastic equations. *Annals of Mathematical Statistics* 21: 570–582.

Basmann, R. L. 1960. On finite sample distributions of generalized classical linear indentifiability test statistics. *Journal of the American Statistical Association* 55: 650–659.

Bound, J., D. A. Jaeger, and R. M. Baker. 1995. Problems with instrumental variables estimation when the correlation between the instruments and the endogenous explanatory variable is weak. *Journal of the American Statistical Association* 90: 443–450.

Cragg, J. G., and S. G. Donald. 1993. Testing identifiability and specification in instrumental variables models. *Econometric Theory* 9: 222–240.

Davidson, R., and J. G. MacKinnon. 1993. *Estimation and Inference in Econometrics*. New York: Oxford University Press.

Hahn, J., and J. Hausman. 2003. Weak instruments: Diagnosis and cures in empirical econometrics. *American Economic Review Papers and Proceedings* 93: 118–125.

Hall, A. R., G. D. Rudebusch, and D. W. Wilcox. 1996. Judging instrument relevance in instrumental variables estimation. *International Economic Review* 37: 283–298.

Hansen, L. P. 1982. Large sample properties of generalized method of moments estimators. *Econometrica* 50: 1029–1054.

Judge, G. G., W. E. Griffiths, R. C. Hill, H. Lütkepohl, and T.-C. Lee. 1985. *The Theory and Practice of Econometrics*. 2nd ed. New York: Wiley.

Nelson, C. R. and R. Startz. 1990. The distribution of the instrumental variable estimator and its t ratio when the instrument is a poor one. *Journal of Business* 63: S125–S140.

Poi, B. P. 2006. Jackknife instrumental variables estimation in Stata. *Stata Journal* 6: 364–376.

Sargan, J. D. 1958. The estimation of economic relationships using instrumental variables. *Econometrica* 26: 393–415.

Shea, J. 1997. Instrument relevance in multivariate linear models: A simple measure. *Review of Economics and Statistics* 79: 348–352.

Staiger, D., and J. H. Stock. 1997. Instrumental variables regression with weak instruments. *Econometrica* 65: 557–586.

Stock, J. H., J. H. Wright, and M. Yogo. 2002. A survey of weak instruments and weak identification in generalized method of moments. *Journal of Business and Economic Statistics* 20: 518–529.

Stock, J. H. and M. Yogo. 2005. Testing for weak instruments in linear IV regression. In *Identification and Inference for Econometric Models: Essays in Honor of Thomas Rothenberg*, ed. D. W. K. Andrews and J. H. Stock, 80–108. New York: Cambridge University Press.

Wooldridge, J. M. 1995. Score diagnostics for linear models estimated by two stage least squares. In *Advances in Econometrics and Quantitative Economics: Essays in Honor of Professor C. R. Rao*, ed. G. S. Maddala, P. C. B. Phillips, and T. N. Srinivasan, 66–87. Oxford: Blackwell.

Also See

[R] **ivregress** — Single-equation instrumental-variables regression

[U] **20 Estimation and postestimation commands**

Title

ivtobit — Tobit model with endogenous regressors

Syntax

Maximum likelihood estimator

`ivtobit` *depvar* [*varlist*$_1$] (*varlist*$_2$ = *varlist*$_{iv}$) [*if*] [*in*] [*weight*] ,
`ll`[(*#*)] `ul`[(*#*)] [*mle_options*]

Two-step estimator

`ivtobit` *depvar* [*varlist*$_1$] (*varlist*$_2$ = *varlist*$_{iv}$) [*if*] [*in*] [*weight*] , `twostep`
`ll`[(*#*)] `ul`[(*#*)] [*tse_options*]

mle_options	description
Model	
* `ll`[(*#*)]	lower limit for left censoring
* `ul`[(*#*)]	upper limit for right censoring
`mle`	use conditional maximum-likelihood estimator; the default
`constraints(`*constraints*`)`	apply specified linear constraints
SE/Robust	
`vce(`*vcetype*`)`	*vcetype* may be `oim`, `robust`, `cluster` *clustvar*, `opg`, `bootstrap`, or `jackknife`
Reporting	
`level(`*#*`)`	set confidence level; default is `level(95)`
`first`	report first-stage regression
Max options	
maximize_options	control the maximization process

Must specify at least one of `ll`[(#*)] and `ul`[(*#*)].

(*Continued on next page*)

tse_options	description
Model	
*twostep	use Newey's two-step estimator; the default is mle
*ll[(#)]	lower limit for left censoring
*ul[(#)]	upper limit for right censoring
SE	
vce(*vcetype*)	*vcetype* may be standard, bootstrap, or jackknife
Reporting	
level(#)	set confidence level; default is level(95)
first	report first-stage regression

*twostep is required. Must specify at least one of ll[(#)] and ul[(#)].

depvar, $varlist_1$, $varlist_2$, and $varlist_{iv}$ may contain time-series operators; see [U] **11.4.3 Time-series varlists**.

bootstrap, by, jackknife, rolling, statsby, svy, and xi are allowed; see [U] **11.1.10 Prefix commands**.

Weights are not allowed with the bootstrap prefix.

vce(), first, twostep, and weights are not allowed with the svy prefix.

fweights, iweights, and pweights are allowed with the maximum likelihood estimator. fweights are allowed with Newey's two-step estimator. See [U] **11.1.6 weight**.

See [U] **20 Estimation and postestimation commands** for more capabilities of estimation commands.

Description

ivtobit fits tobit models where one or more of the regressors is endogenously determined. By default, ivtobit uses maximum likelihood estimation. Alternatively, Newey's (1987) minimum chi-squared estimator can be invoked with the twostep option. See [R] **ivprobit** for probit estimation with endogenous regressors and [R] **tobit** for tobit estimation when the model contains no endogenous regressors.

Options for ML estimator

Model

ll(#) and ul(#) indicate the lower and upper limits for censoring, respectively. You may specify one or both. Observations with *depvar* $\leq$ ll() are left-censored; observations with *depvar* $\geq$ ul() are right-censored; and remaining observations are not censored. You do not have to specify the censoring values at all. It is enough to type ll, ul, or both. When you do not specify a censoring value, ivtobit assumes that the lower limit is the minimum observed in the data (if ll is specified) and that the upper limit is the maximum (if ul is specified).

mle requests that the conditional maximum-likelihood estimator be used. This is the default.

constraints(*constraints*); see [R] **estimation options**.

SE/Robust

vce(*vcetype*) specifies the type of standard error reported, which includes types that are derived from asymptotic theory, that are robust to some kinds of misspecification, that allow for intragroup correlation, and that use bootstrap or jackknife methods; see [R] ***vce_option***.

Reporting

level(#); see [R] **estimation options**.

first requests that the parameters for the reduced-form equations showing the relationships between the endogenous variables and instruments be displayed. For the two-step estimator, first shows the first-stage regressions. For the maximum likelihood estimator, these parameters are estimated jointly with the parameters of the tobit equation. The default is not to show these parameter estimates.

Max options

maximize_options: difficult, technique(*algorithm_spec*), iterate(#), [no]log, trace, gradient, showstep, hessian, shownrtolerance, tolerance(#), ltolerance(#), gtolerance(#), nrtolerance(#), nonrtolerance, from(*init_specs*); see [R] **maximize**. This model's likelihood function can be difficult to maximize, especially with multiple endogenous variables. The difficult and technique(bfgs) options may be helpful in achieving convergence.

Setting the optimization type to technique(bhhh) resets the default *vcetype* to vce(opg).

Options for two-step estimator

Model

twostep is required and requests that Newey's (1987) efficient two-step estimator be used to obtain the coefficient estimates.

ll(#) and ul(#) indicate the lower and upper limits for censoring, respectively. You may specify one or both. Observations with *depvar* ≤ ll() are left-censored; observations with *depvar* ≥ ul() are right-censored; and remaining observations are not censored. You do not have to specify the censoring values at all. It is enough to type ll, ul, or both. When you do not specify a censoring value, ivtobit assumes that the lower limit is the minimum observed in the data (if ll is specified) and that the upper limit is the maximum (if ul is specified).

SE

vce(*vcetype*) specifies the type of standard error reported, which includes types that are derived from asymptotic theory and that use bootstrap or jackknife methods; see [R] ***vce_option***.

Reporting

level(#); see [R] **estimation options**.

first requests that the parameters for the reduced-form equations showing the relationships between the endogenous variables and instruments be displayed. For the two-step estimator, first shows the first-stage regressions. For the maximum likelihood estimator, these parameters are estimated jointly with the parameters of the tobit equation. The default is not to show these parameter estimates.

Remarks

ivtobit fits models with censored dependent variables and endogenous regressors. You can use it to fit a tobit model when you suspect that one or more of the regressors is correlated with the error term. ivtobit is to tobit what ivregress is to linear regression analysis; see [R] **ivregress** for more information.

Formally, the model is

$$\begin{aligned} y_{1i}^* &= \boldsymbol{y}_{2i}\boldsymbol{\beta} + \boldsymbol{x}_{1i}\boldsymbol{\gamma} + u_i \\ \boldsymbol{y}_{2i} &= \boldsymbol{x}_{1i}\boldsymbol{\Pi}_1 + \boldsymbol{x}_{2i}\boldsymbol{\Pi}_2 + \boldsymbol{v}_i \end{aligned}$$

where $i = 1, \ldots, N$, $\boldsymbol{y}_{2i}$ is a $1 \times p$ vector of endogenous variables, $\boldsymbol{x}_{1i}$ is a $1 \times k_1$ vector of exogenous variables, $\boldsymbol{x}_{2i}$ is a $1 \times k_2$ vector of additional instruments, and the equation for $\boldsymbol{y}_{2i}$ is written in reduced form. By assumption $(u_i, \boldsymbol{v}_i) \sim \mathrm{N}(\boldsymbol{0})$. $\boldsymbol{\beta}$ and $\boldsymbol{\gamma}$ are vectors of structural parameters, and $\boldsymbol{\Pi}_1$ and $\boldsymbol{\Pi}_2$ are matrices of reduced-form parameters. We do not observe y_{1i}^*; instead, we observe

$$y_{1i} = \begin{cases} a & y_{1i}^* < a \\ y_{1i}^* & a \le y_{1i}^* \le b \\ b & y_{1i}^* > b \end{cases}$$

The order condition for identification of the structural parameters is that $k_2 \geq p$. Presumably $\boldsymbol{\Sigma}$ is not block diagonal between u_i and $\boldsymbol{v}_i$; otherwise, $\boldsymbol{y}_{2i}$ would not be endogenous.

❑ Technical Note

This model is derived under the assumption that $(u_i, \boldsymbol{v}_i)$ is independently and identically distributed multivariate normal for all i. The `vce(cluster` *clustvar*`)` option can be used to control for a lack of independence. As with the standard tobit model without endogeneity, if u_i is heteroskedastic, point estimates will be inconsistent.

❑

▷ Example 1

Using the same dataset as in [R] **ivprobit**, we now want to estimate women's incomes. In our hypothetical world, all women who choose not to work receive \$10,000 in welfare and child-support payments. Therefore, we never observe incomes under \$10,000: a woman offered a job with an annual wage less than that would not accept and instead would collect the welfare payment. We model income as a function of the number of years of schooling completed, the number of children at home, and other household income. We again believe that `other_inc` is endogenous, so we use `male_educ` as an instrument.

```
. use http://www.stata-press.com/data/r10/laborsup
. ivtobit fem_inc fem_educ kids (other_inc = male_educ), ll
Fitting exogenous tobit model
Fitting full model
Iteration 0:   log likelihood = -3332.8771
Iteration 1:   log likelihood = -3303.8855
Iteration 2:   log likelihood = -3300.8595
Iteration 3:   log likelihood = -3300.8376
Iteration 4:   log likelihood = -3300.8376
Tobit model with endogenous regressors                Number of obs   =        500
                                                      Wald chi2(3)    =     122.15
Log likelihood = -3300.8376                           Prob > chi2     =     0.0000

------------------------------------------------------------------------------
     fem_inc |      Coef.   Std. Err.      z    P>|z|     [95% Conf. Interval]
-------------+----------------------------------------------------------------
   other_inc |   -1.20984    .174987    -6.91   0.000    -1.552808   -.8668715
    fem_educ |   4.463251   .5220069     8.55   0.000     3.440136    5.486365
        kids |  -4.267294   .9498655    -4.49   0.000    -6.128996   -2.405592
       _cons |   13.72174   9.705962     1.41   0.157    -5.301592    32.74508
-------------+----------------------------------------------------------------
      /alpha |   .4183454   .1817899     2.30   0.021     .0620437    .7746471
        /lns |    3.15463   .0520291    60.63   0.000     3.052655    3.256605
        /lnv |   2.813383   .0316228    88.97   0.000     2.751404    2.875363
-------------+----------------------------------------------------------------
           s |   23.44436   1.219788                      21.17148    25.96125
           v |   16.66621   .5270318                      15.66461    17.73186
------------------------------------------------------------------------------
Instrumented:  other_inc
Instruments:   fem_educ kids male_educ
------------------------------------------------------------------------------
Wald test of exogeneity (/alpha = 0): chi2(1) =     5.30  Prob > chi2 = 0.0214
  Obs. summary:          272  left-censored observations at fem_inc<=10
                         228     uncensored observations
                           0 right-censored observations
```

Since we did not specify `mle` or `twostep`, `ivtobit` used the maximum likelihood estimator by default. `ivtobit` fits a tobit model, ignoring endogeneity, to get starting values for the full model. The header of the output contains the maximized log likelihood, the number of observations, and a Wald statistic and p-value for the test of the hypothesis that all the slope coefficients are jointly zero. At the end of the output, we see a count of the censored and uncensored observations.

Near the bottom of the output is a Wald test of the exogeneity of the instrumented variables. If the test statistic is not significant, there is not sufficient information in the sample to reject the null hypothesis of no endogeneity. Then the point estimates from `ivtobit` are consistent, although those from `tobit` are likely to have smaller standard errors.

◁

Various two-step estimators have also been proposed for the endogenous tobit model, and Newey's (1987) minimum chi-squared estimator is available with the `twostep` option.

(*Continued on next page*)

▷ Example 2

Refitting our labor-supply model with the two-step estimator yields

```
. ivtobit fem_inc fem_educ kids (other_inc = male_educ), ll twostep
Two-step tobit with endogenous regressors         Number of obs   =        500
                                                  Wald chi2(3)    =     117.40
                                                  Prob > chi2     =     0.0000

------------------------------------------------------------------------------
     fem_inc |      Coef.   Std. Err.      z    P>|z|     [95% Conf. Interval]
-------------+----------------------------------------------------------------
   other_inc |  -.9045397   .1329947    -6.80   0.000    -1.165205   -.6438749
    fem_educ |    3.27239   .3969189     8.24   0.000     2.494444    4.050337
        kids |  -3.312356   .7219712    -4.59   0.000    -4.727394   -1.897319
       _cons |   19.24735   7.373561     2.61   0.009     4.795433    33.69926
------------------------------------------------------------------------------
Instrumented:  other_inc
Instruments:   fem_educ kids male_educ
------------------------------------------------------------------------------
Wald test of exogeneity:      chi2(1) =     4.64          Prob > chi2 = 0.0312
  Obs. summary:         272  left-censored observations at fem_inc<=10
                        228     uncensored observations
                          0 right-censored observations
```

All the coefficients have the same signs as their counterparts in the maximum likelihood model. The Wald test at the bottom of the output confirms our earlier finding of endogeneity.

◁

❑ Technical Note

In the tobit model with endogenous regressors, we assume that $(u_i, \boldsymbol{v}_i)$ is multivariate normal with covariance matrix

$$\mathrm{Var}(u_i, \boldsymbol{v}_i) = \boldsymbol{\Sigma} = \begin{bmatrix} \sigma_u^2 & \boldsymbol{\Sigma}_{21}' \\ \boldsymbol{\Sigma}_{21} & \boldsymbol{\Sigma}_{22} \end{bmatrix}$$

Using the properties of the multivariate normal distribution, $\mathrm{Var}(u_i|\boldsymbol{v}_i) \equiv \sigma_{u|v}^2 = \sigma_u^2 - \boldsymbol{\Sigma}_{21}'\boldsymbol{\Sigma}_{22}^{-1}\boldsymbol{\Sigma}_{21}$. Calculating the marginal effects on the conditional expected values of the observed and latent dependent variables and on the probability of censoring requires an estimate of σ_u^2. The two-step estimator identifies only $\sigma_{u|v}^2$, not σ_u^2, so only the linear prediction and its standard error are available after you have used the `twostep` option. However, unlike the two-step probit estimator described in [R] **ivprobit**, the two-step tobit estimator does identify $\boldsymbol{\beta}$ and $\boldsymbol{\gamma}$. See Wooldridge (2002, 532) for more information.

❑

Saved Results

`ivtobit` saves the following in `e()`:

Scalars

`e(N)`	# of observations	`e(ll)`	log likelihood
`e(k)`	# of parameters	`e(N_clust)`	# of clusters
`e(k_eq)`	# of equations	`e(endog_ct)`	number of endogenous regressors
`e(k_eq_model)`	# of equations in model Wald test	`e(p)`	model Wald p-value
`e(k_aux)`	# of auxiliary parameters	`e(p_exog)`	exogeneity test Wald p-value
`e(k_dv)`	# of dependent variables	`e(chi2)`	model Wald χ^2
`e(N_unc)`	# of uncensored observations	`e(chi2_exog)`	Wald χ^2 test of exogeneity
`e(N_lc)`	# of left-censored observations	`e(rc)`	return code
`e(N_rc)`	# of right-censored observations	`e(ic)`	# of iterations
`e(llopt)`	contents of `ll()`	`e(rank)`	rank of `e(V)`
`e(ulopt)`	contents of `ul()`	`e(converged)`	`1` if converged, `0` otherwise
`e(df_m)`	model degrees of freedom		

Macros

`e(cmd)`	`ivtobit`	`e(diparm#)`	display transformed parameter #
`e(cmdline)`	command as typed	`e(opt)`	type of optimization
`e(depvar)`	name of dependent variable	`e(method)`	`ml` or `twostep`
`e(instd)`	instrumented variables	`e(ml_method)`	type of `ml` method
`e(insts)`	instruments	`e(user)`	name of likelihood-evaluator program
`e(wtype)`	weight type	`e(technique)`	maximization technique
`e(wexp)`	weight expression	`e(crittype)`	optimization criterion
`e(title)`	title in estimation output	`e(properties)`	`b V`
`e(clustvar)`	name of cluster variable	`e(footnote)`	program used to implement the footnote display
`e(chi2type)`	`Wald`; type of model χ^2 test		
`e(vce)`	*vcetype* specified in `vce()`	`e(predict)`	program used to implement `predict`
`e(vcetype)`	title used to label Std. Err.		

Matrices

`e(b)`	coefficient vector	`e(V)`	variance–covariance matrix of the estimators
`e(ilog)`	iteration log (up to 20 iterations)		
`e(Sigma)`	$\widehat{\boldsymbol{\Sigma}}$	`e(gradient)`	gradient vector

Functions

`e(sample)`	marks estimation sample

Methods and Formulas

`ivtobit` is implemented as an ado-file.

The estimation procedure used by `ivtobit` is similar to that used by `ivprobit`. For compactness, we write the model as

$$y_{1i}^* = \boldsymbol{z}_i\boldsymbol{\delta} + u_i \tag{1a}$$

$$\boldsymbol{y}_{2i} = \boldsymbol{x}_i\boldsymbol{\Pi} + \boldsymbol{v}_i \tag{1b}$$

where $\boldsymbol{z}_i = (\boldsymbol{y}_{2i}, \boldsymbol{x}_{1i})$, $\boldsymbol{x}_i = (\boldsymbol{x}_{1i}, \boldsymbol{x}_{2i})$, $\boldsymbol{\delta} = (\boldsymbol{\beta}', \boldsymbol{\gamma}')'$, and $\boldsymbol{\Pi} = (\boldsymbol{\Pi}_1', \boldsymbol{\Pi}_2')'$. We do not observe y_{1i}^*; instead, we observe

$$y_{1i} = \begin{cases} a & y_{1i}^* < a \\ y_{1i}^* & a \leq y_{1i}^* \leq b \\ b & y_{1i}^* > b \end{cases}$$

$(u_i, \boldsymbol{v}_i)$ is distributed multivariate normal with mean zero and covariance matrix

$$\boldsymbol{\Sigma} = \begin{bmatrix} \sigma_u^2 & \boldsymbol{\Sigma}_{21}' \\ \boldsymbol{\Sigma}_{21} & \boldsymbol{\Sigma}_{22} \end{bmatrix}$$

Using the properties of the multivariate normal distribution, we can write $u_i = \boldsymbol{v}_i'\boldsymbol{\alpha} + \epsilon_i$, where $\boldsymbol{\alpha} = \boldsymbol{\Sigma}_{22}^{-1}\boldsymbol{\Sigma}_{21}$; $\epsilon_i \sim \mathrm{N}(0; \sigma_{u|v}^2)$, where $\sigma_{u|v}^2 = \sigma_u^2 - \boldsymbol{\Sigma}_{21}'\boldsymbol{\Sigma}_{22}^{-1}\boldsymbol{\Sigma}_{21}$; and ϵ_i is independent of $\boldsymbol{v}_i$, $\boldsymbol{z}_i$, and $\boldsymbol{x}_i$.

The likelihood function is straightforward to derive since we can write the joint density $f(y_{1i}, \boldsymbol{y}_{2i}|\boldsymbol{x}_i)$ as $f(y_{1i}|\boldsymbol{y}_{2i}, \boldsymbol{x}_i)\, f(\boldsymbol{y}_{2i}|\boldsymbol{x}_i)$. With one endogenous regressor,

$$\ln f(y_{2i}|\boldsymbol{x}_i) = -\frac{1}{2}\left\{ \ln 2\pi + \ln\sigma_v^2 + \frac{(y_{2i} - \boldsymbol{x}_i\boldsymbol{\Pi})^2}{\sigma_v^2} \right\}$$

and

$$\ln f(y_{1i}|y_{2i}, \boldsymbol{x}_i) = \begin{cases} \ln\left\{1 - \Phi(m_i/\sigma_{u|v})\right\} & y_{1i} = a \\ -\frac{1}{2}\left\{ \ln 2\pi + \ln\sigma_{u|v}^2 + \frac{(y_{1i} - m_i)^2}{\sigma_{u|v}^2} \right\} & a < y_{1i} < b \\ \ln\Phi(m_i/\sigma_{u|v}) & y_{1i} = b \end{cases}$$

where

$$m_i = \boldsymbol{z}_i\boldsymbol{\delta} + \alpha\,(y_{2i} - \boldsymbol{x}_i\boldsymbol{\Pi})$$

and $\Phi(\cdot)$ is the normal distribution function so that the log likelihood for observation i is

$$\ln L_i = w_i\left\{ \ln f(y_{1i}|y_{2i}, \boldsymbol{x}_i) + \ln f(y_{2i}|\boldsymbol{x}_i) \right\}$$

where w_i is the weight for observation i or one if no weights were specified. Instead of estimating $\sigma_{u|v}$ and σ_v directly, we estimate $\ln\sigma_{u|v}$ and $\ln\sigma_v$.

For multiple endogenous regressors, we have

$$\ln f(\boldsymbol{y}_{2i}|\boldsymbol{x}_i) = -\frac{1}{2}\left(\ln 2\pi + \ln|\boldsymbol{\Sigma}_{22}| + \boldsymbol{v}_i'\boldsymbol{\Sigma}_{22}^{-1}\boldsymbol{v}_i \right)$$

and $\ln f(y_{1i}|\boldsymbol{y}_{2i}, \boldsymbol{x}_i)$ is the same as before, except that now

$$m_i = \boldsymbol{z}_i\boldsymbol{\delta} + (\boldsymbol{y}_{2i} - \boldsymbol{x}_i\boldsymbol{\Pi})\boldsymbol{\Sigma}_{22}^{-1}\boldsymbol{\Sigma}_{21}$$

Instead of maximizing the log-likelihood function with respect to $\boldsymbol{\Sigma}$, we maximize with respect to the Cholesky decomposition $\boldsymbol{S}$ of $\boldsymbol{\Sigma}$; that is, there exists a lower triangular matrix $\boldsymbol{S}$ such that $\boldsymbol{S}\boldsymbol{S}' = \boldsymbol{\Sigma}$. This maximization ensures that $\boldsymbol{\Sigma}$ is positive definite, as a covariance matrix must be. Let

$$
S = \begin{bmatrix} s_{11} & 0 & 0 & \ldots & 0 \\ s_{21} & s_{22} & 0 & \ldots & 0 \\ s_{31} & s_{32} & s_{33} & \ldots & 0 \\ \vdots & \vdots & \vdots & \ddots & \vdots \\ s_{p+1,1} & s_{p+1,2} & s_{p+1,3} & \ldots & s_{p+1,p+1} \end{bmatrix}
$$

The two-step estimates are obtained using Newey's (1987) minimum chi-squared estimator. The procedure is identical to the one described in [R] **ivprobit**, except that `tobit` is used instead of `probit`.

Acknowledgments

The two-step estimator is based on the `tobitiv` command written by Jonah Gelbach, College of Law, Florida State University, and the `ivtobit` command written by Joe Harkness, Institute of Policy Studies, Johns Hopkins University.

References

Miranda, A., and S. Rabe-Hesketh. 2006. Maximum likelihood estimation of endogenous switching and sample selection models for binary, ordinal, and count variables. *Stata Journal* 6: 285–308.

Newey, W. K. 1987. Efficient estimation of limited dependent variable models with endogenous explanatory variables. *Journal of Econometrics* 36: 231–250.

Wooldridge, J. M. 2002. *Econometric Analysis of Cross Section and Panel Data.* Cambridge, MA: MIT Press.

Also See

[R] **ivtobit postestimation** — Postestimation tools for ivtobit

[R] **ivprobit** — Probit model with endogenous regressors

[R] **ivregress** — Single-equation instrumental-variables regression

[R] **regress** — Linear regression

[R] **tobit** — Tobit regression

[XT] **xtintreg** — Random-effects interval-data regression models

[XT] **xttobit** — Random-effects tobit models

[U] **20 Estimation and postestimation commands**

Title

ivtobit postestimation — Postestimation tools for ivtobit

Description

The following postestimation commands are available for `ivtobit`:

command	description
`adjust`[1]	adjusted predictions of $\mathbf{x}\beta$
`estat`[2]	AIC, BIC, VCE, and estimation sample summary
`estat` (svy)	postestimation statistics for survey data
`estimates`	cataloging estimation results
`hausman`	Hausman's specification test
`lincom`	point estimates, standard errors, testing, and inference for linear combinations of coefficients
`lrtest`[3]	likelihood-ratio test; not available with two-step estimator
`mfx`	marginal effects or elasticities
`nlcom`	point estimates, standard errors, testing, and inference for nonlinear combinations of coefficients
`predict`	predictions, residuals, influence statistics, and other diagnostic measures
`predictnl`	point estimates, standard errors, testing, and inference for generalized predictions
`suest`[2]	seemingly unrelated estimation
`test`	Wald tests for simple and composite linear hypotheses
`testnl`	Wald tests of nonlinear hypotheses

[1] `adjust` is not appropriate with time-series operators.
[2] `estat ic` and `suest` are not appropriate after `ivtobit, twostep`.
[3] `lrtest` is not appropriate with `svy` estimation results.

See the corresponding entries in the *Stata Base Reference Manual* for details, but see [SVY] **estat** for details about `estat` (svy).

Syntax for predict

After ML or twostep

`predict` [*type*] *newvar* [*if*] [*in*] [, *statistic*]

After ML

`predict` [*type*] {*stub** | *newvarlist*} [*if*] [*in*] , `scores`

statistic	description
Main	
`xb`	linear prediction; the default
`stdp`	standard error of the linear prediction
`stdf`	standard error of the forecast; not available with two-step estimator
`pr(`*a*`,`*b*`)`	$\Pr(a < y_j < b)$; not available with two-step estimator
`e(`*a*`,`*b*`)`	$E(y_j \vert a < y_j < b)$; not available with two-step estimator
`ystar(`*a*`,`*b*`)`	$E(y_j^*)$, $y_j = \max\{a, \min(y_j, b)\}$; not available with two-step estimator

These statistics are available both in and out of sample; type `predict ... if e(sample) ...` if wanted only for the estimation sample.

`stdf` is not allowed with `svy` estimation results.

where *a* and *b* may be numbers or variables; *a* missing ($a \geq .$) means $-\infty$, and *b* missing ($b \geq .$) means $+\infty$; see [U] **12.2.1 Missing values**.

Options for predict

Main

`xb`, the default, calculates the linear prediction.

`stdp` calculates the standard error of the linear prediction. It can be thought of as the standard error of the predicted expected value or mean for the observation's covariate pattern. The standard error of the prediction is also referred to as the standard error of the fitted value.

`stdf` calculates the standard error of the forecast, which is the standard error of the point prediction for 1 observation. It is commonly referred to as the standard error of the future or forecast value. By construction, the standard errors produced by `stdf` are always larger than those produced by `stdp`; see *Methods and Formulas* in [R] **regress**. `stdf` is not available with the two-step estimator.

`pr(`*a*`,`*b*`)` calculates $\Pr(a < \boldsymbol{x}_j\boldsymbol{b} + u_j < b)$, the probability that $y_j|\boldsymbol{x}_j$ would be observed in the interval (a, b).

a and *b* may be specified as numbers or variable names; *lb* and *ub* are variable names;
`pr(20,30)` calculates $\Pr(20 < \boldsymbol{x}_j\boldsymbol{b} + u_j < 30)$;
`pr(`*lb*`,`*ub*`)` calculates $\Pr(lb < \boldsymbol{x}_j\boldsymbol{b} + u_j < ub)$; and
`pr(20,`*ub*`)` calculates $\Pr(20 < \boldsymbol{x}_j\boldsymbol{b} + u_j < ub)$.

a missing ($a \geq .$) means $-\infty$; `pr(.,30)` calculates $\Pr(-\infty < \boldsymbol{x}_j\boldsymbol{b} + u_j < 30)$;
`pr(`*lb*`,30)` calculates $\Pr(-\infty < \boldsymbol{x}_j\boldsymbol{b} + u_j < 30)$ in observations for which $lb \geq .$
and calculates $\Pr(lb < \boldsymbol{x}_j\boldsymbol{b} + u_j < 30)$ elsewhere.

b missing ($b \geq .$) means $+\infty$; `pr(20,.)` calculates $\Pr(+\infty > \boldsymbol{x}_j\boldsymbol{b} + u_j > 20)$;
`pr(20,`*ub*`)` calculates $\Pr(+\infty > \boldsymbol{x}_j\boldsymbol{b} + u_j > 20)$ in observations for which $ub \geq .$
and calculates $\Pr(20 < \boldsymbol{x}_j\boldsymbol{b} + u_j < ub)$ elsewhere.

`pr(`*a*`,`*b*`)` is not available with the two-step estimator.

`e(`*a*`,`*b*`)` calculates $E(\boldsymbol{x}_j\boldsymbol{b} + u_j \mid a < \boldsymbol{x}_j\boldsymbol{b} + u_j < b)$, the expected value of $y_j|\boldsymbol{x}_j$ conditional on $y_j|\boldsymbol{x}_j$ being in the interval (a, b), meaning that $y_j|\boldsymbol{x}_j$ is censored. *a* and *b* are specified as they are for `pr()`. `e(`*a*`,`*b*`)` is not available with the two-step estimator.

`ystar(`*a*`,`*b*`)` calculates $E(y_j^*)$, where $y_j^* = a$ if $\boldsymbol{x}_j\boldsymbol{b} + u_j \leq a$, $y_j^* = b$ if $\boldsymbol{x}_j\boldsymbol{b} + u_j \geq b$, and $y_j^* = \boldsymbol{x}_j\boldsymbol{b} + u_j$ otherwise, meaning that y_j^* is truncated. *a* and *b* are specified as they are for `pr()`. `ystar(`*a*`,`*b*`)` is not available with the two-step estimator.

`scores`, not available with `twostep`, calculates equation-level score variables.

For models with one endogenous regressor, five new variables are created.

The first new variable will contain $\partial \ln L / \partial (\boldsymbol{z}_i \boldsymbol{\delta})$.

The second new variable will contain $\partial \ln L / \partial (\boldsymbol{x}_i \boldsymbol{\Pi})$.

The third new variable will contain $\partial \ln L / \partial \alpha$.

The fourth new variable will contain $\partial \ln L / \partial \ln \sigma_{u|v}$.

The fifth new variable will contain $\partial \ln L / \partial \ln \sigma_v$.

For models with p endogenous regressors, $p + \{(p+1)(p+2)\}/2 + 1$ new variables are created.

The first new variable will contain $\partial \ln L / \partial (\boldsymbol{z}_i \boldsymbol{\delta})$.

The second through $(p+1)$th new score variables will contain $\partial \ln L / \partial (\boldsymbol{x}_i \boldsymbol{\Pi}_k)$, $k = 1, \ldots, p$, where $\boldsymbol{\Pi}_k$ is the kth column of $\boldsymbol{\Pi}$.

The remaining score variables will contain the partial derivatives of $\ln L$ with respect to s_{11}, s_{21}, ..., $s_{p+1,1}$, s_{22}, ..., $s_{p+1,2}$, ..., $s_{p+1,p+1}$, where $s_{m,n}$ denotes the (m, n) element of the Cholesky decomposition of the error covariance matrix.

Remarks

Remarks are presented under the following headings:

Marginal effects
Obtaining predicted values

Marginal effects

▷ Example 1

We can obtain marginal effects by using the `mfx` command after `ivtobit`. For the labor-supply model of example 1 in [R] **ivtobit**, suppose that we wanted to know the marginal effects on the woman's expected income, conditional on her income being greater than $10,000.

```
. mfx compute, predict(e(10, .)) eqlist(fem_inc) force
Marginal effects after ivtobit
      y  = E(fem_inc|fem_inc>10) (predict, e(10, .))
         =  26.027323
```

variable	dy/dx	Std. Err.	z	P>|z|	[95% C.I.]		X
other_~c	-.3352235	.04722	-7.10	0.000	-.427772	-.242675	49.6023
fem_educ	1.236682	.13825	8.95	0.000	.965714	1.50765	12.046
kids	-1.182386	.26181	-4.52	0.000	-1.69553	-.669241	1.976

Increasing the number of children in the family by one decreases the conditional expected wage by $1,182 (wages in our dataset are measured in thousands of dollars). If we had not used the `eqlist()` option, `mfx` would have included a row in the table for `male_educ`, for which the marginal effect is identically zero since that variable is an instrument. We used the `force` option so that `mfx` would calculate standard errors. Otherwise, `mfx` would refuse to do so because `other_inc` is an endogenous variable.

◁

Obtaining predicted values

After fitting your model using `ivtobit`, you can obtain the linear prediction and its standard error for both the estimation sample and other samples using the `predict` command. If you used the maximum likelihood estimator, you can also obtain conditional expected values of the observed and latent dependent variables, the standard error of the forecast, and the probability of observing the dependent variable in a specified interval. See [U] **20 Estimation and postestimation commands** and [R] **predict**.

Methods and Formulas

All postestimation commands listed above are implemented as ado-files.

The linear prediction is calculated as $z_i\widehat{\boldsymbol{\delta}}$, where $\widehat{\boldsymbol{\delta}}$ is the estimated value of $\boldsymbol{\delta}$, and z_i and $\boldsymbol{\delta}$ are defined in (1*a*) of [R] **ivtobit**. Expected values and probabilities are calculated using the same formulas as those used by the standard exogenous tobit model.

Also See

[R] **ivtobit** — Tobit model with endogenous regressors

[U] **20 Estimation and postestimation commands**

Title

jackknife — Jackknife estimation

Syntax

jackknife *exp_list* [, *options eform_option*] : *command*

options	description
Main	
eclass	number of observations used is stored in e(N)
rclass	number of observations used is stored in r(N)
n(*exp*)	specify *exp* that evaluates to the number of observations used
Options	
cluster(*varlist*)	variables identifying sample clusters
idcluster(*newvar*)	create new cluster ID variable
saving(*filename*, ...)	save results to *filename*; save statistics in double precision; save results to *filename* every # replications
keep	keep pseudovalues
mse	use MSE formula for variance estimation
Reporting	
level(#)	set confidence level; default is level(95)
notable	suppress table of results
noheader	suppress table header
nolegend	suppress table legend
verbose	display the full table legend
nodots	suppress the replication dots
noisily	display any output from *command*
trace	trace the *command*
title(*text*)	use *text* as title for jackknife results
Advanced	
nodrop	do not drop observations
reject(*exp*)	identify invalid results
† *eform_option*	display coefficient table in exponentiated form

† *eform_option* does not appear in the dialog.

svy is allowed; see [SVY] **svy jackknife**.

All weight types supported by *command* are allowed except aweights; see [U] **11.1.6 weight**.

See [U] **20 Estimation and postestimation commands** for more capabilities of estimation commands.

exp_list contains (*name*: *elist*)
elist
eexp

elist contains	*newvar* = (*exp*)
	(*exp*)
eexp is	*specname*
	[*eqno*]*specname*
specname is	_b
	_b[]
	_se
	_se[]
eqno is	##
	name

exp is a standard Stata expression; see [U] **13 Functions and expressions**.

Distinguish between [], which are to be typed, and [ ], which indicate optional arguments.

Description

jackknife performs jackknife estimation. Typing

```
. jackknife exp_list: command
```

executes *command* once for each observation in the dataset, leaving the associated observation out of the calculations that make up *exp_list*.

command defines the statistical command to be executed. Most Stata commands and user-written programs can be used with jackknife, as long as they follow standard Stata syntax and allow the *if* qualifier; see [U] **11 Language syntax**. The by prefix may not be part of *command*.

exp_list specifies the statistics to be collected from the execution of *command*. If *command* changes the contents in e(b), *exp_list* is optional and defaults to _b.

Many estimation commands allow the vce(jackknife) option. For those commands, we recommend using vce(jackknife) over jackknife since the estimation command already handles clustering and other model-specific details for you. The jackknife prefix command is intended for use with nonestimation commands, such as summarize, user-written commands, or functions of coefficients.

jknife is a synonym for jackknife.

Options

Main

eclass, rclass, and n(*exp*) specify where *command* saves the number of observations on which it based the calculated results. We strongly advise you to specify one of these options.

eclass specifies that *command* save the number of observations in e(N).

rclass specifies that *command* save the number of observations in r(N).

n(*exp*) specifies an expression that evaluates to the number of observations used. Specifying n(r(N)) is equivalent to specifying option rclass. Specifying n(e(N)) is equivalent to specifying option eclass. If *command* saves the number of observations in r(N1), specify n(r(N1)).

If you specify no options, jackknife will assume eclass or rclass, depending upon which of e(N) and r(N) is not missing (in that order). If both e(N) and r(N) are missing, jackknife assumes that all observations in the dataset contribute to the calculated result. If that assumption is incorrect, the reported standard errors will be incorrect. For instance, say that you specify

```
. jackknife coef=_b[x2]: myreg y x1 x2 x3
```

where myreg uses e(n) instead of e(N) to identify the number of observations used in calculations. Further assume that observation 42 in the dataset has x3 equal to missing. The 42nd observation plays no role in obtaining the estimates, but jackknife has no way of knowing that and will use the wrong N. If, on the other hand, you specify

```
. jackknife coef=_b[x2], n(e(n)): myreg y x1 x2 x3
```

jackknife will notice that observation 42 plays no role. Option n(e(n)) is specified because myreg is an estimation command but it saves the number of observations used in e(n) (instead of the standard e(N)). When jackknife runs the regression omitting the 42nd observation, jackknife will observe that e(n) has the same value as when jackknife previously ran the regression using all the observations. Thus jackknife will know that myreg did not use the observation.

Options

cluster(*varlist*) specifies the variables identifying sample clusters. If cluster() is specified, one cluster is left out of each call to *command*, instead of 1 observation.

idcluster(*newvar*) creates a new variable containing a unique integer identifier for each resampled cluster, starting at 1 and leading up to the number of clusters. This option may be specified only when the cluster() option is specified. idcluster() helps identify the cluster to which a pseudovalue belongs.

saving(*filename* [, *suboptions*]) creates a Stata data file (.dta file) consisting of (for each statistic in *exp_list*) a variable containing the jackknife replicates.

double specifies that the results for each replication be stored as doubles, meaning 8-byte reals. By default, they are stored as floats, meaning 4-byte reals. This option may be used without the saving() option to compute the variance estimates using double precision.

every(#) specifies that results be written to disk every #th replication. every() should be specified only in conjunction with saving() when *command* takes a long time for each replication. This will allow recovery of partial results should some other software crash your computer. See [P] **postfile**.

replace specifies that *filename* be overwritten, if it exists. This option does not appear in the dialog.

keep specifies that new variables be added to the dataset containing the pseudovalues of the requested statistics. For instance, if you typed

```
. jackknife coef=_b[x2], eclass keep: regress y x1 x2 x3
```

new variable coef would be added to the dataset containing the pseudovalues for _b[x2]. Let b be the value of _b[x2] when all observations are used to fit the model, and let $b(j)$ be the value when the jth observation is omitted. The pseudovalues are defined as

$$\text{pseudovalue}_j = N\{b - b(j)\} + b(j)$$

where N is the number of observations used to produce b.

`mse` specifies that `jackknife` compute the variance by using deviations of the replicates from the observed value of the statistics based on the entire dataset. By default, `jackknife` computes the variance by using deviations of the pseudovalues from their mean.

Reporting

`level(#)`; see [R] **estimation options**.

`notable` suppresses the display of the table of results.

`noheader` suppresses display of the table header. This option implies `nolegend`.

`nolegend` suppresses display of the table legend. The table legend identifies the rows of the table with the expressions they represent.

`verbose` specifies that the full table legend be displayed. By default, coefficients and standard errors are not displayed.

`nodots` suppresses display of the replication dots. By default, one dot character is displayed for each successful replication. A red 'x' is displayed if *command* returns an error or if one of the values in *exp_list* is missing.

`noisily` specifies that any output from *command* be displayed. This option implies the `nodots` option.

`trace` causes a trace of the execution of *command* to be displayed. This option implies the `noisily` option.

`title(`*text*`)` specifies a title to be displayed above the table of jackknife results; the default title is `Jackknife results` or what is produced in `e(title)` by an estimation command.

Advanced

`nodrop` prevents observations outside `e(sample)` and the *if* and *in* qualifiers from being dropped before the data are resampled.

`reject(`*exp*`)` identifies an expression that indicates when results should be rejected. When *exp* is true, the resulting values are reset to missing values.

The following option is available with `jackknife` but is not shown in the dialog:

eform_option causes the coefficient table to be displayed in exponentiated form: for each coefficient, exp(b) rather than b is displayed. Standard errors and confidence intervals are also transformed. Display of the intercept, if any, is suppressed.

command determines which of the following are allowed (`eform(`*string*`)` and `eform` are always allowed).

eform_option	description
`eform(`*string*`)`	use *string* for the column title
`eform`	exponentiated coefficient; *string* is "`exp(b)`"
`hr`	hazard ratio; *string* is "`Haz. Ratio`"
`irr`	incidence-rate ratio; *string* is "`IRR`"
`or`	odds ratio; *string* is "`Odds ratio`"
`rrr`	relative-risk ratio; *string* is "`RRR`"

Remarks

Remarks are presented under the following headings:

Introduction
Jackknifed standard deviation
Collecting multiple statistics
Collecting coefficients

Introduction

Although the jackknife—developed in the late 1940s and early 1950s—is of largely historical interest today, it is still useful in searching for overly influential observations. This feature is often forgotten. In any case, the jackknife is

1. an alternative, first-order unbiased estimator for a statistic;
2. a data-dependent way to calculate the standard error of the statistic and to obtain significance levels and confidence intervals; and
3. a way of producing measures called pseudovalues for each observation, reflecting the observation's influence on the overall statistic.

The idea behind the simplest form of the jackknife—the one implemented here—is to repeatedly calculate the statistic in question, each time omitting just one of the dataset's observations. Assume that our statistic of interest is the sample mean. Let y_j be the jth observation of our data on some measurement y, where $j = 1, \ldots, N$ and N is the sample size. If $\overline{y}$ is the sample mean of y using the entire dataset and $\overline{y}_{(j)}$ is the mean when the jth observation is omitted, then

$$\overline{y} = \frac{(N-1)\,\overline{y}_{(j)} + y_j}{N}$$

Solving for y_j, we obtain

$$y_j = N\,\overline{y} - (N-1)\,\overline{y}_{(j)}$$

These are the pseudovalues that `jackknife` calculates. To move this discussion beyond the sample mean, let $\widehat{\theta}$ be the value of our statistic (not necessarily the sample mean) using the entire dataset, and let $\widehat{\theta}_{(j)}$ be the computed value of our statistic with the jth observation omitted. The pseudovalue for the jth observation is

$$\widehat{\theta}_j^* = N\,\widehat{\theta} - (N-1)\,\widehat{\theta}_{(j)}$$

The mean of the pseudovalues is the alternative, first-order unbiased estimator mentioned above, and the standard error of the mean of the pseudovalues is an estimator for the standard error of $\widehat{\theta}$ (Tukey 1958).

When the `cluster()` option is given, clusters are omitted instead of observations, and N is the number of clusters instead of the sample size.

The jackknife estimate of variance has been largely replaced by the bootstrap estimate (see [R] **bootstrap**), which is widely viewed as more efficient and robust. The use of jackknife pseudovalues to detect outliers is too often forgotten and is something the bootstrap does not provide. See Mosteller and Tukey (1977, 133–163) and Mooney and Duval (1993, 22–27) for more information.

▷ Example 1

As our first example, we will show that the jackknife standard error of the sample mean is equivalent to the standard error of the sample mean computed using the classical formula in the `ci` command. We use the `double` option to compute the standard errors with the same precision as the `ci` command.

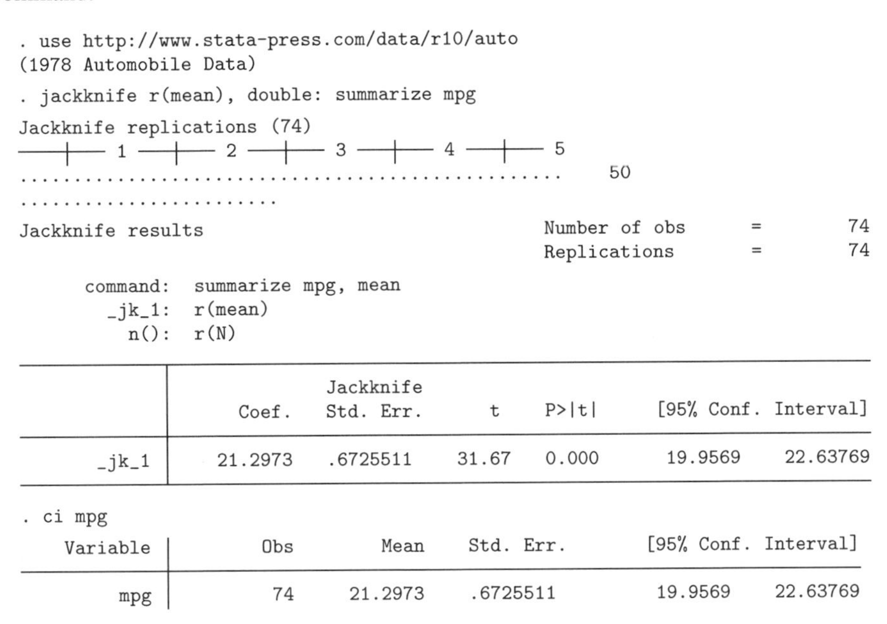

```
. use http://www.stata-press.com/data/r10/auto
(1978 Automobile Data)

. jackknife r(mean), double: summarize mpg
Jackknife replications (74)
----+--- 1 ---+--- 2 ---+--- 3 ---+--- 4 ---+--- 5
..................................................    50
........................
Jackknife results                               Number of obs    =        74
                                                Replications     =        74

      command:  summarize mpg, mean
        _jk_1:  r(mean)
          n():  r(N)

------------------------------------------------------------------------------
             |              Jackknife
             |      Coef.   Std. Err.      t    P>|t|     [95% Conf. Interval]
-------------+----------------------------------------------------------------
       _jk_1 |    21.2973    .6725511    31.67   0.000     19.9569     22.63769
------------------------------------------------------------------------------

. ci mpg
    Variable |        Obs        Mean    Std. Err.       [95% Conf. Interval]
-------------+---------------------------------------------------------------
         mpg |         74     21.2973    .6725511        19.9569     22.63769
```

◁

Jackknifed standard deviation

▷ Example 2

Mosteller and Tukey (1977, 139–140) request a 95% confidence interval for the standard deviation of the 11 values:

$$0.1,\quad 0.1,\quad 0.1,\quad 0.4,\quad 0.5,\quad 1.0,\quad 1.1,\quad 1.3,\quad 1.9,\quad 1.9,\quad 4.7$$

Stata's `summarize` command calculates the mean and standard deviation and saves them as `r(mean)` and `r(sd)`. To obtain the jackknifed standard deviation of the 11 values and save the pseudovalues as a new variable `sd`, we would type

```
. input x

             x
  1. 0.1
  2. 0.1
  3. 0.1
  4. 0.4
  5. 0.5
  6. 1.0
  7. 1.1
  8. 1.3
  9. 1.9
 10. 1.9
 11. 4.7
 12. end
```

```
. jackknife sd=r(sd), rclass keep: summarize x
(running summarize on estimation sample)
Jackknife replications (11)
----+--- 1 ---+--- 2 ---+--- 3 ---+--- 4 ---+--- 5
...........
Jackknife results                          Number of obs      =         11
                                           Replications       =         11

      command:  summarize x
           sd:  r(sd)
          n():  r(N)

------------------------------------------------------------------------------
             |              Jackknife
             |      Coef.   Std. Err.      t    P>|t|     [95% Conf. Interval]
-------------+----------------------------------------------------------------
          sd |   1.343469    .624405     2.15   0.057     -.047792     2.73473
------------------------------------------------------------------------------
```

Interpreting the output, the standard deviation reported by `summarize mpg` is 1.34. The jackknife standard error is 0.62. The 95% confidence interval for the standard deviation is −.048 to 2.73.

By specifying `keep`, `jackknife` creates in our dataset a new variable, `sd`, for the pseudovalues.

```
. list, sep(4)

     +----------------+
     |   x         sd |
     |----------------|
  1. |  .1   1.139977 |
  2. |  .1   1.139977 |
  3. |  .1   1.139977 |
  4. |  .4   .8893147 |
     |----------------|
  5. |  .5    .824267 |
  6. |   1    .632489 |
  7. | 1.1   .6203189 |
  8. | 1.3   .6218889 |
     |----------------|
  9. | 1.9    .835419 |
 10. | 1.9    .835419 |
 11. | 4.7   7.703949 |
     +----------------+
```

The jackknife estimate is the average of variable `sd`, so `sd` contains the individual values of our statistic. We can see that the last observation is substantially larger than the others. The last observation is certainly an outlier, but whether that reflects the considerable information it contains or indicates that it should be excluded from analysis depends on the context of the problem. Here Mosteller and Tukey created the dataset by sampling from an exponential distribution, so the observation is informative.

◁

▷ Example 3

Let us repeat the example above using the automobile dataset, obtaining the standard error of the standard deviation of mpg.

```
. use http://www.stata-press.com/data/r10/auto
(1978 Automobile Data)
. jackknife sd=r(sd), rclass keep: summarize mpg
(running summarize on estimation sample)
Jackknife replications (74)
```

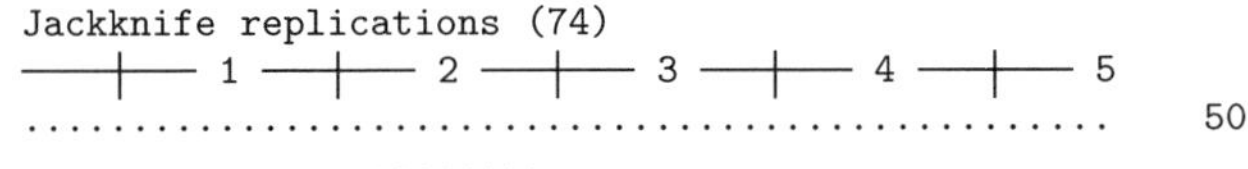


```
..................................................     50
........................

Jackknife results                          Number of obs      =        74
                                           Replications       =        74

      command:  summarize mpg
           sd:  r(sd)
          n():  r(N)

------------------------------------------------------------------------------
             |              Jackknife
             |      Coef.   Std. Err.      t    P>|t|     [95% Conf. Interval]
-------------+----------------------------------------------------------------
          sd |   5.785503   .6072509     9.53   0.000     4.575254    6.995753
------------------------------------------------------------------------------
```

Let's look at sd more carefully:

```
. summarize sd, detail
                    pseudovalues: r(sd)
-------------------------------------------------------------
      Percentiles      Smallest
 1%     2.870471       2.870471
 5%     2.870471       2.870471
10%     2.906255       2.870471       Obs                  74
25%     3.328489       2.870471       Sum of Wgt.          74

50%     3.948335                      Mean           5.817374
                        Largest       Std. Dev.       5.22377
75%     6.844418       17.34316
90%     9.597018        19.7617       Variance       27.28777
95%     17.34316        19.7617       Skewness        4.07202
99%     38.60905       38.60905       Kurtosis       23.37823

. list make mpg sd if sd > 30

     +-------------------------------+
     |      make   mpg          sd   |
     |-------------------------------|
71.  | VW Diesel    41    38.60905   |
     +-------------------------------+
```

Here the VW Diesel is the only diesel car in our dataset. ◁

(Continued on next page)

Collecting multiple statistics

▷ Example 4

`jackknife` is not limited to collecting just one statistic. For instance, we can use `summarize, detail` and then obtain the jackknife estimate of the standard deviation and skewness. `summarize, detail` saves the standard deviation in `r(sd)` and the skewness in `r(skewness)`, so we might type

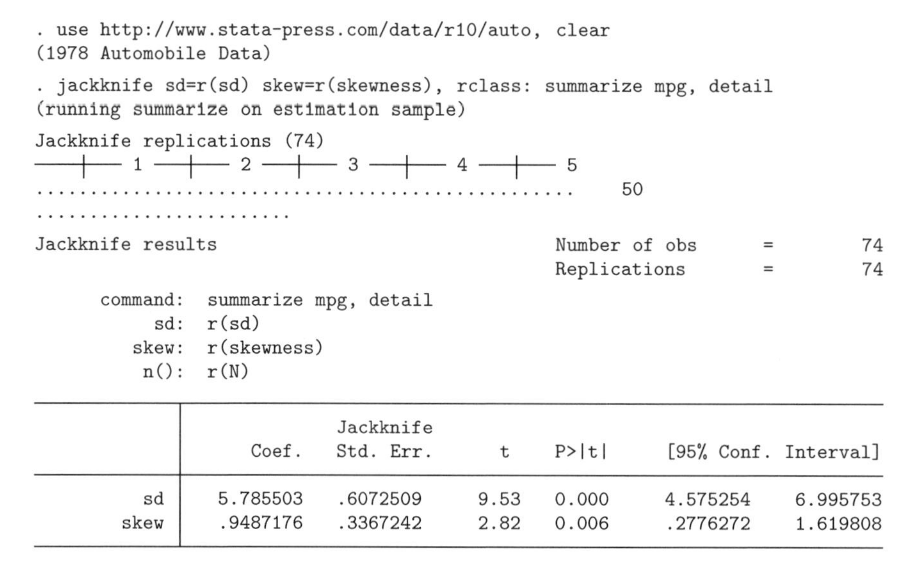

```
. use http://www.stata-press.com/data/r10/auto, clear
(1978 Automobile Data)

. jackknife sd=r(sd) skew=r(skewness), rclass: summarize mpg, detail
(running summarize on estimation sample)

Jackknife replications (74)
----+--- 1 ---+--- 2 ---+--- 3 ---+--- 4 ---+--- 5
..................................................    50
........................

Jackknife results                             Number of obs      =        74
                                              Replications       =        74

      command:  summarize mpg, detail
           sd:  r(sd)
         skew:  r(skewness)
          n():  r(N)
```

	Coef.	Jackknife Std. Err.	t	P>|t|	[95% Conf.	Interval]
sd	5.785503	.6072509	9.53	0.000	4.575254	6.995753
skew	.9487176	.3367242	2.82	0.006	.2776272	1.619808

◁

Collecting coefficients

▷ Example 5

`jackknife` can also collect coefficients from estimation commands. For instance, using `auto.dta`, we might wish to obtain the jackknife standard errors of the coefficients from a regression in which we model the mileage of a car by its weight and trunk space. To do this, we could refer to the coefficients as `_b[weight]`, `_b[trunk]`, `_se[weight]`, and `_se[trunk]` in the *exp_list*, or we could simply use the extended expressions `_b`. In fact, `jackknife` assumes `_b` by default when used with estimation commands.

```
. use http://www.stata-press.com/data/r10/auto, clear
(1978 Automobile Data)

. jackknife: regress mpg weight trunk
(running regress on estimation sample)

Jackknife replications (74)
----+--- 1 ---+--- 2 ---+--- 3 ---+--- 4 ---+--- 5
..................................................    50
........................

Linear regression                               Number of obs      =        74
                                                Replications       =        74
                                                F(   2,      73)   =     78.10
                                                Prob > F           =    0.0000
                                                R-squared          =    0.6543
                                                Adj R-squared      =    0.6446
                                                Root MSE           =    3.4492

------------------------------------------------------------------------------
             |              Jackknife
         mpg |      Coef.   Std. Err.      t    P>|t|     [95% Conf. Interval]
-------------+----------------------------------------------------------------
      weight |  -.0056527   .0010216    -5.53   0.000    -.0076887   -.0036167
       trunk |   -.096229   .1486236    -0.65   0.519    -.3924354    .1999773
       _cons |   39.68913   1.873324    21.19   0.000      35.9556    43.42266
------------------------------------------------------------------------------
```

If you are going to use `jackknife` to estimate standard errors of model coefficients, we recommend using the `vce(jackknife)` option when it is allowed with the estimation command; see [R] ***vce_option***.

```
. regress mpg weight trunk, vce(jackknife, nodots)

Linear regression                               Number of obs      =        74
                                                Replications       =        74
                                                F(   2,      73)   =     78.10
                                                Prob > F           =    0.0000
                                                R-squared          =    0.6543
                                                Adj R-squared      =    0.6446
                                                Root MSE           =    3.4492

------------------------------------------------------------------------------
             |              Jackknife
         mpg |      Coef.   Std. Err.      t    P>|t|     [95% Conf. Interval]
-------------+----------------------------------------------------------------
      weight |  -.0056527   .0010216    -5.53   0.000    -.0076887   -.0036167
       trunk |   -.096229   .1486236    -0.65   0.519    -.3924354    .1999773
       _cons |   39.68913   1.873324    21.19   0.000      35.9556    43.42266
------------------------------------------------------------------------------
```

◁

John Wilder Tukey (1915–2000) was born in Massachusetts. He studied chemistry at Brown and mathematics at Princeton and afterward worked at both Princeton and Bell Labs, as well as being involved in a great many government projects, consultancies, and committees. He made outstanding contributions to several areas of statistics, including time series, multiple comparisons, robust statistics, and exploratory data analysis. Tukey was extraordinarily energetic and inventive, not least in his use of terminology: he is credited with the terms bit and software, in addition to ANOVA, boxplot, data analysis, hat matrix, jackknife, stem-and-leaf plot, trimming, and winsorizing, among many others. Tukey's direct and indirect impacts mark him as one the greatest statisticians of all time.

Saved Results

`jknife` saves the following in `e()`:

Scalars

`e(N_reps)`	number of complete replications	`e(N)`	sample size
`e(N_misreps)`	number of incomplete replications	`e(N_clust)`	number of clusters
`e(df_r)`	degrees of freedom		

Macros

`e(cmdname)`	command name from *command*	`e(pseudo)`	new variables containing pseudovalues
`e(cmd)`	same as `e(cmdname)` or `jackknife`		
`e(command)`	*command*	`e(prefix)`	`jackknife`
`e(cmdline)`	command as typed	`e(cluster)`	cluster variables
`e(nfunction)`	`e(N)`, `r(N)`, `n()` option, or empty	`e(exp#)`	expression for the #th statistic
`e(wtype)`	weight type	`e(mse)`	from `mse` option
`e(wexp)`	weight expression	`e(vce)`	`jackknife`
`e(title)`	title in estimation output	`e(vcetype)`	title used to label Std. Err.
		`e(properties)`	`b V`

Matrices

`e(b)`	observed statistics	`e(b_jk)`	jackknife estimates
`e(V)`	jackknife variance–covariance matrix		

When *exp_list* is `_b`, `jackknife` will also carry forward most of the results already in `e()` from *command*.

Methods and Formulas

`jackknife` is implemented as an ado-file.

Let $\widehat{\theta}$ be the observed value of the statistic, that is, the value of the statistic calculated using the original dataset. Let $\widehat{\theta}_{(j)}$ be the value of the statistic computed by leaving out the *j*th observation (or cluster); thus $j = 1, 2, \dots, N$ identifies an individual observation (or cluster), and N is the total number of observations (or clusters). The *j*th pseudovalue is given by

$$\widehat{\theta}_j^* = \widehat{\theta}_{(j)} + N\{\widehat{\theta} - \widehat{\theta}_{(j)}\}$$

When the `mse` option is specified, the standard error is estimated as

$$\widehat{\text{se}} = \left\{ \frac{N-1}{N} \sum_{j=1}^{N} (\widehat{\theta}_{(j)} - \widehat{\theta})^2 \right\}^{1/2}$$

and the jackknife estimate is

$$\bar{\theta}_{(.)} = \frac{1}{N} \sum_{j=1}^{N} \widehat{\theta}_{(j)}$$

Otherwise, the standard error is estimated as

$$\widehat{\text{se}} = \left\{ \frac{1}{N(N-1)} \sum_{j=1}^{N} (\widehat{\theta}_j^* - \bar{\theta}^*)^2 \right\}^{1/2} \qquad \bar{\theta}^* = \frac{1}{N} \sum_{j=1}^{N} \widehat{\theta}_j^*$$

where $\bar{\theta}^*$ is the jackknife estimate. The variance–covariance matrix is similarly computed.

References

Brillinger, D. R. 2002. John W. Tukey: His life and professional contributions. *Annals of Statistics* 30: 1535–1575.

Gould, W. 1995. sg34: Jackknife estimation. *Stata Technical Bulletin* 24: 25–29. Reprinted in *Stata Technical Bulletin Reprints*, vol. 4, pp. 165–170.

Mooney, C. Z., and R. D. Duval. 1993. *Bootstrapping: A Nonparametric Approach to Statistical Inference.* Newbury Park, CA: Sage.

Mosteller, F., and J. W. Tukey. 1977. *Data Analysis and Regression: A Second Course in Statistics.* Reading, MA: Addison–Wesley.

Tukey, J. W. 1958. Bias and confidence in not-quite large samples. Abstract in *Annals of Mathematical Statistics* 29: 614.

Also See

[R] **jackknife postestimation** — Postestimation tools for jackknife

[R] **bootstrap** — Bootstrap sampling and estimation

[R] **permute** — Monte Carlo permutation tests

[R] **simulate** — Monte Carlo simulations

[SVY] **svy jackknife** — Jackknife estimation for survey data

[U] **13.5 Accessing coefficients and standard errors**

[U] **13.6 Accessing results from Stata commands**

[U] **20 Estimation and postestimation commands**

Title

jackknife postestimation — Postestimation tools for jackknife

Description

The following postestimation commands are available for `jackknife`:

command	description
* `adjust`	adjusted predictions of $\mathbf{x}\beta$ or $\exp(\mathbf{x}\beta)$
`estat`	AIC, BIC, VCE, and estimation sample summary
`estimates`	cataloging estimation results
`lincom`	point estimates, standard errors, testing, and inference for linear combinations of coefficients
* `mfx`	marginal effects or elasticities
`nlcom`	point estimates, standard errors, testing, and inference for nonlinear combinations of coefficients
* `predict`	predictions, residuals, influence statistics, and other diagnostic measures
* `predictnl`	point estimates, standard errors, testing, and inference for nonlinear combinations of coefficients
`test`	Wald tests for simple and composite linear hypotheses
`testnl`	Wald tests of nonlinear hypotheses

*This postestimation command is allowed only if it may be used after *command*.
See the corresponding entries in the *Stata Base Reference Manual* for details.

Syntax for predict

The syntax of `predict` (and even if `predict` is allowed) following `jackknife` depends upon the *command* used with `jackknife`.

Methods and Formulas

All postestimation commands listed above are implemented as ado-files.

Also See

[R] **jackknife** — Jackknife estimation

[U] **20 Estimation and postestimation commands**

Title

kappa — Interrater agreement

Syntax

Interrater agreement, two unique raters

`kap` *varname*$_1$ *varname*$_2$ [*if*] [*in*] [*weight*] [, *options*]

Weights for weighting disagreements

`kapwgt` *wgtid* [1 \ # 1 [\ # # 1 . . .]]

Interrater agreement, nonunique raters, variables record ratings for each rater

`kap` *varname*$_1$ *varname*$_2$ *varname*$_3$ [. . .] [*if*] [*in*] [*weight*]

Interrater agreement, nonunique raters, variables record frequency of ratings

`kappa` *varlist* [*if*] [*in*]

options	description
Main	
`tab`	display table of assessments
`wgt(`*wgtid*`)`	specify how to weight disagreements; see *Options* for alternatives
`absolute`	treat rating categories as absolute

`fweight`s are allowed; see [U] **11.1.6 weight**.

Description

`kap` (first syntax) calculates the kappa-statistic measure of interrater agreement when there are two unique raters and two or more ratings.

`kapwgt` defines weights for use by `kap` in measuring the importance of disagreements.

`kap` (second syntax) and `kappa` calculate the kappa-statistic measure when there are two or more (nonunique) raters and two outcomes, more than two outcomes when the number of raters is fixed, and more than two outcomes when the number of raters varies. `kap` (second syntax) and `kappa` produce the same results; they merely differ in how they expect the data to be organized.

`kap` assumes that each observation is a subject. *varname*$_1$ contains the ratings by the first rater, *varname*$_2$ by the second rater, and so on.

`kappa` also assumes that each observation is a subject. The variables, however, record the frequencies with which ratings were assigned. The first variable records the number of times the first rating was assigned, the second variable records the number of times the second rating was assigned, and so on.

Options

Main

`tab` displays a tabulation of the assessments by the two raters.

`wgt(`*wgtid*`)` specifies that *wgtid* be used to weight disagreements. You can define your own weights by using `kapwgt`; `wgt()` then specifies the name of the user-defined matrix. For instance, you might define

```
. kapwgt mine 1 \ .8 1 \ 0 .8 1 \ 0 0 .8 1
```

and then

```
. kap rata ratb, wgt(mine)
```

Also two prerecorded weights are available.

`wgt(w)` specifies weights $1 - |i - j|/(k - 1)$, where i and j index the rows and columns of the ratings by the two raters and k is the maximum number of possible ratings.

`wgt(w2)` specifies weights $1 - \{(i - j)/(k - 1)\}^2$.

`absolute` is relevant only if `wgt()` is also specified; see `wgt()` above. Option `absolute` modifies how i, j, and k are defined and how corresponding entries are found in a user-defined weighting matrix. When `absolute` is not specified, i and j refer to the row and column index, not to the ratings themselves. Say that the ratings are recorded as $\{0, 1, 1.5, 2\}$. There are four ratings; $k = 4$, and i and j are still 1, 2, 3, and 4 in the formulas above. Index 3, for instance, corresponds to rating = 1.5. This system is convenient but can, with some data, lead to difficulties.

When `absolute` is specified, all ratings must be integers, and they must be coded from the set $\{1, 2, 3, \ldots\}$. Not all values need be used; integer values that do not occur are simply assumed to be unobserved.

Remarks

Remarks are presented under the following headings:

Two raters
More than two raters

The kappa-statistic measure of agreement is scaled to be 0 when the amount of agreement is what would be expected to be observed by chance and 1 when there is perfect agreement. For intermediate values, Landis and Koch (1977a, 165) suggest the following interpretations:

below 0.0	Poor
0.00 – 0.20	Slight
0.21 – 0.40	Fair
0.41 – 0.60	Moderate
0.61 – 0.80	Substantial
0.81 – 1.00	Almost Perfect

Two raters

▷ Example 1

Consider the classification by two radiologists of 85 xeromammograms as normal, benign disease, suspicion of cancer, or cancer (a subset of the data from Boyd et al. 1982 and discussed in the context of kappa in Altman 1991, 403–405).

```
. use http://www.stata-press.com/data/r10/rate2
(Altman p. 403)
. tabulate rada radb
```

Radiologist A's assessment	Radiologist B's assessment normal	benign	suspect	cancer	Total
normal	21	12	0	0	33
benign	4	17	1	0	22
suspect	3	9	15	2	29
cancer	0	0	0	1	1
Total	28	38	16	3	85

Our dataset contains two variables: `rada`, radiologist A's assessment, and `radb`, radiologist B's assessment. Each observation is a patient.

We can obtain the kappa measure of interrater agreement by typing

```
. kap rada radb
```

Agreement	Expected Agreement	Kappa	Std. Err.	Z	Prob>Z
63.53%	30.82%	0.4728	0.0694	6.81	0.0000

If each radiologist had made his determination randomly (but with probabilities equal to the overall proportions), we would expect the two radiologists to agree on 30.8% of the patients. In fact, they agreed on 63.5% of the patients, or 47.3% of the way between random agreement and perfect agreement. The amount of agreement indicates that we can reject the hypothesis that they are making their determinations randomly.

◁

▷ Example 2: Weighted kappa, prerecorded weight w

There is a difference between two radiologists disagreeing about whether a xeromammogram indicates cancer or the suspicion of cancer and disagreeing about whether it indicates cancer or is normal. The weighted kappa attempts to deal with this. `kap` provides two "prerecorded" weights, `w` and `w2`:

```
. kap rada radb, wgt(w)
Ratings weighted by:
   1.0000  0.6667  0.3333  0.0000
   0.6667  1.0000  0.6667  0.3333
   0.3333  0.6667  1.0000  0.6667
   0.0000  0.3333  0.6667  1.0000
```

Agreement	Expected Agreement	Kappa	Std. Err.	Z	Prob>Z
86.67%	69.11%	0.5684	0.0788	7.22	0.0000

The `w` weights are given by $1 - |i - j|/(k - 1)$, where i and j index the rows of columns of the ratings by the two raters and k is the maximum number of possible ratings. The weighting matrix is printed above the table. Here the rows and columns of the 4×4 matrix correspond to the ratings normal, benign, suspicious, and cancerous.

A weight of 1 indicates that an observation should count as perfect agreement. The matrix has 1s down the diagonals—when both radiologists make the same assessment, they are in agreement. A weight of, say, 0.6667 means that they are in two-thirds agreement. In our matrix, they get that score if they are "one apart"—one radiologist assesses cancer and the other is merely suspicious, or one is suspicious and the other says benign, and so on. An entry of 0.3333 means that they are in one-third agreement, or, if you prefer, two-thirds disagreement. That is the score attached when they are "two apart". Finally, they are in complete disagreement when the weight is zero, which happens only when they are three apart—one says cancer and the other says normal. ◁

▷ Example 3: Weighted kappa, prerecorded weight w2

The other prerecorded weight is `w2`, where the weights are given by $1 - \{(i - j)/(k - 1)\}^2$:

```
. kap rada radb, wgt(w2)
Ratings weighted by:
   1.0000   0.8889   0.5556   0.0000
   0.8889   1.0000   0.8889   0.5556
   0.5556   0.8889   1.0000   0.8889
   0.0000   0.5556   0.8889   1.0000

             Expected
Agreement   Agreement     Kappa   Std. Err.         Z      Prob>Z
-----------------------------------------------------------------
  94.77%      84.09%     0.6714     0.1079       6.22      0.0000
```

The `w2` weight makes the categories even more alike and is probably inappropriate here. ◁

▷ Example 4: Weighted kappa, user-defined weights

In addition to using prerecorded weights, we can define our own weights with the `kapwgt` command. For instance, we might feel that suspicious and cancerous are reasonably similar, that benign and normal are reasonably similar, but that the suspicious/cancerous group is nothing like the benign/normal group:

```
. kapwgt xm 1 \ .8 1 \ 0 0 1 \ 0 0 .8 1
. kapwgt xm
 1.0000
 0.8000 1.0000
 0.0000 0.0000 1.0000
 0.0000 0.0000 0.8000 1.0000
```

We name the weights `xm`, and after the weight name, we enter the lower triangle of the weighting matrix, using \ to separate rows. We have four outcomes, so we continued entering numbers until we had defined the fourth row of the weighting matrix. If we type `kapwgt` followed by a name and nothing else, it shows us the weights recorded under that name. Satisfied that we have entered them correctly, we now use the weights to recalculate kappa:

```
. kap rada radb, wgt(xm)
Ratings weighted by:
   1.0000   0.8000   0.0000   0.0000
   0.8000   1.0000   0.0000   0.0000
   0.0000   0.0000   1.0000   0.8000
   0.0000   0.0000   0.8000   1.0000

             Expected
Agreement   Agreement     Kappa   Std. Err.         Z      Prob>Z

  80.47%      52.67%     0.5874     0.0865       6.79      0.0000
```

◁

❑ Technical Note

In addition to using weights for weighting the differences in categories, you can specify Stata's traditional weights for weighting the data. In the examples above, we have 85 observations in our dataset—one for each patient. If we only knew the table of outcomes—that there were 21 patients rated normal by both radiologists, etc.—it would be easier to enter the table into Stata and work from it. The easiest way to enter the data is with `tabi`; see [R] **tabulate twoway**.

```
. tabi 21 12 0 0 \ 4 17 1 0 \ 3 9 15 2 \ 0 0 0 1, replace

           |                    col
       row |         1          2          3          4 |     Total
-----------+--------------------------------------------+----------
         1 |        21         12          0          0 |        33
         2 |         4         17          1          0 |        22
         3 |         3          9         15          2 |        29
         4 |         0          0          0          1 |         1
-----------+--------------------------------------------+----------
     Total |        28         38         16          3 |        85

          Pearson chi2(9) =  77.8111   Pr = 0.000
```

`tabi` reported the Pearson χ^2 for this table, but we do not care about it. The important thing is that, with the `replace` option, `tabi` left the table in memory:

```
. list in 1/5

     +-----------------+
     | row   col   pop |
     |-----------------|
  1. |   1     1    21 |
  2. |   1     2    12 |
  3. |   1     3     0 |
  4. |   1     4     0 |
  5. |   2     1     4 |
     +-----------------+
```

The variable `row` is radiologist A's assessment, `col` is radiologist B's assessment, and `pop` is the number so assessed by both. Thus

```
. kap row col [freq=pop]

             Expected
Agreement   Agreement     Kappa   Std. Err.         Z      Prob>Z

  63.53%      30.82%     0.4728     0.0694       6.81      0.0000
```

If we are going to keep these data, the names `row` and `col` are not indicative of what the data reflect. We could (see [U] **12.6 Dataset, variable, and value labels**)

```
. rename row rada
. rename col radb
. label var rada "Radiologist A's assessment"
. label var radb "Radiologist B's assessment"
. label define assess 1 normal 2 benign 3 suspect 4 cancer
. label values rada assess
. label values radb assess
. label data "Altman p. 403"
```

kap's tab option, which can be used with or without weighted data, shows the table of assessments:

```
. kap rada radb [freq=pop], tab
Radiologist
        A's          Radiologist B's assessment
 assessment     normal     benign    suspect     cancer        Total

     normal         21         12          0          0           33
     benign          4         17          1          0           22
    suspect          3          9         15          2           29
     cancer          0          0          0          1            1

      Total         28         38         16          3           85

             Expected
Agreement   Agreement     Kappa   Std. Err.         Z      Prob>Z

  63.53%      30.82%     0.4728     0.0694       6.81      0.0000
```

❑

❑ Technical Note

You have data on individual patients. There are two raters, and the possible ratings are 1, 2, 3, and 4, but neither rater ever used rating 3:

```
. use http://www.stata-press.com/data/r10/rate2no3, clear
. tabulate ratera raterb

                        raterb
    ratera          1          2          4        Total

         1          6          4          3           13
         2          5          3          3           11
         4          1          1         26           28

     Total         12          8         32           52
```

Here kap would determine that the ratings are from the set $\{1, 2, 4\}$ because those were the only values observed. kap would expect a user-defined weighting matrix to be 3×3, and if it were not, kap would issue an error message. In the formula-based weights, the calculation would be based on $i, j = 1, 2, 3$ corresponding to the three observed ratings $\{1, 2, 4\}$.

Specifying the absolute option would clarify that the ratings are 1, 2, 3, and 4; it just so happens that rating $= 3$ was never assigned. If a user-defined weighting matrix were also specified, kap would expect it to be 4×4 or larger (larger because we can think of the ratings being 1, 2, 3, 4, 5, ... and it just so happens that ratings 5, 6, ... were never observed, just as rating $= 3$ was not observed). In the formula-based weights, the calculation would be based on $i, j = 1, 2, 4$.

```
. kap ratera raterb, wgt(w)
Ratings weighted by:
   1.0000   0.5000   0.0000
   0.5000   1.0000   0.5000
   0.0000   0.5000   1.0000

             Expected
Agreement   Agreement     Kappa   Std. Err.         Z      Prob>Z
-----------------------------------------------------------------
  79.81%      57.17%     0.5285     0.1169       4.52      0.0000
. kap ratera raterb, wgt(w) absolute
Ratings weighted by:
   1.0000   0.6667   0.0000
   0.6667   1.0000   0.3333
   0.0000   0.3333   1.0000

             Expected
Agreement   Agreement     Kappa   Std. Err.         Z      Prob>Z
-----------------------------------------------------------------
  81.41%      55.08%     0.5862     0.1209       4.85      0.0000
```

If all conceivable ratings are observed in the data, specifying **absolute** makes no difference. For instance, if rater A assigns ratings $\{1,2,4\}$ and rater B assigns $\{1,2,3,4\}$, the complete set of assigned ratings is $\{1,2,3,4\}$, the same that **absolute** would specify. Without **absolute**, it makes no difference whether the ratings are coded $\{1,2,3,4\}$, $\{0,1,2,3\}$, $\{1,7,9,100\}$, $\{0,1,1.5,2.0\}$, or otherwise.

❑

More than two raters

For more than two raters, the mathematics are such that the two raters are not considered unique. For instance, if there are three raters, there is no assumption that the three raters who rate the first subject are the same as the three raters that rate the second. Although we call this the "more than two raters" case, it can be used with two raters when the raters' identities vary.

The nonunique rater case can be usefully broken down into three subcases: (a) there are two possible ratings, which we will call positive and negative; (b) there are more than two possible ratings, but the number of raters per subject is the same for all subjects; and (c) there are more than two possible ratings, and the number of raters per subject varies. **kappa** handles all these cases. To emphasize that there is no assumption of constant identity of raters across subjects, the variables specified contain counts of the number of raters rating the subject into a particular category.

> Jacob Cohen (1923–1998) was born in New York City. After studying psychology at City College of New York and New York University, he worked as a medical psychologist and then from 1959 at New York University. He made many contributions to research methods, including the kappa measure. He persistently emphasized the value of multiple regression and the importance of power and of measuring effects rather than testing significance.

(Continued on next page)

▷ Example 5: Two ratings

Fleiss, Levin, and Paik (2003, 612) offers the following hypothetical ratings by different sets of raters on 25 subjects:

Subject	No. of raters	No. of pos. ratings	Subject	No. of raters	No. of pos. ratings
1	2	2	14	4	3
2	2	0	15	2	0
3	3	2	16	2	2
4	4	3	17	3	1
5	3	3	18	2	1
6	4	1	19	4	1
7	3	0	20	5	4
8	5	0	21	3	2
9	2	0	22	4	0
10	4	4	23	3	0
11	5	5	24	3	3
12	3	3	25	2	2
13	4	4			

We have entered these data into Stata, and the variables are called `subject`, `raters`, and `pos`. `kappa`, however, requires that we specify variables containing the number of positive ratings and negative ratings, that is, `pos` and `raters-pos`:

```
. use http://www.stata-press.com/data/r10/p612
. gen neg = raters-pos
. kappa pos neg
Two-outcomes, multiple raters:
         Kappa          Z        Prob>Z
       ----------------------------------
        0.5415        5.28       0.0000
```

We would have obtained the same results if we had typed `kappa neg pos`.

◁

▷ Example 6: More than two ratings, constant number of raters, kappa

Each of 10 subjects is rated into one of three categories by five raters (Fleiss, Levin, and Paik 2003, 615):

```
. use http://www.stata-press.com/data/r10/p615, clear
. list
```

	subject	cat1	cat2	cat3
1.	1	1	4	0
2.	2	2	0	3
3.	3	0	0	5
4.	4	4	0	1
5.	5	3	0	2
6.	6	1	4	0
7.	7	5	0	0
8.	8	0	4	1
9.	9	1	0	4
10.	10	3	0	2

We obtain the kappa statistic:

```
. kappa cat1-cat3
     Outcome │   Kappa          Z     Prob>Z
─────────────┼──────────────────────────────
        cat1 │   0.2917       2.92    0.0018
        cat2 │   0.6711       6.71    0.0000
        cat3 │   0.3490       3.49    0.0002
─────────────┼──────────────────────────────
    combined │   0.4179       5.83    0.0000
```

The first part of the output shows the results of calculating kappa for each of the categories separately against an amalgam of the remaining categories. For instance, the `cat1` line is the two-rating kappa, where positive is `cat1` and negative is `cat2` or `cat3`. The test statistic, however, is calculated differently (see *Methods and Formulas*). The combined kappa is the appropriately weighted average of the individual kappas. There is considerably less agreement about the rating of subjects into the first category than there is for the second.

◁

▷ Example 7: More than two ratings, constant number of raters, kap

Now suppose that we have the same data as in the previous example but that the data are organized differently:

```
. use http://www.stata-press.com/data/r10/p615b
. list

     ┌─────────────────────────────────────────────────────────┐
     │ subject   rater1   rater2   rater3   rater4   rater5    │
     ├─────────────────────────────────────────────────────────┤
  1. │       1        1        2        2        2        2    │
  2. │       2        1        1        3        3        3    │
  3. │       3        3        3        3        3        3    │
  4. │       4        1        1        1        1        3    │
  5. │       5        1        1        1        3        3    │
     ├─────────────────────────────────────────────────────────┤
  6. │       6        1        2        2        2        2    │
  7. │       7        1        1        1        1        1    │
  8. │       8        2        2        2        2        3    │
  9. │       9        1        3        3        3        3    │
 10. │      10        1        1        1        3        3    │
     └─────────────────────────────────────────────────────────┘
```

Here we would use `kap` rather than `kappa` since the variables record ratings for each rater.

```
. kap rater1 rater2 rater3 rater4 rater5
There are 5 raters per subject:
     Outcome │   Kappa          Z     Prob>Z
─────────────┼──────────────────────────────
           1 │   0.2917       2.92    0.0018
           2 │   0.6711       6.71    0.0000
           3 │   0.3490       3.49    0.0002
─────────────┼──────────────────────────────
    combined │   0.4179       5.83    0.0000
```

It does not matter which rater is which when there are more than two raters.

◁

▷ Example 8: More than two ratings, varying number of raters, kappa

In this unfortunate case, kappa can be calculated, but there is no test statistic for testing against $\kappa > 0$. We do nothing differently—kappa calculates the total number of raters for each subject, and, if it is not a constant, kappa suppresses the calculation of test statistics.

```
. use http://www.stata-press.com/data/r10/rvary
. list

     +-----------------------------------+
     | subject   cat1   cat2   cat3      |
     |-----------------------------------|
  1. |       1      1      3      0      |
  2. |       2      2      0      3      |
  3. |       3      0      0      5      |
  4. |       4      4      0      1      |
  5. |       5      3      0      2      |
     |-----------------------------------|
  6. |       6      1      4      0      |
  7. |       7      5      0      0      |
  8. |       8      0      4      1      |
  9. |       9      1      0      2      |
 10. |      10      3      0      2      |
     +-----------------------------------+

. kappa cat1-cat3
            Outcome |    Kappa          Z     Prob>Z
--------------------+---------------------------------
               cat1 |    0.2685         .          .
               cat2 |    0.6457         .          .
               cat3 |    0.2938         .          .
--------------------+---------------------------------
           combined |    0.3816         .          .
Note:  number of ratings per subject vary; cannot calculate test
       statistics.
```

◁

▷ Example 9: More than two ratings, varying number of raters, kap

This case is similar to the previous example, but the data are organized differently:

```
. use http://www.stata-press.com/data/r10/rvary2
. list

     +---------------------------------------------------------+
     | subject   rater1   rater2   rater3   rater4   rater5    |
     |---------------------------------------------------------|
  1. |       1        1        2        2        .        2    |
  2. |       2        1        1        3        3        3    |
  3. |       3        3        3        3        3        3    |
  4. |       4        1        1        1        1        3    |
  5. |       5        1        1        1        3        3    |
     |---------------------------------------------------------|
  6. |       6        1        2        2        2        2    |
  7. |       7        1        1        1        1        1    |
  8. |       8        2        2        2        2        3    |
  9. |       9        1        3        .        .        3    |
 10. |      10        1        1        1        3        3    |
     +---------------------------------------------------------+
```

Here we specify kap instead of kappa since the variables record ratings for each rater.

```
. kap rater1-rater5
There are between 3 and 5 (median = 5.00) raters per subject:

         Outcome |    Kappa          Z     Prob>Z
-----------------+-------------------------------
               1 |    0.2685         .         .
               2 |    0.6457         .         .
               3 |    0.2938         .         .
-----------------+-------------------------------
        combined |    0.3816         .         .
Note:  number of ratings per subject vary; cannot calculate test
       statistics.
```

◁

Saved Results

kap and kappa save the following in r():

Scalars

r(N)	number of subjects (kap only)	r(kappa)	kappa
r(prop_o)	observed proportion of agreement (kap only)	r(z)	z statistic
r(prop_e)	expected proportion of agreement (kap only)	r(se)	standard error for kappa statistic

Methods and Formulas

kap, kapwgt, and kappa are implemented as ado-files.

The kappa statistic was first proposed by Cohen (1960). The generalization for weights reflecting the relative seriousness of each possible disagreement is due to Cohen (1968). The analysis-of-variance approach for $k = 2$ and $m \geq 2$ is due to Landis and Koch (1977b). See Altman (1991, 403–409) or Dunn (2000, chap. 2) for an introductory treatment and Fleiss, Levin, and Paik (2003, chap. 18) for a more detailed treatment. All formulas below are as presented in Fleiss, Levin, and Paik (2003). Let m be the number of raters, and let k be the number of rating outcomes.

kap: m = 2

Define w_{ij} ($i = 1, \dots, k$, $j = 1, \dots, k$) as the weights for agreement and disagreement (wgt()), or, if the data are not weighted, define $w_{ii} = 1$ and $w_{ij} = 0$ for $i \neq j$. If wgt(w) is specified, $w_{ij} = 1 - |i - j|/(k - 1)$. If wgt(w2) is specified, $w_{ij} = 1 - \left\{(i - j)/(k - 1)\right\}^2$.

The observed proportion of agreement is

$$p_o = \sum_{i=1}^{k} \sum_{j=1}^{k} w_{ij} p_{ij}$$

where p_{ij} is the fraction of ratings i by the first rater and j by the second. The expected proportion of agreement is

$$p_e = \sum_{i=1}^{k}\sum_{j=1}^{k} w_{ij} p_{i\cdot} p_{\cdot j}$$

where $p_{i\cdot} = \sum_j p_{ij}$ and $p_{\cdot j} = \sum_i p_{ij}$.

Kappa is given by $\widehat{\kappa} = (p_o - p_e)/(1 - p_e)$.

The standard error of $\widehat{\kappa}$ for testing against 0 is

$$\widehat{s}_0 = \frac{1}{(1-p_e)\sqrt{n}} \left(\left[\sum_i \sum_j p_{i\cdot} p_{\cdot j} \{ w_{ij} - (\overline{w}_{i\cdot} + \overline{w}_{\cdot j}) \}^2 \right] - p_e^2 \right)^{1/2}$$

where n is the number of subjects being rated, $\overline{w}_{i\cdot} = \sum_j p_{\cdot j} w_{ij}$, and $\overline{w}_{\cdot j} = \sum_i p_{i\cdot} w_{ij}$. The test statistic $Z = \kappa / s_0$ is assumed to be distributed $N(0,1)$.

kappa: m > 2, k = 2

Each subject i, $i = 1, \ldots, n$ is found by x_i of m_i raters to be positive (the choice as to what is labeled positive is arbitrary).

The overall proportion of positive ratings is $\overline{p} = \sum_i x_i/(n\overline{m})$, where $\overline{m} = \sum_i m_i/n$. The between-subjects mean square is (approximately)

$$B = \frac{1}{n} \sum_i \frac{(x_i - m_i \overline{p})^2}{m_i}$$

and the within-subject mean square is

$$W = \frac{1}{n(\overline{m} - 1)} \sum_i \frac{x_i(m_i - x_i)}{m_i}$$

Kappa is then defined as

$$\widehat{\kappa} = \frac{B - W}{B + (\overline{m} - 1)W}$$

The standard error for testing against 0 (Fleiss and Cuzick 1979) is approximately equal to and is calculated as

$$\widehat{s}_0 = \frac{1}{(\overline{m} - 1)\sqrt{n \overline{m}_H}} \left\{ 2(\overline{m}_H - 1) + \frac{(\overline{m} - \overline{m}_H)(1 - 4\overline{pq})}{\overline{mpq}} \right\}^{1/2}$$

where $\overline{m}_H$ is the harmonic mean of m_i and $\overline{q} = 1 - \overline{p}$.

The test statistic $Z = \widehat{\kappa}/\widehat{s}_0$ is assumed to be distributed $N(0,1)$.

kappa: m > 2, k > 2

Let x_{ij} be the number or ratings on subject i, $i = 1, \ldots, n$ into category j, $j = 1, \ldots, k$. Define $\overline{p}_j$ as the overall proportion of ratings in category j, $\overline{q}_j = 1 - \overline{p}_j$, and let $\widehat{\kappa}_j$ be the kappa statistic given above for $k = 2$ when category j is compared with the amalgam of all other categories. Kappa is (Landis and Koch 1977b)

$$\overline{\kappa} = \frac{\sum\limits_j \overline{p}_j \overline{q}_j \widehat{\kappa}_j}{\sum\limits_j \overline{p}_j \overline{q}_j}$$

In the case where the number of raters per subject $\sum_j x_{ij}$ is a constant m for all i, Fleiss, Nee, and Landis (1979) derived the following formulas for the approximate standard errors. The standard error for testing $\widehat{\kappa}_j$ against 0 is

$$\widehat{s}_j = \left\{ \frac{2}{nm(m-1)} \right\}^{1/2}$$

and the standard error for testing $\overline{\kappa}$ is

$$\overline{s} = \frac{\sqrt{2}}{\sum\limits_j \overline{p}_j \overline{q}_j \sqrt{nm(m-1)}} \left\{ \Big(\sum_j \overline{p}_j \overline{q}_j \Big)^2 - \sum_j \overline{p}_j \overline{q}_j (\overline{q}_j - \overline{p}_j) \right\}^{1/2}$$

References

Abramson, J. H., and Z. H. Abramson. 2001. *Making Sense of Data: A Self-Instruction Manual on the Interpretation of Epidemiological Data.* 3rd ed. New York: Oxford University Press.

Altman, D. G. 1991. *Practical Statistics for Medical Research.* London: Chapman & Hall/CRC.

Boyd, N. F., C. Wolfson, M. Moskowitz, T. Carlile, M. Petitclerc, H. A. Ferri, E. Fishell, A. Gregoire, M. Kiernan, J. D. Longley, I. S. Simor, and A. B. Miller. 1982. Observer variation in the interpretation of xeromammograms. *Journal of the National Cancer Institute* 68: 357–63.

Campbell, M. J., and D. Machin. 1999. *Medical Statistics: A Commonsense Approach.* 3rd ed. New York: Wiley.

Cohen, J. 1960. A coefficient of agreement for nominal scales. *Educational and Psychological Measurement* 20: 37–46.

——. 1968. Weighted kappa: Nominal scale agreement with provision for scaled disagreement or partial credit. *Psychological Bulletin* 70: 213–220.

Cox, N. J. 2006. Assessing agreement of measurements and predictions in geomorphology. *Geomorphology* 76: 332–346.

Dunn, G. 2000. *Statistics in Psychiatry.* London: Arnold.

Fleiss, J. L., and J. Cuzick. 1979. The reliability of dichotomous judgments: Unequal numbers of judges per subject. *Applied Psychological Measurement* 3: 537–542.

Fleiss, J. L., B. Levin, and M. C. Paik. 2003. *Statistical Methods for Rates and Proportions.* 3rd ed. New York: Wiley.

Fleiss, J. L., J. C. M. Nee, and J. R. Landis. 1979. Large sample variance of kappa in the case of different sets of raters. *Psychological Bulletin* 86: 974–977.

Gould, W. W. 1997. stata49: Interrater agreement. *Stata Technical Bulletin* 40: 2–8. Reprinted in *Stata Technical Bulletin Reprints*, vol. 7, pp. 20–28.

Landis, J. R., and G. G. Koch. 1977a. The measurement of observer agreement for categorical data. *Biometrics* 33: 159–174.

——. 1977b. A one-way components of variance model for categorical data. *Biometrics* 33: 671–679.

Reichenheim, M. E. 2000. sxd3: Sample size for the kappa statistic of interrater agreement. *Stata Technical Bulletin* 58: 41–45. Reprinted in *Stata Technical Bulletin Reprints*, vol. 10, pp. 382–387.

——. 2004. Confidence intervals for the kappa statistic. *Stata Journal* 4: 421–428.

Shrout, P. E. 2001. Jacob Cohen (1923–1998). *American Psychologist* 56: 166.

Steichen, T. J. and N. J. Cox. 1998a. sg84: Concordance correlation coefficient. *Stata Technical Bulletin* 43: 35–39. Reprinted in *Stata Technical Bulletin Reprints*, vol. 8, pp. 137–143.

——. 1998b. sg84.1: Concordance correlation coefficient, revisited. *Stata Technical Bulletin* 45: 21–23. Reprinted in *Stata Technical Bulletin Reprints*, vol. 8, pp. 143–145.

——. 2000a. sg84.2: Concordance correlation coefficient: Update for Stata 6. *Stata Technical Bulletin* 54: 25–26. Reprinted in *Stata Technical Bulletin Reprints*, vol. 9, pp. 169–170.

——. 2000b. sg84.3: Concordance correlation coefficient: Minor corrections. *Stata Technical Bulletin* 58: 9. Reprinted in *Stata Technical Bulletin Reprints*, vol. 10, p. 137.

——. 2002. A note on the concordance correlation coefficient. *Stata Journal* 2: 183–189.

Title

kdensity — Univariate kernel density estimation

Syntax

`kdensity` *varname* [*if*] [*in*] [*weight*] [, *options*]

options	description
Main	
`kernel(`*kernel*`)`	specify kernel function; default is `kernel(epanechnikov)`
`bwidth(`#`)`	half-width of kernel
`generate(`$newvar_x$ $newvar_d$`)`	store the estimation points in $newvar_x$ and the density estimate in $newvar_d$
`n(`#`)`	estimate density using # points; default is min(N, 50)
`at(`var_x`)`	estimate density using the values specified by var_x
`nograph`	suppress graph
Kernel plot	
cline_options	affect rendition of the plotted kernel density estimate
Density plots	
`normal`	add normal density to the graph
`normopts(`*cline_options*`)`	affect rendition of normal density
`student(`#`)`	add Student's t density with # degrees of freedom to the graph
`stopts(`*cline_options*`)`	affect rendition of the Student's t density
Add plots	
`addplot(`*plot*`)`	add other plots to the generated graph
Y axis, X axis, Titles, Legend, Overall	
twoway_options	any options other than `by()` documented in [G] ***twoway_options***

kernel	description
`epanechnikov`	Epanechnikov kernel function; the default
`epan2`	alternative Epanechnikov kernel function
`biweight`	biweight kernel function
`cosine`	cosine trace kernel function
`gaussian`	Gaussian kernel function
`parzen`	Parzen kernel function
`rectangle`	rectangle kernel function
`triangle`	triangle kernel function

`fweight`s, `aweight`s, and `iweight`s are allowed; see [U] **11.1.6 weight**.

Description

kdensity produces kernel density estimates and graphs the result.

Options

Main

kernel(*kernel*) specifies the kernel function for use in calculating the kernel density estimate. The default kernel is the Epanechnikov kernel (epanechnikov).

bwidth(#) specifies the half-width of the kernel, the width of the density window around each point. If bwidth() is not specified, the "optimal" width is calculated and used. The optimal width is the width that would minimize the mean integrated squared error if the data were Gaussian and a Gaussian kernel were used, so it is not optimal in any global sense. In fact, for multimodal and highly skewed densities, this width is usually too wide and oversmooths the density (Silverman 1992).

generate($newvar_x$ $newvar_d$) stores the results of the estimation. $newvar_x$ will contain the points at which the density is estimated. $newvar_d$ will contain the density estimate.

n(#) specifies the number of points at which the density estimate is to be evaluated. The default is $\min(N, 50)$, where N is the number of observations in memory.

at(var_x) specifies a variable that contains the values at which the density should be estimated. This option allows you to more easily obtain density estimates for different variables or different subsamples of a variable and then overlay the estimated densities for comparison.

nograph suppresses the graph. This option is often used with the generate() option.

Kernel plot

cline_options affect the rendition of the plotted kernel density estimate. See [G] ***cline_options***.

Density plots

normal requests that a normal density be overlaid on the density estimate for comparison.

normopts(*cline_options*) specifies details about the rendition of the normal curve, such as the color and style of line used. See [G] ***cline_options***.

student(#) specifies that a Student's t density with # degrees of freedom be overlaid on the density estimate for comparison.

stopts(*cline_options*) affects the rendition of the Student's t density. See [G] ***cline_options***.

Add plots

addplot(*plot*) provides a way to add other plots to the generated graph. See [G] ***addplot_option***.

Y axis, X axis, Titles, Legend, Overall

twoway_options are any of the options documented in [G] ***twoway_options***, excluding by(). These include options for titling the graph (see [G] ***title_options***) and for saving the graph to disk (see [G] ***saving_option***).

Remarks

Kernel density estimators approximate the density $f(x)$ from observations on x. Histograms do this, too, and the histogram itself is a kind of kernel density estimate. The data are divided into nonoverlapping intervals, and counts are made of the number of data points within each interval. Histograms are bar graphs that depict these frequency counts—the bar is centered at the midpoint of each interval—and its height reflects the average number of data points in the interval.

In more general kernel density estimates, the range is still divided into intervals, and estimates of the density at the center of intervals are produced. One difference is that the intervals are allowed to overlap. We can think of sliding the interval—called a window—along the range of the data and collecting the center-point density estimates. The second difference is that, rather than merely counting the number of observations in a window, a kernel density estimator assigns a weight between 0 and 1—based on the distance from the center of the window—and sums the weighted values. The function that determines these weights is called the kernel.

Kernel density estimates have the advantages of being smooth and of being independent of the choice of origin (corresponding to the location of the bins in a histogram).

See Salgado-Ugarte, Shimizu, and Taniuchi (1993) and Fox (1990) for discussions of kernel density estimators that stress their use as exploratory data-analysis tools.

▷ Example 1: Histogram and kernel density estimate

Goeden (1978) reports data consisting of 316 length observations of coral trout. We wish to investigate the underlying density of the lengths. To begin on familiar ground, we might draw a histogram. In [R] **histogram**, we suggest setting the bins to $\min(\sqrt{n}, 10 \cdot \log_{10} n)$, which for $n = 316$ is roughly 18:

```
. use http://www.stata-press.com/data/r10/trocolen
. histogram length, bin(18)
(bin=18, start=226, width=19.777778)
```

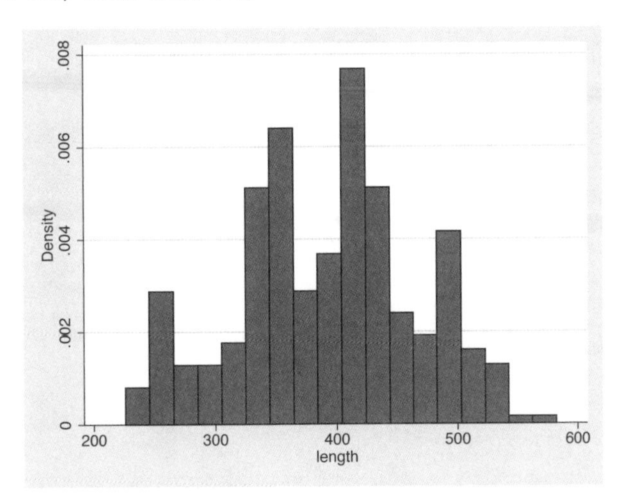

The kernel density estimate, on the other hand, is smooth.

```
. kdensity length
```

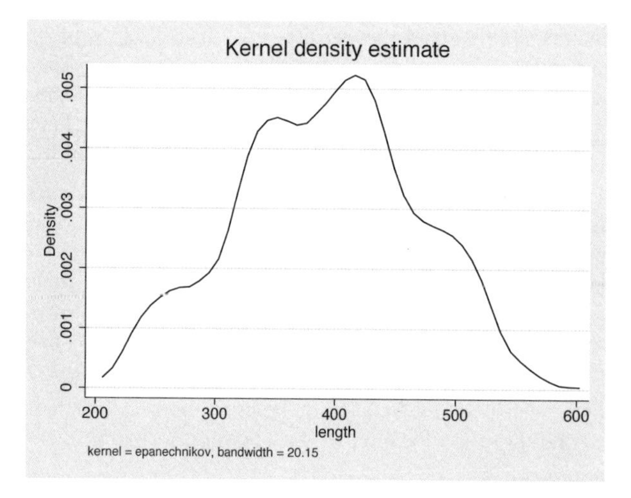

Kernel density estimators are, however, sensitive to an assumption, just as are histograms. In histograms, we specify a number of bins. For kernel density estimators, we specify a width. In the graph above, we used the default width. **kdensity** is smarter than **twoway histogram** in that its default width is not a fixed constant. Even so, the default width is not necessarily best.

kdensity saves the width in the returned scalar **bwidth**, so typing **display r(bwidth)** reveals it. Doing this, we discover that the width is approximately 20.

Widths are similar to the inverse of the number of bins in a histogram in that smaller widths provide more detail. The units of the width are the units of x, the variable being analyzed. The width is specified as a half-width, meaning that the kernel density estimator with half-width 20 corresponds to sliding a window of size 40 across the data.

We can specify half-widths for ourselves by using the **bwidth()** option. Smaller widths do not smooth the density as much:

```
. kdensity length, bwidth(10)
```

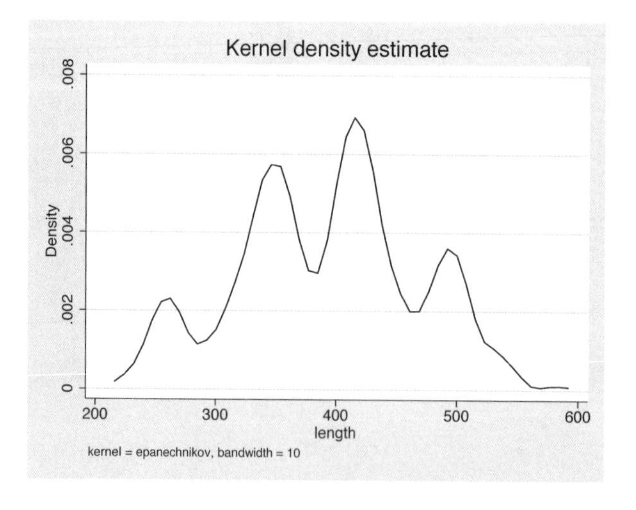

```
. kdensity length, bwidth(15)
```

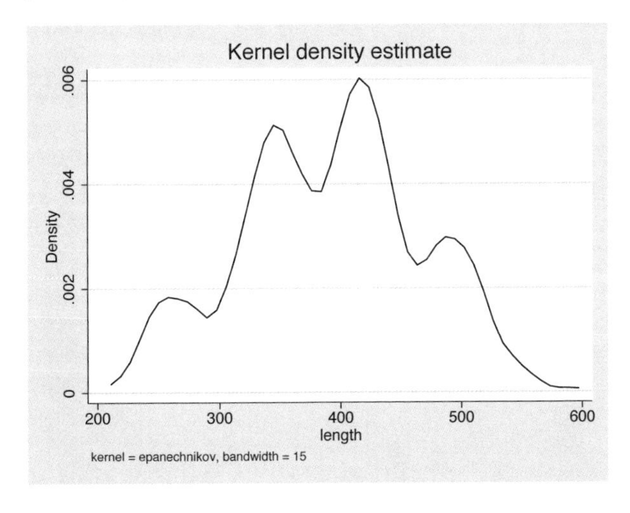

◁

▷ Example 2: Different kernels can produce different results

When widths are held constant, different kernels can produce surprisingly different results. This is really an attribute of the kernel and width combination; for a given width, some kernels are more sensitive than others at identifying peaks in the density estimate.

We can see this when using a dataset with lots of peaks. In the automobile dataset, we characterize the density of `weight`, the weight of the vehicles. Below we compare the Epanechnikov and Parzen kernels.

```
. use http://www.stata-press.com/data/r10/auto, clear
(1978 Automobile Data)
. kdensity weight, kernel(epanechnikov) nograph generate(x epan)
. kdensity weight, kernel(parzen) nograph generate(x2 parzen)
. label var epan "Epanechnikov density estimate"
. label var parzen "Parzen density estimate"
. line epan parzen x, sort ytitle(Density) legend(cols(1))
```

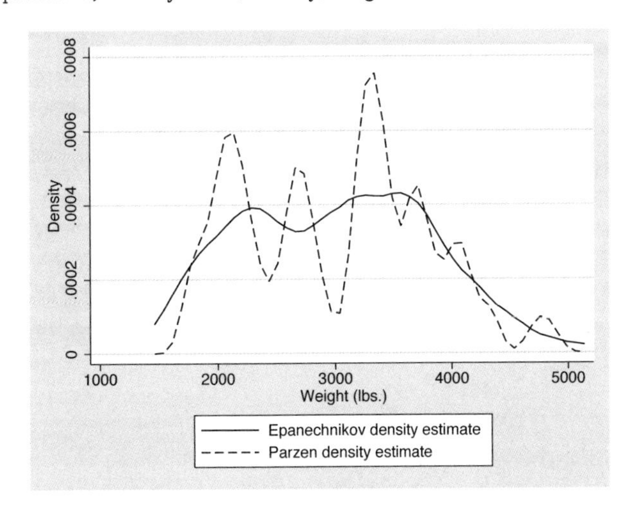

We did not specify a width, so we obtained the default width. That width is not a function of the selected kernel, but of the data. See *Methods and Formulas* for the calculation of the optimal width.

◁

▷ Example 3: Density with overlaid normal density

In examining the density estimates, we may wish to overlay a normal density or a Student's t density for comparison. Using automobile weights, we can get an idea of the distance from normality by using the `normal` option.

```
. kdensity weight, kernel(epanechnikov) normal
```

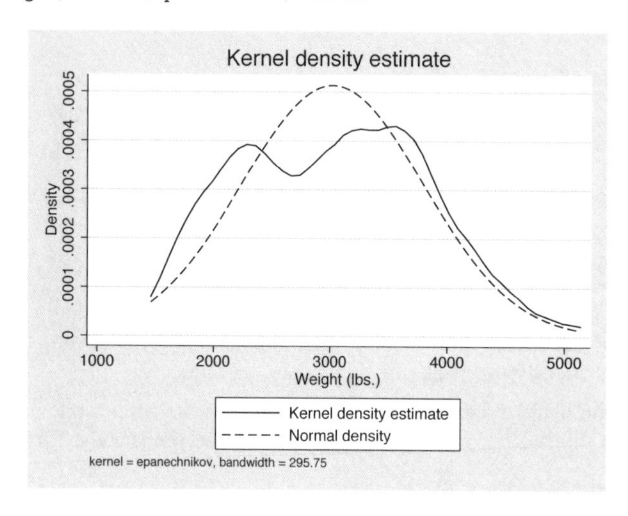

◁

▷ Example 4: Compare two densities

We also may want to compare two or more densities. In this example, we will compare the density estimates of the weights for the foreign and domestic cars.

```
. use http://www.stata-press.com/data/r10/auto, clear
(1978 Automobile Data)
. kdensity weight, nograph generate(x fx)
. kdensity weight if foreign==0, nograph generate(fx0) at(x)
. kdensity weight if foreign==1, nograph generate(fx1) at(x)
. label var fx0 "Domestic cars"
. label var fx1 "Foreign cars"
```

```
. line fx0 fx1 x, sort ytitle(Density)
```

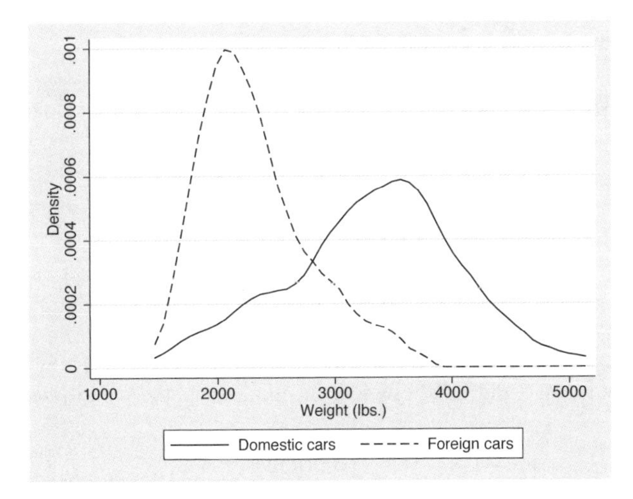

◁

❑ Technical Note

Although all the examples we included had densities of less than 1, the density may exceed 1.

The probability density $f(x)$ of a continuous variable, x, has the units and dimensions of the reciprocal of x. If x is measured in meters, $f(x)$ has units $1/\text{meter}$. Thus the density is not measured on a probability scale, so it is possible for $f(x)$ to exceed 1.

To see this, think of a uniform density on the interval 0 to 1. The area under the density curve is 1: this is the product of the density, which is constant at 1, and the range, which is 1. If the variable is then transformed by doubling, the area under the curve remains 1 and is the product of the density, constant at 0.5, and the range, which is 2. Conversely, if the variable is transformed by halving, the area under the curve also remains at 1 and is the product of the density, constant at 2, and the range, which is 0.5. (Strictly, the range is measured in certain units, and the density is measured in the reciprocal of those units, so the units cancel on multiplication.)

❑

Saved Results

`kdensity` saves the following in `r()`:

Scalars

`r(bwidth)`	kernel bandwidth
`r(n)`	number of points at which the estimate was evaluated
`r(scale)`	density bin width

Macros

`r(kernel)`	name of kernel

Methods and Formulas

`kdensity` is implemented as an ado-file.

A kernel density estimate is formed by summing the weighted values calculated with the kernel function K, as in

$$\widehat{f}_K = \frac{1}{qh}\sum_{i=1}^{n} w_i K\left(\frac{x - X_i}{h}\right)$$

where $q = \sum_i w_i$ if weights are frequency weights (`fweight`) or analytic weights (`aweight`), and $q = 1$ if weights are importance weights (`iweights`). Anyalytical weights are rescaled so that $\sum_i w_i = n$ (see [U] **11 Language syntax**). If weights are not used, then $w_i = 1$, for $i = 1, \ldots, n$. `kdensity` includes seven different kernel functions. The Epanechnikov is the default function if no other kernel is specified and is the most efficient in minimizing the mean integrated squared error.

Kernel	Formula	
Biweight	$K[z] = \begin{cases} \frac{15}{16}(1 - z^2)^2 \\ 0 \end{cases}$	if $\lvert z\rvert < 1$ otherwise
Cosine	$K[z] = \begin{cases} 1 + \cos(2\pi z) \\ 0 \end{cases}$	if $\lvert z\rvert < 1/2$ otherwise
Epanechnikov	$K[z] = \begin{cases} \frac{3}{4}(1 - \frac{1}{5}z^2)/\sqrt{5} \\ 0 \end{cases}$	if $\lvert z\rvert < \sqrt{5}$ otherwise
Epan2	$K[z] = \begin{cases} \frac{3}{4}(1 - z^2) \\ 0 \end{cases}$	if $\lvert z\rvert < 1$ otherwise
Gaussian	$K[z] = \frac{1}{\sqrt{2\pi}}e^{-z^2/2}$	
Parzen	$K[z] = \begin{cases} \frac{4}{3} - 8z^2 + 8\lvert z\rvert^3 \\ 8(1 - \lvert z\rvert)^3/3 \\ 0 \end{cases}$	if $\lvert z\rvert \le 1/2$ if $1/2 < \lvert z\rvert \le 1$ otherwise
Rectangular	$K[z] = \begin{cases} 1/2 \\ 0 \end{cases}$	if $\lvert z\rvert < 1$ otherwise
Triangular	$K[z] = \begin{cases} 1 - \lvert z\rvert \\ 0 \end{cases}$	if $\lvert z\rvert < 1$ otherwise

From the definitions given in the table, we can see that the choice of h will drive how many values are included in estimating the density at each point. This value is called the *window width* or *bandwidth*. If the window width is not specified, it is determined as

$$m = \min\left(\sqrt{\text{variance}_x},\ \frac{\text{interquartile range}_x}{1.349}\right)$$

$$h = \frac{0.9m}{n^{1/5}}$$

where x is the variable for which we wish to estimate the kernel and n is the number of observations.

Most researchers agree that the choice of kernel is not as important as the choice of bandwidth. There is a great deal of literature on choosing bandwidths under various conditions; see, for example, Parzen (1962) or Tapia and Thompson (1978). Also see Newton (1988) for a comparison with sample spectral density estimation in time-series applications.

Acknowledgments

We gratefully acknowledge the previous work by Isaías H. Salgado-Ugarte of Universidad Nacional Autónoma de México, and Makoto Shimizu and Toru Taniuchi of the University of Tokyo; see Salgado-Ugarte, Shimizu, and Taniuchi (1993). Their article provides a good overview of the subject of univariate kernel density estimation and presents arguments for its use in exploratory data analysis.

References

Cox, N. J. 2005. Speaking Stata: Density probability plots. *Stata Journal* 5: 259–273.

Fiorio, C. V. 2004. Confidence intervals for kernel density estimation. *Stata Journal* 4: 168–179.

Fox, J. 1990. Describing univariate distributions. In *Modern Methods of Data Analysis*, ed. J. Fox and J. S. Long, 58–125. Newbury Park, CA: Sage.

Goeden, G. B. 1978. A monograph of the coral trout, *Plectropomus leopardus* (Lacépède). *Research Bulletin Fisheries Service Queensland* 1: 1–42.

Kohler, U., and F. Kreuter. 2005. *Data Analysis Using Stata*. College Station, TX: Stata Press.

Newton, H. J. 1988. *TIMESLAB: A Time Series Analysis Laboratory*. Belmont, CA: Wadsworth & Brooks/Cole.

Parzen, E. 1962. On estimation of a probability density function and mode. *Annals of Mathematical Statistics* 32: 1065–1076.

Royston, J. P., and N. J. Cox. 2005. A multivariable scatterplot smoother. *Stata Journal* 5: 405–412.

Salgado-Ugarte, I. H., and M. A. Pérez-Hernández. 2003. Exploring the use of variable bandwidth kernel density estimators. *Stata Journal* 3: 133–147.

Salgado-Ugarte, I. H., M. Shimizu, and T. Taniuchi. 1993. snp6: Exploring the shape of univariate data using kernel density estimators. *Stata Technical Bulletin* 16: 8–19. Reprinted in *Stata Technical Bulletin Reprints*, vol. 3, pp. 155–173.

——. 1995a. snp6.1: ASH, WARPing, and kernel density estimation for univariate data. *Stata Technical Bulletin* 26: 23–31. Reprinted in *Stata Technical Bulletin Reprints*, vol. 5, pp. 161–172.

——. 1995b. snp6.2: Practical rules for bandwidth selection in univariate density estimation. *Stata Technical Bulletin* 27: 5–19. Reprinted in *Stata Technical Bulletin Reprints*, vol. 5, pp. 172–190.

——. 1997. snp13: Nonparametric assessment of multimodality for univariate data. *Stata Technical Bulletin* 38: 27–35. Reprinted in *Stata Technical Bulletin Reprints*, vol. 7, pp. 232–243.

Scott, D. W. 1992. *Multivariate Density Estimation: Theory, Practice, and Visualization*. New York: Wiley.

Silverman, B. W. 1992. *Density Estimation for Statistics and Data Analysis*. London: Chapman & Hall.

Simonoff, J. S. 1996. *Smoothing Methods in Statistics*. New York: Springer.

Steichen, T. J. 1998. gr33: Violin plots. *Stata Technical Bulletin* 46: 13–18. Reprinted in *Stata Technical Bulletin Reprints*, vol. 8, pp. 57–65.

Tapia, R. A., and J. R. Thompson. 1978. *Nonparametric Probability Density Estimation*. Baltimore: Johns Hopkins University Press.

Van Kerm, P. 2003. Adaptive kernel density estimation. *Stata Journal* 3: 148–156.

Wand, M. P., and M. C. Jones. 1995. *Kernel Smoothing*. London: Chapman & Hall.

Also See

[R] **histogram** — Histograms for continuous and categorical variables

Title

ksmirnov — Kolmogorov–Smirnov equality-of-distributions test

Syntax

One-sample Kolmogorov–Smirnov test

`ksmirnov` *varname* = *exp* [*if*] [*in*]

Two-sample Kolmogorov–Smirnov test

`ksmirnov` *varname* [*if*] [*in*] , `by(`*groupvar*`)` [`exact`]

Description

`ksmirnov` performs one- and two-sample Kolmogorov–Smirnov tests of the equality of distributions. In the first syntax, *varname* is the variable whose distribution is being tested, and *exp* must evaluate to the corresponding (theoretical) cumulative. In the second syntax, *groupvar* must take on two distinct values. The distribution of *varname* for the first value of *groupvar* is compared with that of the second value.

When testing for normality, please see [R] **sktest** and [R] **swilk**.

Options for two-sample test

Main

`by(`*groupvar*`)` is required. It specifies a binary variable that identifies the two groups.

`exact` specifies that the exact p-value be computed. This may take a long time if $n > 50$.

Remarks

▷ Example 1: Two-sample test

Say that we have data on `x` that resulted from two different experiments, labeled as `group==1` and `group==2`. Our data contain

```
. use http://www.stata-press.com/data/r10/ksxmpl
. list

     +------------+
     | group    x |
     |------------|
  1. |     2    2 |
  2. |     1    0 |
  3. |     2    3 |
  4. |     1    4 |
  5. |     1    5 |
     |------------|
  6. |     2    8 |
  7. |     2   10 |
     +------------+
```

We wish to use the two-sample Kolmogorov–Smirnov test to determine if there are any differences in the distribution of x for these two groups:

```
. ksmirnov x, by(group)
Two-sample Kolmogorov-Smirnov test for equality of distribution functions
 Smaller group           D       P-value  Corrected
 ---------------------------------------------------
 1:                 0.5000        0.424
 2:                -0.1667        0.909
 Combined K-S:      0.5000        0.785      0.735
```

The first line tests the hypothesis that x for group 1 contains *smaller* values than group 2. The largest difference between the distribution functions is 0.5. The approximate p-value for this is 0.424, which is not significant.

The second line tests the hypothesis that x for group 1 contains *larger* values than group 2. The largest difference between the distribution functions in this direction is 0.1667. The approximate p-value for this small difference is 0.909.

Finally, the approximate p-value for the combined test is 0.785, corrected to 0.735. The p-values **ksmirnov** calculates are based on the asymptotic distributions derived by Smirnov (1939). These approximations are not good for small samples ($n < 50$). They are too conservative—real p-values tend to be substantially smaller. We have also included a less conservative approximation for the nondirectional hypothesis based on an empirical continuity correction—the 0.735 reported in the third column.

That number, too, is only an approximation. An exact value can be calculated using the **exact** option:

```
. ksmirnov x, by(group) exact
Two-sample Kolmogorov-Smirnov test for equality of distribution functions
 Smaller group           D       P-value      Exact
 ---------------------------------------------------
 1:                 0.5000        0.424
 2:                -0.1667        0.909
 Combined K-S:      0.5000        0.785      0.657
```

◁

▷ Example 2: One-sample test

Let's now test whether x in the example above is distributed normally. Kolmogorov–Smirnov is not a particularly powerful test in testing for normality, and we do not endorse such use of it; see [R] **sktest** and [R] **swilk** for better tests.

In any case, we will test against a normal distribution with the same mean and standard deviation:

```
. summarize x
    Variable |       Obs        Mean    Std. Dev.       Min        Max
-------------+--------------------------------------------------------
           x |         7    4.571429    3.457222          0         10
. ksmirnov x = normal((x-4.571429)/3.457222)
One-sample Kolmogorov-Smirnov test against theoretical distribution
           normal((x-4.571429)/3.457222)
 Smaller group           D       P-value  Corrected
 ---------------------------------------------------
 x:                 0.1650        0.683
 Cumulative:       -0.1250        0.803
 Combined K-S:      0.1650        0.991      0.978
```

Since Stata has no way of knowing that we based this calculation on the calculated mean and standard deviation of `x`, the test statistics will be slightly conservative in addition to being approximations. Nevertheless, they clearly indicate that the data cannot be distinguished from normally distributed data.

◁

Saved Results

`ksmirnov` saves the following in `r()`:

Scalars

`r(D_1)`	D from line 1	`r(D)`	combined D
`r(p_1)`	p-value from line 1	`r(p)`	combined p-value
`r(D_2)`	D from line 2	`r(p_cor)`	corrected combined p-value
`r(p_2)`	p-value from line 2	`r(p_exact)`	exact combined p-value

Macros

`r(group1)`	name of group from line 1	`r(group2)`	name of group from line 2

Methods and Formulas

`ksmirnov` is implemented as an ado-file.

In general, the Kolmogorov–Smirnov test (Kolmogorov 1933; Smirnov 1939; also see Conover 1999, 428–465) is not very powerful against differences in the tails of distributions. In return for this, it is fairly powerful for alternative hypotheses that involve lumpiness or clustering in the data.

The directional hypotheses are evaluated with the statistics

$$D^+ = \max_x \left\{ F(x) - G(x) \right\}$$

$$D^- = \min_x \left\{ F(x) - G(x) \right\}$$

where $F(x)$ and $G(x)$ are the empirical distribution functions for the sample being compared. The combined statistic is

$$D = \max\Big(|D^+|\,,\,|D^-| \Big)$$

The p-value for this statistic may be obtained by evaluating the asymptotic limiting distribution. Let m be the sample size for the first sample, and let n be the sample size for the second sample. Smirnov (1939) shows that

$$\lim_{m,n\to\infty} \Pr\left\{ \sqrt{mn/(m+n)}D_{m,n} \le z \right\} = 1 - 2\sum_{i=1}^{\infty} \big(-1\big)^{i-1} \exp\big(-2i^2z^2\big)$$

The first five terms form the approximation P_a used by Stata. The exact p-value is calculated by a counting algorithm; see Gibbons (1971, 127–131). A corrected p-value was obtained by modifying the asymptotic p-value using a numerical approximation technique

$$Z = \Phi^{-1}(P_a) + 1.04/\min(m,n) + 2.09/\max(m,n) - 1.35/\sqrt{mn/(m+n)}$$

$$p\text{-value} = \Phi(Z)$$

where $\Phi()$ is the cumulative normal distribution.

Andrei Nikolayevich Kolmogorov (1903–1987), of Russia, was one of the great mathematicians of the 20th century, making outstanding contributions in many different branches, including set theory, measure theory, probability and statistics, approximation theory, functional analysis, classical dynamics, and theory of turbulence. He was a faculty member at Moscow State University for more than 60 years.

Nikolai Vasilyevich Smirnov (1900–1966) was a Russian statistician whose work included contributions in nonparametric statistics, order statistics, and goodness of fit. After army service and the study of philosophy and philology, he turned to mathematics and eventually rose to be head of mathematical statistics at the Steklov Mathematical Institute in Moscow.

References

Aivazian, S. A. 1997. Smirnov, Nikolai Vasilyevich. In *Leading Personalities in Statistical Sciences from the Seventeenth Century to the Present*, ed. N. L. Johnson and S. Kotz, 208–210. New York: Wiley.

Conover, W. J. 1999. *Practical Nonparametric Statistics*. 3rd ed. New York: Wiley.

Gibbons, J. D. 1971. *Nonparametric Statistical Inference*. New York: McGraw–Hill.

Johnson, N. L., and S. Kotz. 1997. Kolmogorov, Andrei Nikolayevich. In *Leading Personalities in Statistical Sciences from the Seventeenth Century to the Present*, ed. N. L. Johnson and S. Kotz, 255–256. New York: Wiley.

Kolmogorov, A. N. 1933. Sulla determinazione empirica di una legge di distribuzione. *Giornale dell' Istituto Italiano degli Attuari* 4: 83–91.

Riffenburgh, R. H. 2005. *Statistics in Medicine*. 2nd ed. New York: Elsevier.

Smirnov, N. V. 1939. Estimate of deviation between empirical distribution functions in two independent samples (in Russian). *Bulletin Moscow University* 2: 3–16.

Also See

[R] **runtest** — Test for random order

[R] **sktest** — Skewness and kurtosis test for normality

[R] **swilk** — Shapiro–Wilk and Shapiro–Francia tests for normality

Title

kwallis — Kruskal–Wallis equality-of-populations rank test

Syntax

`kwallis` *varname* [*if*] [*in*] , `by(`*groupvar*`)`

Description

`kwallis` tests the hypothesis that several samples are from the same population. In the syntax diagram above, *varname* refers to the variable recording the outcome, and *groupvar* refers to the variable denoting the population. `by()` is required.

Option

`by(`*groupvar*`)` is required. It specifies a binary variable that identifies the two groups.

Remarks

▷ Example 1

We have data on the 50 states. The data contain the median age of the population `medage` and the region of the country `region` for each state. We wish to test for the equality of the median age distribution across all four regions simultaneously:

```
. use http://www.stata-press.com/data/r10/census
(1980 Census data by state)
. kwallis medage, by(region)
Kruskal-Wallis equality-of-populations rank test
```

region	Obs	Rank Sum
NE	9	376.50
N Cntrl	12	294.00
South	16	398.00
West	13	206.50

```
chi-squared =    17.041 with 3 d.f.
probability =     0.0007
chi-squared with ties =    17.062 with 3 d.f.
probability =     0.0007
```

From the output, we see that we can reject the hypothesis that the populations are the same at any level below 0.07%. ◁

Saved Results

`kwallis` saves the following in `r()`:

Scalars

`r(df)`	degrees of freedom	`r(chi2)`	χ^2
		`r(chi2_adj)`	χ^2 adjusted for ties

Methods and Formulas

kwallis is implemented as an ado-file.

The Kruskal–Wallis test (Kruskal and Wallis 1952a,b; also see Altman 1991, 213–215, Conover 1999, 288–297, and Riffenburgh 2005, 287–291) is a multiple-sample generalization of the two-sample Wilcoxon (also called Mann–Whitney) rank sum test (Wilcoxon 1945; Mann and Whitney 1947). Samples of sizes n_j, $j = 1, \dots, m$ are combined and ranked in ascending order of magnitude. Tied values are assigned the average ranks. Let n denote the overall sample size, and let $R_j = \sum_{i=1}^{n_j} R(X_{ji})$ denote the sum of the ranks for the jth sample. The Kruskal–Wallis one-way analysis-of-variance test H is defined as

$$H = \frac{1}{S^2} \left\{ \sum_{j=1}^{m} \frac{R_j^2}{n_j} - \frac{n(n+1)^2}{4} \right\}$$

where

$$S^2 = \frac{1}{n-1} \left\{ \sum_{\text{all ranks}} R(X_{ji})^2 - \frac{n(n+1)^2}{4} \right\}$$

If there are no ties, this equation simplifies to

$$H = \frac{12}{n(n+1)} \sum_{j=1}^{m} \frac{R_j^2}{n_j} - 3(n+1)$$

The sampling distribution of H is approximately χ^2 with $m - 1$ degrees of freedom.

William Henry Kruskal (1919–2005) was born in New York City. He studied mathematics and statistics at Antioch College, Harvard, and Columbia, and joined the University of Chicago in 1951. He has made many outstanding contributions to linear models, nonparametric statistics, government statistics, and the history and methodology of statistics.

Wilson Allen Wallis (1912–1998) was born in Philadelphia. He studied psychology and economics at the Universities of Minnesota and Chicago and at Columbia. He taught at Yale, Stanford, and Chicago, before moving as President (later Chancellor) to the University of Rochester in 1962. He also served in several Republican administrations. Wallis served as editor of the *Journal of the American Statistical Association*, coauthored a popular introduction to statistics, and contributed to nonparametric statistics.

References

Altman, D. G. 1991. *Practical Statistics for Medical Research.* London: Chapman & Hall/CRC.

Conover, W. J. 1999. *Practical Nonparametric Statistics.* 3rd ed. New York: Wiley.

Kruskal, W. H., and W. A. Wallis. 1952a. Use of ranks in one-criterion variance analysis. *Journal of the American Statistical Association* 47: 583–621.

——. 1952b. Errata for use of ranks in one-criterion variance analysis. *Journal of the American Statistical Association* 48: 907–911.

Mann, H. B., and D. R. Whitney. 1947. On a test of whether one of two random variables is stochastically larger than the other. *Annals of Mathematical Statistics* 18: 50–60.

Newson, R. 2006. Confidence intervals for rank statistics: Somers' D and extensions. *Stata Journal* 6: 309–334.

Olkin, I. 1991. A conversation with W. Allen Wallis. *Statistical Science* 6: 121–140.

Riffenburgh, R. H. 2005. *Statistics in Medicine*. 2nd ed. New York: Academic Press.

Wilcoxon, F. 1945. Individual comparisons by ranking methods. *Biometrics* 1: 80–83.

Zabell, S. 1994. A conversation with William Kruskal. *Statistical Science* 9: 285–303.

Also See

[R] **nptrend** — Test for trend across ordered groups

[R] **oneway** — One-way analysis of variance

[R] **sdtest** — Variance-comparison tests

[R] **signrank** — Equality tests on matched data

Title

ladder — Ladder of powers

Syntax

Ladder of powers

`ladder` *varname* [*if*] [*in*] [, `generate(`*newvar*`)` `noadjust`]

Ladder-of-powers histograms

`gladder` *varname* [*if*] [*in*] [, *histogram_options combine_options*]

Ladder-of-powers quantile–normal plots

`qladder` *varname* [*if*] [*in*] [, *qnorm_options combine_options*]

`by` is allowed with `ladder`; see [D] **by**.

Description

`ladder` searches a subset of the ladder of powers (Tukey 1977) for a transform that converts *varname* into a normally distributed variable. `sktest` tests for normality; see [R] **sktest**. Also see [R] **boxcox**.

`gladder` displays nine histograms of transforms of *varname* according to the ladder of powers. `gladder` is useful pedagogically, but we do not advise looking at histograms for research work; `ladder` or `qnorm` (see [R] **diagnostic plots**) is preferred.

`qladder` displays the quantiles of transforms of *varname* according to the ladder of powers against the quantiles of a normal distribution.

Options for ladder

Main

`generate(`*newvar*`)` saves the transformed values corresponding to the minimum chi-squared value from the table. We do not recommend using `generate()` because it is literal in interpreting the minimum, thus ignoring nearly equal but perhaps more interpretable transforms.

`noadjust` is the `noadjust` option to `sktest`; see [R] **sktest**.

Options for gladder

histogram_options affect the rendition of the histograms across all relevant transformations. See [R] **histogram**. Here the `normal` option is assumed, so you must supply the `nonormal` option to suppress the overlaid normal density. Also, `gladder` does not allow the `width(#)` option of `histogram`.

combine_options are any of the options documented in [G] **graph combine**. These include options for titling the graph (see [G] ***title_options***) and for saving the graph to disk (see [G] ***saving_option***).

Options for qladder

qnorm_options affect the rendition of the quantile–normal plots across all relevant transformations. See [R] **diagnostic plots**.

combine_options are any of the options documented in [G] **graph combine**. These include options for titling the graph (see [G] ***title_options***) and for saving the graph to disk (see [G] ***saving_option***).

Remarks

▷ Example 1: ladder

We have data on the mileage rating of 74 automobiles and wish to find a transform that makes the variable normally distributed:

```
. use http://www.stata-press.com/data/r10/auto
(1978 Automobile Data)
. ladder mpg
Transformation          formula                 chi2(2)       P(chi2)
---------------------------------------------------------------------
cubic                   mpg^3                     43.59         0.000
square                  mpg^2                     27.03         0.000
identity                mpg                       10.95         0.004
square root             sqrt(mpg)                  4.94         0.084
log                     log(mpg)                   0.87         0.647
1/(square root)         1/sqrt(mpg)                0.20         0.905
inverse                 1/mpg                      2.36         0.307
1/square                1/(mpg^2)                 11.99         0.002
1/cubic                 1/(mpg^3)                 24.30         0.000
```

If we had typed `ladder mpg, gen(mpgx)`, the variable `mpgx` containing $1/\sqrt{\texttt{mpg}}$ would have been automatically generated for us. This is the perfect example of why you should not, in general, specify the `generate()` option. We also cannot reject the hypothesis that the inverse of `mpg` is normally distributed and that 1/`mpg`—gallons per mile—has a better interpretation. It is a measure of energy consumption.
◁

▷ Example 2: gladder

`gladder` explores the same transforms as `ladder` but presents results graphically:

(Continued on next page)

```
. gladder mpg, fraction
```

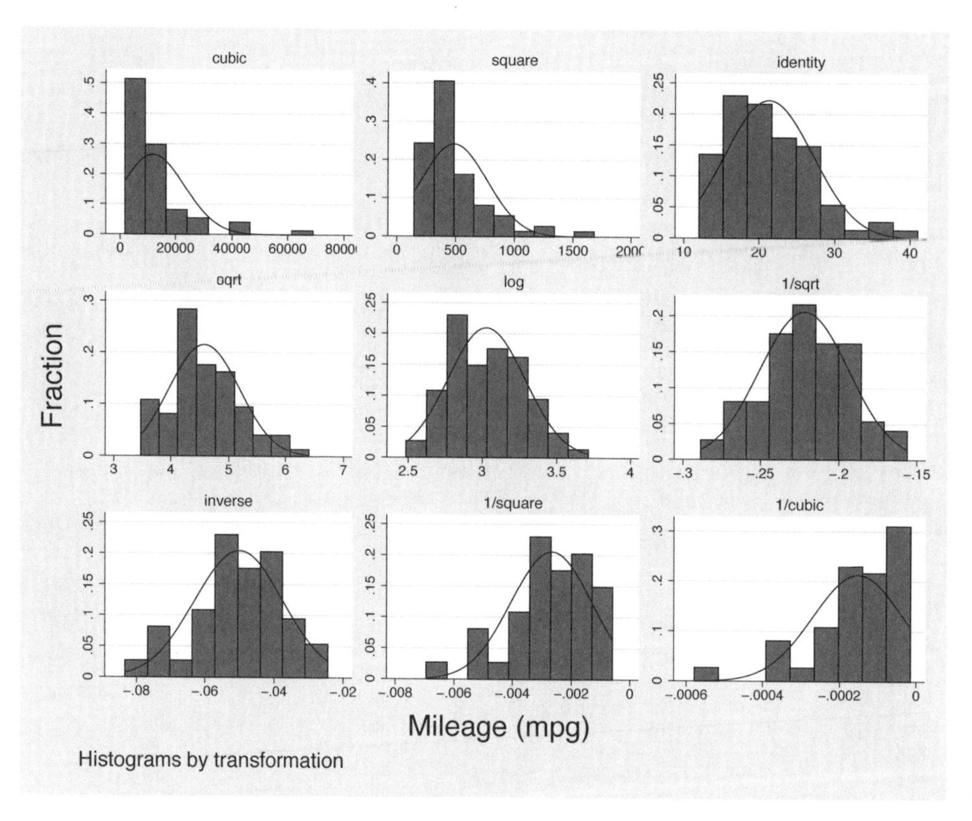

◁

❑ Technical Note

gladder is useful pedagogically, but be careful when using it for research work, especially with many observations. For instance, consider the following data on the average July temperature in degrees Fahrenheit for 954 U.S. cities:

```
. use http://www.stata-press.com/data/r10/citytemp
(City Temperature Data)

. ladder tempjuly
Transformation        formula                   chi2(2)      P(chi2)

cubic                 tempjuly^3                  47.49        0.000
square                tempjuly^2                  19.70        0.000
identity              tempjuly                     3.83        0.147
square root           sqrt(tempjuly)               1.83        0.400
log                   log(tempjuly)                5.40        0.067
1/(square root)       1/sqrt(tempjuly)            13.72        0.001
inverse               1/tempjuly                  26.36        0.000
1/square              1/(tempjuly^2)              64.43        0.000
1/cubic               1/(tempjuly^3)                  .        0.000
```

The period in the last line indicates that the χ^2 is very large; see [R] **sktest**.

From the table, we see that there is certainly a difference in normality between the square and square-root transform. If, however, you can see the difference between the transforms in the diagram below, you have better eyes than we do:

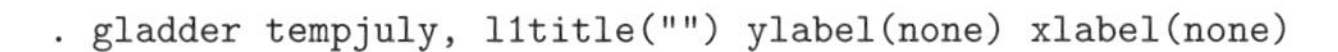

```
. gladder tempjuly, l1title("") ylabel(none) xlabel(none)
```

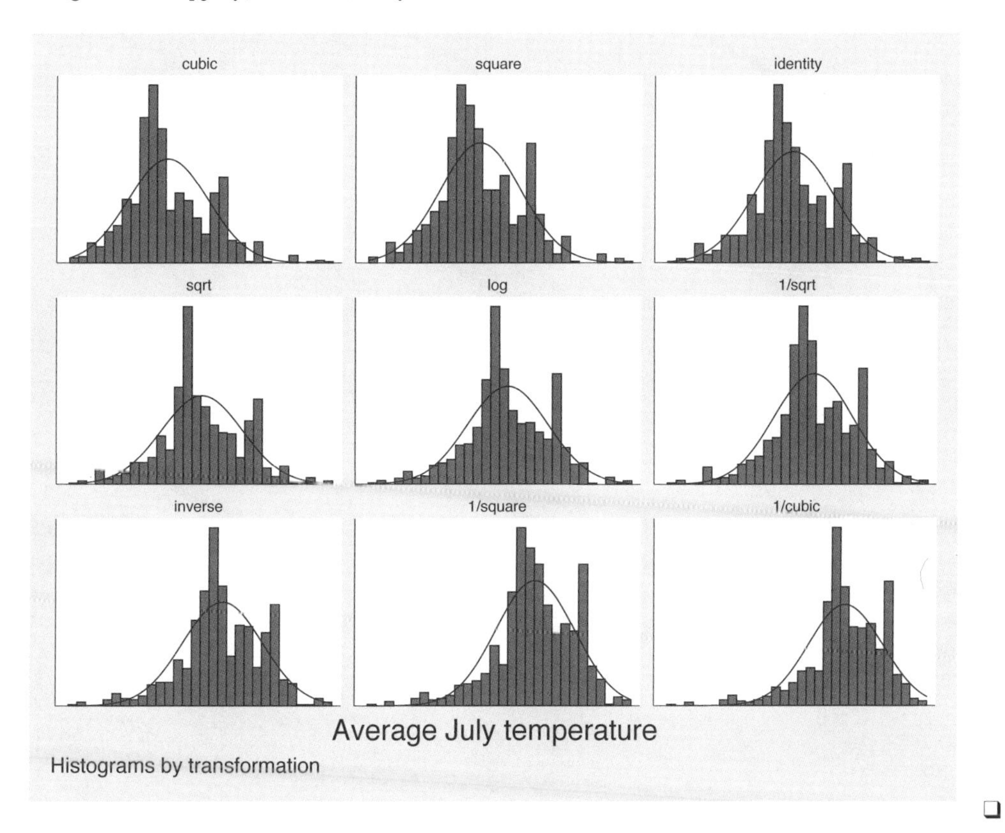

❑

▷ Example 3: qladder

A better graph for seeing normality is the quantile–normal graph, which can be produced by `qladder`.

(Continued on next page)

```
. qladder tempjuly, ylabel(none) xlabel(none)
```

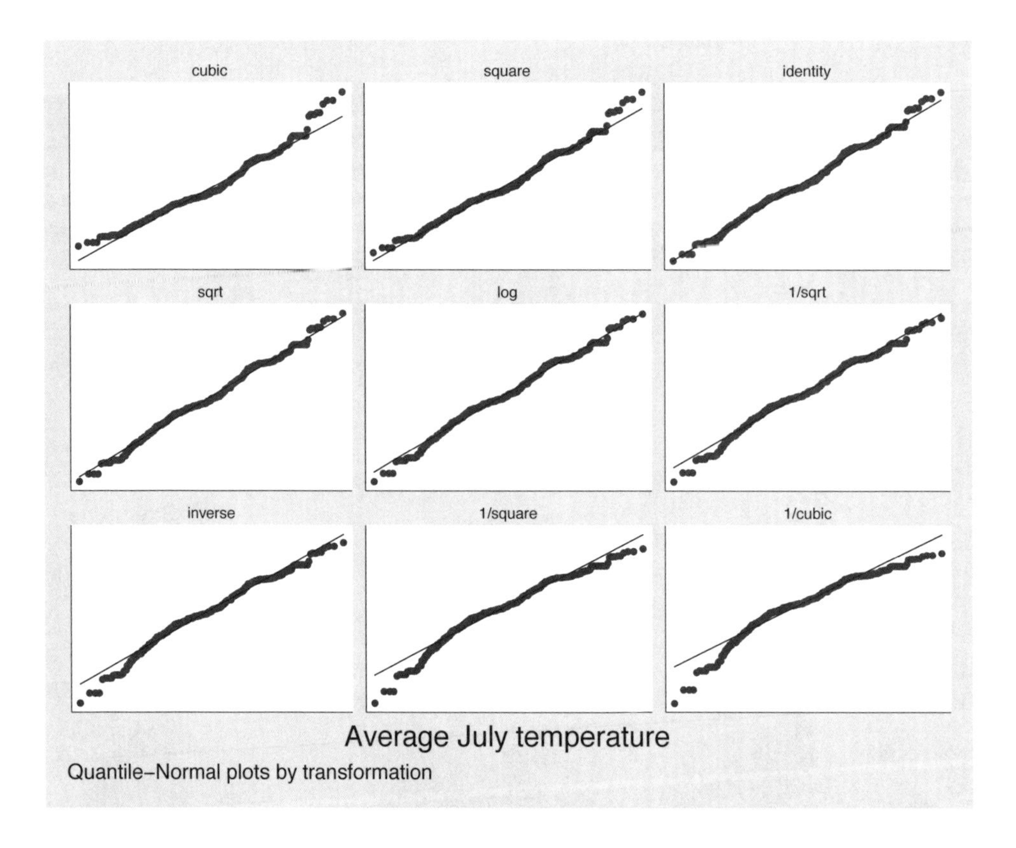

This graph shows that for the square transform, the upper tail—and only the upper tail—diverges from what would be expected. This divergence is detected by `sktest` as a problem with skewness, as we would learn from using `sktest` to examine `tempjuly` squared and square rooted.

◁

Saved Results

ladder saves the following in r():

Scalars

r(N)	number of observations
r(invcube)	χ^2 for inverse-cubic transformation
r(P_invcube)	significance level for inverse-cubic transformation
r(invsq)	χ^2 for inverse-square transformation
r(P_invsq)	significance level for inverse-square transformation
r(inv)	χ^2 for inverse transformation
r(P_inv)	significance level for inverse transformation
r(invsqrt)	χ^2 for inverse-root transformation
r(P_invsqrt)	significance level for inverse-root transformation
r(log)	χ^2 for log transformation
r(P_log)	significance level for log transformation
r(sqrt)	χ^2 for square-root transformation
r(P_sqrt)	significance level for square-root transformation
r(ident)	χ^2 for untransformed data
r(P_ident)	significance level for untransformed data
r(square)	χ^2 for square transformation
r(P_square)	significance level for square transformation
r(cube)	χ^2 for cubic transformation
r(P_cube)	significance level for cubic transformation

Methods and Formulas

ladder, gladder, and qladder are implemented as ado-files.

For ladder, results are as reported by sktest; see [R] **sktest**. If generate() is specified, the transform with the minimum χ^2 value is chosen.

gladder sets the number of bins to $\min(\sqrt{n}, 10\log_{10} n)$, rounded to the closest integer, where n is the number of unique values of *varname*. See [R] **histogram** for a discussion of the optimal number of bins.

Also see Findley (1990) for a ladder-of-powers variable transformation program that produces one-way graphs with overlaid box plots, in addition to histograms with overlaid normals. Buchner and Findley (1990) discuss ladder-of-powers transformations as one aspect of preliminary data analysis. Also see Hamilton (1992, 18–23) and Hamilton (2006, 127–129).

Acknowledgment

qladder was written by Jeroen Weesie, Utrecht University, Utrecht, The Netherlands.

References

Buchner, D. M., and T. W. Findley. 1990. Research in physical medicine and rehabilitation: VIII preliminary data analysis. *American Journal of Physical Medicine and Rehabilitation* 69: 154–169.

Cox, N. J. 2005. Speaking Stata: Density probability plots. *Stata Journal* 5: 259–273.

Findley, T. W. 1990. sed3: Variable transformation and evaluation. *Stata Technical Bulletin* 2: 15. Reprinted in *Stata Technical Bulletin Reprints*, vol. 1, pp. 85–86.

Hamilton, L. C. 1992. *Regression with Graphics: A Second Course in Applied Statistics*. Belmont, CA: Duxbury.

——. 2006. *Statistics with Stata (Updated for Version 9)*. Belmont, CA: Duxbury.

Tukey, J. W. 1977. *Exploratory Data Analysis*. Reading, MA: Addison–Wesley.

Also See

[R] **diagnostic plots** — Distributional diagnostic plots

[R] **lnskew0** — Find zero-skewness log or Box–Cox transform

[R] **lv** — Letter-value displays

[R] **sktest** — Skewness and kurtosis test for normality

Title

level — Set default confidence level

Syntax

```
set level # [, permanently]
```

Description

`set level` specifies the default confidence level for confidence intervals for all commands that report confidence intervals. The initial value is 95, meaning 95% confidence intervals. # may be between 10.00 and 99.99, and # can have at most two digits after the decimal point.

Option

`permanently` specifies that, in addition to making the change right now, the `level` setting be remembered and become the default setting when you invoke Stata.

Remarks

To change the level of confidence intervals reported by a particular command, you need not reset the default confidence level. All commands that report confidence intervals have a `level(#)` option. When you do not specify the option, the confidence intervals are calculated for the default level set by `set level`, or 95% if you have not reset it.

▷ Example 1

We use the `ci` command to obtain the confidence interval for the mean of `mpg`:

```
. use http://www.stata-press.com/data/r10/auto
(1978 Automobile Data)

. ci mpg

    Variable |        Obs        Mean    Std. Err.       [95% Conf. Interval]
-------------+---------------------------------------------------------------
         mpg |         74     21.2973    .6725511        19.9569    22.63769
```

To obtain 90% confidence intervals, we would type

```
. ci mpg, level(90)

    Variable |        Obs        Mean    Std. Err.       [90% Conf. Interval]
-------------+---------------------------------------------------------------
         mpg |         74     21.2973    .6725511       20.17683    22.41776
```

or

```
. set level 90

. ci mpg

    Variable |        Obs        Mean    Std. Err.       [90% Conf. Interval]
-------------+---------------------------------------------------------------
         mpg |         74     21.2973    .6725511       20.17683    22.41776
```

If we opt for the second alternative, the next time that we fit a model (say with `regress`), 90% confidence intervals will be reported. If we wanted 95% confidence intervals, we could specify `level(95)` on the estimation command, or we could reset the default by typing `set level 95`.

The current setting of `level()` is stored as the c-class value `c(level)`; see [P] **creturn**.

◁

Also See

[R] **query** — Display system parameters

[P] **creturn** — Return c-class values

[U] **20 Estimation and postestimation commands**

[U] **20.7 Specifying the width of confidence intervals**

Title

lincom — Linear combinations of estimators

Syntax

`lincom` *exp* [, *options*]

options	description
`eform`	generic label; `exp(b)`; the default
`or`	odds ratio
`hr`	hazard ratio
`irr`	incidence-rate ratio
`rrr`	relative-risk ratio
`level(#)`	set confidence level; default is `level(95)`

where *exp* is any linear combination of coefficients that is a valid syntax for `test`; see [R] **test**. *exp* must not contain an equal sign.

Description

`lincom` computes point estimates, standard errors, t or z statistics, p-values, and confidence intervals for linear combinations of coefficients after any estimation command. Results can optionally be displayed as odds ratios, hazard ratios, incidence-rate ratios, or relative risk ratios.

`lincom` can be used with `svy` estimation results; see [SVY] **svy postestimation**.

Options

`eform`, `or`, `hr`, `irr`, and `rrr` all report coefficient estimates as $\exp(\widehat{\beta})$ rather than $\widehat{\beta}$. Standard errors and confidence intervals are similarly transformed. `or` is the default after `logistic`. The only difference in these options is how the output is labeled.

Option	Label	Explanation	Example commands
`eform`	`exp(b)`	Generic label	
`or`	`Odds Ratio`	Odds ratio	`logistic, logit`
`hr`	`Haz. Ratio`	Hazard ratio	`stcox, streg`
`irr`	`IRR`	Incidence-rate ratio	`poisson`
`rrr`	`RRR`	Relative risk ratio	`mlogit`

exp may not contain any additive constants when you use the `eform`, `or`, `hr`, `irr`, or `rrr` option.

`level(#)` specifies the confidence level, as a percentage, for confidence intervals. The default is `level(95)` or as set by `set level`; see [U] **20.7 Specifying the width of confidence intervals**.

Remarks

Remarks are presented under the following headings:

Using lincom

After fitting a model and obtaining estimates for coefficients $\beta_1, \beta_2, \ldots, \beta_k$, you may want to view estimates for linear combinations of the β_i, such as $\beta_1 - \beta_2$. `lincom` can display estimates for any linear combination of the form $c_0 + c_1\beta_1 + c_2\beta_2 + \cdots + c_k\beta_k$.

`lincom` works after any estimation command for which `test` works. Any valid expression for `test` syntax 1 (see [R] **test**) is a valid expression for `lincom`.

`lincom` is useful for viewing odds ratios, hazard ratios, etc., for one group (i.e., one set of covariates) relative to another group (i.e., another set of covariates). See the examples below.

▷ Example 1

We perform a linear regression:

```
. use http://www.stata-press.com/data/r10/regress
. regress y x1 x2 x3
      Source |       SS       df       MS              Number of obs =     148
-------------+------------------------------           F(  3,   144) =   96.12
       Model |  3259.3561     3  1086.45203           Prob > F      =  0.0000
    Residual |  1627.56282   144  11.3025196           R-squared     =  0.6670
-------------+------------------------------           Adj R-squared =  0.6600
       Total |  4886.91892   147  33.2443464           Root MSE      =  3.3619

------------------------------------------------------------------------------
           y |      Coef.   Std. Err.      t    P>|t|     [95% Conf. Interval]
-------------+----------------------------------------------------------------
          x1 |   1.457113    1.07461     1.36   0.177     -.666934    3.581161
          x2 |   2.221682   .8610358     2.58   0.011     .5197797    3.923583
          x3 |   -.006139   .0005543   -11.08   0.000    -.0072345   -.0050435
       _cons |   36.10135   4.382693     8.24   0.000     27.43863    44.76407
------------------------------------------------------------------------------
```

To see the difference of the coefficients of `x2` and `x1`, we type

```
. lincom x2 - x1
 ( 1)  - x1 + x2 = 0

------------------------------------------------------------------------------
           y |      Coef.   Std. Err.      t    P>|t|     [95% Conf. Interval]
-------------+----------------------------------------------------------------
         (1) |   .7645682   .9950282     0.77   0.444     -1.20218    2.731316
------------------------------------------------------------------------------
```

The expression can be any linear combination.

```
. lincom 3*x1 + 500*x3
 ( 1)  3 x1 + 500 x3 = 0

------------------------------------------------------------------------------
           y |      Coef.   Std. Err.      t    P>|t|     [95% Conf. Interval]
-------------+----------------------------------------------------------------
         (1) |   1.301825   3.396624     0.38   0.702    -5.411858    8.015507
------------------------------------------------------------------------------
```

Nonlinear expressions are not allowed.

```
. lincom x2/x1
not possible with test
r(131);
```

For information about estimating nonlinear expressions, see [R] **nlcom**.

◁

❑ Technical Note

`lincom` uses the same shorthands for coefficients as `test` (see [R] **test**). When you type `x1`, for instance, `lincom` knows that you mean the coefficient of `x1`. The formal syntax for referencing this coefficient is actually `_b[x1]`, or alternatively, `_coef[x1]`. So, more formally, in the last example we could have typed

```
. lincom 3*_b[x1] + 500*_b[x3]
  (output omitted)
```

❑

Odds ratios and incidence-rate ratios

After logistic regression, the `or` option can be specified with `lincom` to display odds ratios for any effect. Incidence-rate ratios after commands such as `poisson` can be similarly obtained by specifying the `irr` option.

▷ Example 2

Consider the low birthweight dataset from Hosmer and Lemeshow (2000, 25). We fit a logistic regression model of low birthweight (variable `low`) on the following variables:

Variable	Description	Coding
`age`	age in years	
`black`	race black	1 if black, 0 otherwise
`other`	race other	1 if race other, 0 otherwise
`smoke`	smoking status	1 if smoker, 0 if nonsmoker
`ht`	history of hypertension	1 if yes, 0 if no
`ui`	uterine irritability	1 if yes, 0 if no
`lwd`	maternal weight before pregnancy	1 if weight $<$ 110 lb., 0 otherwise
`ptd`	history of premature labor	1 if yes, 0 if no
`agelwd`	`age` $\times$ `lwd`	
`smokelwd`	`smoke` $\times$ `lwd`	

We first fit a model without the interaction terms `agelwd` and `smokelwd` (Hosmer and Lemeshow 1989, table 4.8) by using `logit`.

```
. use http://www.stata-press.com/data/r10/lbw3
(Hosmer & Lemeshow data)

. logit low age lwd black other smoke ptd ht ui

Iteration 0:   log likelihood =   -117.336
Iteration 1:   log likelihood = -99.431174
Iteration 2:   log likelihood = -98.785718
Iteration 3:   log likelihood =    -98.778
Iteration 4:   log likelihood = -98.777998

Logistic regression                               Number of obs   =        189
                                                  LR chi2(8)      =      37.12
                                                  Prob > chi2     =     0.0000
Log likelihood = -98.777998                       Pseudo R2       =     0.1582

------------------------------------------------------------------------------
         low |      Coef.   Std. Err.      z    P>|z|     [95% Conf. Interval]
-------------+----------------------------------------------------------------
         age |  -.0464796   .0373888    -1.24   0.214    -.1197603    .0268011
         lwd |   .8420615   .4055338     2.08   0.038     .0472299    1.636893
       black |   1.073456   .5150752     2.08   0.037     .0639273    2.082985
       other |    .815367   .4452979     1.83   0.067    -.0574008    1.688135
       smoke |   .8071996    .404446     2.00   0.046     .0145001    1.599899
         ptd |   1.281678   .4621157     2.77   0.006     .3759478    2.187408
          ht |   1.435227   .6482699     2.21   0.027     .1646415    2.705813
          ui |   .6576256   .4666192     1.41   0.159    -.2569313    1.572182
       _cons |  -1.216781   .9556797    -1.27   0.203    -3.089878     .656317
------------------------------------------------------------------------------
```

To get the odds ratio for black smokers relative to white nonsmokers (the reference group), we type

```
. lincom black + smoke, or

 ( 1)  black + smoke = 0

------------------------------------------------------------------------------
         low | Odds Ratio   Std. Err.      z    P>|z|     [95% Conf. Interval]
-------------+----------------------------------------------------------------
         (1) |   6.557805   4.744692     2.60   0.009     1.588176    27.07811
------------------------------------------------------------------------------
```

`lincom` computed $\exp(\beta_{\texttt{black}} + \beta_{\texttt{smoke}}) = 6.56$. To see the odds ratio for white smokers relative to black nonsmokers, we type

```
. lincom smoke - black, or

 ( 1) - black + smoke = 0

------------------------------------------------------------------------------
         low | Odds Ratio   Std. Err.      z    P>|z|     [95% Conf. Interval]
-------------+----------------------------------------------------------------
         (1) |   .7662425   .4430176    -0.46   0.645     .2467334    2.379603
------------------------------------------------------------------------------
```

Now let's add the interaction terms to the model (Hosmer and Lemeshow 1989, table 4.10). This time, we will use `logistic` rather than `logit`. By default, `logistic` displays odds ratios.

```
. generate agelwd = age*lwd

. generate smokelwd = smoke*lwd
```

```
. logistic low age black other smoke ht ui lwd ptd agelwd smokelwd
Logistic regression                               Number of obs   =        189
                                                  LR chi2(10)     =      42.66
                                                  Prob > chi2     =     0.0000
Log likelihood =  -96.00616                       Pseudo R2       =     0.1818
```

low	Odds Ratio	Std. Err.	z	P>|z|	[95% Conf.	Interval]
age	.9194513	.041896	-1.84	0.065	.8408967	1.005344
black	2.95383	1.532788	2.09	0.037	1.068277	8.167462
other	2.137589	.9919132	1.64	0.102	.8608713	5.307749
smoke	3.168096	1.452377	2.52	0.012	1.289956	7.780755
ht	3.893141	2.5752	2.05	0.040	1.064768	14.2346
ui	2.071284	.9931385	1.52	0.129	.8092928	5.301191
lwd	.1772934	.3312383	-0.93	0.354	.0045539	6.902359
ptd	3.426633	1.615282	2.61	0.009	1.360252	8.632086
agelwd	1.15883	.09602	1.78	0.075	.9851216	1.36317
smokelwd	.2447849	.2003996	-1.72	0.086	.0491956	1.217988

Hosmer and Lemeshow (1989, table 4.13) consider the effects of smoking (smoke = 1) and low maternal weight before pregnancy (lwd = 1). The effect of smoking among non-low-weight mothers (lwd = 0) is given by the odds ratio 3.17 for smoke in the logistic output. The effect of smoking among low-weight mothers is given by

```
. lincom smoke + smokelwd
 ( 1)  smoke + smokelwd = 0
```

low	Odds Ratio	Std. Err.	z	P>|z|	[95% Conf.	Interval]
(1)	.7755022	.5749508	-0.34	0.732	.1813465	3.316322

We did not have to specify the or option. After logistic, lincom assumes or by default.

The effect of low weight (lwd = 1) is more complicated since we fit an age × lwd interaction. We must specify the age of mothers for the effect. The effect among 30-year-old nonsmokers is given by

```
. lincom lwd + 30*agelwd
 ( 1)  lwd + 30.0 agelwd = 0
```

low	Odds Ratio	Std. Err.	z	P>|z|	[95% Conf.	Interval]
(1)	14.7669	13.56689	2.93	0.003	2.439266	89.39625

lincom computed $\exp(\beta_{\texttt{lwd}} + 30\beta_{\texttt{agelwd}}) = 14.8$. It may seem odd that we entered it as lwd + 30*agelwd, but remember that lwd and agelwd are just lincom's (and test's) shorthand for _b[lwd] and _b[agelwd]. We could have typed

```
. lincom _b[lwd] + 30*_b[agelwd]
 ( 1)  lwd + 30.0 agelwd = 0
```

low	Odds Ratio	Std. Err.	z	P>|z|	[95% Conf.	Interval]
(1)	14.7669	13.56689	2.93	0.003	2.439266	89.39625

◁

Multiple-equation models

lincom also works with multiple-equation models. The only difference is how you refer to the coefficients. Recall that for multiple-equation models, coefficients are referenced using the syntax

[*eqno*] *varname*

where *eqno* is the equation number or equation name and *varname* is the corresponding variable name for the coefficient; see [U] **13.5 Accessing coefficients and standard errors** and [R] **test** for details.

▷ Example 3

Let's consider the example from [R] **mlogit** (Tarlov et al. 1989; Wells et al. 1989).

```
. use http://www.stata-press.com/data/r10/sysdsn2, clear
(Health insurance data)
. mlogit insure age male nonwhite site2 site3, nolog
Multinomial logistic regression                   Number of obs   =        615
                                                  LR chi2(10)     =      42.99
                                                  Prob > chi2     =     0.0000
Log likelihood = -534.36165                       Pseudo R2       =     0.0387
```

insure	Coef.	Std. Err.	z	P>|z|	[95% Conf.	Interval]
Prepaid						
age	-.011745	.0061946	-1.90	0.058	-.0238862	.0003962
male	.5616934	.2027465	2.77	0.006	.1643175	.9590693
nonwhite	.9747768	.2363213	4.12	0.000	.5115955	1.437958
site2	.1130359	.2101903	0.54	0.591	-.2989296	.5250013
site3	-.5879879	.2279351	-2.58	0.010	-1.034733	-.1412433
_cons	.2697127	.3284422	0.82	0.412	-.3740222	.9134476
Uninsure						
age	-.0077961	.0114418	-0.68	0.496	-.0302217	.0146294
male	.4518496	.3674867	1.23	0.219	-.268411	1.17211
nonwhite	.2170589	.4256361	0.51	0.610	-.6171725	1.05129
site2	-1.211563	.4705127	-2.57	0.010	-2.133751	-.2893747
site3	-.2078123	.3662926	-0.57	0.570	-.9257327	.510108
_cons	-1.286943	.5923219	-2.17	0.030	-2.447872	-.1260135

```
(insure==Indemnity is the base outcome)
```

To see the estimate of the sum of the coefficient of male and the coefficient of nonwhite for the Prepaid outcome, we type

```
. lincom [Prepaid]male + [Prepaid]nonwhite
 ( 1)  [Prepaid]male + [Prepaid]nonwhite = 0
```

insure	Coef.	Std. Err.	z	P>|z|	[95% Conf.	Interval]
(1)	1.53647	.3272489	4.70	0.000	.8950741	2.177866

To view the estimate as a ratio of relative risks (see [R] **mlogit** for the definition and interpretation), we specify the rrr option.

```
. lincom [Prepaid]male + [Prepaid]nonwhite, rrr
 ( 1)  [Prepaid]male + [Prepaid]nonwhite = 0
```

insure	RRR	Std. Err.	z	P>|z|	[95% Conf. Interval]	
(1)	4.648154	1.521103	4.70	0.000	2.447517	8.827451

◁

Saved Results

lincom saves the following in r():

Scalars

r(estimate)	point estimate
r(se)	estimate of standard error
r(df)	degrees of freedom

Methods and Formulas

lincom is implemented as an ado-file.

References

Hosmer, D. W., Jr., and S. Lemeshow. 1989. *Applied Logistic Regression*. New York: Wiley.

——. 2000. *Applied Logistic Regression*. 2nd ed. New York: Wiley.

Tarlov, A. R., J. E. Ware, Jr., S. Greenfield, E. C. Nelson, E. Perrin, and M. Zubkoff. 1989. The medical outcomes study. *Journal of the American Medical Association* 262: 925–930.

Wells, K. E., R. D. Hays, M. A. Burnam, W. H. Rogers, S. Greenfield, and J. E. Ware, Jr. 1989. Detection of depressive disorder for patients receiving prepaid or fee-for-service care. *Journal of the American Medical Association* 262: 3298–3302.

Also See

[R] **nlcom** — Nonlinear combinations of estimators

[R] **test** — Test linear hypotheses after estimation

[R] **testnl** — Test nonlinear hypotheses after estimation

[U] **13.5 Accessing coefficients and standard errors**

[U] **20 Estimation and postestimation commands**

Title

linktest — Specification link test for single-equation models

Syntax

`linktest` [*if*] [*in*] [, *cmd_options*]

When *if* and *in* are not specified, the link test is performed on the same sample as the previous estimation.

Description

`linktest` performs a link test for model specification after any single-equation estimation command, such as `logistic`, `regress`, `stcox`, etc.

Option

Main

cmd_options must be the same options specified with the underlying estimation command.

Remarks

The form of the link test implemented here is based on an idea of Tukey (1949), which was further described by Pregibon (1980), elaborating on work in his unpublished thesis (Pregibon 1979). See *Methods and Formulas* below for more details.

▷ Example 1

We want to explain the mileage ratings of cars in our automobile dataset by using the weight, engine displacement, and whether the car is manufactured outside the United States:

```
. use http://www.stata-press.com/data/r10/auto
(1978 Automobile Data)

. regress mpg weight displ foreign
      Source |       SS       df       MS              Number of obs =      74
-------------+------------------------------           F(  3,    70) =   45.88
       Model |  1619.71935     3  539.906448           Prob > F      =  0.0000
    Residual |  823.740114    70  11.7677159           R-squared     =  0.6629
-------------+------------------------------           Adj R-squared =  0.6484
       Total |  2443.45946    73  33.4720474           Root MSE      =  3.4304

------------------------------------------------------------------------------
         mpg |      Coef.   Std. Err.      t    P>|t|     [95% Conf. Interval]
-------------+----------------------------------------------------------------
      weight |  -.0067745   .0011665    -5.81   0.000    -.0091011   -.0044479
displacement |   .0019286   .0100701     0.19   0.849    -.0181556    .0220129
     foreign |  -1.600631   1.113648    -1.44   0.155    -3.821732    .6204699
       _cons |   41.84795   2.350704    17.80   0.000     37.15962    46.53628
------------------------------------------------------------------------------
```

On the basis of the R^2, we are reasonably pleased with this model.

If our model really is specified correctly, then if we were to regress `mpg` on the prediction and the prediction squared, the prediction squared would have no explanatory power. This is what `linktest` does:

```
. linktest
      Source |       SS       df       MS              Number of obs =      74
-------------+------------------------------           F(  2,    71) =   76.75
       Model |  1670.71514     2  835.357572           Prob > F      =  0.0000
    Residual |  772.744316    71  10.8837228           R-squared     =  0.6837
-------------+------------------------------           Adj R-squared =  0.6748
       Total |  2443.45946    73  33.4720474           Root MSE      =   3.299

------------------------------------------------------------------------------
         mpg |      Coef.   Std. Err.      t    P>|t|     [95% Conf. Interval]
-------------+----------------------------------------------------------------
        _hat |  -.4127198   .6577736    -0.63   0.532    -1.724283    .8988434
      _hatsq |   .0338198    .015624     2.16   0.034     .0026664    .0649732
       _cons |   14.00705   6.713276     2.09   0.041     .6211539    27.39294
------------------------------------------------------------------------------
```

We find that the prediction squared does have explanatory power, so our specification is not as good as we thought.

Although `linktest` is formally a test of the specification of the dependent variable, it is often interpreted as a test that, conditional on the specification, the independent variables are specified incorrectly. We will follow that interpretation and now include weight squared in our model:

```
. generate weight2 = weight*weight
. regress mpg weight weight2 displ foreign
      Source |       SS       df       MS              Number of obs =      74
-------------+------------------------------           F(  4,    69) =   39.37
       Model |  1699.02634     4  424.756584           Prob > F      =  0.0000
    Residual |  744.433124    69  10.7888859           R-squared     =  0.6953
-------------+------------------------------           Adj R-squared =  0.6777
       Total |  2443.45946    73  33.4720474           Root MSE      =  3.2846

------------------------------------------------------------------------------
         mpg |      Coef.   Std. Err.      t    P>|t|     [95% Conf. Interval]
-------------+----------------------------------------------------------------
      weight |  -.0173257   .0040488    -4.28   0.000    -.0254028   -.0092486
     weight2 |   1.87e-06   6.89e-07     2.71   0.008     4.93e-07    3.24e-06
displacement |  -.0101625   .0106236    -0.96   0.342     -.031356     .011031
     foreign |  -2.560016   1.123506    -2.28   0.026    -4.801349   -.3186832
       _cons |   58.23575   6.449882     9.03   0.000     45.36859    71.10291
------------------------------------------------------------------------------
```

Now we perform the link test on our new model:

(*Continued on next page*)

```
. linktest
      Source |       SS       df       MS              Number of obs =      74
-------------+------------------------------           F(  2,    71) =   81.08
       Model |  1699.39489     2  849.697445           Prob > F      =  0.0000
    Residual |   744.06457    71  10.4797827           R-squared     =  0.6955
-------------+------------------------------           Adj R-squared =  0.6869
       Total |  2443.45946    73  33.4720474           Root MSE      =  3.2372

------------------------------------------------------------------------------
         mpg |      Coef.   Std. Err.      t    P>|t|     [95% Conf. Interval]
-------------+----------------------------------------------------------------
        _hat |   1.141987   .7612218     1.50   0.138    -.3758456    2.659821
      _hatsq |  -.0031916   .0170194    -0.19   0.852    -.0371272    .0307441
       _cons |   -1.50305   8.196444    -0.18   0.855    -17.84629    14.84019
------------------------------------------------------------------------------
```

We now pass the link test.

◁

▷ Example 2

Above we followed a standard misinterpretation of the link test—when we discovered a problem, we focused on the explanatory variables of our model. We might consider varying exactly what the link test tests. The link test told us that our dependent variable was misspecified. For those with an engineering background, mpg is indeed a strange measure. It would make more sense to model energy consumption—gallons per mile—in terms of weight and displacement:

```
. gen gpm = 1/mpg
. regress gpm weight displ foreign
      Source |       SS       df       MS              Number of obs =      74
-------------+------------------------------           F(  3,    70) =   76.33
       Model |  .009157962     3  .003052654           Prob > F      =  0.0000
    Residual |  .002799666    70  .000039995           R-squared     =  0.7659
-------------+------------------------------           Adj R-squared =  0.7558
       Total |  .011957628    73  .000163803           Root MSE      =  .00632

------------------------------------------------------------------------------
         gpm |      Coef.   Std. Err.      t    P>|t|     [95% Conf. Interval]
-------------+----------------------------------------------------------------
      weight |   .0000144   2.15e-06     6.72   0.000     .0000102    .0000187
displacement |   .0000186   .0000186     1.00   0.319    -.0000184    .0000557
     foreign |   .0066981   .0020531     3.26   0.002     .0026034    .0107928
       _cons |   .0008917   .0043337     0.21   0.838    -.0077515     .009535
------------------------------------------------------------------------------
```

This model looks every bit as reasonable as our original model.

```
. linktest
      Source |       SS       df       MS              Number of obs =      74
-------------+------------------------------           F(  2,    71) =  117.06
       Model |  .009175219     2  .004587609           Prob > F      =  0.0000
    Residual |  .002782409    71  .000039189           R-squared     =  0.7673
-------------+------------------------------           Adj R-squared =  0.7608
       Total |  .011957628    73  .000163803           Root MSE      =  .00626

------------------------------------------------------------------------------
         gpm |      Coef.   Std. Err.      t    P>|t|     [95% Conf. Interval]
-------------+----------------------------------------------------------------
        _hat |   .6608413    .515275     1.28   0.204    -.3665877     1.68827
      _hatsq |   3.275857   4.936655     0.66   0.509    -6.567553    13.11927
       _cons |    .008365   .0130468     0.64   0.523    -.0176496    .0343795
------------------------------------------------------------------------------
```

Specifying the model in terms of gallons per mile also solves the specification problem and results in a more parsimonious specification.

◁

▷ Example 3

The link test can be used with any single-equation estimation procedure, not solely regression. Let's turn our problem around and attempt to explain whether a car is manufactured outside the United States by its mileage rating and weight. To save paper, we will specify `logit`'s `nolog` option, which suppresses the iteration log:

```
. logit foreign mpg weight, nolog
Logistic regression                               Number of obs   =         74
                                                  LR chi2(2)      =      35.72
                                                  Prob > chi2     =     0.0000
Log likelihood = -27.175156                       Pseudo R2       =     0.3966

------------------------------------------------------------------------------
     foreign |      Coef.   Std. Err.      z    P>|z|     [95% Conf. Interval]
-------------+----------------------------------------------------------------
         mpg |  -.1685869   .0919174    -1.83   0.067    -.3487418     .011568
      weight |  -.0039067   .0010116    -3.86   0.000    -.0058894    -.001924
       _cons |   13.70837   4.518707     3.03   0.002     4.851864    22.56487
------------------------------------------------------------------------------
```

When we run `linktest` after `logit`, the result is another logit specification:

```
. linktest, nolog
Logistic regression                               Number of obs   =         74
                                                  LR chi2(2)      =      36.83
                                                  Prob > chi2     =     0.0000
Log likelihood = -26.615714                       Pseudo R2       =     0.4090

------------------------------------------------------------------------------
     foreign |      Coef.   Std. Err.      z    P>|z|     [95% Conf. Interval]
-------------+----------------------------------------------------------------
        _hat |   .8438531   .2738759     3.08   0.002     .3070661     1.38064
      _hatsq |  -.1559115   .1568642    -0.99   0.320    -.4633596    .1515366
       _cons |   .2630557   .4299598     0.61   0.541      -.57965    1.105761
------------------------------------------------------------------------------
```

The link test reveals no problems with our specification.

If there had been a problem, we would have been virtually forced to accept the misinterpretation of the link test—we would have reconsidered our specification of the independent variables. When using `logit`, we have no control over the specification of the dependent variable other than to change likelihood functions.

We admit to having seen a dataset once for which the link test rejected the logit specification. We did change the likelihood function, refitting the model using `probit`, and satisfied the link test. Probit has thinner tails than logit. In general, however, you will not be so lucky.

◁

(Continued on next page)

❑ Technical Note

You should specify the same options with `linktest` that you do with the estimation command, although you do not have to follow this advice as literally as we did in the preceding example. `logit`'s `nolog` option merely suppresses a part of the output, not what is estimated. We specified `nolog` both times to save paper.

If you are testing a tobit model, you must specify the censoring points just as you do with the `tobit` command.

If you are not sure which options are important, duplicate exactly what you specified on the estimation command.

If you do not specify `if` *exp* or `in` *range* with `linktest`, Stata will by default perform the link test on the same sample as the previous estimation. Suppose that you omitted some data when performing your estimation, but want to calculate the link test on all the data, which you might do if you believe the model is appropriate for all the data. You would type `linktest if e(sample) < .` to do this.

❑

Saved Results

`linktest` saves the following in `r()`:

Scalars

`r(t)`	t statistic on `_hatsq`	`r(df)`	degrees of freedom

`linktest` is *not* an estimation command in the sense that it leaves previous estimation results unchanged. For instance, after running a regression and performing the link test, typing `regress` without arguments after the link test still replays the original regression.

For integrating an estimation command with `linktest`, `linktest` assumes that the name of the estimation command is stored in `e(cmd)` and that the name of the dependent variable is stored in `e(depvar)`. After estimation, it assumes that the number of degrees of freedom for the t test is given by `e(df_m)` if the macro is defined.

If the estimation command reports Z statistics instead of t statistics, `linktest` will also report Z statistics. The Z statistic, however, is still returned in `r(t)`, and `r(df)` is set to a missing value.

Methods and Formulas

`linktest` is implemented as an ado-file.

The link test is based on the idea that if a regression or regression-like equation is properly specified, you should be able to find no additional independent variables that are significant except by chance. One kind of specification error is called a link error. In regression, this means that the dependent variable needs a transformation or "link" function to properly relate to the independent variables. The idea of a link test is to add an independent variable to the equation that is especially likely to be significant if there is a link error.

Let

$$\mathbf{y} = f(\mathbf{X}\beta)$$

be the model and $\widehat{\beta}$ be the parameter estimates. `linktest` calculates

$$\texttt{\_hat} = \mathbf{X}\widehat{\beta}$$

and

$$\texttt{\_hatsq} = \texttt{\_hat}^2$$

The model is then refitted with these two variables, and the test is based on the significance of `_hatsq`. This is the form suggested by Pregibon (1979) based on an idea of Tukey (1949). Pregibon (1980) suggests a slightly different method that has come to be known as "Pregibon's goodness-of-link test". We prefer the older version because it is universally applicable, straightforward, and a good second-order approximation. It can be applied to any single-equation estimation technique, whereas Pregibon's more recent tests are estimation-technique specific.

References

Pregibon, D. 1979. *Data Analytic Methods for Generalized Linear Models*. Ph.D. Dissertation. University of Toronto.

——. 1980. Goodness of link tests for generalized linear models. *Applied Statistics* 29: 15–24.

Tukey, J. W. 1949. One degree of freedom for nonadditivity. *Biometrics* 5: 232–242.

Also See

[R] **regress postestimation** — Postestimation tools for regress

Title

lnskew0 — Find zero-skewness log or Box–Cox transform

Syntax

Zero-skewness log

`lnskew0` *newvar* = *exp* [*if*] [*in*] [, *options*]

Zero-skewness Box–Cox transform

`bcskew0` *newvar* = *exp* [*if*] [*in*] [, *options*]

options	description
Main	
`delta(#)`	increment for derivative of skewness function; default is `delta(0.02)` for `lnskew0` and `delta(0.01)` for `bcskew0`
`zero(#)`	value for determining convergence; default is `zero(0.001)`
`level(#)`	set confidence level; default is `level(95)`

Description

`lnskew0` creates *newvar* $= \ln(\pm exp - k)$, choosing k and the sign of *exp* so that the skewness of *newvar* is zero.

`bcskew0` creates *newvar* $= (exp^{\lambda} - 1)/\lambda$, the Box–Cox power transformation (Box and Cox 1964), choosing λ so that the skewness of *newvar* is zero. *exp* must be strictly positive. Also see [R] **boxcox** for maximum likelihood estimation of λ.

Options

Main

`delta(#)` specifies the increment used for calculating the derivative of the skewness function with respect to k (`lnskew0`) or λ (`bcskew0`). The default values are 0.02 for `lnskew0` and 0.01 for `bcskew0`.

`zero(#)` specifies a value for skewness to determine convergence that is small enough to be considered zero and is, by default, 0.001.

`level(#)` specifies the confidence level for the confidence interval for k (`lnskew0`) or λ (`bcskew0`). The confidence interval is calculated only if `level()` is specified. # is specified as an integer; 95 means 95% confidence intervals. The `level()` option is honored only if the number of observations exceeds 7.

Remarks

▷ Example 1: lnskew0

Using our automobile dataset (see [U] **1.2.1 Sample datasets**), we want to generate a new variable equal to ln(`mpg` − k) to be approximately normally distributed. `mpg` records the miles per gallon for each of our cars. One feature of the normal distribution is that it has skewness 0.

```
. use http://www.stata-press.com/data/r10/auto
(1978 Automobile Data)
. lnskew0 lnmpg = mpg
       Transform |         k     [95% Conf. Interval]       Skewness
-----------------+---------------------------------------------------
       ln(mpg-k) |  5.383659      (not calculated)         -7.05e-06
```

This created the new variable `lnmpg` = ln(`mpg` − 5.384):

```
. describe lnmpg
              storage  display     value
variable name   type   format      label      variable label
------------------------------------------------------------------------
lnmpg           float  %9.0g                  ln(mpg-5.383659)
```

Since we did not specify the `level()` option, no confidence interval was calculated. At the outset, we could have typed

```
. use http://www.stata-press.com/data/r10/auto
(Automobile Data)
. lnskew0 lnmpg = mpg, level(95)
       Transform |         k     [95% Conf. Interval]       Skewness
-----------------+---------------------------------------------------
       ln(mpg-k) |  5.383659    -17.12339    9.892416      -7.05e-06
```

The confidence interval is calculated under the assumption that ln(`mpg` − k) really does have a normal distribution. It would be perfectly reasonable to use `lnskew0`, even if we did not believe that the transformed variable would have a normal distribution—if we literally wanted the zero-skewness transform—although, then the confidence interval would be an approximation of unknown quality to the true confidence interval. If we now wanted to test the believability of the confidence interval, we could also test our new variable `lnmpg` by using `swilk` with the `lnnormal` option.

◁

❑ Technical Note

`lnskew0` and `bcskew0` report the resulting skewness of the variable merely to reassure you of the accuracy of its results. In our example above, `lnskew0` found k such that the resulting skewness was $-7 \cdot 10^{-6}$, a number close enough to zero for all practical purposes. If we wanted to make it even smaller, we could specify the `zero()` option. Typing `lnskew0 new=mpg, zero(1e-8)` changes the estimated k to 5.383552 from 5.383659 and reduces the calculated skewness to $-2 \cdot 10^{-11}$.

When you request a confidence interval, `lnskew0` may report the lower confidence interval as '`.`', which should be taken as indicating the lower confidence limit $k_L = -\infty$. (This cannot happen with `bcskew0`.)

As an example, consider a sample of size n on x and assume that the skewness of x is positive, but not significantly so, at the desired significance level—say, 5%. Then no matter how large and negative you make k_L, there is no value extreme enough to make the skewness of $\ln(x - k_L)$ equal the corresponding percentile (97.5 for a 95% confidence interval) of the distribution of skewness in a normal distribution of the same sample size. You cannot do this because the distribution of $\ln(x - k_L)$ tends to that of x—apart from location and scale shift—as $x \to \infty$. This "problem" never applies to the upper confidence limit, k_U, because the skewness of $\ln(x - k_U)$ tends to $-\infty$ as k tends upwards to the minimum value of x.

❑

▷ Example 2: bcskew0

In example 1, using `lnskew0` with a variable such as `mpg` is probably undesirable. `mpg` has a natural zero, and we are shifting that zero arbitrarily. On the other hand, use of `lnskew0` with a variable such as temperature measured in Fahrenheit or Celsius would be more appropriate, as the zero is indeed arbitrary.

For a variable like `mpg`, it makes more sense to use the Box–Cox power transform (Box and Cox 1964):

$$y^{(\lambda)} = \frac{y^\lambda - 1}{\lambda}$$

λ is free to take on any value, but $y^{(1)} = y - 1$, $y^{(0)} = \ln(y)$, and $y^{(-1)} = 1 - 1/y$.

`bcskew0` works like `lnskew0`:

```
. bcskew0 bcmpg = mpg, level(95)
      Transform |         L        [95% Conf. Interval]         Skewness
----------------+----------------------------------------------------------
    (mpg^L-1)/L |  -.3673283     -1.212752    .4339645          .0001898
```

The 95% confidence interval includes $\lambda = -1$ (λ is labeled `L` in the output), which has a rather more pleasing interpretation—gallons per mile—than $(\texttt{mpg}^{-.3673} - 1)/(-.3673)$. The confidence interval, however, is calculated assuming that the power transformed variable is normally distributed. It makes perfect sense to use `bcskew0`, even when you do not believe that the transformed variable will be normally distributed, but then the confidence interval is an approximation of unknown quality. If you believe that the transformed data are normally distributed, you can alternatively use `boxcox` to estimate λ; see [R] **boxcox**.

◁

Saved Results

`lnskew0` and `bcskew0` save the following in `r()`:

Scalars

`r(gamma)`	k (`lnskew0`)
`r(lambda)`	λ (`bcskew0`)
`r(lb)`	lower bound of confidence interval
`r(ub)`	upper bound of confidence interval
`r(skewness)`	resulting skewness of transformed variable

Methods and Formulas

lnskew0 and bcskew0 are implemented as ado-files.

Skewness is as calculated by summarize; see [R] **summarize**. Newton's method with numeric, uncentered derivatives is used to estimate k (lnskew0) and λ (bcskew0). For lnskew0, the initial value is chosen so that the minimum of $x - k$ is 1, and thus $\ln(x - k)$ is 0. bcskew0 starts with $\lambda = 1$.

Acknowledgment

lnskew0 and bcskew0 were written by Patrick Royston of the MRC Clinical Trials Unit, London.

Reference

Box, G. E. P., and D. R. Cox. 1964. An analysis of transformations. *Journal of the Royal Statistical Society, Series B* 26: 211–243.

Also See

[R] **ladder** — Ladder of powers

[R] **boxcox** — Box–Cox regression models

[R] **swilk** — Shapiro–Wilk and Shapiro–Francia tests for normality

Title

log — Echo copy of session to file or device

Syntax

Report status of log file

log

log query [*logname*]

Open log file

log using *filename* [, append replace [text | smcl] name(*logname*)]

Close log, temporarily suspend logging, or resume logging

log {close | off | on} [*logname*]

Report status of command log file

cmdlog

Open command log file

cmdlog using *filename* [, append replace]

Close command log, temporarily suspend logging, or resume logging

cmdlog {close | on | off}

Set default format for logs

set logtype {text | smcl} [, permanently]

Specify screen width

set linesize #

In addition to using the log command, you may access the capabilities of log by selecting **File > Log** from the menu and choosing one of the options in the list.

Description

log allows you to make a full record of your Stata session. A log is a file containing what you type and Stata's output. You may start multiple log files at the same time, and you may refer to them with a *logname*.

`cmdlog` allows you to make a record of what you type during your Stata session. A command log contains only what you type, so it is a subset of a full log.

You can make full logs, command logs, or both simultaneously. Neither is produced until you tell Stata to start logging.

Command logs are always ASCII text files, making them easy to convert into do-files. (In this respect, it would make more sense if the default extension of a command log file was `.do` because command logs are do-files. The default is `.txt`, not `.do`, however, to keep you from accidentally overwriting your important do-files.)

Full logs are recorded in one of two formats: SMCL (Stata Markup and Control Language) or text (meaning ASCII). The default is SMCL, but you can use `set logtype` to change that, or you can specify an option to state the format you wish. We recommend SMCL because it preserves fonts and colors. SMCL logs can be converted to ASCII text or to other formats by using the `translate` command; see [R] **translate**. You can also use `translate` to produce printable versions of SMCL logs. SMCL logs can be viewed and printed from the Viewer, as can any text file; see [R] **view**.

When using multiple log files, you may have up to five SMCL logs and five text logs open at the same time.

`log` or `cmdlog`, typed without arguments, reports the status of logging. `log query`, when passed an optional *logname*, reports the status of that log.

`log using` and `cmdlog using` open a log file. `log close` and `cmdlog close` close the file. Between times, `log off` and `cmdlog off`, and `log on` and `cmdlog on`, can temporarily suspend and resume logging.

If *filename* is specified without an extension, one of the suffixes `.smcl`, `.log`, or `.txt` is added. The extension `.smcl` or `.log` is added by `log`, depending on whether the file format is SMCL or ASCII text. The extension `.txt` is added by `cmdlog`. If *filename* contains embedded spaces, remember to enclose it in double quotes.

`set logtype` specifies the default format in which full logs are to be recorded. Initially, full logs are recorded in SMCL format.

`set linesize` specifies the width of the screen currently being used. Not all Stata commands respect `linesize`. Also, note that there is no `permanently` option allowed with `set linesize`.

If you resize your Results window, `set linesize` will automatically be reset.

Options for use with both log and cmdlog

`append` specifies that results be appended to an existing file. If the file does not already exist, a new file is created.

`replace` specifies that *filename*, if it already exists, be overwritten. When you do not specify either `replace` or `append`, the file is assumed to be new. If the specified file already exists, an error message is issued and logging is not started.

Options for use with log

`text` and `smcl` specify the format in which the log is to be recorded. The default is complicated to describe but is what you would expect:

If you specify the file as *filename*`.smcl`, the default is to write the log in SMCL format (regardless of the value of `set logtype`).

If you specify the file as *filename*.log, the default is to write the log in text format (regardless of the value of the set logtype).

If you type *filename* without an extension and specify neither the smcl option nor the text option, the default is to write the file according to the value of set logtype. If you have not set logtype, then that default is SMCL. Also the *filename* you specified will be fixed to read *filename*.smcl if a SMCL log is being created or *filename*.log if a text log is being created.

If you specify either of the options text or smcl, then what you specify determines how the log is written. If *filename* was specified without an extension, the appropriate extension is added for you.

If you open multiple log files, you may choose a different format for each file.

name(*logname*) specifies an optional name you may use to refer to the log while it is open. You can start multiple log files, give each a different *logname*, and then close, temporarily suspend, or resume them each individually.

Option for use with set logtype

permanently specifies that, in addition to making the change right now, the logtype setting be remembered and become the default setting when you invoke Stata.

Remarks

For a detailed explanation of logs, see [U] **15 Printing and preserving output**.

When you open a full log, the default is to show the name of the file and a time and date stamp:

```
. log using myfile
-------------------------------------------------------------------------------
      log:  C:\data\proj1\myfile.smcl
 log type:  smcl
opened on:  12 Jan 2007, 12:28:23

.
```

The above information will appear in the log. If you do not want this information to appear, precede the command by quietly:

```
. quietly log using myfile
```

quietly will not suppress any error messages or anything else you need to know.

Similarly, when you close a full log, the default is to show the full information,

```
. log close
      log:  C:\data\proj1\myfile.smcl
 log type:  smcl
closed on:  12 Jan 2007, 12:32:41
-------------------------------------------------------------------------------
```

and that information will also appear in the log. If you want to suppress that, type quietly log close.

Saved Results

log and cmdlog save the following in r():

Macros	
r(filename)	name of file
r(status)	on or off
r(type)	text or smcl

Also See

[R] **translate** — Print and translate logs

[R] **query** — Display system parameters

[GSM] **17 Saving and printing results by using logs**

[GSW] **17 Saving and printing results by using logs**

[GSU] **17 Saving and printing results by using logs**

[U] **15 Printing and preserving output**

Title

logistic — Logistic regression, reporting odds ratios

Syntax

`logistic` *depvar indepvars* [*if*] [*in*] [*weight*] [, *options*]

options	description
Model	
`offset(`*varname*`)`	include *varname* in model with coefficient constrained to 1
`asis`	retain perfect predictor variables
SE/Robust	
`vce(`*vcetype*`)`	*vcetype* may be `oim`, `robust`, `cluster` *clustvar*, `bootstrap`, or `jackknife`
Reporting	
`level(`*#*`)`	set confidence level; default is `level(95)`
`coef`	report estimated coefficients
Max options	
maximize_options	control the maximization process; seldom used

depvar and *indepvars* may contain time-series operators; see [U] **11.4.3 Time-series varlists**.

`bootstrap`, `by`, `jackknife`, `nestreg`, `rolling`, `statsby`, `stepwise`, `svy`, and `xi` are allowed; see [U] **11.1.10 Prefix commands**.

Weights are not allowed with the `bootstrap` prefix.

`vce()` and weights are not allowed with the `svy` prefix.

`fweight`s, `iweight`s, and `pweight`s are allowed; see [U] **11.1.6 weight**.

See [U] **20 Estimation and postestimation commands** for more capabilities of estimation commands.

Description

`logistic` fits a logistic regression model of *depvar* on *varlist*, where *depvar* is a 0/1 variable (or, more precisely, a 0/non-0 variable). Without arguments, `logistic` redisplays the last `logistic` estimates. `logistic` displays estimates as odds ratios; to view coefficients, type `logit` after running `logistic`. To obtain odds ratios for any covariate pattern relative to another, see [R] **lincom**.

Options

Model

`offset(`*varname*`)`; see [R] **estimation options**.

`asis` forces retention of perfect predictor variables and their associated perfectly predicted observations and may produce instabilities in maximization; see [R] **probit**.

SE/Robust

`vce(`*vcetype*`)` specifies the type of standard error reported, which includes types that are derived from asymptotic theory, that are robust to some kinds of misspecification, that allow for intragroup correlation, and that use bootstrap or jackknife methods; see [R] ***vce_option***.

Reporting

`level(`*#*`)`; see [R] **estimation options**.

`coef` causes `logistic` to report the estimated coefficients rather than the odds ratios (exponentiated coefficients). `coef` may be specified when the model is fitted or may be used later to redisplay results. `coef` affects only how results are displayed and not how they are estimated.

Max options

maximize_options: `iterate(`*#*`)`, `tolerance(`*#*`)`, `ltolerance(`*#*`)`; see [R] **maximize**. These options are seldom used.

Remarks

Remarks are presented under the following headings:

logistic and logit
Robust estimate of variance

logistic and logit

`logistic` provides an alternative and preferred way to fit maximum-likelihood logit models, the other choice being `logit` ([R] **logit**).

First, let us dispose of some confusing terminology. We use the words logit and logistic to mean the same thing: maximum likelihood estimation. To some, one or the other of these words connotes transforming the dependent variable and using weighted least squares to fit the model, but that is not how we use either word here. Thus the `logit` and `logistic` commands produce the same results.

The `logistic` command is generally preferred to `logit` because `logistic` presents the estimates in terms of odds ratios rather than coefficients. To a few people, this may seem a disadvantage, but you can type `logit` without arguments after `logistic` to see the underlying coefficients.

Nevertheless, [R] **logit** is still worth reading because `logistic` shares the same features as `logit`, including omitting variables due to collinearity or one-way causation.

For an introduction to logistic regression, see Lemeshow and Hosmer (1998), Pagano and Gauvreau (2000, 470–487), or Pampel (2000); for a complete but nonmathematical treatment, see Kleinbaum and Klein (2002); and for a thorough discussion, see Hosmer and Lemeshow (2000). See Gould (2000) for a discussion of the interpretation of logistic regression. See Dupont (2002) for a discussion of logistic regression with examples using Stata. For a discussion using Stata with an emphasis on model specification, see Vittinghoff et al. (2005).

Stata has a variety of commands for performing estimation when the dependent variable is dichotomous or polychotomous. See Long and Freese (2006) for a book devoted to fitting these models with Stata. Here is a list of some estimation commands that may be of interest. See [I] **estimation commands** for a complete list of all of Stata's estimation commands.

`asclogit`	[R] **asclogit**	Alternative-specific conditional logit (McFadden's choice) model
`asmprobit`	[R] **asmprobit**	Alternative-specific multinomial probit regression
`asroprobit`	[R] **asroprobit**	Alternative-specific rank-ordered probit regression
`binreg`	[R] **binreg**	Generalized linear models for the binomial family
`biprobit`	[R] **biprobit**	Bivariate probit regression
`blogit`	[R] **glogit**	Logit regression for grouped data
`bprobit`	[R] **glogit**	Probit regression for grouped data
`clogit`	[R] **clogit**	Conditional (fixed-effects) logistic regression
`cloglog`	[R] **cloglog**	Complementary log-log regression
`exlogistic`	[R] **exlogistic**	Exact logistic regression
`glm`	[R] **glm**	Generalized linear models
`glogit`	[R] **glogit**	Weighted least-squares logistic regression for grouped data
`gprobit`	[R] **glogit**	Weighted least-squares probit regression for grouped data
`heckprob`	[R] **heckprob**	Probit model with selection
`hetprob`	[R] **hetprob**	Heteroskedastic probit model
`ivprobit`	[R] **ivprobit**	Probit model with endogenous regressors
`logit`	[R] **logit**	Logistic regression, reporting coefficients
`mlogit`	[R] **mlogit**	Multinomial (polytomous) logistic regression
`mprobit`	[R] **mprobit**	Multinomial probit regression
`nlogit`	[R] **nlogit**	Nested logit regression (RUM-consistent and nonnormalized)
`ologit`	[R] **ologit**	Ordered logistic regression
`oprobit`	[R] **oprobit**	Ordered probit regression
`probit`	[R] **probit**	Probit regression
`rologit`	[R] **rologit**	Rank-ordered logistic regression
`scobit`	[R] **scobit**	Skewed logistic regression
`slogit`	[R] **slogit**	Stereotype logistic regression
`svy:` *cmd*	[SVY] **svy estimation**	Survey versions of many of these commands are available; see [SVY] **svy estimation**
`xtcloglog`	[XT] **xtcloglog**	Random-effects and population-averaged cloglog models
`xtgee`	[XT] **xtgee**	GEE population-averaged generalized linear models
`xtlogit`	[XT] **xtlogit**	Fixed-effects, random-effects, and population-averaged logit models
`xtprobit`	[XT] **xtprobit**	Random-effects and population-averaged probit models

▷ Example 1

Consider the following dataset from a study of risk factors associated with low birthweight described in Hosmer and Lemeshow (2000, 25).

```
. use http://www.stata-press.com/data/r10/lbw
(Hosmer & Lemeshow data)

. describe

Contains data from http://www.stata-press.com/data/r10/lbw.dta
  obs:           189                          Hosmer & Lemeshow data
 vars:            11                          15 Jan 2007 05:01
 size:         3,402 (95.1% of memory free)
-------------------------------------------------------------------------------
              storage  display     value
variable name   type   format      label      variable label
-------------------------------------------------------------------------------
id              int    %8.0g                  identification code
low             byte   %8.0g                  birthweight<2500g
age             byte   %8.0g                  age of mother
lwt             int    %8.0g                  weight at last menstrual period
race            byte   %8.0g       race       race
smoke           byte   %8.0g                  smoked during pregnancy
ptl             byte   %8.0g                  premature labor history (count)
ht              byte   %8.0g                  has history of hypertension
ui              byte   %8.0g                  presence, uterine irritability
ftv             byte   %8.0g                  number of visits to physician
                                                during 1st trimester
bwt             int    %8.0g                  birthweight (grams)
-------------------------------------------------------------------------------
Sorted by:
```

We want to investigate the causes of low birthweight. Here `race` is a categorical variable indicating whether a person is white (`race` = 1), black (`race` = 2), or other (`race` = 3). We want indicator (dummy) variables for `race` included in the regression. (One of the dummies, of course, must be omitted.) Thus before we can fit the model, we must create the dummy variables for `race`.

There are several ways we could do this, but the easiest is to let another Stata command, `xi`, do it for us. We type `xi:` in front of our `logistic` command and in our *varlist* include not `race` but `i.race` to indicate that we want the indicator variables for this categorical variable; see [R] **xi** for the full details.

```
. xi: logistic low age lwt i.race smoke ptl ht ui
i.race            _Irace_1-3          (naturally coded; _Irace_1 omitted)

Logistic regression                               Number of obs   =        189
                                                  LR chi2(8)      =      33.22
                                                  Prob > chi2     =     0.0001
Log likelihood =   -100.724                       Pseudo R2       =     0.1416

------------------------------------------------------------------------------
         low | Odds Ratio   Std. Err.      z    P>|z|     [95% Conf. Interval]
-------------+----------------------------------------------------------------
         age |   .9732636   .0354759    -0.74   0.457     .9061578    1.045339
         lwt |   .9849634   .0068217    -2.19   0.029     .9716834    .9984249
    _Irace_2 |   3.534767   1.860737     2.40   0.016     1.259736    9.918406
    _Irace_3 |   2.368079   1.039949     1.96   0.050     1.001356    5.600207
       smoke |   2.517698    1.00916     2.30   0.021     1.147676    5.523162
         ptl |   1.719161   .5952579     1.56   0.118     .8721455    3.388787
          ht |   6.249602   4.322408     2.65   0.008     1.611152    24.24199
          ui |     2.1351   .9808153     1.65   0.099     .8677528      5.2534
------------------------------------------------------------------------------
```

The odds ratios are for a one-unit change in the variable. If we wanted the odds ratio for `age` to be in terms of 4-year intervals, we would type

```
. gen age4 = age/4
. xi: logistic low age4 lwt i.race smoke ptl ht ui
```
(output omitted)

After `logistic`, we can type `logit` to see the model in terms of coefficients and standard errors:

```
. logit
Logistic regression                               Number of obs   =        189
                                                  LR chi2(8)      =      33.22
                                                  Prob > chi2     =     0.0001
Log likelihood =   -100.724                       Pseudo R2       =     0.1416
```

low	Coef.	Std. Err.	z	P>|z|	[95% Conf.	Interval]
age	-.0271003	.0364504	-0.74	0.457	-.0985418	.0443412
lwt	-.0151508	.0069259	-2.19	0.029	-.0287253	-.0015763
_Irace_2	1.262647	.5264101	2.40	0.016	.2309024	2.294392
_Irace_3	.8620792	.4391531	1.96	0.050	.0013548	1.722804
smoke	.9233448	.4008266	2.30	0.021	.1377391	1.708951
ptl	.5418366	.346249	1.56	0.118	-.136799	1.220472
ht	1.832518	.6916292	2.65	0.008	.4769494	3.188086
ui	.7585135	.4593768	1.65	0.099	-.1418484	1.658875
_cons	.4612239	1.20459	0.38	0.702	-1.899729	2.822176

If we wanted to see the `logistic` output again, we would type `logistic` without arguments.

◁

▷ Example 2

We can specify the confidence interval for the odds ratios with the `level()` option, and we can do this either at estimation time or when replaying the model. For instance, to see our previous models with narrower, 90% confidence intervals, we might type

```
. logistic, level(90)
Logistic regression                               Number of obs   =        189
                                                  LR chi2(8)      =      33.22
                                                  Prob > chi2     =     0.0001
Log likelihood =   -100.724                       Pseudo R2       =     0.1416
```

low	Odds Ratio	Std. Err.	z	P>|z|	[90% Conf.	Interval]
age	.9732636	.0354759	-0.74	0.457	.9166258	1.033401
lwt	.9849634	.0068217	-2.19	0.029	.9738063	.9962483
_Irace_2	3.534767	1.860737	2.40	0.016	1.487028	8.402379
_Irace_3	2.368079	1.039949	1.96	0.050	1.149971	4.876471
smoke	2.517698	1.00916	2.30	0.021	1.302185	4.867819
ptl	1.719161	.5952579	1.56	0.118	.9726876	3.038505
ht	6.249602	4.322408	2.65	0.008	2.003487	19.49478
ui	2.1351	.9808153	1.65	0.099	1.00291	4.545424

◁

Robust estimate of variance

If you specify `vce(robust)`, Stata reports the robust estimate of variance described in [U] **20.15 Obtaining robust variance estimates**. Here is the model previously fitted with the robust estimate of variance:

```
. xi: logistic low age lwt i.race smoke ptl ht ui, vce(robust)
i.race            _Irace_1-3          (naturally coded; _Irace_1 omitted)

Logistic regression                               Number of obs   =        189
                                                  Wald chi2(8)    =      29.02
                                                  Prob > chi2     =     0.0003
Log pseudolikelihood =   -100.724                 Pseudo R2       =     0.1416

------------------------------------------------------------------------------
             |               Robust
         low | Odds Ratio   Std. Err.      z    P>|z|     [95% Conf. Interval]
-------------+----------------------------------------------------------------
         age |   .9732636   .0329376    -0.80   0.423     .9108015    1.040009
         lwt |   .9849634   .0070209    -2.13   0.034     .9712984    .9988206
    _Irace_2 |   3.534767   1.793616     2.49   0.013     1.307504    9.556051
    _Irace_3 |   2.368079   1.026563     1.99   0.047     1.012512    5.538501
       smoke |   2.517698   .9736416     2.39   0.017     1.179852    5.372537
         ptl |   1.719161   .7072902     1.32   0.188     .7675715    3.850476
          ht |   6.249602   4.102026     2.79   0.005     1.726445     22.6231
          ui |     2.1351   1.042775     1.55   0.120     .8197749    5.560858
------------------------------------------------------------------------------
```

Also you can specify `vce(cluster` *clustvar*`)` and then, within cluster, relax the assumption of independence. To illustrate this, we have made some fictional additions to the low-birthweight data.

Say that these data are not a random sample of mothers but instead are a random sample of mothers from a random sample of hospitals. In fact, that may be true—we do not know the history of these data.

Hospitals specialize, and it would not be too incorrect to say that some hospitals specialize in more difficult cases. We are going to show two extremes. In one, all hospitals are alike, but we are going to estimate under the possibility that they might differ. In the other, hospitals are strikingly different. In both cases, we assume that patients are drawn from 20 hospitals.

In both examples, we will fit the same model, and we will type the same command to fit it. Below are the same data we have been using but with a new variable `hospid`, which identifies from which of the 20 hospitals each patient was drawn (and which we have made up):

(Continued on next page)

```
. use http://www.stata-press.com/data/r10/hospid1, clear
. xi: logistic low age lwt i.race smoke ptl ht ui, vce(cluster hospid)
i.race            _Irace_1-3          (naturally coded; _Irace_1 omitted)
Logistic regression                               Number of obs   =        189
                                                  Wald chi2(8)    =      49.67
                                                  Prob > chi2     =     0.0000
Log pseudolikelihood =   -100.724                 Pseudo R2       =     0.1416
                                 (Std. Err. adjusted for 20 clusters in hospid)
```

low	Odds Ratio	Robust Std. Err.	z	P>|z|	[95% Conf.	Interval]
age	.9732636	.0397476	-0.66	0.507	.898396	1.05437
lwt	.9849634	.0057101	-2.61	0.009	.9738352	.9962187
_Irace_2	3.534767	2.013285	2.22	0.027	1.157563	10.79386
_Irace_3	2.368079	.8451325	2.42	0.016	1.176562	4.766257
smoke	2.517698	.8284259	2.81	0.005	1.321062	4.79826
ptl	1.719161	.6676221	1.40	0.163	.8030814	3.680219
ht	6.249602	4.066275	2.82	0.005	1.745911	22.37086
ui	2.1351	1.093144	1.48	0.138	.7827337	5.824014

The standard errors are similar to the standard errors we have previously obtained, whether we used the robust or conventional estimators. In this example, we invented the hospital IDs randomly.

Here are the results of the estimation with the same data but with a different set of hospital IDs:

```
. use http://www.stata-press.com/data/r10/hospid2
. xi: logistic low age lwt i.race smoke ptl ht ui, vce(cluster hospid)
i.race            _Irace_1-3          (naturally coded; _Irace_1 omitted)
Logistic regression                               Number of obs   =        189
                                                  Wald chi2(8)    =       7.19
                                                  Prob > chi2     =     0.5167
Log pseudolikelihood =   -100.724                 Pseudo R2       =     0.1416
                                 (Std. Err. adjusted for 20 clusters in hospid)
```

low	Odds Ratio	Robust Std. Err.	z	P>|z|	[95% Conf.	Interval]
age	.9732636	.0293064	-0.90	0.368	.9174862	1.032432
lwt	.9849634	.0106123	-1.41	0.160	.9643817	1.005984
_Irace_2	3.534767	3.120338	1.43	0.153	.6265521	19.9418
_Irace_3	2.368079	1.297738	1.57	0.116	.8089594	6.932114
smoke	2.517698	1.570287	1.48	0.139	.7414969	8.548654
ptl	1.719161	.6799153	1.37	0.171	.7919046	3.732161
ht	6.249602	7.165454	1.60	0.110	.660558	59.12808
ui	2.1351	1.411977	1.15	0.251	.5841231	7.804266

Note the strikingly larger standard errors. What happened? In these data, women most likely to have low-birthweight babies are sent to certain hospitals, and the decision on likeliness is based not just on age, smoking history, etc., but on other things that doctors can see but that are not recorded in our data. Thus merely because a woman is at one of the centers identifies her to be more likely to have a low-birthweight baby.

Saved Results

`logistic` saves the following in `e()`:

Scalars

`e(N)`	number of observations	`e(r2_p)`	pseudo-R-squared
`e(N_cds)`	number of completely determined successes	`e(ll)`	log likelihood
`e(N_cdf)`	number of completely determined failures	`e(ll_0)`	log likelihood, constant-only model
`e(df_m)`	model degrees of freedom	`e(N_clust)`	number of clusters
		`e(chi2)`	χ^2

Macros

`e(cmd)`	`logistic`	`e(chi2type)`	`Wald` or `LR`; type of model χ^2 test
`e(cmdline)`	command as typed	`e(vce)`	*vcetype* specified in `vce()`
`e(depvar)`	name of dependent variable	`e(vcetype)`	title used to label Std. Err.
`e(wtype)`	weight type	`e(crittype)`	optimization criterion
`e(wexp)`	weight expression	`e(properties)`	`b V`
`e(title)`	title in estimation output	`e(estat_cmd)`	program used to implement `estat`
`e(clustvar)`	name of cluster variable	`e(predict)`	program used to implement `predict`
`e(offset)`	offset		

Matrices

`e(b)`	coefficient vector	`e(V)`	variance–covariance matrix of the estimators

Functions

`e(sample)`	marks estimation sample

Methods and Formulas

`logistic` is implemented as an ado-file.

Define $\mathbf{x}_j$ as the (row) vector of independent variables, augmented by 1, and $\mathbf{b}$ as the corresponding estimated parameter (column) vector. The logistic regression model is fitted by `logit`; see [R] **logit** for details of estimation.

The odds ratio corresponding to the ith coefficient is $\psi_i = \exp(b_i)$. The standard error of the odds ratio is $s_i^{\psi} = \psi_i s_i$, where s_i is the standard error of b_i estimated by `logit`.

Define $I_j = \mathbf{x}_j\mathbf{b}$ as the predicted index of the jth observation. The predicted probability of a positive outcome is

$$p_j = \frac{\exp(I_j)}{1+\exp(I_j)}$$

References

Archer, K. J., and S. Lemeshow. 2006. Goodness-of-fit test for a logistic regression model fitted using survey sample data. *Stata Journal* 6: 97–105.

Brady, A. R. 1998. sbe21: Adjusted population attributable fractions from logistic regression. *Stata Technical Bulletin* 42: 8–12. Reprinted in *Stata Technical Bulletin Reprints*, vol. 7, pp. 137–143.

Cleves, M. A., and A. Tosetto. 2000. sg139: Logistic regression when binary outcome is measured with uncertainty. *Stata Technical Bulletin* 55: 20–23.

Collett, D. 2003. *Modelling Binary Data*. 2nd ed. London: Chapman & Hall/CRC.

Dupont, W. D. 2002. *Statistical Modeling for Biomedical Researchers: A Simple Introduction to the Analysis of Complex Data*. Cambridge: Cambridge University Press.

Freese, J. 2002. Least likely observations in regression models for categorical outcomes. *Stata Journal* 2: 296–300.

Garrett, J. M. 1997. sbe14: Odds ratios and confidence intervals for logistic regression models with effect modification. *Stata Technical Bulletin* 36: 15–22. Reprinted in *Stata Technical Bulletin Reprints*, vol. 6, pp. 104–114.

Gould, W. W. 2000. sg124: Interpreting logistic regression in all its forms. *Stata Technical Bulletin* 53: 19–29. Reprinted in *Stata Technical Bulletin Reprints*, vol. 9, pp. 257–270.

Hilbe, J. M. 1997. sg63: Logistic regression: Standardized coefficients and partial correlations. *Stata Technical Bulletin* 35: 21–22. Reprinted in *Stata Technical Bulletin Reprints*, vol. 6, pp. 162–163.

Hosmer, D. W., Jr., and S. Lemeshow. 2000. *Applied Logistic Regression*. 2nd ed. New York: Wiley.

Irala-Estévez, J. de, and M. A. Martínez. 2000. sg125: Automatic estimation of interaction effects and their confidence intervals. *Stata Technical Bulletin* 53: 29–31. Reprinted in *Stata Technical Bulletin Reprints*, vol. 9, pp. 270–273.

Kleinbaum, D. G., and M. Klein. 2002. *Logistic Regression: A Self-Learning Text*. 2nd ed. New York: Springer.

Lemeshow, S., and D. W. Hosmer, Jr. 1998. Logistic regression. In *Encyclopedia of Biostatistics*, ed. P. Armitage and T. Colton, 2316–2327. New York: Wiley.

Lemeshow, S., and J.-R. Le Gall. 1994. Modeling the severity of illness of ICU patients: A systems update. *Journal of the American Medical Association* 272: 1049–1055.

Long, J. S., and J. Freese. 2006. *Regression Models for Categorical Dependent Variables Using Stata*. 2nd ed. College Station, TX: Stata Press.

Miranda, A., and S. Rabe-Hesketh. 2006. Maximum likelihood estimation of endogenous switching and sample selection models for binary, ordinal, and count variables. *Stata Journal* 6: 285–308.

Mitchell, M. N., and X. Chen. 2005. Visualizing main effects and interactions for binary logit models. *Stata Journal* 5: 64–82.

Pagano, M., and K. Gauvreau. 2000. *Principles of Biostatistics*. 2nd ed. Belmont, CA: Duxbury.

Pampel, F. C. 2000. *Logistic Regression: A Primer*. Thousand Oaks, CA: Sage.

Paul, C. 1998. sg92: Logistic regression for data including multiple imputations. *Stata Technical Bulletin* 45: 28–30. Reprinted in *Stata Technical Bulletin Reprints*, vol. 8, pp. 180–183.

Pearce, M. S. 2000. sg148: Profile likelihood confidence intervals for explanatory variables in logistic regression. *Stata Technical Bulletin* 56: 45–47.

Pregibon, D. 1981. Logistic regression diagnostics. *Annals of Statistics* 9: 705–724.

Reilly, M., and A. Salim. 2000. sg156: Mean score method for missing covariate data in logistic regression models. *Stata Technical Bulletin* 58: 25–27. Reprinted in *Stata Technical Bulletin Reprints*, vol. 10, pp. 256–258.

Schonlau, M. 2005. Boosted regression (boosting): An introductory tutorial and a Stata plugin. *Stata Journal* 5: 330–354.

Vittinghoff, E., D. V. Glidden, S. C. Shiboski, and C. E. McCulloch. 2005. *Regression Methods in Biostatistics: Linear, Logistic, Survival, and Repeated Measures Models*. New York: Springer.

Xu, J., and J. S. Long. 2005. Confidence intervals for predicted outcomes in regression models for categorical outcomes. *Stata Journal* 5: 537–559.

Also See

[R] **logistic postestimation** — Postestimation tools for logistic

[R] **roc** — Receiver operating characteristic (ROC) analysis

[R] **brier** — Brier score decomposition

[R] **exlogistic** — Exact logistic regression

[R] **logit** — Logistic regression, reporting coefficients

[SVY] **svy estimation** — Estimation commands for survey data

[XT] **xtlogit** — Fixed-effects, random-effects, and population-averaged logit models

[U] **20 Estimation and postestimation commands**

Title

logistic postestimation — Postestimation tools for logistic

Description

The following postestimation commands are of special interest after `logistic`:

command	description
`estat clas`	`estat classification` reports various summary statistics, including the classification table
`estat gof`	Pearson or Hosmer–Lemeshow goodness-of-fit test
`lroc`	graphs the ROC curve and calculates the area under the curve
`lsens`	graphs sensitivity and specificity versus probability cutoff

These commands are not appropriate after the `svy` prefix.

For information about these commands, see below.

The following standard postestimation commands are also available:

command	description
`adjust`[1]	adjusted predictions of $\mathbf{x}\beta$, probabilities, or $\exp(\mathbf{x}\beta)$
`estat`	AIC, BIC, VCE, and estimation sample summary
`estat` (svy)	postestimation statistics for survey data
`estimates`	cataloging estimation results
`lincom`	point estimates, standard errors, testing, and inference for linear combinations of coefficients
`linktest`	link test for model specification
`lrtest`[2]	likelihood-ratio test
`mfx`	marginal effects or elasticities
`nlcom`	point estimates, standard errors, testing, and inference for nonlinear combinations of coefficients
`predict`	predictions, residuals, influence statistics, and other diagnostic measures
`predictnl`	point estimates, standard errors, testing, and inference for generalized predictions
`suest`	seemingly unrelated estimation
`test`	Wald tests for simple and composite linear hypotheses
`testnl`	Wald tests of nonlinear hypotheses

[1] `adjust` is not appropriate with time-series operators.

[2] `lrtest` is not appropriate with `svy` estimation results.

See the corresponding entries in the *Stata Base Reference Manual* for details, but see [SVY] **estat** for details about `estat` (svy).

Special-interest postestimation commands

estat classification reports various summary statistics, including the classification table.

estat gof reports the Pearson goodness-of-fit test or the Hosmer–Lemeshow goodness-of-fit test.

lroc graphs the ROC curve and calculates the area under the curve.

lsens graphs sensitivity and specificity versus probability cutoff and optionally creates new variables containing these data.

estat classification, estat gof, lroc, and lsens produce statistics and graphs either for the estimation sample or for any set of observations. However, they always use the estimation sample by default. When weights, if, or in is used with logistic, it is not necessary to repeat the qualifier with these commands when you want statistics computed for the estimation sample. Specify if, in, or the all option only when you want statistics computed for a set of observations other than the estimation sample. Specify weights (only fweights are allowed with these commands) only when you want to use a different set of weights.

By default, estat classification, estat gof, lroc, and lsens use the last model fitted by logistic. You may also directly specify the model to these postestimation commands by inputting a vector of coefficients with the beta() option and passing the name of the dependent variable *depvar* to these commands.

estat classification and estat gof require that the current estimation results be from logistic, logit, or probit. lroc and lsens commands may also be used after logit or probit.

Syntax for predict

predict [*type*] *newvar* [*if*] [*in*] [, *statistic* nooffset rules asif]

statistic	description
Main	
pr	probability of a positive outcome; the default
xb	linear prediction
stdp	standard error of the linear prediction
*dbeta	Pregibon (1981) $\Delta\widehat{\beta}$ influence statistic
*deviance	deviance residual
*dx2	Hosmer and Lemeshow (2000, 174) $\Delta\chi^2$ influence statistic
*ddeviance	Hosmer and Lemeshow (2000, 174) ΔD influence statistic
*hat	Pregibon (1981) leverage
*number	sequential number of the covariate pattern
*residuals	Pearson residuals; adjusted for number sharing covariate pattern
*rstandard	standardized Pearson residuals; adjusted for number sharing covariate pattern
score	first derivative of the log likelihood with respect to $\mathbf{x}_j\boldsymbol{\beta}$

Unstarred statistics are available both in and out of sample; type predict ... if e(sample) ... if wanted only for the estimation sample. Starred statistics are calculated only for the estimation sample, even when if e(sample) is not specified.

pr, xb, stdp, and score are the only options allowed with svy estimation results.

Options for predict

Main

`pr`, the default, calculates the probability of a positive outcome.

`xb` calculates the linear prediction.

`stdp` calculates the standard error of the linear prediction.

`dbeta` calculates the Pregibon (1981) $\Delta\widehat{\beta}$ influence statistic, a standardized measure of the difference in the coefficient vector that is due to deletion of the observation along with all others that share the same covariate pattern. In Hosmer and Lemeshow (2000, 144–145) jargon, this statistic is M-asymptotic; that is, it is adjusted for the number of observations that share the same covariate pattern.

`deviance` calculates the deviance residual.

`dx2` calculates the Hosmer and Lemeshow (2000, 174) $\Delta\chi^2$ influence statistic, reflecting the decrease in the Pearson χ^2 because of deletion of the observation and all others that share the same covariate pattern.

`ddeviance` calculates the Hosmer and Lemeshow (2000, 174) ΔD influence statistic, which is the change in the deviance residual that is due to deletion of the observation and all others that share the same covariate pattern.

`hat` calculates the Pregibon (1981) leverage or the diagonal elements of the hat matrix adjusted for the number of observations that share the same covariate pattern.

`number` numbers the covariate patterns—observations with the same covariate pattern have the same `number`. Observations not used in estimation have `number` set to missing. The first covariate pattern is numbered 1, the second 2, and so on.

`residuals` calculates the Pearson residual as given by Hosmer and Lemeshow (2000, 145) and adjusted for the number of observations that share the same covariate pattern.

`rstandard` calculates the standardized Pearson residual as given by Hosmer and Lemeshow (2000, 173) and adjusted for the number of observations that share the same covariate pattern.

`score` calculates the equation-level score, $\partial \ln L/\partial(\mathbf{x}_j\boldsymbol{\beta})$.

Options

`nooffset` is relevant only if you specified `offset(`*varname*`)` for `logistic`. It modifies the calculations made by `predict` so that they ignore the offset variable; the linear prediction is treated as $\mathbf{x}_j\mathbf{b}$ rather than as $\mathbf{x}_j\mathbf{b} + \text{offset}_j$.

`rules` requests that Stata use any rules that were used to identify the model when making the prediction. By default, Stata calculates missing for excluded observations. See [R] **logit** for an example.

`asif` requests that Stata ignore the rules and the exclusion criteria and calculate predictions for all observations possible using the estimated parameter from the model. See [R] **logit** for an example.

Syntax for estat classification

`estat classification` [*if*] [*in*] [*weight*] [, *class_options*]

class_options	description
Main	
`all`	display summary statistics for all observations in the data
`cutoff(#)`	positive outcome threshold; default is `cutoff(0.5)`
`beta(`*matname*`)`	row vector containing coefficients for the model

`fweight`s are allowed; see [U] **11.1.6 weight**.

Options for estat classification

Main

`all` requests that the statistic be computed for all observations in the data, ignoring any `if` or `in` restrictions specified by `logistic`.

`cutoff(#)` specifies the value for determining whether an observation has a predicted positive outcome. An observation is classified as positive if its predicted probability is $\geq$ #. The default is 0.5.

`beta(`*matname*`)` specifies a row vector containing coefficients for a logistic model. The columns of the row vector must be labeled with the corresponding names of the independent variables in the data. The dependent variable *depvar* must be specified immediately after the command name. See *Models other than last fitted model* later in this entry.

Syntax for estat gof

`estat gof` [*if*] [*in*] [*weight*] [, *gof_options*]

gof_options	description
Main	
`group(#)`	perform Hosmer–Lemeshow goodness-of-fit test using # quantiles
`all`	execute test for all observations in the data
`outsample`	adjust degrees of freedom for samples outside estimation sample
`table`	display table of groups used for test
`beta(`*matname*`)`	row vector containing coefficients for the model

`fweight`s are allowed; see [U] **11.1.6 weight**.

Options for estat gof

Main

`group(#)` specifies the number of quantiles to be used to group the data for the Hosmer–Lemeshow goodness-of-fit test. `group(10)` is typically specified. If this option is not given, the Pearson goodness-of-fit test is computed using the covariate patterns in the data as groups.

`all` requests that the statistic be computed for all observations in the data, ignoring any `if` or `in` restrictions specified with `logistic`.

`outsample` adjusts the degrees of freedom for the Pearson and Hosmer–Lemeshow goodness-of-fit tests for samples outside the estimation sample. See *Samples other than estimation sample* later in this entry.

`table` displays a table of the groups used for the Hosmer–Lemeshow or Pearson goodness-of-fit test with predicted probabilities, observed and expected counts for both outcomes, and totals for each group.

`beta(`*matname*`)` specifies a row vector containing coefficients for a logistic model. The columns of the row vector must be labeled with the corresponding names of the independent variables in the data. The dependent variable *depvar* must be specified immediately after the command name. See *Models other than last fitted model* later in this entry.

Syntax for lroc

`lroc` [*depvar*] [*if*] [*in*] [*weight*] [, *lroc_options*]

lroc_options	description
Main	
`all`	compute area under ROC curve and graph curve for all observations
`nograph`	suppress graph
Advanced	
`beta(`*matname*`)`	row vector containing coefficients for a logistic model
Plot	
cline_options	change the look of the line
marker_options	change look of markers (color, size, etc.)
marker_label_options	add marker labels; change look or position
Reference line	
`rlopts(`*cline_options*`)`	affect rendition of the reference line
Add plots	
`addplot(`*plot*`)`	add other plots to the generated graph
Y axis, X axis, Titles, Legend, Overall	
twoway_options	any options other than `by()` documented in [G] ***twoway_options***

`fweight`s are allowed; see [U] **11.1.6 weight**.

Options for lroc

Main

`all` requests that the statistic be computed for all observations in the data, ignoring any `if` or `in` restrictions specified by `logistic`.

`nograph` suppresses graphical output.

Advanced

beta(*matname*) specifies a row vector containing coefficients for a logistic model. The columns of the row vector must be labeled with the corresponding names of the independent variables in the data. The dependent variable *depvar* must be specified immediately after the command name. See *Models other than last fitted model* later in this entry.

Plot

cline_options, *marker_options*, and *marker_label_options* affect the rendition of the ROC curve—the plotted points connected by lines. These options affect the size and color of markers, whether and how the markers are labeled, and whether and how the points are connected; see [G] ***cline_options***, [G] ***marker_options***, and [G] ***marker_label_options***.

Reference line

rlopts(*cline_options*) affects the rendition of the reference line; see [G] ***cline_options***.

Add plots

addplot(*plot*) provides a way to add other plots to the generated graph. See [G] ***addplot_option***.

Y axis, X axis, Titles, Legend, Overall

twoway_options are any of the options documented in [G] ***twoway_options***, excluding by(). These include options for titling the graph (see [G] ***title_options***) and for saving the graph to disk (see [G] ***saving_option***).

Syntax for lsens

lsens [*depvar*] [*if*] [*in*] [*weight*] [, *lsens_options*]

lsens_options	description
Main	
all	graph all observations in the data
genprob(*varname*)	create variable containing probability cutoffs
gensens(*varname*)	create variable containing sensitivity
genspec(*varname*)	create variable containing specificity
replace	overwrite existing variables
nograph	suppress the graph
Advanced	
beta(*matname*)	row vector containing coefficients for the model
Plot	
connect_options	affect rendition of the plotted points connected by lines
Add plots	
addplot(*plot*)	add other plots to the generated graph
Y axis, X axis, Titles, Legend, Overall	
twoway_options	any options other than by() documented in [G] ***twoway_options***

fweights are allowed; see [U] **11.1.6 weight**.

Options for lsens

Main

`all` requests that the statistic be computed for all observations in the data, ignoring any `if` or `in` restrictions specified with `logistic`.

`genprob(`*varname*`)`, `gensens(`*varname*`)`, and `genspec(`*varname*`)` specify the names of new variables created to contain, respectively, the probability cutoffs and the corresponding sensitivity and specificity. These new variables will be added to the dataset and, in the process, may result in up to two new observations being added to the dataset in memory. Values are recorded for $p = 0$, for each of the observed predicted probabilities, and for $p = 1$. The total number of observations required to do this can be fewer than `_N`, the same as `_N`, or `_N`+1, or `_N`+2. If more observations are added, they are added at the end of the dataset and the values of the original variables are set to missing in the added observations. How the values added align with existing observations is irrelevant.

`replace` requests that existing variables specified for `genprob()`, `gensens()`, or `genspec()` be overwritten.

`nograph` suppresses graphical output.

Advanced

`beta(`*matname*`)` specifies a row vector containing coefficients for a logistic model. The columns of the row vector must be labeled with the corresponding names of the independent variables in the data. The dependent variable *depvar* must be specified immediately after the command name. See *Models other than last fitted model* later in this entry.

Plot

connect_options affect the rendition of the plotted points connected by lines; see *connect_options* in [G] **graph twoway scatter**.

Add plots

`addplot(`*plot*`)` provides a way to add other plots to the generated graph. See [G] ***addplot_option***.

Y axis, X axis, Titles, Legend, Overall

twoway_options are any of the options documented in [G] ***twoway_options***, excluding `by()`. These include options for titling the graph (see [G] ***title_options***) and for saving the graph to disk (see [G] ***saving_option***).

(*Continued on next page*)

Remarks

Remarks are presented under the following headings:

predict after logistic

predict is used after logistic to obtain predicted probabilities, residuals, and influence statistics for the estimation sample. The suggested diagnostic graphs below are from Hosmer and Lemeshow (2000), where they are more elaborately explained. Also see Collett (2003, 129–168) for a thorough discussion of model checking.

predict without options

Typing predict *newvar* after estimation calculates the predicted probability of a positive outcome.

We previously ran the model logistic low age lwt _Irace_2 _Irace_3 smoke ptl ht ui. We obtain the predicted probabilities of a positive outcome by typing

```
. use http://www.stata-press.com/data/r10/lbw
(Hosmer & Lemeshow data)
. xi: logistic low age lwt i.race smoke ptl ht ui
  (output omitted)
. predict p
(option pr assumed; Pr(low))
. summarize p low
```

Variable	Obs	Mean	Std. Dev.	Min	Max
p	189	.3121693	.1913915	.0272559	.8391283
low	189	.3121693	.4646093	0	1

predict with the xb and stdp options

predict with the xb option calculates the linear combination $x_j\mathbf{b}$, where x_j are the independent variables in the jth observation and $\mathbf{b}$ is the estimated parameter vector. This is sometimes known as the index function since the cumulative distribution function indexed at this value is the probability of a positive outcome.

With the stdp option, predict calculates the standard error of the prediction, which is *not* adjusted for replicated covariate patterns in the data. The influence statistics described below are adjusted for replicated covariate patterns in the data.

predict with the residuals option

`predict` can calculate more than predicted probabilities. The Pearson residual is defined as the square root of the contribution of the covariate pattern to the Pearson χ^2 goodness-of-fit statistic, signed according to whether the observed number of positive responses within the covariate pattern is less than or greater than expected. For instance,

```
. predict r, residuals
. summarize r, detail
                        Pearson residual
-------------------------------------------------------------
      Percentiles      Smallest
 1%    -1.750923      -2.283885
 5%    -1.129907      -1.750923
10%    -.9581174      -1.636279       Obs                 189
25%    -.6545911      -1.636279       Sum of Wgt.         189

50%    -.3806923                      Mean          -.0242299
                        Largest       Std. Dev.      .9970949
75%     .8162894        2.23879
90%     1.510355       2.317558       Variance       .9941981
95%     1.747948       3.002206       Skewness       .8618271
99%     3.002206       3.126763       Kurtosis       3.038448
```

We notice the prevalence of a few large positive residuals:

```
. sort r
. list id r low p age race in -5/1

     +-----------------------------------------------+
     | id          r   low          p   age    race |
     |-----------------------------------------------|
185. | 33   2.224501     1   .1681123    19   white |
186. | 57    2.23879     1    .166329    15   white |
187. | 16   2.317558     1   .1569594    27   other |
188. | 77   3.002206     1   .0998678    26   white |
189. | 36   3.126763     1   .0927932    24   white |
     +-----------------------------------------------+
```

predict with the number option

Covariate patterns play an important role in logistic regression. Two observations are said to share the same covariate pattern if the independent variables for the two observations are identical. Although we might think of having individual observations, the statistical information in the sample can be summarized by the covariate patterns, the number of observations with that covariate pattern, and the number of positive outcomes within the pattern. Depending on the model, the number of covariate patterns can approach or be equal to the number of observations, or it can be considerably less.

Stata calculates all the residual and diagnostic statistics in terms of covariate patterns, not observations. That is, all observations with the same covariate pattern are given the same residual and diagnostic statistics. Hosmer and Lemeshow (2000, 144–145) argue that such "M-asymptotic" statistics are more useful than "N-asymptotic" statistics.

To understand the difference, think of an observed positive outcome with predicted probability of 0.8. Taking the observation in isolation, the residual must be positive—we expected 0.8 positive responses and observed 1. This may indeed be the correct residual, but not necessarily. Under the M-asymptotic definition, we ask how many successes we observed across all observations with this covariate pattern. If that number were, say, six, and there were a total of 10 observations with this covariate pattern, then the residual is negative for the covariate pattern—we expected eight positive outcomes but observed six. `predict` makes this kind of calculation and then attaches the same residual to all observations in the covariate pattern.

Occasionally you might want to find all observations sharing a covariate pattern. `number` allows you to do this:

```
. predict pattern, number
. summarize pattern
    Variable |       Obs        Mean    Std. Dev.       Min        Max
-------------+--------------------------------------------------------
     pattern |       189     89.2328    53.16573          1        182
```

We previously fitted the model `logistic low age lwt _Irace_2 _Irace_3 smoke ptl ht ui` over 189 observations. There are 182 covariate patterns in our data.

predict with the deviance option

The deviance residual is defined as the square root of the contribution to the likelihood-ratio test statistic of a saturated model versus the fitted model. It has slightly different properties from the Pearson residual (see Hosmer and Lemeshow 2000, 145–147):

```
. predict d, deviance
. summarize d, detail
                      deviance residual
-------------------------------------------------------------
      Percentiles      Smallest
 1%    -1.843472      -1.911621
 5%     -1.33477      -1.843472
10%    -1.148316      -1.843472       Obs                 189
25%    -.8445325      -1.674869       Sum of Wgt.         189

50%    -.5202702                      Mean          -.1228811
                        Largest       Std. Dev.      1.049237
75%     .9129041       1.894089
90%     1.541558       1.924457       Variance       1.100898
95%     1.673338       2.146583       Skewness       .6598857
99%     2.146583       2.180542       Kurtosis       2.036938
```

predict with the rstandard option

Pearson residuals do not have a standard deviation equal to 1. `rstandard` generates Pearson residuals normalized to have an *expected* standard deviation equal to 1.

```
. predict rs, rstandard
. summarize r rs
    Variable |       Obs        Mean    Std. Dev.       Min        Max
-------------+--------------------------------------------------------
           r |       189   -.0242299    .9970949  -2.283885   3.126763
          rs |       189   -.0279135    1.026406    -2.4478   3.149081

. correlate r rs
(obs=189)

             |        r       rs
-------------+------------------
           r |   1.0000
          rs |   0.9998   1.0000
```

Remember that we previously created `r` containing the (unstandardized) Pearson residuals. In these data, whether we use standardized or unstandardized residuals does not matter much.

predict with the hat option

`hat` calculates the leverage of a covariate pattern—a scaled measure of distance in terms of the independent variables. Large values indicate covariate patterns far from the average covariate pattern that can have a large effect on the fitted model even if the corresponding residual is small. Consider the following graph:

```
. predict h, hat
. scatter h r, xline(0)
```

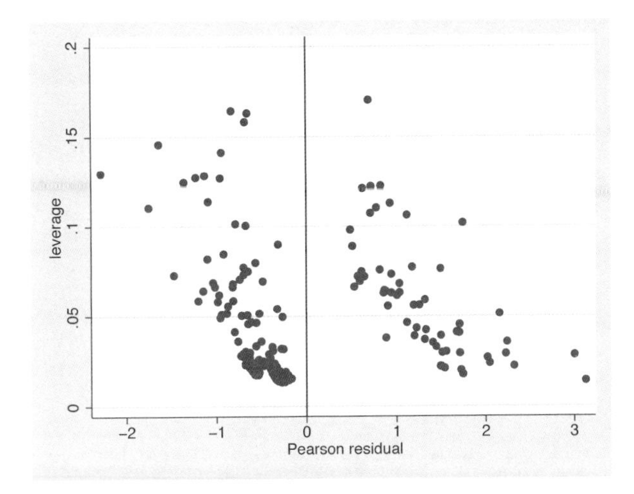

The points to the left of the vertical line are observed negative outcomes; here our data contain almost as many covariate patterns as observations, so most covariate patterns are unique. In such unique patterns, we observe either 0 or 1 success and expect p, thus forcing the sign of the residual. If we had fewer covariate patterns—if we did not have continuous variables in our model—there would be no such interpretation, and we would not have drawn the vertical line at 0.

Points on the left and right edges of the graph represent large residuals—covariate patterns that are not fitted well by our model. Points at the top of our graph represent high leverage patterns. When analyzing the influence of observations on the model, we are most interested in patterns with high leverage and small residuals—patterns that might otherwise escape our attention.

predict with the dx2 option

There are many ways to measure influence, and `hat` is one example. `dx2` measures the decrease in the Pearson χ^2 goodness-of-fit statistic that would be caused by deleting an observation (and all others sharing the covariate pattern):

```
. predict dx2, dx2
. scatter dx2 p
```

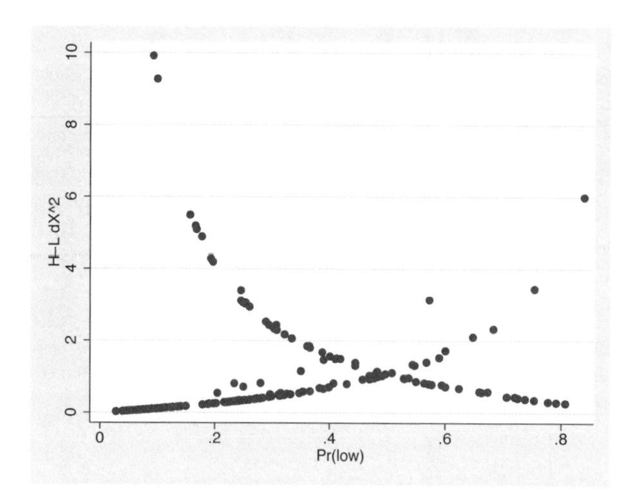

Paraphrasing Hosmer and Lemeshow (2000, 178–179), the points going from the top left to the bottom right correspond to covariate patterns with the number of positive outcomes equal to the number in the group; the points on the other curve correspond to 0 positive outcomes. In our data, most of the covariate patterns are unique, so the points tend to lie along one or the other curves; the points that are off the curves correspond to the few repeated covariate patterns in our data in which all the outcomes are not the same.

We examine this graph for large values of `dx2`—there are two at the top left.

predict with the ddeviance option

Another measure of influence is the change in the deviance residuals due to deletion of a covariate pattern:

```
. predict dd, ddeviance
```

As with `dx2`, we typically graph `ddeviance` against the probability of a positive outcome. We direct you to Hosmer and Lemeshow (2000, 178) for an example and for the interpretation of this graph.

predict with the dbeta option

One of the more direct measures of influence of interest to model fitters is the Pregibon (1981) dbeta measure, a measure of the change in the coefficient vector that would be caused by deleting an observation (and all others sharing the covariate pattern):

```
. predict db, dbeta
. scatter db p
```

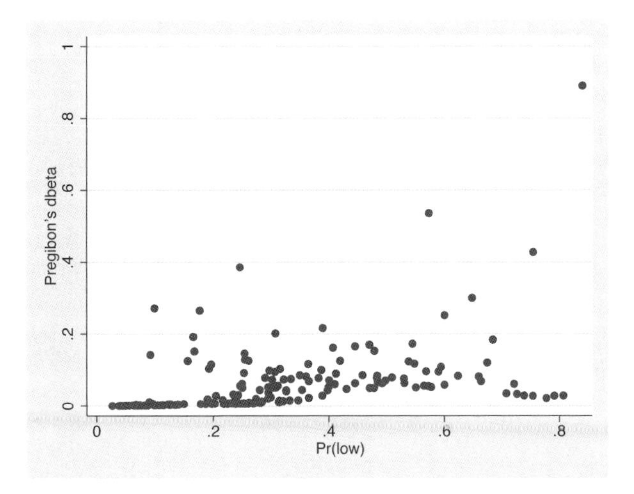

One observation has a large effect on the estimated coefficients. We can easily find this point:

```
. sort db
. list in 1
```

189.

id	low	age	lwt	race	smoke	ptl	ht	ui	ftv	bwt
188	0	25	95	white	1	3	0	1	0	3637

_Irace_2	_Irace_3	p	r	pattern	d
0	0	.8391283	-2.283885	117	-1.911621

rs	h	dx2	dd	db
-2.4478	.1294439	5.991726	4.197658	.8909163

Hosmer and Lemeshow (2000, 180) suggest a graph that combines two of the influence measures:

(*Continued on next page*)

```
. scatter dx2 p [w=db], title("Symbol size proportional to dBeta") mfcolor(none)
(analytic weights assumed)
```

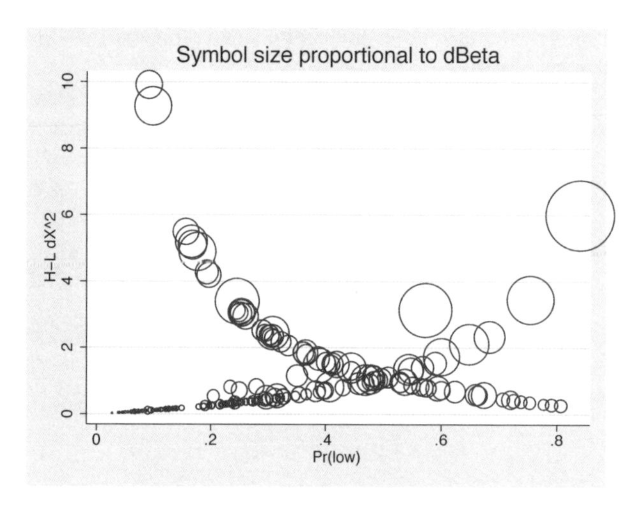

We can easily spot the most influential points by the `dbeta` and `dx2` measures.

estat classification

▷ Example 1

`estat classification` presents the classification statistics and classification table after `logistic`.

```
. use http://www.stata-press.com/data/r10/lbw
(Hosmer & Lemeshow data)
. xi: logistic low age lwt i.race smoke ptl ht ui
  (output omitted)
. estat classification
Logistic model for low
```

Classified	True D	True ~D	Total
+	21	12	33
-	38	118	156
Total	59	130	189

```
Classified + if predicted Pr(D) >= .5
True D defined as low != 0

Sensitivity                     Pr( +| D)   35.59%
Specificity                     Pr( -|~D)   90.77%
Positive predictive value       Pr( D| +)   63.64%
Negative predictive value       Pr(~D| -)   75.64%

False + rate for true ~D        Pr( +|~D)    9.23%
False - rate for true D         Pr( -| D)   64.41%
False + rate for classified +   Pr(~D| +)   36.36%
False - rate for classified -   Pr( D| -)   24.36%

Correctly classified                        73.54%
```

By default, estat classification uses a cutoff of 0.5, although you can vary this with the cutoff() option. You can use the lsens command to review the potential cutoffs; see lsens below.

◁

estat gof

estat gof computes goodness-of-fit tests: either the Pearson χ^2 test or the Hosmer–Lemeshow test.

By default, estat classification, estat gof, lroc, and lsens compute statistics for the estimation sample by using the last model fitted by logistic. However, samples other than the estimation sample can be specified; see *Samples other than the estimation sample* later in this entry.

▷ Example 2

estat gof, typed without options, presents the Pearson χ^2 goodness-of-fit test for the fitted model. The Pearson χ^2 goodness-of-fit test is a test of the observed against expected number of responses using cells defined by the covariate patterns; see *predict with the number option* earlier in this entry for the definition of covariate patterns.

```
. estat gof
Logistic model for low, goodness-of-fit test
       number of observations =       189
 number of covariate patterns =       182
             Pearson chi2(173) =       179.24
                   Prob > chi2 =         0.3567
```

Our model fits reasonably well. However, the number of covariate patterns is close to the number of observations, making the applicability of the Pearson χ^2 test questionable but not necessarily inappropriate. Hosmer and Lemeshow (2000, 147–150) suggest regrouping the data by ordering on the predicted probabilities and then forming, say, 10 nearly equal-sized groups. estat gof with the group() option does this:

```
. estat gof, group(10)
Logistic model for low, goodness-of-fit test
  (Table collapsed on quantiles of estimated probabilities)
       number of observations =       189
             number of groups =        10
      Hosmer-Lemeshow chi2(8) =         9.65
                  Prob > chi2 =         0.2904
```

Again we cannot reject our model. If we specify the table option, estat gof displays the groups along with the expected and observed number of positive responses (low-birthweight babies):

(*Continued on next page*)

```
. estat gof, group(10) table
Logistic model for low, goodness-of-fit test
  (Table collapsed on quantiles of estimated probabilities)
```

Group	Prob	Obs_1	Exp_1	Obs_0	Exp_0	Total
1	0.0827	0	1.2	19	17.8	19
2	0.1276	2	2.0	17	17.0	19
3	0.2015	6	3.2	13	15.8	19
4	0.2432	1	4.3	18	14.7	19
5	0.2792	7	4.9	12	14.1	19
6	0.3138	7	5.6	12	13.4	19
7	0.3872	6	6.5	13	12.5	19
8	0.4828	7	8.2	12	10.8	19
9	0.5941	10	10.3	9	8.7	19
10	0.8391	13	12.8	5	5.2	18

```
      number of observations =       189
            number of groups =        10
      Hosmer-Lemeshow chi2(8) =         9.65
                 Prob > chi2 =         0.2904
```

◁

❑ Technical Note

`estat gof` with the `group()` option puts all observations with the same predicted probabilities into the same group. If, as in the previous example, we request 10 groups, the groups that `estat gof` makes are $[p_0, p_{10}]$, $(p_{10}, p_{20}]$, $(p_{20}, p_{30}]$, ..., $(p_{90}, p_{100}]$, where p_k is the kth percentile of the predicted probabilities, with p_0 the minimum and p_{100} the maximum.

If there are many ties at the quantile boundaries, as will often happen if all independent variables are categorical and there are only a few of them, the sizes of the groups will be uneven. If the totals in some of the groups are small, the χ^2 statistic for the Hosmer–Lemeshow test may be unreliable. In this case, fewer groups should be specified, or the Pearson goodness-of-fit test may be a better choice.

❑

▷ Example 3

The `table` option can be used without the `group()` option. We would not want to specify this for our current model because there were 182 covariate patterns in the data, caused by including the two continuous variables, `age` and `lwt`, in the model. As an aside, we fit a simpler model and specify `table` with `estat gof`:

```
. logistic low _Irace_2 _Irace_3 smoke ui
Logistic regression                               Number of obs   =        189
                                                  LR chi2(4)      =      18.80
                                                  Prob > chi2     =     0.0009
Log likelihood = -107.93404                       Pseudo R2       =     0.0801
```

low	Odds Ratio	Std. Err.	z	P>|z|	[95% Conf.	Interval]
_Irace_2	3.052746	1.498084	2.27	0.023	1.166749	7.987368
_Irace_3	2.922593	1.189226	2.64	0.008	1.31646	6.488269
smoke	2.945742	1.101835	2.89	0.004	1.41517	6.131701
ui	2.419131	1.047358	2.04	0.041	1.03546	5.651783

```
. estat gof, tab
```

Logistic model for low, goodness-of-fit test

Group	Prob	Obs_1	Exp_1	Obs_0	Exp_0	Total
1	0.1230	3	4.9	37	35.1	40
2	0.2533	1	1.0	3	3.0	4
3	0.2907	16	13.7	31	33.3	47
4	0.2923	15	12.6	28	30.4	43
5	0.2997	3	3.9	10	9.1	13
6	0.4978	4	4.0	4	4.0	8
7	0.4998	4	4.5	5	4.5	9
8	0.5087	2	1.5	1	1.5	3
9	0.5469	2	4.4	6	3.6	8
10	0.5577	6	5.6	4	4.4	10
11	0.7449	3	3.0	1	1.0	4

Group	Prob	_Irace_2	_Irace_3	smoke	ui
1	0.1230	0	0	0	0
2	0.2533	0	0	0	1
3	0.2907	0	1	0	0
4	0.2923	0	0	1	0
5	0.2997	1	0	0	0
6	0.4978	0	1	0	1
7	0.4998	0	0	1	1
8	0.5087	1	0	0	1
9	0.5469	0	1	1	0
10	0.5577	1	0	1	0
11	0.7449	0	1	1	1

```
      number of observations =       189
number of covariate patterns =        11
             Pearson chi2(6) =         5.71
                 Prob > chi2 =         0.4569
```

◁

❑ Technical Note

`logistic` and `estat gof` keep track of the estimation sample. If you type `logistic ... if x==1`, then when you type `estat gof`, the statistics will be calculated on the `x==1` subsample of the data automatically.

You should specify `if` or `in` with `estat gof` only when you wish to calculate statistics for a set of observations other than the estimation sample. See *Samples other than the estimation sample* later in this entry.

If the `logistic` model was fitted with `fweights`, `estat gof` properly accounts for the weights in its calculations. (`estat gof` does not allow `pweights`.) You do not have to specify the weights when you run `estat gof`. Weights should be specified with `estat gof` only when you wish to use a different set of weights.

❑

lroc

For other receiver operating characteristic (ROC) commands and a complete description, see [R] **roc**, [R] **rocfit**, and [R] **rocfit postestimation**.

`lroc` graphs the ROC curve—a graph of sensitivity versus one minus specificity as the cutoff c is varied—and calculates the area under it. Sensitivity is the fraction of observed positive-outcome cases that are correctly classified; specificity is the fraction of observed negative-outcome cases that are correctly classified. When the purpose of the analysis is classification, you must choose a cutoff.

The curve starts at $(0,0)$, corresponding to $c=1$, and continues to $(1,1)$, corresponding to $c=0$. A model with no predictive power would be a 45° line. The greater the predictive power, the more bowed the curve, and hence the area beneath the curve is often used as a measure of the predictive power. A model with no predictive power has area 0.5; a perfect model has area 1.

The ROC curve was first discussed in signal detection theory (Peterson, Birdsall, and Fox 1954) and then was quickly introduced into psychology (Tanner and Swets 1954). It has since been applied in other fields, particularly medicine (for instance, Metz 1978). For a classic text on ROC techniques, see Green and Swets (1966).

▷ Example 4

ROC curves are typically used when the point of the analysis is classification—which it is not in our low-birthweight model. Nevertheless, the ROC curve is

```
. lroc
Logistic model for low
number of observations =      189
area under ROC curve   =   0.6658
```

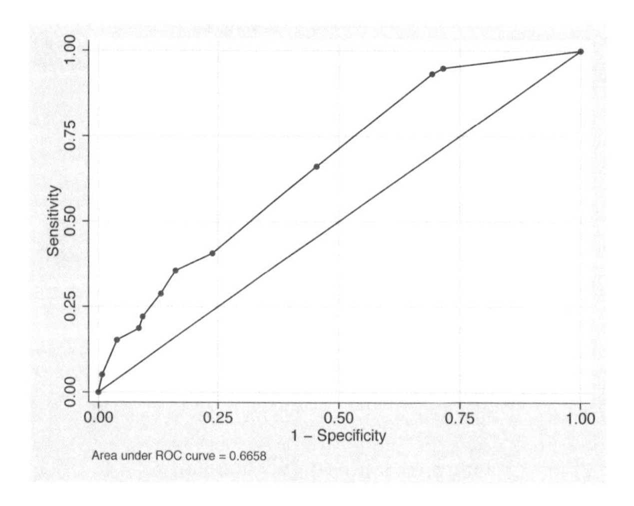

We see that the area under the curve is 0.6658.

◁

lsens

`lsens` also plots sensitivity and specificity; it plots both sensitivity and specificity versus probability cutoff c. The graph is equivalent to what you would get from `estat classification` if you varied the cutoff probability from 0 to 1.

```
. lsens
```

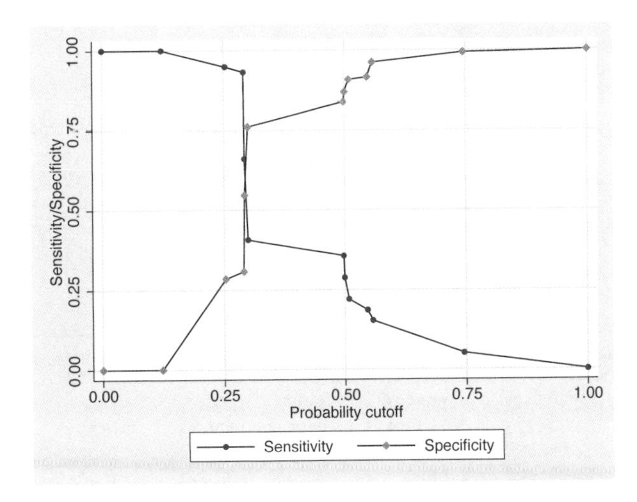

`lsens` optionally creates new variables containing the probability cutoff, sensitivity, and specificity.

```
. lsens, genprob(p) gensens(sens) genspec(spec) nograph
```

The variables created will have $M + 2$ distinct nonmissing values: one for each of the M covariate patterns, one for $c = 0$, and another for $c = 1$.

Samples other than the estimation sample

`estat gof`, `estat classification`, `lroc`, and `lsens` can be used with samples other than the estimation sample. By default, these commands remember the estimation sample used with the last `logistic` command. To override this, simply use an `if` or `in` restriction to select another set of observations, or specify the `all` option to force the command to use all the observations in the dataset.

If you use `estat gof` with a sample that is completely different from the estimation sample (i.e., no overlap), you should also specify the `outsample` option so that the χ^2 statistic properly adjusts the degrees of freedom upward. For an overlapping sample, the conservative thing to do is to leave the degrees of freedom the same as they are for the estimation sample.

▷ Example 5

We want to develop a model for predicting low-birthweight babies. One approach would be to divide our data into two groups, a developmental sample and a validation sample. See Lemeshow and Le Gall (1994) and Tilford, Roberson, and Fiser (1995) for more information on developing prediction models and severity-scoring systems.

We will do this with the low-birthweight data that we considered previously. First, we randomly divide the data into two samples.

```
. use http://www.stata-press.com/data/r10/lbw, clear
(Hosmer & Lemeshow data)
. set seed 1
. gen r = uniform()
. sort r
. gen group = 1 if _n <= _N/2
(95 missing values generated)
. replace group = 2 if group >=.
(95 real changes made)
```

Then we fit a model using the first sample (`group==1`), which is our developmental sample.

```
. xi: logistic low age lwt i.race smoke ptl ht ui if group==1
i.race            _Irace_1-3          (naturally coded; _Irace_1 omitted)
Logistic regression                               Number of obs   =         94
                                                  LR chi2(8)      =      29.14
                                                  Prob > chi2     =     0.0003
Log likelihood = -44.293342                       Pseudo R2       =     0.2475
```

low	Odds Ratio	Std. Err.	z	P>|z|	[95% Conf.	Interval]
age	.91542	.0553937	-1.46	0.144	.8130414	1.03069
lwt	.9744276	.0112295	-2.25	0.025	.9526649	.9966874
_Irace_2	5.063678	3.78442	2.17	0.030	1.170327	21.90913
_Irace_3	2.606209	1.657608	1.51	0.132	.7492483	9.065522
smoke	.909912	.5252898	-0.16	0.870	.2934966	2.820953
ptl	3.033543	1.507048	2.23	0.025	1.145718	8.03198
ht	21.07656	22.64788	2.84	0.005	2.565304	173.1652
ui	.988479	.6699458	-0.02	0.986	.2618557	3.731409

To test calibration in the developmental sample, we calculate the Hosmer–Lemeshow goodness-of-fit test by using `estat gof`.

```
. estat gof, group(10)
Logistic model for low, goodness-of-fit test
  (Table collapsed on quantiles of estimated probabilities)
       number of observations =        94
             number of groups =        10
      Hosmer-Lemeshow chi2(8) =         6.67
                  Prob > chi2 =         0.5721
```

We did not specify an `if` statement with `estat gof` since we wanted to use the estimation sample. Since the test is not significant, we are satisfied with the fit of our model.

Running `lroc` gives a measure of the discrimination:

```
. lroc, nograph
Logistic model for low
number of observations =       94
area under ROC curve   =   0.8156
```

Now we test the calibration of our model by performing a goodness-of-fit test on the validation sample. We specify the `outsample` option so that the number of degrees of freedom is 10 rather than 8.

```
. estat gof if group==2, group(10) table outsample
Logistic model for low, goodness-of-fit test
  (Table collapsed on quantiles of estimated probabilities)
```

Group	Prob	Obs_1	Exp_1	Obs_0	Exp_0	Total
1	0.0725	1	0.4	9	9.6	10
2	0.1202	4	0.8	5	8.2	9
3	0.1549	3	1.3	7	8.7	10
4	0.1888	1	1.5	8	7.5	9
5	0.2609	3	2.2	7	7.8	10
6	0.3258	4	2.7	5	6.3	9
7	0.4217	2	3.7	8	6.3	10
8	0.4915	3	4.1	6	4.9	9
9	0.6265	4	5.5	6	4.5	10
10	0.9737	4	7.1	5	1.9	9

```
      number of observations =        95
            number of groups =        10
     Hosmer-Lemeshow chi2(10) =        28.03
                 Prob > chi2 =         0.0018
```

We must acknowledge that our model does not fit well on the validation sample. The model's discrimination in the validation sample is appreciably lower, as well.

```
. lroc if group==2, nograph
Logistic model for low
number of observations =       95
area under ROC curve   =   0.5839
```

◁

Models other than the last fitted model

By default, `estat classification`, `estat gof`, `lroc`, and `lsens` use the last model fitted by `logistic`. You may also directly specify the model to `lroc` and `lsens` by inputting a vector of coefficients with the `beta()` option and passing the name of the dependent variable *depvar* to these commands.

▷ Example 6

Suppose that someone publishes the following logistic model of low birthweight,

$$\Pr(\texttt{low}=1) = F(-0.02\,\texttt{age} - 0.01\,\texttt{lwt} + 1.3\,\texttt{black} + 1.1\,\texttt{smoke} + 0.5\,\texttt{ptl} + 1.8\,\texttt{ht} + 0.8\,\texttt{ui} + 0.5)$$

where F is the cumulative logistic distribution. These coefficients are not odds ratios; they are the equivalent of what `logit` produces.

We can see whether this model fits our data. First we enter the coefficients as a row vector and label its columns with the names of the independent variables plus `_cons` for the constant (see [P] **matrix define** and [P] **matrix rownames**).

```
. use http://www.stata-press.com/data/r10/lbw3, clear
(Hosmer & Lemeshow data)
. matrix input b = (-.02, -.01, 1.3, 1.1, .5, 1.8, .8, .5)
. matrix colnames b = age lwt black smoke ptl ht ui _cons
```

Here we use `lroc` to examine the predictive ability of the model:

```
. lroc low, beta(b) nograph
Logistic model for low
number of observations =      189
area under ROC curve   =   0.7275
```

The area under the curve indicates that this model does have some predictive power. We could also obtain a graph of sensitivity and specificity as a function of the cutoff probability by typing

```
. lsens low, beta(b) nograph
```

◁

Saved Results

`estat classification` saves the following in `r()`:

Scalars

`r(P_corr)`	percent correctly classified	`r(P_1p)`	positive predictive value
`r(P_p1)`	sensitivity	`r(P_0n)`	negative predictive value
`r(P_n0)`	specificity	`r(P_0p)`	false positive rate given classified positive
`r(P_p0)`	false positive rate given true negative	`r(P_1n)`	false negative rate given classified negative
`r(P_n1)`	false negative rate given true positive		

`estat gof` saves the following in `r()`:

Scalars

`r(N)`	number of observations	`r(df)`	degrees of freedom
`r(m)`	number of covariate patterns or groups	`r(chi2)`	χ^2

`lroc` saves the following in `r()`:

Scalars

`r(N)`	number of observations	`r(area)`	area under the ROC curve

`lsens` saves the following in `r()`:

Scalars

`r(N)`	number of observations

Methods and Formulas

All postestimation commands listed above are implemented as ado-files.

estat gof

Let M be the total number of covariate patterns among the N observations. View the data as collapsed on covariate patterns $j = 1, 2, \ldots, M$, and define m_j as the total number of observations having covariate pattern j and y_j as the total number of positive responses among observations with covariate pattern j. Define p_j as the predicted probability of a positive outcome in covariate pattern j.

The Pearson χ^2 goodness-of-fit statistic is

$$\chi^2 = \sum_{j=1}^{M} \frac{(y_j - m_j p_j)^2}{m_j p_j (1 - p_j)}$$

This χ^2 statistic has approximately $M - k$ degrees of freedom for the estimation sample, where k is the number of independent variables, including the constant. For a sample outside the estimation sample, the statistic has M degrees of freedom.

The Hosmer–Lemeshow goodness-of-fit χ^2 (Hosmer and Lemeshow 1980; Lemeshow and Hosmer 1982; Hosmer, Lemeshow, and Klar 1988) is calculated similarly, except that rather than using the M covariate patterns as the group definition, the quantiles of the predicted probabilities are used to form groups. Let $G = \#$ be the number of quantiles requested with `group(#)`. The smallest index $1 \le q(i) \le M$, such that

$$W_{q(i)} = \sum_{j=1}^{q(i)} m_j \ge \frac{N}{G}$$

gives $p_{q(i)}$ as the upper boundary of the ith quantile for $i = 1, 2, \dots, G$. Let $q(0) = 1$ denote the first index.

The groups are then

$$[p_{q(0)}, p_{q(1)}], (p_{q(1)}, p_{q(2)}], \dots, (p_{q(G-1)}, p_{q(G)}]$$

If the `table` option is given, the upper boundaries $p_{q(1)}, \dots, p_{q(G)}$ of the groups appear next to the group number on the output.

The resulting χ^2 statistic has approximately $G - 2$ degrees of freedom for the estimation sample. For a sample outside the estimation sample, the statistic has G degrees of freedom.

predict after logistic

Index j will now be used to index observations, not covariate patterns. Define M_j for each observation as the total number of observations sharing j's covariate pattern. Define Y_j as the total number of positive responses among observations sharing j's covariate pattern.

The Pearson residual for the jth observation is defined as

$$r_j = \frac{Y_j - M_j p_j}{\sqrt{M_j p_j (1 - p_j)}}$$

For $M_j > 1$, the deviance residual d_j is defined as

$$d_j = \pm \left[2 \left\{ Y_j \ln\left(\frac{Y_j}{M_j p_j} \right) + (M_j - Y_j) \ln\left(\frac{M_j - Y_j}{M_j (1 - p_j)} \right) \right\} \right]^{1/2}$$

where the sign is the same as the sign of $(Y_j - M_j p_j)$. In the limiting cases, the deviance residual is given by

$$d_j = \begin{cases} -\sqrt{2M_j|\ln(1-p_j)|} & \text{if } Y_j = 0 \\ \sqrt{2M_j|\ln p_j|} & \text{if } Y_j = M_j \end{cases}$$

The *unadjusted* diagonal elements of the hat matrix h_{Uj} are given by $h_{Uj} = (\mathbf{XVX}')_{jj}$, where V is the estimated covariance matrix of parameters. The adjusted diagonal elements h_j created by `hat` are then $h_j = M_j p_j(1-p_j)h_{Uj}$.

The standardized Pearson residual r_{Sj} is $r_j/\sqrt{1-h_j}$.

The Pregibon (1981) $\Delta\widehat{\beta}_j$ influence statistic is

$$\Delta\widehat{\beta}_j = \frac{r_j^2 h_j}{(1-h_j)^2}$$

The corresponding change in the Pearson χ^2 is r_{Sj}^2. The corresponding change in the deviance residual is $\Delta D_j = d_j^2/(1-h_j)$.

estat classification and lsens

Again let j index observations. Define c as the `cutoff()` specified by the user or, if not specified, as 0.5. Let p_j be the predicted probability of a positive outcome and y_j be the actual outcome, which we will treat as 0 or 1, although Stata treats it as 0 and non-0, excluding missing observations.

A prediction is classified as *positive* if $p_j \geq c$ and otherwise is classified as *negative*. The classification is *correct* if it is *positive* and $y_j = 1$ or if it is *negative* and $y_j = 0$.

Sensitivity is the fraction of $y_j = 1$ observations that are correctly classified. *Specificity* is the percentage of $y_j = 0$ observations that are correctly classified.

lroc

The ROC curve is a graph of *specificity* against (1 − *sensitivity*). This is guaranteed to be a monotone nondecreasing function since the number of correctly predicted successes increases and the number of correctly predicted failures decreases as the classification cutoff c decreases.

The area under the ROC curve is the area on the bottom of this graph and is determined by integrating the curve. The vertices of the curve are determined by sorting the data according to the predicted index, and the integral is computed using the trapezoidal rule.

References

Archer, K. J., and S. Lemeshow. 2006. Goodness-of-fit test for a logistic regression model fitted using survey sample data. *Stata Journal* 6: 97–105.

Collett, D. 2003. *Modelling Binary Data*. 2nd ed. London: Chapman & Hall/CRC.

Garrett, J. M. 2000. sg157: Predicted values calculated from linear or logistic regression models. *Stata Technical Bulletin* 58: 27–30. Reprinted in *Stata Technical Bulletin Reprints*, vol. 10, pp. 258–261.

Green, D. M., and J. A. Swets. 1966. *Signal Detection Theory and Psychophysics*. New York: Wiley.

Hosmer, D. W., Jr., and S. Lemeshow. 1980. Goodness-of-fit tests for the multiple logistic regression model. *Communications in Statistics* A9: 1043–1069.

——. 2000. *Applied Logistic Regression*. 2nd ed. New York: Wiley.

Hosmer, D. W., Jr., S. Lemeshow, and J. Klar. 1988. Goodness-of-fit testing for the logistic regression model when the estimated probabilities are small. *Biometric Journal* 30: 911–924.

Lemeshow, S., and D. W. Hosmer, Jr. 1982. A review of goodness of fit statistics for use in the development of logistic regression models. *American Journal of Epidemiology* 115: 92–106.

Lemeshow, S., and J.-R. Le Gall. 1994. Modeling the severity of illness of ICU patients: A systems update. *Journal of the American Medical Association* 272: 1049–1055.

Metz, C. E. 1978. Basic principles of ROC analysis. *Seminars in Nuclear Medicine* 8: 283–298.

Mitchell, M. N., and X. Chen. 2005. Visualizing main effects and interactions for binary logit models. *Stata Journal* 5: 64–82.

Peterson, W. W., T. G. Birdsall, and W. C. Fox. 1954. The theory of signal detection. *Transactions IRE Professional Group on Information Theory*, PGIT-4: 171–212.

Pregibon, D. 1981. Logistic regression diagnostics. *Annals of Statistics* 9: 705–724.

Seed, P. T., and A. Tobias. 2001. sbe36.1: Summary statistics for diagnostic tests. *Stata Technical Bulletin* 59: 25–27. Reprinted in *Stata Technical Bulletin Reprints*, vol. 10, pp. 90–93.

Tanner, W. P., Jr., and J. A. Swets. 1954. A decision-making theory of visual detection. *Psychological Review* 61: 401–409.

Tilford, J. M., P. K. Roberson, and D. H. Fiser. 1995. sbe12: Using lfit and lroc to evaluate mortality prediction models. *Stata Technical Bulletin* 28: 14–18. Reprinted in *Stata Technical Bulletin Reprints*, vol. 5, pp. 77–81.

Tobias, A. 2000. sbe36: Summary statistics report for diagnostic tests. *Stata Technical Bulletin* 56: 16–18. Reprinted in *Stata Technical Bulletin Reprints*, vol. 10, pp. 87–90.

Tobias, A., and M. J. Campbell. 1998. sg90: Akaike's information criterion and Schwarz's criterion. *Stata Technical Bulletin* 45: 23–25. Reprinted in *Stata Technical Bulletin Reprints*, vol. 8, pp. 174–177.

Weesie, J. 1998. sg87: Windmeijer's goodness-of-fit test for logistic regression. *Stata Technical Bulletin* 44: 22–27. Reprinted in *Stata Technical Bulletin Reprints*, vol. 8, pp. 153–160.

Also See

[R] **logistic** — Logistic regression, reporting odds ratios

[U] **20 Estimation and postestimation commands**

Title

logit — Logistic regression, reporting coefficients

Syntax

`logit` *depvar* [*indepvars*] [*if*] [*in*] [*weight*] [, *options*]

options	description
Model	
`noconstant`	suppress constant term
`offset(`*varname*`)`	include *varname* in model with coefficient constrained to 1
`asis`	retain perfect predictor variables
SE/Robust	
`vce(`*vcetype*`)`	*vcetype* may be `oim`, `robust`, `cluster` *clustvar*, `bootstrap`, or `jackknife`
Reporting	
`level(`*#*`)`	set confidence level; default is `level(95)`
`or`	report odds ratios
Max options	
maximize_options	control the maximization process; seldom used
† `nocoef`	do not display coefficient table; seldom used

depvar and *indepvars* may contain time-series operators; see [U] **11.4.3 Time-series varlists**.

`bootstrap`, `by`, `jackknife`, `nestreg`, `rolling`, `statsby`, `stepwise`, `svy`, and `xi` are allowed; see [U] **11.1.10 Prefix commands**.

Weights are not allowed with the `bootstrap` prefix.

`vce()`, `nocoef`, and weights are not allowed with the `svy` prefix.

† `nocoef` does not appear in the dialog box.

`fweight`s, `iweight`s, and `pweight`s are allowed; see [U] **11.1.6 weight**.

See [U] **20 Estimation and postestimation commands** for more capabilities of estimation commands.

Description

`logit` fits a maximum-likelihood logit model. *depvar* $=$ 0 indicates a negative outcome; *depvar* $\neq$ 0 and *depvar* $\neq$. (typically *depvar* $=$ 1) indicates a positive outcome.

Also see [R] **logistic**; `logistic` displays estimates as odds ratios. Many users prefer the `logistic` command to `logit`. Results are the same regardless of which you use—both are the maximum-likelihood estimator. Several auxiliary commands that can be run after `logit`, `probit`, or `logistic` estimation are described in [R] **logistic postestimation**. A list of related estimation commands is given in [R] **logistic**.

If estimating on grouped data, see [R] **glogit**.

Options

Model

`noconstant`, `offset(`*varname*`)`; see [R] **estimation options**.

`asis` forces retention of perfect predictor variables and their associated perfectly predicted observations and may produce instabilities in maximization; see [R] **probit**.

SE/Robust

`vce(`*vcetype*`)` specifies the type of standard error reported, which includes types that are derived from asymptotic theory, that are robust to some kinds of misspecification, that allow for intragroup correlation, and that use bootstrap or jackknife methods; see [R] ***vce_option***.

Reporting

`level(`*#*`)`; see [R] **estimation options**.

`or` reports the estimated coefficients transformed to odds ratios, i.e., e^b rather than b. Standard errors and confidence intervals are similarly transformed. This option affects how results are displayed, not how they are estimated. `or` may be specified at estimation or when replaying previously estimated results.

Max options

maximize_options: `iterate(`*#*`)`, `[no]log`, `trace`, `tolerance(`*#*`)`, `ltolerance(`*#*`)`; see [R] **maximize**. These options are seldom used.

The following option is available with `logit` but is not shown in the dialog box:

`nocoef` specifies that the coefficient table not be displayed. This option is sometimes used by program writers but is of no use interactively.

Remarks

Remarks are presented under the following headings:

Basic usage
Model identification

Basic usage

`logit` fits maximum likelihood models with dichotomous dependent (left-hand-side) variables coded as 0/1 (or, more precisely, coded as 0 and not-0).

▷ Example 1

We have data on the make, weight, and mileage rating of 22 foreign and 52 domestic automobiles. We wish to fit a logit model explaining whether a car is foreign on the basis of its weight and mileage. Here is an overview of our data:

```
. use http://www.stata-press.com/data/r10/auto
(1978 Automobile Data)

. keep make mpg weight foreign

. describe

Contains data from http://www.stata-press.com/data/r10/auto.dta
  obs:            74                          1978 Automobile Data
 vars:             4                          13 Apr 2007 17:45
 size:         1,998 (99.7% of memory free)  (_dta has notes)
-------------------------------------------------------------------------------
              storage  display     value
variable name   type   format      label      variable label
-------------------------------------------------------------------------------
make            str18  %-18s                  Make and Model
mpg             int    %8.0g                  Mileage (mpg)
weight          int    %8.0gc                 Weight (lbs.)
foreign         byte   %8.0g       origin     Car type
-------------------------------------------------------------------------------
Sorted by:  foreign
     Note:  dataset has changed since last saved

. inspect foreign

foreign:  Car type                             Number of Observations
------------------                         ------------------------------
                                            Total   Integers   Nonintegers
|  #                        Negative            -          -             -
|  #                        Zero               52         52             -
|  #                        Positive           22         22             -
|  #                                        -----      -----         -----
|  #    #                   Total              74         74             -
|  #    #                   Missing             -
+----------------------                     -----
0                      1                       74
   (2 unique values)

      foreign is labeled and all values are documented in the label.
```

The variable `foreign` takes on two unique values, 0 and 1. The value 0 denotes a domestic car, and 1 denotes a foreign car.

The model that we wish to fit is

$$\Pr(\texttt{foreign} = 1) = F(\beta_0 + \beta_1 \texttt{weight} + \beta_2 \texttt{mpg})$$

where $F(z) = e^z/(1 + e^z)$ is the cumulative logistic distribution.

To fit this model, we type

```
. logit foreign weight mpg
Iteration 0:   log likelihood =  -45.03321
Iteration 1:   log likelihood = -29.898968
Iteration 2:   log likelihood = -27.495771
Iteration 3:   log likelihood = -27.184006
Iteration 4:   log likelihood = -27.175166
Iteration 5:   log likelihood = -27.175156
Logistic regression                               Number of obs   =         74
                                                  LR chi2(2)      =      35.72
                                                  Prob > chi2     =     0.0000
Log likelihood = -27.175156                       Pseudo R2       =     0.3966

------------------------------------------------------------------------------
     foreign |      Coef.   Std. Err.      z    P>|z|     [95% Conf. Interval]
-------------+----------------------------------------------------------------
      weight |  -.0039067   .0010116    -3.86   0.000    -.0058894    -.001924
         mpg |  -.1685869   .0919174    -1.83   0.067    -.3487418     .011568
       _cons |   13.70837   4.518707     3.03   0.002     4.851864    22.56487
------------------------------------------------------------------------------
```

We find that heavier cars are less likely to be foreign and that cars yielding better gas mileage are also less likely to be foreign, at least holding the weight of the car constant.

See [R] **maximize** for an explanation of the output.

◁

❑ Technical Note

Stata interprets a value of 0 as a negative outcome (failure) and treats all other values (except missing) as positive outcomes (successes). Thus if your dependent variable takes on the values 0 and 1, 0 is interpreted as failure and 1 as success. If your dependent variable takes on the values 0, 1, and 2, 0 is still interpreted as failure, but both 1 and 2 are treated as successes.

If you prefer a more formal mathematical statement, when you type `logit` y x, Stata fits the model

$$\Pr(y_j \neq 0 \mid \mathbf{x}_j) = \frac{\exp(\mathbf{x}_j\boldsymbol{\beta})}{1 + \exp(\mathbf{x}_j\boldsymbol{\beta})}$$

❑

Model identification

The `logit` command has one more feature, and it is probably the most useful. `logit` automatically checks the model for identification, and, if it is underidentified, drops whatever variables and observations are necessary for estimation to proceed. (`logistic`, `probit`, and `ivprobit` do this as well.)

▷ Example 2

Have you ever fitted a logit model where one or more of your independent variables perfectly predicted one or the other outcome?

For instance, consider the following data:

Outcome y	Independent Variable x
0	1
0	1
0	0
1	0

Say that we wish to predict the outcome on the basis of the independent variable. The outcome is always zero whenever the independent variable is one. In our data, $\Pr(y = 0 \mid x = 1) = 1$, which means that the logit coefficient on x must be minus infinity with a corresponding infinite standard error. At this point, you may suspect that we have a problem.

Unfortunately, not all such problems are so easily detected, especially if you have a lot of independent variables in your model. If you have ever had such difficulties, you have experienced one of the more unpleasant aspects of computer optimization. The computer has no idea that it is trying to solve for an infinite coefficient as it begins its iterative process. All it knows is that, at each step, making the coefficient a little bigger, or a little smaller, works wonders. It continues on its merry way until either (1) the whole thing comes crashing to the ground when a numerical overflow error occurs or (2) it reaches some predetermined cutoff that stops the process. In the meantime, you have been waiting. The estimates that you finally receive, if you receive any at all, may be nothing more than numerical roundoff.

Stata watches for these sorts of problems, alerts us, fixes them, and properly fits the model.

Let's return to our automobile data. Among the variables we have in the data is one called `repair`, which takes on three values. A value of 1 indicates that the car has a poor repair record, 2 indicates an average record, and 3 indicates a better-than-average record. Here is a tabulation of our data:

```
. use http://www.stata-press.com/data/r10/repair, clear
(1978 Automobile Data)

. tabulate foreign repair

           |              repair
   foreign |         1          2          3 |     Total
-----------+---------------------------------+----------
  Domestic |        10         27          9 |        46
   Foreign |         0          3          9 |        12
-----------+---------------------------------+----------
     Total |        10         30         18 |        58
```

All the cars with poor repair records (`repair==1`) are domestic. If we were to attempt to predict `foreign` on the basis of the repair records, the predicted probability for the `repair==1` category would have to be zero. This in turn means that the logit coefficient must be minus infinity, and that would set most computer programs buzzing.

Let's try Stata on this problem. First, we make up two new variables, `rep_is_1` and `rep_is_2`, which indicate the `repair` category.

```
. generate rep_is_1 = (repair==1)

. generate rep_is_2 = (repair==2)
```

The statement `generate rep_is_1 = (repair==1)` creates a new variable, `rep_is_1`, which takes on the value 1 when `repair` is 1, and zero otherwise. Similarly, the next `generate` statement creates `rep_is_2` that takes on the value 1 when `repair` is 2, and zero otherwise. We are now ready to fit our logit model. See [R] **probit** for the corresponding probit model.

```
. logit foreign rep_is_1 rep_is_2
Note: rep_is_1 != 0 predicts failure perfectly
      rep_is_1 dropped and 10 obs not used
Iteration 0:   log likelihood = -26.992087
Iteration 1:   log likelihood = -22.483187
Iteration 2:   log likelihood = -22.230498
Iteration 3:   log likelihood = -22.229139
Iteration 4:   log likelihood = -22.229138
Logistic regression                               Number of obs   =         48
                                                  LR chi2(1)      =       9.53
                                                  Prob > chi2     =     0.0020
Log likelihood = -22.229138                       Pseudo R2       =     0.1765

------------------------------------------------------------------------------
     foreign |      Coef.   Std. Err.      z    P>|z|     [95% Conf. Interval]
-------------+----------------------------------------------------------------
    rep_is_2 |  -2.197225   .7698003    -2.85   0.004    -3.706006   -.6884436
       _cons |   3.89e-16   .4714045     0.00   1.000    -.9239359    .9239359
------------------------------------------------------------------------------
```

Remember that all the cars with poor repair records (`rep_is_1`) are domestic, so the model cannot be fitted, or at least it cannot be fitted if we restrict ourselves to finite coefficients. Stata noted that fact: "Note: rep_is_1 != 0 predicts failure perfectly". This is Stata's mathematically precise way of saying what we said in English. When `rep_is_1` is not equal to 0, the car is domestic.

Stata then went on to say, "rep_is_1 dropped and 10 obs not used". This is Stata eliminating the problem. First, the variable `rep_is_1` had to be removed from the model because it would have an infinite coefficient. Then the 10 observations that led to the problem had to be eliminated, as well, so as not to bias the remaining coefficients in the model. The 10 observations that are not used are the 10 domestic cars that have poor repair records.

Finally, Stata fitted what was left of the model, using the remaining observations.

◁

❑ Technical Note

Stata is pretty smart about catching problems like this. It will catch "one-way causation by a dummy variable", as we demonstrated above.

Stata also watches for "two-way causation", that is, a variable that perfectly determines the outcome, both successes and failures. Here Stata says, "so-and-so predicts outcome perfectly" and stops. Statistics dictates that no model can be fitted.

Stata also checks your data for collinear variables; it will say, "so-and-so dropped because of collinearity". No observations need to be eliminated in this case, and model fitting will proceed without the offending variable.

It will also catch a subtle problem that can arise with continuous data. For instance, if we were estimating the chances of surviving the first year after an operation, and if we included in our model `age`, and if all the persons over 65 died within the year, Stata would say, "age > 65 predicts failure perfectly". It would then inform us about the fixup it takes and fit what can be fitted of our model.

`logit` (and `logistic`, `probit`, and `ivprobit`) will also occasionally display messages such as

```
Note: 4 failures and 0 successes completely determined.
```

There are two causes for a message like this. The first—and most unlikely—case occurs when a continuous variable (or a combination of a continuous variable with other continuous or dummy variables) is simply a great predictor of the dependent variable. Consider Stata's `auto.dta` dataset with 6 observations removed.

```
. use http://www.stata-press.com/data/r10/auto, clear
(1978 Automobile Data)

. drop if foreign==0 & gear_ratio>3.1
(6 observations deleted)

. logit foreign mpg weight gear_ratio, nolog

Logistic regression                               Number of obs   =         68
                                                  LR chi2(3)      =      72.64
                                                  Prob > chi2     =     0.0000
Log likelihood = -6.4874814                       Pseudo R2       =     0.8484

------------------------------------------------------------------------------
     foreign |      Coef.   Std. Err.      z    P>|z|     [95% Conf. Interval]
-------------+----------------------------------------------------------------
         mpg |  -.4944907   .2655508    -1.86   0.063    -1.014961    .0259792
      weight |  -.0060919    .003101    -1.96   0.049    -.0121698    -.000014
  gear_ratio |   15.70509   8.166234     1.92   0.054    -.3004359    31.71061
       _cons |  -21.39527   25.41486    -0.84   0.400    -71.20747    28.41694
------------------------------------------------------------------------------
Note: 4 failures and 0 successes completely determined.
```

There are no missing standard errors in the output. If you receive the "completely determined" message and have one or more missing standard errors in your output, see the second case discussed below.

Note `gear_ratio`'s large coefficient. `logit` thought that the 4 observations with the smallest predicted probabilities were essentially predicted perfectly.

```
. predict p
(option pr assumed; Pr(foreign))

. sort p

. list p in 1/4

     +----------+
     |        p |
     |----------|
  1. | 1.34e-10 |
  2. | 6.26e-09 |
  3. | 7.84e-09 |
  4. | 1.49e-08 |
     +----------+
```

If this happens to you, you don't have to do anything. Computationally, the model is sound. The second case discussed below requires careful examination.

The second case occurs when the independent terms are all dummy variables or continuous ones with repeated values (e.g., age). Here one or more of the estimated coefficients will have missing standard errors. For example, consider this dataset consisting of 5 observations.

```
. list

     +----------------+
     | y   x1   x2    |
     |----------------|
  1. | 0    0    0    |
  2. | 0    1    0    |
  3. | 1    1    0    |
  4. | 0    0    1    |
  5. | 1    0    1    |
     +----------------+
```

```
. logit y x1 x2, nolog
Logistic regression                               Number of obs   =          5
                                                  LR chi2(2)      =       1.18
                                                  Prob > chi2     =     0.5530
Log likelihood = -2.7725887                       Pseudo R2       =     0.1761

------------------------------------------------------------------------------
           y |      Coef.   Std. Err.      z    P>|z|     [95% Conf. Interval]
-------------+----------------------------------------------------------------
          x1 |   18.26157          2     9.13   0.000     14.34164     22.1815
          x2 |   18.26157          .        .       .            .           .
       _cons |  -18.26157   1.414214   -12.91   0.000    -21.03338   -15.48976
------------------------------------------------------------------------------
Note: 1 failure and 0 successes completely determined.
. predict p
(option pr assumed; Pr(y))
. list

     +----------------------+
     | y   x1   x2        p |
     |----------------------|
  1. | 0    0    0   1.17e-08 |
  2. | 0    1    0         .5 |
  3. | 1    1    0         .5 |
  4. | 0    0    1         .5 |
  5. | 1    0    1         .5 |
     +----------------------+
```

Two things are happening here. First, **logit** can fit the outcome (y = 0) for the covariate pattern x1 = 0 and x2 = 0 (i.e., the first observation) perfectly. This observation is the "1 failure ... completely determined". Second, if this observation is dropped, then x1, x2, and the constant are collinear.

This is the cause of the message "completely determined" and the missing standard errors. It happens when you have a covariate pattern (or patterns) with only one outcome and there is collinearity when the observations corresponding to this covariate pattern are dropped.

If this happens to you, confirm the causes. First, identify the covariate pattern with only one outcome. (For your data, replace x1 and x2 with the independent variables of your model.)

```
. drop p
. egen pattern = group(x1 x2)
. quietly logit y x1 x2
. predict p
(option pr assumed; Pr(y))
. summarize p
    Variable |       Obs        Mean    Std. Dev.       Min        Max
-------------+--------------------------------------------------------
           p |         5          .4    .2236068   1.17e-08         .5
```

If successes were completely determined, that means that there are predicted probabilities that are almost 1. If failures were completely determined, that means that there are predicted probabilities that are almost 0. The latter is the case here, so we locate the corresponding value of pattern:

```
. tabulate pattern if p < 1e-7
   group(x1 |
        x2) |      Freq.     Percent        Cum.
------------+-----------------------------------
          1 |          1      100.00      100.00
------------+-----------------------------------
      Total |          1      100.00
```

Once we omit this covariate pattern from the estimation sample, `logit` can deal with the collinearity:

```
. logit y x1 x2 if pattern !=1, nolog
note: x2 dropped because of collinearity
Logistic regression                               Number of obs   =          4
                                                  LR chi2(1)      =       0.00
                                                  Prob > chi2     =     1.0000
Log likelihood = -2.7725887                       Pseudo R2       =     0.0000

------------------------------------------------------------------------------
           y |      Coef.   Std. Err.      z    P>|z|     [95% Conf. Interval]
-------------+----------------------------------------------------------------
          x1 |          0          2     0.00   1.000    -3.919928    3.919928
       _cons |          0   1.414214     0.00   1.000    -2.771808    2.771808
------------------------------------------------------------------------------
```

We omit the collinear variable. Then we must decide whether to include or omit the observations with `pattern` = 1. We could include them,

```
. logit y x1, nolog
Logistic regression                               Number of obs   =          5
                                                  LR chi2(1)      =       0.14
                                                  Prob > chi2     =     0.7098
Log likelihood = -3.2958369                       Pseudo R2       =     0.0206

------------------------------------------------------------------------------
           y |      Coef.   Std. Err.      z    P>|z|     [95% Conf. Interval]
-------------+----------------------------------------------------------------
          x1 |   .6931472   1.870827     0.37   0.711    -2.973605      4.3599
       _cons |  -.6931472   1.224742    -0.57   0.571    -3.093597    1.707302
------------------------------------------------------------------------------
```

or exclude them,

```
. logit y x1 if pattern !=1, nolog
Logistic regression                               Number of obs   =          4
                                                  LR chi2(1)      =       0.00
                                                  Prob > chi2     =     1.0000
Log likelihood = -2.7725887                       Pseudo R2       =     0.0000

------------------------------------------------------------------------------
           y |      Coef.   Std. Err.      z    P>|z|     [95% Conf. Interval]
-------------+----------------------------------------------------------------
          x1 |          0          2     0.00   1.000    -3.919928    3.919928
       _cons |          0   1.414214     0.00   1.000    -2.771808    2.771808
------------------------------------------------------------------------------
```

If the covariate pattern that predicts outcome perfectly is meaningful, you may want to exclude these observations from the model. Here you would report that covariate pattern such and such predicted outcome perfectly and that the best model for the rest of the data is .... But, more likely, the perfect prediction was simply the result of having too many predictors in the model. Then you would omit the extraneous variables from further consideration and report the best model for all the data.

❑

Saved Results

logit saves the following in e():

Scalars

e(N)	number of observations	e(r2_p)	pseudo-R-squared
e(N_cds)	number of completely determined successes	e(ll)	log likelihood
e(N_cdf)	number of completely determined failures	e(ll_0)	log likelihood, constant-only model
e(df_m)	model degrees of freedom	e(N_clust)	number of clusters
		e(chi2)	χ^2

Macros

e(cmd)	logit	e(chi2type)	Wald or LR; type of model χ^2 test
e(cmdline)	command as typed	e(vce)	*vcetype* specified in vce()
e(depvar)	name of dependent variable	e(vcetype)	title used to label Std. Err.
e(wtype)	weight type	e(crittype)	optimization criterion
e(wexp)	weight expression	e(properties)	b V
e(title)	title in estimation output	e(estat_cmd)	program used to implement estat
e(clustvar)	name of cluster variable	e(predict)	program used to implement predict
e(offset)	offset		

Matrices

e(b)	coefficient vector	e(V)	variance–covariance matrix of the estimators

Functions

e(sample)	marks estimation sample

Methods and Formulas

logit is implemented as an ado-file.

Cramer (2003, chap. 9) surveys the prehistory and history of the logit model. The word "logit" was coined by Berkson (1944) and is analogous to the word "probit". For an introduction to probit and logit, see, for example, Aldrich and Nelson (1984), Johnston and DiNardo (1997), Long (1997), Long and Freese (2006), Pampel (2000), or Powers and Xie (2000).

The likelihood function for logit is

$$\ln L = \sum_{j \in S} w_j \ln F(\mathbf{x}_j \mathbf{b}) + \sum_{j \notin S} w_j \ln\{1 - F(\mathbf{x}_j \mathbf{b})\}$$

where S is the set of all observations j, such that $y_j \neq 0$, $F(z) = e^z/(1+e^z)$, and w_j denotes the optional weights. $\ln L$ is maximized as described in [R] **maximize**.

If robust standard errors are requested, the calculation described in *Methods and Formulas* of [R] **regress** is carried forward with $\mathbf{u}_j = \{1 - F(\mathbf{x}_j \mathbf{b})\}\mathbf{x}_j$ for the positive outcomes and $-F(\mathbf{x}_j \mathbf{b})\mathbf{x}_j$ for the negative outcomes. q_c is given by its asymptotic-like formula.

Joseph Berkson (1899–1982) was born in New York City and studied at the College of the City of New York, Columbia, and Johns Hopkins, earning both an M.D. and a doctorate in statistics. He then worked at Johns Hopkins before moving to the Mayo Clinic in 1931 as a biostatistician. Among many other contributions, his most influential drew upon a long-sustained interest in the logistic function, especially his 1944 paper on bioassay, in which he introduced the term "logit". Berkson was a frequent participant in controversy, sometimes humorous, sometimes bitter, on subjects such as the evidence for links between smoking and various diseases and the relative merits of probit and logit methods and of different calculation methods.

References

Aldrich, J. H., and F. D. Nelson. 1984. *Linear Probability, Logit, and Probit Models*. Newbury Park, CA: Sage.

Archer, K. J., and S. Lemeshow. 2006. Goodness-of-fit test for a logistic regression model fitted using survey sample data. *Stata Journal* 6: 97–105.

Berkson, J. 1944. Application of the logistic function to bio-assay. *Journal of the American Statistical Association* 39: 357–365.

Cleves, M., and A. Tosetto. 2000. sg139: Logistic regression when binary outcome is measured with uncertainty. *Stata Technical Bulletin* 55: 20–23. Reprinted in *Stata Technical Bulletin Reprints*, vol. 10, pp. 152–156.

Cramer, J. S. 2003. *Logit Models from Economics and Other Fields*. Cambridge: Cambridge University Press.

Hosmer, D. W., Jr., and S. Lemeshow. 2000. *Applied Logistic Regression*. 2nd ed. New York: Wiley.

Johnston, J., and J. DiNardo. 1997. *Econometric Methods*. 4th ed. New York: McGraw–Hill.

Judge, G. G., W. E. Griffiths, R. C. Hill, H. Lütkepohl, and T.-C. Lee. 1985. *The Theory and Practice of Econometrics*. 2nd ed. New York: Wiley.

Long, J. S. 1997. *Regression Models for Categorical and Limited Dependent Variables*. Thousand Oaks, CA: Sage.

Long, J. S., and J. Freese. 2006. *Regression Models for Categorical Dependent Variables Using Stata*. 2nd ed. College Station, TX: Stata Press.

Miranda, A., and S. Rabe-Hesketh. 2006. Maximum likelihood estimation of endogenous switching and sample selection models for binary, ordinal, and count variables. *Stata Journal* 6: 285–308.

Mitchell, M. N., and X. Chen. 2005. Visualizing main effects and interactions for binary logit models. *Stata Journal* 5: 64–82.

O'Fallon, W. M. 1998. Berkson, Joseph. In *Encyclopedia of Biostatistics*, vol. 1, ed. P. Armitage and T. Colton, 290–295. Chichester, UK: Wiley.

Pampel, F. C. 2000. *Logistic Regression: A Primer*. Thousand Oaks, CA: Sage.

Powers, D. A., and Y. Xie. 2000. *Statistical Methods for Categorical Data Analysis*. San Diego, CA: Academic Press.

Pregibon, D. 1981. Logistic regression diagnostics. *Annals of Statistics* 9: 705–724.

Schonlau, M. 2005. Boosted regression (boosting): An introductory tutorial and a Stata plugin. *Stata Journal* 5: 330–354.

Xu, J., and J. S. Long. 2005. Confidence intervals for predicted outcomes in regression models for categorical outcomes. *Stata Journal* 5: 537–559.

Also See

[R] **logit postestimation** — Postestimation tools for logit

[R] **roc** — Receiver operating characteristic (ROC) analysis

[R] **brier** — Brier score decomposition

[R] **exlogistic** — Exact logistic regression

[R] **glogit** — Logit and probit regression for grouped data

[R] **logistic** — Logistic regression, reporting odds ratios

[R] **probit** — Probit regression

[SVY] **svy estimation** — Estimation commands for survey data

[XT] **xtlogit** — Fixed-effects, random-effects, and population-averaged logit models

[U] **20 Estimation and postestimation commands**

Title

logit postestimation — Postestimation tools for logit

Description

The following postestimation commands are of special interest after `logit`:

command	description
`estat clas`	`estat classification` reports various summary statistics, including the classification table
`estat gof`	Pearson or Hosmer–Lemeshow goodness-of-fit test
`lroc`	graphs the ROC curve and calculates the area under the curve
`lsens`	graphs sensitivity and specificity versus probability cutoff

These commands are not appropriate after the `svy` prefix.

For information about these commands, see [R] **logistic postestimation**.

The following standard postestimation commands are also available:

command	description
`adjust`[1]	adjusted predictions of $\mathbf{x}\beta$, probabilities, or $\exp(\mathbf{x}\beta)$
`estat`	AIC, BIC, VCE, and estimation sample summary
`estat` (svy)	postestimation statistics for survey data
`estimates`	cataloging estimation results
`lincom`	point estimates, standard errors, testing, and inference for linear combinations of coefficients
`linktest`	link test for model specification
`lrtest`[2]	likelihood-ratio test
`mfx`	marginal effects or elasticities
`nlcom`	point estimates, standard errors, testing, and inference for nonlinear combinations of coefficients
`predict`	predictions, residuals, influence statistics, and other diagnostic measures
`predictnl`	point estimates, standard errors, testing, and inference for generalized predictions
`suest`	seemingly unrelated estimation
`test`	Wald tests for simple and composite linear hypotheses
`testnl`	Wald tests of nonlinear hypotheses

[1] `adjust` is not appropriate with time-series operators.

[2] `lrtest` is not appropriate with `svy` estimation results.

See the corresponding entries in the *Stata Base Reference Manual* for details, but see [SVY] **estat** for details about `estat` (svy).

Syntax for predict

predict [*type*] *newvar* [*if*] [*in*] [, *statistic* nooffset rules asif]

statistic	description
Main	
pr	probability of a positive outcome; the default
xb	$\mathbf{x}_j\mathbf{b}$, fitted values
stdp	standard error of the prediction
*dbeta	Pregibon (1981) $\Delta\widehat{\beta}$ influence statistic
*deviance	deviance residual
*dx2	Hosmer and Lemeshow (2000) $\Delta\,\chi^2$ influence statistic
*ddeviance	Hosmer and Lemeshow (2000) $\Delta\,D$ influence statistic
*hat	Pregibon (1981) leverage
*number	sequential number of the covariate pattern
*residuals	Pearson residuals; adjusted for number sharing covariate pattern
*rstandard	standardized Pearson residuals; adjusted for number sharing covariate pattern
score	first derivative of the log likelihood with respect to $\mathbf{x}_j\boldsymbol{\beta}$

Unstarred statistics are available both in and out of sample; type predict ... if e(sample) ... if wanted only for the estimation sample. Starred statistics are calculated only for the estimation sample, even when if e(sample) is not specified.

pr, xb, stdp, and score are the only options allowed with svy estimation results.

Options for predict

Main

pr, the default, calculates the probability of a positive outcome.

xb calculates the linear prediction.

stdp calculates the standard error of the linear prediction.

dbeta calculates the Pregibon (1981) $\Delta\widehat{\beta}$ influence statistic, a standardized measure of the difference in the coefficient vector that is due to deletion of the observation along with all others that share the same covariate pattern. In Hosmer and Lemeshow (2000) jargon, this statistic is M-asymptotic; that is, it is adjusted for the number of observations that share the same covariate pattern.

deviance calculates the deviance residual.

dx2 calculates the Hosmer and Lemeshow (2000) $\Delta\chi^2$ influence statistic, reflecting the decrease in the Pearson χ^2 that is due to the deletion of the observation and all others that share the same covariate pattern.

ddeviance calculates the Hosmer and Lemeshow (2000) ΔD influence statistic, which is the change in the deviance residual that is due to deletion of the observation and all others that share the same covariate pattern.

hat calculates the Pregibon (1981) leverage or the diagonal elements of the hat matrix adjusted for the number of observations that share the same covariate pattern.

number numbers the covariate patterns—observations with the same covariate pattern have the same number. Observations not used in estimation have number set to missing. The first covariate pattern is numbered 1, the second 2, and so on.

`residuals` calculates the Pearson residual as given by Hosmer and Lemeshow (2000) and adjusted for the number of observations that share the same covariate pattern.

`rstandard` calculates the standardized Pearson residual as given by Hosmer and Lemeshow (2000) and adjusted for the number of observations that share the same covariate pattern.

`score` calculates the equation-level score, $\partial \ln L/\partial(\mathbf{x}_j\boldsymbol{\beta})$.

Options

`nooffset` is relevant only if you specified `offset(`*varname*`)` for `logit`. It modifies the calculations made by `predict` so that they ignore the offset variable; the linear prediction is treated as $\mathbf{x}_j\mathbf{b}$ rather than as $\mathbf{x}_j\mathbf{b} + \text{offset}_j$.

`rules` requests that Stata use any rules that were used to identify the model when making the prediction. By default, Stata calculates missing for excluded observations.

`asif` requests that Stata ignore the rules and exclusion criteria and calculate predictions for all observations possible using the estimated parameter from the model.

Remarks

Once you have fitted a logit model, you can obtain the predicted probabilities by using the `predict` command for both the estimation sample and other samples; see [U] **20 Estimation and postestimation commands** and [R] **predict**. Here we will make only a few more comments.

`predict` without arguments calculates the predicted probability of a positive outcome, i.e., $\Pr(y_j = 1) = F(\mathbf{x}_j\mathbf{b})$. With the `xb` option, `predict` calculates the linear combination $\mathbf{x}_j\mathbf{b}$, where $\mathbf{x}_j$ are the independent variables in the jth observation and $\mathbf{b}$ is the estimated parameter vector. This is sometimes known as the index function since the cumulative distribution function indexed at this value is the probability of a positive outcome.

In both cases, Stata remembers any rules used to identify the model and calculates missing for excluded observations, unless `rules` or `asif` is specified. For information about the other statistics available after `predict`, see [R] **logistic postestimation**.

▷ Example 1

In example 2 of [R] **logit**, we fitted the logit model `logit foreign rep_is_1 rep_is_2`. To obtain predicted probabilities, type

```
. use http://www.stata-press.com/data/r10/repair
(1978 Automobile Data)

. generate rep_is_1 = (repair==1)

. generate rep_is_2 = (repair==2)

. logit foreign rep_is_1 rep_is_2
  (output omitted)
. predict p
(option pr assumed; Pr(foreign))
(10 missing values generated)

. summarize foreign p

    Variable |       Obs        Mean    Std. Dev.       Min        Max
-------------+--------------------------------------------------------
     foreign |        58    .2068966    .4086186          0          1
           p |        48         .25    .1956984         .1         .5
```

Stata remembers any rules used to identify the model and sets predictions to missing for any excluded observations. In the previous example, `logit` dropped the variable `rep_is_1` from our model and excluded 10 observations. Thus when we typed `predict p`, those same 10 observations were again excluded, and their predictions were set to missing.

`predict`'s `rules` option uses the rules in the prediction. During estimation, we were told "rep_is_1 != 0 predicts failure perfectly", so the rule is that when `rep_is_1` is not zero, we should predict 0 probability of success or a positive outcome:

```
. predict p2, rules
. summarize foreign p p2
    Variable |       Obs        Mean    Std. Dev.       Min        Max
-------------+--------------------------------------------------------
     foreign |        58    .2068966    .4086186          0          1
           p |        48         .25    .1956984         .1         .5
          p2 |        58    .2068966    .2016268          0         .5
```

`predict`'s `asif` option ignores the rules and exclusion criteria and calculates predictions for all observations possible using the estimated parameters from the model:

```
. predict p3, asif
. summarize foreign p p2 p3
    Variable |       Obs        Mean    Std. Dev.       Min        Max
-------------+--------------------------------------------------------
     foreign |        58    .2068966    .4086186          0          1
           p |        48         .25    .1956984         .1         .5
          p2 |        58    .2068966    .2016268          0         .5
          p3 |        58    .2931034    .2016268         .1         .5
```

Which is right? What `predict` does by default is the most conservative approach. If many observations had been excluded because of a simple rule, we could be reasonably certain that the `rules` prediction is correct. The `asif` prediction is correct only if the exclusion is a fluke, and we would be willing to exclude the variable from the analysis anyway. Then, however, we would refit the model to include the excluded observations.

◁

Methods and Formulas

All postestimation commands listed above are implemented as ado-files.

See *Methods and Formulas* of [R] **logistic postestimation** for details.

References

Archer, K. J., and S. Lemeshow. 2006. Goodness-of-fit test for a logistic regression model fitted using survey sample data. *Stata Journal* 6: 97–105.

Hosmer, D. W., Jr., and S. Lemeshow. 2000. *Applied Logistic Regression*. 2nd ed. New York: Wiley.

Pregibon, D. 1981. Logistic regression diagnostics. *Annals of Statistics* 9: 705–724.

Also See

[R] **logit** — Logistic regression, reporting coefficients

[R] **logistic postestimation** — Postestimation tools for logistic

[U] **20 Estimation and postestimation commands**

Title

loneway — Large one-way ANOVA, random effects, and reliability

Syntax

`loneway` *response_var group_var* [*if*] [*in*] [*weight*] [, *options*]

options	description
Main	
`mean`	expected value of F distribution; default is 1
`median`	median of F distribution; default is 1
`exact`	exact confidence intervals (groups must be equal with no weights)
`level(#)`	set confidence level; default is `level(95)`

by is allowed; see [D] **by**.

`aweight`s are allowed; see [U] **11.1.6 weight**.

Description

`loneway` fits one-way analysis-of-variance (ANOVA) models on datasets with many levels of *group_var* and presents different ancillary statistics from `oneway` (see [R] **oneway**):

Feature	`oneway`	`loneway`
Fit one-way model	x	x
on fewer than 376 levels	x	x
on more than 376 levels		x
Bartlett's test for equal variance	x	
Multiple-comparison tests	x	
Intragroup correlation and S.E.		x
Intragroup correlation confidence interval		x
Est. reliability of group-averaged score		x
Est. S.D. of group effect		x
Est. S.D. within group		x

Options

Main

`mean` specifies that the expected value of the $F_{k-1,N-k}$ distribution be used as the reference point F_m in the estimation of ρ instead of the default value of 1.

`median` specifies that the median of the $F_{k-1,N-k}$ distribution be used as the reference point F_m in the estimation of ρ instead of the default value of 1.

`exact` requests that exact confidence intervals be computed, as opposed to the default asymptotic confidence intervals. This option is allowed only if the groups are equal in size and weights are not used.

`level(#)` specifies the confidence level, as a percentage, for confidence intervals of the coefficients. The default is `level(95)` or as set by `set level`; see [U] **20.7 Specifying the width of confidence intervals**.

Remarks

Remarks are presented under the following headings:

The one-way ANOVA model

▷ Example 1

`loneway`'s output looks like that of `oneway`, except that `loneway` presents more information at the end. Using our automobile dataset, we have created a (numeric) variable called `manufacturer_grp` identifying the manufacturer of each car, and within each manufacturer we have retained a maximum of four models, selecting those with the lowest mpg. We can compute the intraclass correlation of mpg for all manufacturers with at least four models as follows:

```
. use http://www.stata-press.com/data/r10/auto7
(1978 Automobile Data)
. loneway mpg manufacturer_grp if nummake == 4
                 One-way Analysis of Variance for mpg: Mileage (mpg)

                                                Number of obs =        36
                                                    R-squared =    0.5228

    Source                    SS         df      MS            F     Prob > F
-------------------------------------------------------------------------------
Between manufactur~p      621.88889       8   77.736111      3.70     0.0049
Within manufactur~p          567.75      27   21.027778
-------------------------------------------------------------------------------
Total                     1189.6389      35   33.989683

         Intraclass          Asy.
         correlation         S.E.       [95% Conf. Interval]
         ------------------------------------------------
            0.40270        0.18770      0.03481     0.77060

         Estimated SD of manufactur~p effect         3.765247
         Estimated SD within manufactur~p            4.585605
         Est. reliability of a manufactur~p mean      .72950
              (evaluated at n=4.00)
```

◁

In addition to the standard one-way ANOVA output, `loneway` produces the R-squared, the estimated standard deviation of the group effect, the estimated standard deviation within group, the intragroup correlation, the estimated reliability of the group-averaged mean, and, for unweighted data, the asymptotic standard error and confidence interval for the intragroup correlation.

R-squared

The R-squared is, of course, simply the underlying R^2 for a regression of *response_var* on the levels of *group_var*, or `mpg` on the various manufacturers here.

The random-effects ANOVA model

`loneway` assumes that we observe a variable y_{ij} measured for n_i elements within k groups or classes such that

$$y_{ij} = \mu + \alpha_i + \epsilon_{ij}, \quad i = 1, 2, \dots, k, \quad j = 1, 2, \dots, n_i$$

and α_i and ϵ_{ij} are independent zero-mean random variables with variance σ^2_α and σ^2_ϵ, respectively. This is the random-effects ANOVA model, also known as the components-of-variance model, in which it is typically assumed that the y_{ij} are normally distributed.

The interpretation with respect to our example is that the observed value of our response variable, `mpg`, is created in two steps. First, the ith manufacturer is chosen, and a value α_i is determined—the typical `mpg` for that manufacturer less the overall `mpg` μ. Then a deviation, ϵ_{ij}, is chosen for the jth model within this manufacturer. This is how much that particular automobile differs from the typical `mpg` value for models from this manufacturer.

For our sample of 36 car models, the estimated standard deviations are $\sigma_\alpha = 3.8$ and $\sigma_\epsilon = 4.6$. Thus a little more than half of the variation in `mpg` between cars is attributable to the car model, with the rest attributable to differences between manufacturers. These standard deviations differ from those that would be produced by a (standard) fixed-effects regression in that the regression would require the sum within each manufacturer of the ϵ_{ij}, $\epsilon_{i.}$ for the ith manufacturer, to be zero, whereas these estimates merely impose the constraint that the sum is *expected* to be zero.

Intraclass correlation

There are various estimators of the intraclass correlation, such as the pairwise estimator, which is defined as the Pearson product-moment correlation computed over all possible pairs of observations that can be constructed within groups. For a discussion of various estimators, see Donner (1986). `loneway` computes what is termed the analysis of variance, or ANOVA, estimator. This intraclass correlation is the theoretical upper bound on the variation in *response_var* that is explainable by *group_var*, of which R-squared is an overestimate because of the serendipity of fitting. This correlation is comparable to an R-squared—you do not have to square it.

In our example, the intra-`manu` correlation, the correlation of `mpg` within manufacturer, is 0.40. Since `aweights` were not used and the default correlation was computed (i.e., the `mean` and `median` options were not specified), `loneway` also provided the asymptotic confidence interval and standard error of the intraclass correlation estimate.

Estimated reliability of the group-averaged score

The estimated reliability of the group-averaged score or mean has an interpretation similar to that of the intragroup correlation; it is a comparable number if we average *response_var* by *group_var*, or `mpg` by `manu` in our example. It is the theoretical upper bound of a regression of manufacturer-averaged `mpg` on characteristics of manufacturers. Why would we want to collapse our 36-observation dataset into a 9-observation dataset of manufacturer averages? Because the 36 observations might be a mirage. When General Motors builds cars, do they sometimes put a Pontiac label and sometimes a Chevrolet label on them, so that it appears in our data as if we have two cars when we really have only one, replicated? If that were the case, and if it were the case for many other manufacturers, then we would be forced to admit that we do not have data on 36 cars; we instead have data on nine manufacturer-averaged characteristics.

Saved Results

`loneway` saves the following in `r()`:

Scalars

`r(N)`	number of observations	`r(rho_t)`	estimated reliability
`r(rho)`	intraclass correlation	`r(se)`	asymp. SE of intraclass correlation
`r(lb)`	lower bound of 95% CI for rho	`r(sd_w)`	estimated SD within group
`r(ub)`	upper bound of 95% CI for rho	`r(sd_b)`	estimated SD of group effect

Methods and Formulas

`loneway` is implemented as an ado-file.

The mean squares in the `loneway`'s ANOVA table are computed as

$$\text{MS}_\alpha = \sum_i w_{i\cdot}(\overline{y}_{i\cdot} - \overline{y}_{\cdot\cdot})^2/(k-1)$$

and

$$\text{MS}_\epsilon = \sum_i \sum_j w_{ij}(y_{ij} - \overline{y}_{i\cdot})^2/(N-k)$$

in which

$$w_{i\cdot} = \sum_j w_{ij} \quad w_{\cdot\cdot} = \sum_i w_{i\cdot} \quad \overline{y}_{i\cdot} = \sum_j w_{ij} y_{ij}/w_{i\cdot} \quad \text{and} \quad \overline{y}_{\cdot\cdot} = \sum_i w_{i\cdot}\overline{y}_{i\cdot}/w_{\cdot\cdot}$$

The corresponding expected values of these mean squares are

$$E(\text{MS}_\alpha) = \sigma_\epsilon^2 + g\sigma_\alpha^2 \qquad \text{and} \qquad E(\text{MS}_\epsilon) = \sigma_\epsilon^2$$

in which

$$g = \frac{w_{\cdot\cdot} - \sum_i w_{i\cdot}^2/w_{\cdot\cdot}}{k-1}$$

In the unweighted case, we get

$$g = \frac{N - \sum_i n_i^2/N}{k-1}$$

As expected, $g = m$ for the case of no weights and equal group sizes in the data, i.e., $n_i = m$ for all i. Replacing the expected values with the observed values and solving yields the ANOVA estimates of σ_α^2 and σ_ϵ^2. Substituting these into the definition of the intraclass correlation

$$\rho = \frac{\sigma_\alpha^2}{\sigma_\alpha^2 + \sigma_\epsilon^2}$$

yields the ANOVA estimator of the intraclass correlation:

$$\rho_A = \frac{F_{\text{obs}} - 1}{F_{\text{obs}} - 1 + g}$$

F_{obs} is the observed value of the F statistic from the ANOVA table. For no weights and equal n_i, ρ_A = roh, which is the intragroup correlation defined by Kish (1965). Two slightly different estimators are available through the `mean` and `median` options (Gleason 1997). If either of these options is specified, the estimate of ρ becomes

$$\rho = \frac{F_{\text{obs}} - F_m}{F_{\text{obs}} + (g-1)F_m}$$

For the `mean` option, $F_m = E(F_{k-1,N-K}) = (N-k)/(N-k-2)$, i.e., the expected value of the ANOVA table's F statistic. For the `median` option, F_m is simply the median of the F statistic. Setting F_m to 1 gives ρ_A, so for large samples, these different point estimators are essentially the same. Also, since the intraclass correlation of the random-effects model is by definition nonnegative, for any of the three possible point estimators, ρ is truncated to zero if F_{obs} is less than F_m.

For no weighting, interval estimators for ρ_A are computed. If the groups are equal sized (all n_i equal) and the `exact` option is specified, the following exact (assuming that the y_{ij} are normally distributed) $100(1-\alpha)\%$ confidence interval is computed,

$$\left\{ \frac{F_{\text{obs}} - F_m F_u}{F_{\text{obs}} + (g-1)F_m F_u}, \frac{F_{\text{obs}} - F_m F_l}{F_{\text{obs}} + (g-1)F_m F_l} \right\}$$

with $F_m = 1$, $F_l = F_{\alpha/2,k-1,N-k}$, and $F_u = F_{1-\alpha/2,k-1,N-k}$, $F_{\cdot,k-1,N-k}$ being the cumulative distribution function for the F distribution with $k-1$ and $N-k$ degrees of freedom. If `mean` or `median` is specified, F_m is defined as above. If the groups are equal sized and `exact` is not specified, the following asymptotic $100(1-\alpha)\%$ confidence interval for ρ_A is computed,

$$\left[\rho_A - z_{\alpha/2}\sqrt{V(\rho_A)}, \rho_A + z_{\alpha/2}\sqrt{V(\rho_A)} \right]$$

where $z_{\alpha/2}$ is the $100(1-\alpha/2)$ percentile of the standard normal distribution and $\sqrt{V(\rho_A)}$ is the asymptotic standard error of ρ defined below. This confidence interval is also available for unequal groups. It is not applicable and, therefore, not computed for the estimates of ρ provided by the `mean` and `median` options. Again since the intraclass coefficient is nonnegative, if the lower bound is negative for either confidence interval, it is truncated to zero. As might be expected, the coverage probability of a truncated interval is higher than its nominal value.

The asymptotic standard error of ρ_A, assuming that the y_{ij} are normally distributed, is also computed when appropriate, namely, for unweighted data and when ρ_A is computed (neither the `mean` option nor the `median` option is specified):

$$V(\rho_A) = \frac{2(1-\rho)^2}{g^2}(A+B+C)$$

with

$$A = \frac{\{1+\rho(g-1)\}^2}{N-k}$$

$$B = \frac{(1-\rho)\{1+\rho(2g-1)\}}{k-1}$$

$$C = \frac{\rho^2\{\sum n_i^2 - 2N^{-1}\sum n_i^3 + N^{-2}(\sum n_i^2)^2\}}{(k-1)^2}$$

and ρ_A is substituted for ρ (Donner 1986).

The estimated reliability of the group-averaged score, known as the Spearman–Brown prediction formula in the psychometric literature (Winer, Brown, and Michels 1991, 1014), is

$$\rho_t = \frac{t\rho}{1 + (t-1)\rho}$$

for group size t. `loneway` computes ρ_t for $t = g$.

The estimated standard deviation of the group effect is $\sigma_\alpha = \sqrt{(\mathrm{MS}_\alpha - \mathrm{MS}_\epsilon)/g}$. This deviation comes from the assumption that an observation is derived by adding a group effect to a within-group effect.

The estimated standard deviation within group is the square root of the mean square due to error, or $\sqrt{\mathrm{MS}_\epsilon}$.

Acknowledgment

We thank John Gleason of Syracuse University for his contributions to improving `loneway`.

References

Donner, A. 1986. A review of inference procedures for the intraclass correlation coefficient in the one-way random effects model. *International Statistical Review* 54: 67–82.

Gleason, J. R. 1997. sg65: Computing intraclass correlations and large ANOVAs. *Stata Technical Bulletin* 35: 25–31. Reprinted in *Stata Technical Bulletin Reprints*, vol. 6, pp. 167–176.

Kish, L. 1965. *Survey Sampling*. New York: Wiley.

Marchenko, Y. 2006. Estimating variance components in Stata. *Stata Journal* 6: 1–21.

Winer, B. J., D. R. Brown, and K. M. Michels. 1991. *Statistical Principles in Experimental Design*. 3rd ed. New York: McGraw–Hill.

Also See

[R] **anova** — Analysis of variance and covariance

[R] **oneway** — One-way analysis of variance

Title

lowess — Lowess smoothing

Syntax

`lowess` *yvar xvar* [*if*] [*in*] [, *options*]

options	description
Main	
`mean`	running-mean smooth; default is running-line least squares
`noweight`	suppress weighted regressions; default is tricube weighting function
`bwidth(#)`	use # for the bandwidth; default is `bwidth(0.8)`
`logit`	transform dependent variable to logits
`adjust`	adjust smoothed mean to equal mean of dependent variable
`nograph`	suppress graph
`generate(`*newvar*`)`	create *newvar* containing smoothed values of *yvar*
Plot	
marker_options	change look of markers (color, size, etc.)
marker_label_options	add marker labels; change look or position
Smoothed line	
`lineopts(`*cline_options*`)`	affect rendition of the smoothed line
Add plots	
`addplot(`*plot*`)`	add other plots to generated graph
Y axis, X axis, Titles, Legend, Overall, By	
twoway_options	any of the options documented in [G] ***twoway_options***

yvar and *xvar* may contain time-series operators; see [U] **11.4.3 Time-series varlists**.

Description

`lowess` carries out a locally weighted regression of *yvar* on *xvar*, displays the graph, and optionally saves the smoothed variable.

Warning: `lowess` is computationally intensive and may therefore take a long time to run on a slow computer. Lowess calculations on 1,000 observations, for instance, require performing 1,000 regressions.

Options

Main

`mean` specifies running-mean smoothing; the default is running-line least-squares smoothing.

`noweight` prevents the use of Cleveland's (1979) tricube weighting function; the default is to use this weighting function.

`bwidth(#)` specifies the bandwidth. Centered subsets of `bwidth()` $\times N$ observations are used for calculating smoothed values for each point in the data except for the end points, where smaller, uncentered subsets are used. The greater the `bwidth()`, the greater the smoothing. The default is 0.8.

`logit` transforms the smoothed *yvar* into logits. Predicted values less than .0001 or greater than .9999 are set to $1/N$ and $1 - 1/N$, respectively, before taking logits.

`adjust` adjusts the mean of the smoothed *yvar* to equal the mean of *yvar* by multiplying by an appropriate factor. This option is useful when smoothing binary (0/1) data.

`nograph` suppresses displaying the graph.

`generate(`*newvar*`)` creates *newvar* containing the smoothed values of *yvar*.

Plot

marker_options affect the rendition of markers drawn at the plotted points, including their shape, size, color, and outline; see [G] ***marker_options***.

marker_label_options specify if and how the markers are to be labeled; see [G] ***marker_label_options***.

Smoothed line

`lineopts(`*cline_options*`)` affects the rendition of the lowess-smoothed line; see [G] ***cline_options***.

Add plots

`addplot(`*plot*`)` provides a way to add other plots to the generated graph. See [G] ***addplot_option***.

Y axis, X axis, Titles, Legend, Overall, By

twoway_options are any of the options documented in [G] ***twoway_options***. These include options for titling the graph (see [G] ***title_options***), options for saving the graph to disk (see [G] ***saving_option***), and the `by()` option (see [G] ***by_option***).

Remarks

By default, `lowess` provides locally weighted scatterplot smoothing. The basic idea is to create a new variable (*newvar*) that, for each *yvar* y_i, contains the corresponding smoothed value. The smoothed values are obtained by running a regression of *yvar* on *xvar* by using only the data (x_i, y_i) and a few of the data near this point. In `lowess`, the regression is weighted so that the central point (x_i, y_i) gets the highest weight and points that are farther away (based on the distance $|x_j - x_i|$) receive less weight. The estimated regression line is then used to predict the smoothed value $\widehat{y}_i$ for y_i only. The procedure is repeated to obtain the remaining smoothed values, which means that a separate weighted regression is performed for every point in the data.

Lowess is a desirable smoother because of its locality—it tends to follow the data. Polynomial smoothing methods, for instance, are global in that what happens on the extreme left of a scatterplot can affect the fitted values on the extreme right.

▷ Example 1

The amount of smoothing is affected by `bwidth(#)`. You are warned to experiment with different values. For instance,

```
. use http://www.stata-press.com/data/r10/lowess1
(example data for lowess)
. lowess h1 depth
```

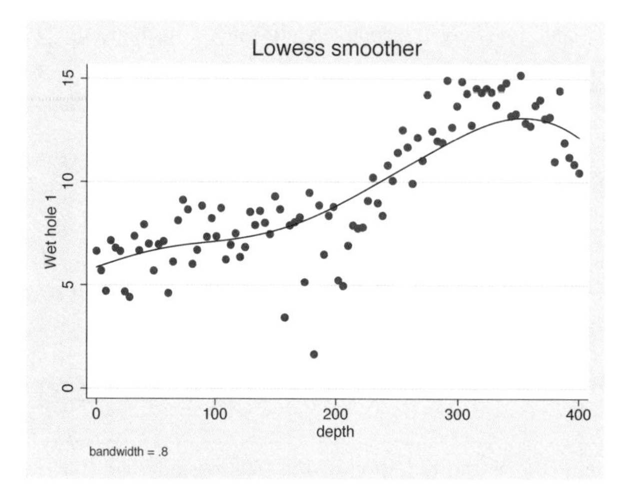

Now compare that with

```
. lowess h1 depth, bwidth(.4)
```

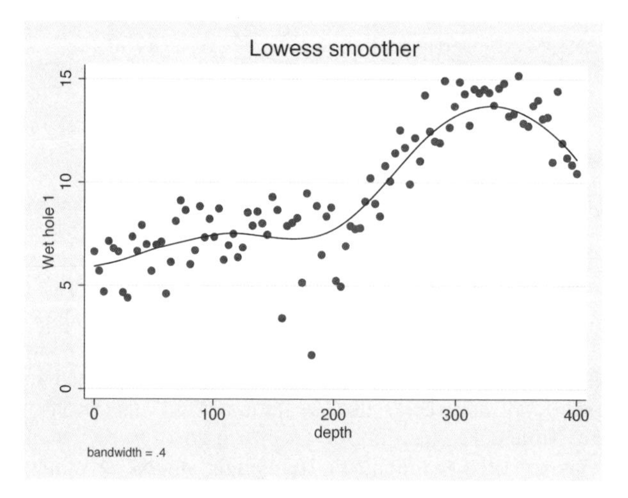

In the first case, the default bandwidth of 0.8 is used, meaning that 80% of the data are used in smoothing each point. In the second case, we explicitly specified a bandwidth of 0.4. Smaller bandwidths follow the original data more closely.

◁

▷ Example 2

Two `lowess` options are especially useful with binary (0/1) data: `adjust` and `logit`. `adjust` adjusts the resulting curve (by multiplication) so that the mean of the smoothed values is equal to the mean of the unsmoothed values. `logit` specifies that the smoothed curve be in terms of the log of the odds ratio:

```
. use http://www.stata-press.com/data/r10/auto
(1978 Automobile Data)
. lowess foreign mpg, ylabel(0 "Domestic" 1 "Foreign") jitter(5) adjust
```

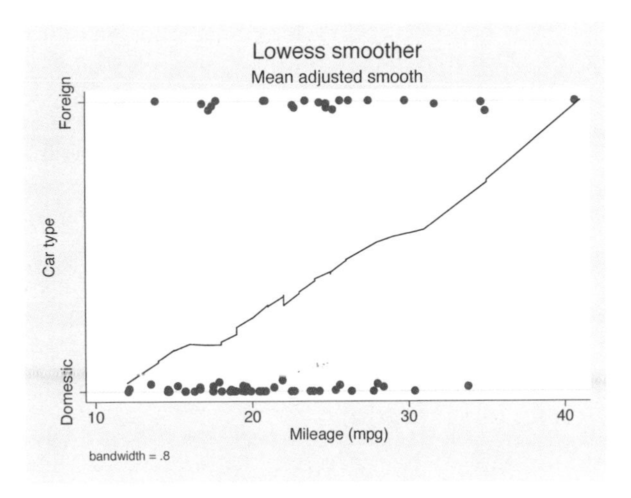


```
. lowess foreign mpg, logit yline(0)
```

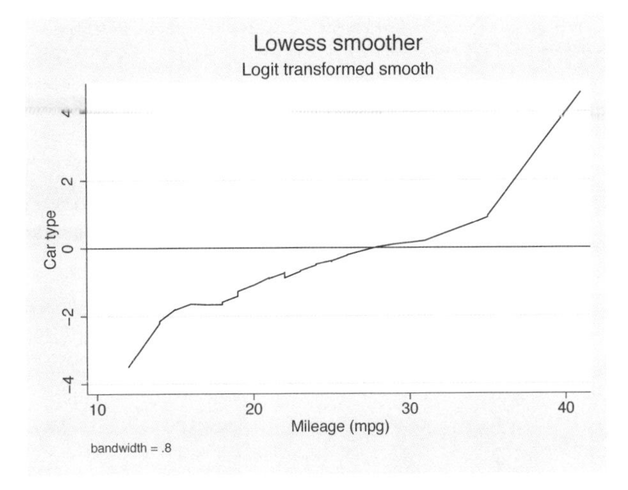

With binary data, if you do not use the `logit` option, it is a good idea to specify `graph`'s `jitter()` option; see [G] **graph twoway scatter**. Since the underlying data (whether the car was manufactured outside the United States in this case) take on only two values, raw data points are more likely to be on top of each other, thus making it impossible to tell how many points there are. `graph`'s `jitter()` option adds some noise to the data to shift the points around. This noise affects only the location of points on the graph, not the lowess curve.

When you specify the `logit` option, the display of the raw data is suppressed.

◁

❑ Technical Note

`lowess` can be used for more than just lowess smoothing. Lowess can be usefully thought of as a combination of two smoothing concepts: the use of predicted values from regression (rather than means) for imputing a smoothed value and the use of the tricube weighting function (rather than a constant weighting function). `lowess` allows you to combine these concepts freely. You can use line smoothing without weighting (specify `noweight`), mean smoothing with tricube weighting (specify `mean`), or mean smoothing without weighting (specify `mean` and `noweight`).

❑

Methods and Formulas

`lowess` is implemented as an ado-file.

Let y_i and x_i be the two variables, and assume that the data are ordered so that $x_i \leq x_{i+1}$ for $i = 1, \dots, N-1$. For each y_i, a smoothed value y_i^s is calculated.

The subset used in calculating y_i^s is indices $i_- = \max(1, i-k)$ through $i_+ = \min(i+k, N)$, where $k = \lfloor (N \cdot \texttt{bwidth} - 0.5)/2 \rfloor$. The weights for each of the observations between $j = i_-, \dots, i_+$ are either 1 (`noweight`) or the tricube (default),

$$w_j = \left\{ 1 - \left(\frac{|x_j - x_i|}{\Delta} \right)^3 \right\}^3$$

where $\Delta = 1.0001 \max(x_{i_+} - x_i, x_i - x_{i_-})$. The smoothed value y_i^s is then the (weighted) mean or the (weighted) regression prediction at x_i.

William S. Cleveland (1943–) studied mathematics and statistics at Princeton and Yale. He worked for several years at Bell Labs in New Jersey and now teaches statistics and computer science at Purdue. He has made key contributions in many areas of statistics, including graphics and data visualization, time series, environmental applications, and analysis of Internet traffic data.

Acknowledgment

`lowess` is a modified version of a command originally written by Patrick Royston of the MRC Clinical Trials Unit, London.

References

Chambers, J. M., W. S. Cleveland, B. Kleiner, and P. A. Tukey. 1983. *Graphical Methods for Data Analysis*. Belmont, CA: Wadsworth.

Cleveland, W. S. 1979. Robust locally weighted regression and smoothing scatterplots. *Journal of the American Statistical Association* 74: 829–836.

——. 1993. *Visualizing Data.* Summit, NJ: Hobart.

——. 1994. *The Elements of Graphing Data.* 2nd ed. Summit, NJ: Hobart.

Cox, N. J. 2005. Speaking Stata: Smoothing in various directions. *Stata Journal* 5: 574–593.

Goodall, C. 1990. A survey of smoothing techniques. In *Modern Methods of Data Analysis*, ed. J. Fox and J. S. Long, 126–176. Newbury Park, CA: Sage.

Royston, J. P. 1991. gr6: Lowess smoothing. *Stata Technical Bulletin* 3: 7–9. Reprinted in *Stata Technical Bulletin Reprints*, vol. 1, pp. 41–44.

Royston, P., and N. J. Cox. 2005. A multivariable scatterplot smoother. *Stata Journal* 5: 405–412.

Salgado-Ugarte, I. H., and M. Shimizu. 1995. snp8: Robust scatterplot smoothing: Enhancements to Stata's ksm. *Stata Technical Bulletin* 25: 23–26. Reprinted in *Stata Technical Bulletin Reprints*, vol. 5, pp. 190–194.

Sasieni, P. 1994. snp7: Natural cubic splines. *Stata Technical Bulletin* 22: 19–22. Reprinted in *Stata Technical Bulletin Reprints*, vol. 4, pp. 171–174.

Also See

[D] **ipolate** — Linearly interpolate (extrapolate) values

[R] **smooth** — Robust nonlinear smoother

[R] **lpoly** — Kernel-weighted local polynomial smoothing

Title

lpoly — Kernel-weighted local polynomial smoothing

Syntax

`lpoly` *yvar xvar* [*if*] [*in*] [*weight*] [, *options*]

options	description
Main	
`kernel(`*kernel*`)`	specify kernel function; default is `kernel(epanechnikov)`
`bwidth(`# | *varname*`)`	specify kernel bandwidth
`degree(`#`)`	specify degree of the polynomial smooth; default is `degree(0)`
`generate(`[*newvar*$_x$] *newvar*$_s$`)`	store smoothing grid in *newvar*$_x$ and smoothed points in *newvar*$_s$
`n(`#`)`	obtain the smooth at # points; default is min(N, 50)
`at(`*varname*`)`	obtain the smooth at the values specified by *varname*
`nograph`	suppress graph
`noscatter`	suppress scatterplot only
SE/CI	
`ci`	plot confidence bands
`level(`#`)`	set confidence level; default is `level(95)`
`se(`*newvar*`)`	store standard errors in *newvar*
`pwidth(`#`)`	specify pilot bandwidth for standard error calculation
`var(`# | *varname*`)`	specify estimates of residual variance
Scatterplot	
marker_options	change look of markers (color, size, etc.)
marker_label_options	add marker labels; change look or position
Smoothed line	
`lineopts(`*cline_options*`)`	affect rendition of the smoothed line
CI plot	
`ciopts(`*cline_options*`)`	affect rendition of the confidence bands
Add plots	
`addplot(`*plot*`)`	add other plots to the generated graph
Y axis, X axis, Titles, Legend, Overall	
twoway_options	any options other than `by()` documented in [G] ***twoway_options***

kernel	description
epanechnikov	Epanechnikov kernel function; the default
epan2	alternative Epanechnikov kernel function
biweight	biweight kernel function
cosine	cosine trace kernel function
gaussian	Gaussian kernel function
parzen	Parzen kernel function
rectangle	rectangle kernel function
triangle	triangle kernel function

fweights and aweights are allowed; see [U] **11.1.6 weight**.

Description

lpoly performs a kernel-weighted local polynomial regression of *yvar* on *xvar* and displays a graph of the smoothed values with (optional) confidence bands.

Options

Main

kernel(*kernel*) specifies the kernel function for use in calculating the weighted local polynomial estimate. The default is kernel(epanechnikov).

bwidth(# | *varname*) specifies the half-width of the kernel—the width of the smoothing window around each point. If bwidth() is not specified, a rule-of-thumb (ROT) bandwidth estimator is calculated and used. A local variable bandwidth may be specified in *varname*, in conjunction with an explicit smoothing grid using option at().

degree(#) specifies the degree of the polynomial to be used in the smoothing. The default is degree(0), meaning local-mean smoothing.

generate([*newvar*$_x$] *newvar*$_s$) stores the smoothing grid in *newvar*$_x$ and the smoothed values in *newvar*$_s$. If at() is not specified, then both *newvar*$_x$ and *newvar*$_s$ must be specified. Otherwise, only *newvar*$_s$ is to be specified.

n(#) specifies the number of points at which the smooth is to be calculated. The default is min(N, 50), where N is the number of observations.

at(*varname*) specifies a variable that contains the values at which the smooth should be calculated. By default, the smoothing is done on an equally spaced grid, but you can use at() to instead perform the smoothing at the observed *x*'s, for example. This option also allows you to more easily obtain smooths for different variables or different subsamples of a variable and then overlay the estimates for comparison.

nograph suppresses drawing the graph of the estimated smooth. This option is often used with the generate() option.

noscatter suppresses superimposing a scatterplot of the observed data over the smooth. This option is useful when the number of resulting points would be so large as to clutter the graph.

SE/CI

`ci` plots confidence bands, using the confidence level specified in `level()`.

`level(#)` specifies the confidence level, as a percentage, for confidence intervals. The default is `level(95)` or as set by `set level`; see [U] **20.7 Specifying the width of confidence intervals**.

`se(`*newvar*`)` stores the estimates of the standard errors in *newvar*. This option requires specifying `generate()` and/or `at()`.

`pwidth(#)` specifies the pilot bandwidth to be used for standard error computations. The default is chosen to be 1.5 times the value of the ROT bandwidth selector. If you specify `pwidth()` without specifying `se()` or `ci`, then option `ci` is assumed.

`var(`# | *varname*`)` specifies an estimate of a constant residual variance or a variable containing estimates of the residual variances at each grid point required for standard error computation. By default, the residual variance at each smoothing point is estimated by the normalized weighted residual sum of squares obtained from locally fitting a polynomial of order $p + 2$, where p is the degree specified in `degree()`. `var(`*varname*`)` is allowed only if `at()` is specified. If you specify `var()` without specifying `se()` or `ci`, then option `ci` is assumed.

Scatterplot

marker_options affect the rendition of markers drawn at the plotted points, including their shape, size, color, and outline; see [G] ***marker_options***.

marker_label_options specify if and how the markers are to be labeled; see [G] ***marker_label_options***.

Smoothed line

`lineopts(`*cline_options*`)` affects the rendition of the smoothed line; see [G] ***cline_options***.

CI plot

`ciopts(`*cline_options*`)` affects the rendition of the confidence bands; see [G] ***cline_options***.

Add plots

`addplot(`*plot*`)` provides a way to add other plots to the generated graph; see [G] ***addplot_option***.

Y axis, X axis, Titles, Legend, Overall

twoway_options are any of the options documented in [G] ***twoway_options***, excluding `by()`. These include options for titling the graph (see [G] ***title_options***) and for saving the graph to disk (see [G] ***saving_option***).

Remarks

Remarks are presented under the following headings:

Introduction
Local polynomial smoothing
Choice of a bandwidth
Confidence bands

Introduction

The last 25 years or so has seen a significant outgrowth in the literature on scatterplot smoothing, otherwise known as univariate nonparametric regression. Of most appeal is the idea of making no assumptions about the functional form for the expected value of a response given a regressor, but instead allowing the data to "speak for themselves." Various methods and estimators fall into the category of nonparametric regression, including local mean smoothing as described independently by Nadaraya (1964) and Watson (1964), the Gasser–Müller (1979) estimator, locally weighted scatterplot smoothing (LOWESS) as described by Cleveland (1979), wavelets (e.g., Donoho 1995), and splines (Eubank 1988), to name a few. Much of the vast literature focuses on automating the amount of smoothing to be performed and dealing with the bias/variance tradeoff inherent to this type of estimation. For example, for Nadaraya–Watson the amount of smoothing is controlled by choosing a *bandwidth*.

Smoothing via local polynomials is by no means a new idea but instead one that has been rediscovered in recent years in articles such as Fan (1992). A natural extension of the local mean smoothing of Nadaraya–Watson, local polynomial regression involves fitting the response to a polynomial form of the regressor via locally weighted least squares. Higher-order polynomials have better bias properties than the zero-degree local polynomials of the Nadaraya–Watson estimator; in general, higher-order polynomials do not require bias adjustment at the boundary of the regression space. For a definitive reference on local polynomial smoothing, see Fan and Gijbels (1996).

Local polynomial smoothing

Consider a set of scatterplot data $\{(x_1, y_1), \dots, (x_n, y_n)\}$ from the model

$$y_i = m(x_i) + \sigma(x_i)\epsilon_i \qquad (1)$$

for some unknown mean and variance functions $m(\cdot)$ and $\sigma^2(\cdot)$, and symmetric errors ϵ_i with $E(\epsilon_i) = 0$ and $\text{Var}(\epsilon_i) = 1$. The goal is to estimate $m(x_0) = E[Y|X = x_0]$, making no assumption about the functional form of $m(\cdot)$.

`lpoly` estimates $m(x_0)$ as the constant term (intercept) of a regression, weighted by the kernel function specified in `kernel()`, of *yvar* on the polynomial terms $(xvar - x_0), (xvar - x_0)^2, \dots, (xvar - x_0)^p$ for each smoothing point x_0. The degree of the polynomial, p, is specified in `degree()`, the amount of smoothing is controlled by the bandwidth specified in `bwidth()`, and the chosen kernel function is specified in `kernel()`.

▷ Example 1

Consider the motorcycle data as examined (among other places) in Fan and Gijbels (1996). The data consist of 133 observations and measure the acceleration (`accel` measured in grams [g]) of a dummy's head during impact over time (`time` measured in milliseconds). For these data, we use `lpoly` to fit a local cubic polynomial with the default bandwidth (obtained using the ROT method) and the default Epanechnikov kernel.

(Continued on next page)

```
. use http://www.stata-press.com/data/r10/motorcycle
(Motorcycle data from Fan & Gijbels (1996))
. lpoly accel time, degree(3)
```

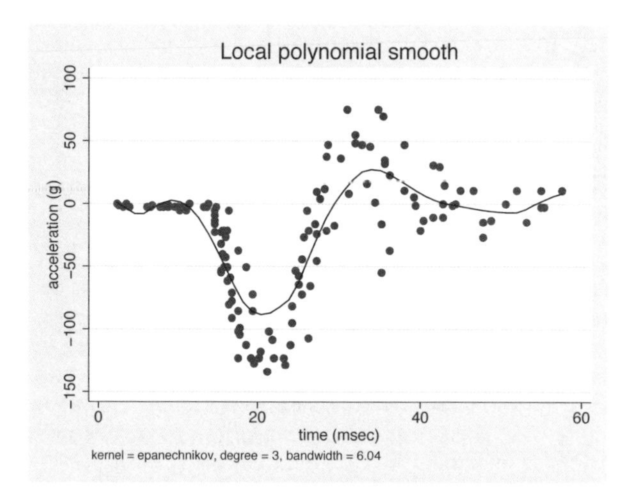

The default bandwidth and kernel settings do not provide a satisfactory fit in this example. To improve the fit we can either supply a different bandwidth by using option `bwidth()` or specify a different kernel by using option `kernel()`. For example, using the alternative Epanechnikov kernel, `kernel(epan2)`, below provides a better fit for these data.

```
. lpoly accel time, degree(3) kernel(epan2)
```

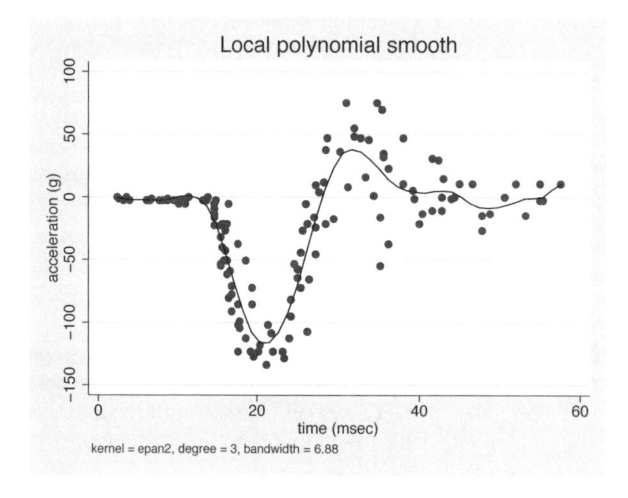

◁

❑ Technical Note

`lpoly` allows specifying in `degree()` both odd and even orders of the polynomial to be used for the smoothing. However, the odd-order, $2k+1$, polynomial approximations are preferable. They have an extra parameter compared with the even-order, $2k$, approximations, which leads to a significant bias reduction and there is no increase of variability associated with adding this extra parameter. Using an odd order when estimating the regression function is therefore usually sufficient. For a more thorough discussion, see Fan and Gijbels (1996).

❑

Choice of a bandwidth

The choice of a bandwidth is crucial for many smoothing techniques, including local polynomial smoothing. In general, using a large bandwidth gives smooths with a large bias, whereas a small bandwidth may result in highly variable smoothed values. Various techniques exist for optimal bandwidth selection. By default, `lpoly` uses the ROT method to estimate the bandwidth used for the smoothing; see *Methods and Formulas* for details.

▷ Example 2

Using the `motorcycle` data, we demonstrate how a local linear polynomial fit changes using different bandwidths.

```
. lpoly accel time, degree(1) kernel(epan2) bwidth(1) generate(at smooth1)
> nograph
. lpoly accel time, degree(1) kernel(epan2) bwidth(7) at(at) generate(smooth2)
> nograph
. label variable smooth1 "smooth: width = 1"
. label variable smooth2 "smooth: width = 7"
. lpoly accel time, degree(1) kernel(epan2) at(at) addplot(line smooth* at)
> legend(label(2 "smooth: width = 3.42 (ROT)")) note("kernel = epan2, degree = 1")
```

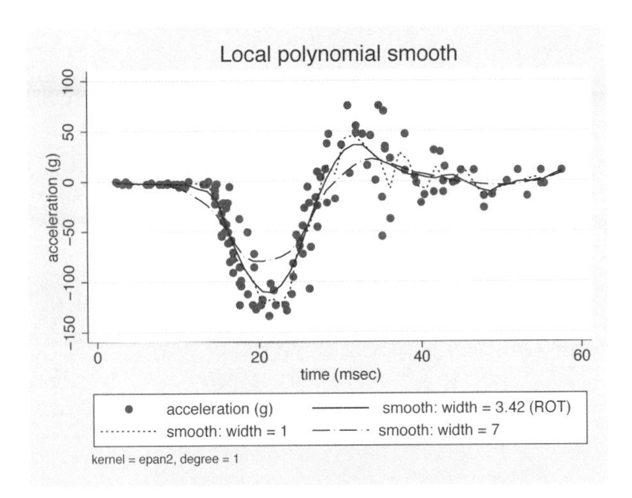

From this graph, we can see that the local linear polynomial fit with larger bandwidth (`width = 7`) corresponds to a smoother line but fails to fit the curvature of the scatterplot data. The smooth obtained using the width equal to one seems to fit most data points, but the corresponding line has several spikes indicating larger variability. The smooth obtained using the ROT bandwidth estimator seems to have a good tradeoff between the fit and variability in this example.

In the above, we also demonstrated how the `generate()` and `addplot()` options may be used to produce overlaid plots obtained from `lpoly` with different options. The `nograph` option saves time when you need to save only results with `generate()`.

However, to avoid generating variables manually, one can use `twoway lpoly` instead; see [G] **graph twoway lpoly** for more details.

```
. twoway scatter accel time ||
>         lpoly accel time, degree(1) kernel(epan2) lpattern(solid) ||
>         lpoly accel time, degree(1) kernel(epan2) bwidth(1)       ||
>         lpoly accel time, degree(1) kernel(epan2) bwidth(7)       ||
>    , legend(label(2 "smooth: width = 3.42 (ROT)") label(3 "smooth: width = 1")
>            label(4 "smooth: width = 7"))
>      title("Local polynomial smooth") note("kernel = epan2, degree = 1")
>      xtitle("time (msec)") ytitle("acceleration (g)")
```

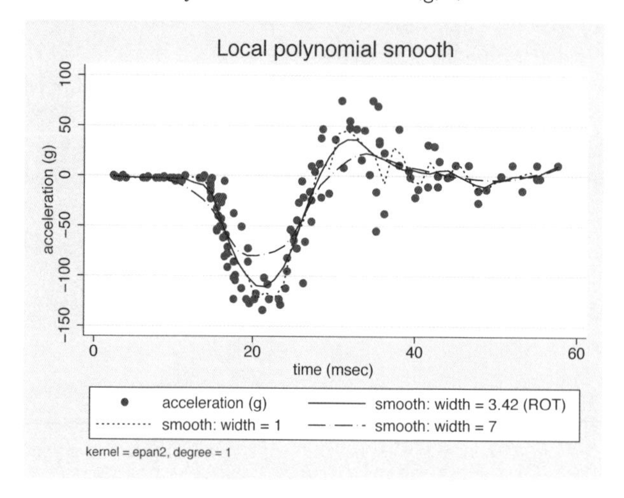

◁

The ROT estimate is commonly used as an initial guess for the amount of smoothing; this approach may be sufficient when the choice of a bandwidth is less important. In other cases, you can pick your own bandwidth.

When the shape of the regression function has a combination of peaked and flat regions, a variable bandwidth may be preferable over the constant bandwidth to allow for different degrees of smoothness in different regions. The `bwidth()` option allows you to specify the values of the local variable bandwidths as those stored in a variable in your data.

Similar issues with bias and variability arise when choosing a pilot bandwidth (option `pwidth()`) used to compute standard errors of the local polynomial smoother. The default value is chosen to be $1.5 \times$ ROT. For a review of methods for pilot bandwidth selection, see Fan and Gijbels (1996).

Confidence bands

The established asymptotic normality of the local polynomial estimators under certain conditions allows the construction of approximate confidence bands. `lpoly` offers the `ci` option to plot these bands.

▷ Example 3

Let us plot the confidence bands for the local polynomial fit from example 1.

```
. lpoly accel time, degree(3) kernel(epan2) ci
```

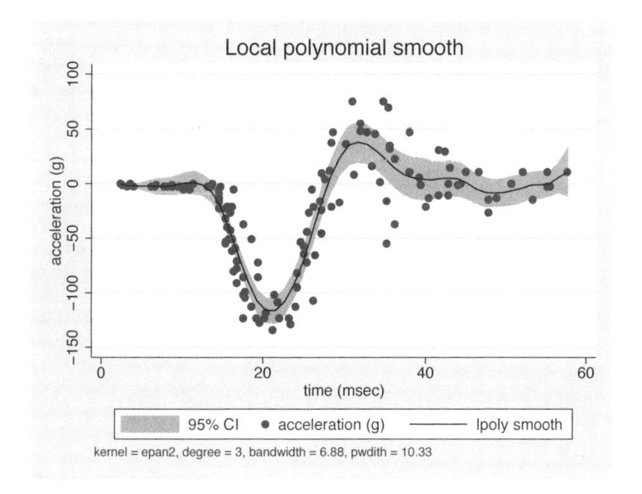

You can obtain graphs with overlaid confidence bands by using `twoway lpolyci`; see [G] **graph twoway lpolyci** for examples.

◁

Constructing the confidence intervals involves computing standard errors obtained by taking a square-root of the estimate of the conditional variance of the local polynomial estimator at each grid point x_0. Estimating the conditional variance requires fitting a polynomial of a higher order locally by using a different bandwidth, the pilot bandwidth. The value of the pilot bandwidth may be supplied by using `pwidth()`. By default, the value of $1.5 \times$ ROT is used. Also, estimates of the residual variance $\sigma^2(x_0)$ at each grid point x_0 are required to obtain the estimates of the conditional variances. These estimates may be supplied by using option `var()`. By default, they are computed using the normalized weighted residual sum of squares from a local polynomial fit of a higher order. See *Methods and Formulas* for details. The standard errors may be saved by using `se()`.

Saved Results

`lpoly` saves the following in `r()`:

Scalars

`r(degree)`	smoothing polynomial degree	`r(bwidth)`	bandwidth of the smooth
`r(ngrid)`	number of successful regressions	`r(pwidth)`	pilot bandwidth
`r(N)`	sample size		

Macros

`r(kernel)`	name of kernel

Methods and Formulas

`lpoly` is implemented as an ado-file.

Consider model (1), written in matrix notation,

$$\mathbf{y} = m(\mathbf{x}) + \epsilon$$

where $\mathbf{y}$ and $\mathbf{x}$ are the $n \times 1$ vectors of scatterplot values, ϵ is the $n \times 1$ vector of errors with zero mean and covariance matrix $\mathbf{\Sigma} = \text{diag}\{\sigma(x_i)\}\mathbf{I}_n$, and $m()$ and $\sigma()$ are some unknown functions. Define $m(x_0) = E[Y|X = x_0]$ and $\sigma^2(x_0) = \text{Var}[Y|X = x_0]$ to be the conditional mean and conditional variance of random variable Y (residual variance), respectively, for some realization x_0 of random variable X.

The method of local polynomial smoothing is based on the approximation of $m(x)$ locally by a pth order polynomial in $(x - x_0)$ for some x in the neighborhood of x_0. For the scatterplot data $\{(x_1, y_1), \ldots, (x_n, y_n)\}$ the pth-order local polynomial smooth $\widehat{m}(x_0)$ is equal to $\widehat{\beta}_0$, an estimate of the intercept of the weighted linear regression,

$$\widehat{\beta} = (\mathbf{X}^T\mathbf{W}\mathbf{X})^{-1}\mathbf{X}^T\mathbf{W}\mathbf{y} \tag{2}$$

where $\widehat{\beta} = (\widehat{\beta}_0, \widehat{\beta}_1, \ldots, \widehat{\beta}_p)^T$ is the vector of estimated regression coefficients (with $\{\widehat{\beta}_j = (j!)^{-1}\widehat{m}^{(j)}(x)|_{x=x_0},\ j = 0, \ldots, p\}$ also representing estimated coefficients from a corresponding Taylor expansion); $\mathbf{X} = \{(x_i - x_0)^j\}_{i,j=1,0}^{n,p}$ is a design matrix; and $\mathbf{W} = \text{diag}\{K_h(x_i - x_0)\}_{n \times n}$ is a weighting matrix with weights $K_h(\cdot)$ defined as $K_h(x) = h^{-1}K(x/h)$, with $K(\cdot)$ being a kernel function and h defining a bandwidth. The kernels are defined in *Methods and Formulas* in [R] **kdensity**.

The default bandwidth is obtained using the ROT method of bandwidth selection. The ROT bandwidth is the plugin estimator of the asymptotically optimal constant bandwidth. This is the bandwidth that minimizes the conditional weighted mean integrated squared error. The ROT plugin bandwidth selector for the smoothing bandwidth h is defined as follows; assuming constant residual variance $\sigma^2(x_0) = \sigma^2$ and odd degree p:

$$\widehat{h} = C_{0,p}(K)\left[\frac{\widehat{\sigma}^2 \int w_0(x)dx}{n\int\{\widehat{m}^{(p+1)}(x)\}^2 w_0(x)f(x)dx}\right]^{1/(2p+3)} \tag{3}$$

where $C_{0,p}(K)$ is a constant, as defined in Fan and Gijbels (1996), that depends on the kernel function $K(\cdot)$, and the degree of a polynomial p and w_0 is chosen to be an indicator function on the interval $[\text{min}_\mathbf{x} + 0.05 \times \text{range}_\mathbf{x}, \text{max}_\mathbf{x} - 0.05 \times \text{range}_\mathbf{x}]$ with $\text{min}_\mathbf{x}$, $\text{max}_\mathbf{x}$, and $\text{range}_\mathbf{x}$ being, respectively, the minimum, maximum, and the range of $\mathbf{x}$. To obtain the estimates of a constant residual variance, $\widehat{\sigma}^2$, and $(p+1)$st order derivative of $m(x)$, denoted as $\widehat{m}^{(p+1)}(x)$, a polynomial in $\mathbf{x}$ of order $(p+3)$ is fitted globally to $\mathbf{y}$. $\widehat{\sigma}^2$ is estimated as a standardized residual sum of squares from this fit.

The expression for the asymptotically optimal constant bandwidth used in constructing the ROT bandwidth estimator is derived for the odd-order polynomial approximations. For even-order polynomial fits the expression would depend not only on $m^{(p+1)}(x)$ but also on $m^{(p+2)}(x)$ and the design density and its derivative, $f(x)$ and $f'(x)$. Therefore, the ROT bandwidth selector would require estimation of these additional quantities. Instead, for an even-degree p of the local polynomial, `lpoly` uses the value of the ROT estimator (3) computed using degree $p + 1$. As such, for even degrees this is not a plugin estimator of the asymptotically optimal constant bandwidth.

The estimates of the conditional variance of local polynomial estimators are obtained using

$$\widehat{\mathrm{Var}}\{\widehat{m}(x_0)|X = x_0\} = \widehat{\sigma}^2_m(x_0) = (\mathbf{X}^T\mathbf{W}\mathbf{X})^{-1}(\mathbf{X}^T\mathbf{W}^2\mathbf{X})(\mathbf{X}^T\mathbf{W}\mathbf{X})^{-1}\widehat{\sigma}^2(x_0) \qquad (4)$$

where $\widehat{\sigma}^2(x_0)$ is estimated by the normalized weighted residual sum of squares from the $(p+2)$nd order polynomial fitted using pilot bandwidth $h^\star$.

When the bias is negligible the normal-approximation method yields a $(1-\alpha)\times 100\%$ confidence interval for $m(x_0)$,

$$\left\{\widehat{m}(x_0) - z_{(1-\alpha/2)}\widehat{\sigma}_m(x_0),\ \widehat{m}(x_0) + z_{(1-\alpha/2)}\widehat{\sigma}_m(x_0)\right\}$$

where $z_{(1-\alpha/2)}$ is the $(1-\alpha/2)$th quantile of the standard Gaussian distribution, and $\widehat{m}(x_0)$ and $\widehat{\sigma}_m(x_0)$ are as defined in (2) and (4), respectively.

References

Cleveland, W. S. 1979. Robust locally weighted regression and smoothing scatterplots. *Journal of the American Statistical Association* 74: 829–836.

Cox, N. J. 2005. Speaking Stata: Smoothing in various directions. *Stata Journal* 5: 574–593.

Donoho, D. L. 1995. Nonlinear solution of linear inverse problems by wavelet–vaguelette decomposition. *Applied and Computational Harmonic Analysis* 2: 101–126.

Eubank, R. L. 1988. *Spline Smoothing and Nonparametric Regression.* New York: Dekker.

Fan, J. 1992. Design-adaptive nonparametric regression. *Journal of the American Statistical Association* 87: 998–1004.

Fan, J., and I. Gijbels. 1996. *Local Polynomial Modelling and Its Applications.* London: Chapman & Hall.

Gasser, T., and H.-G. Müller. 1979. Kernel estimation of regression functions. In *Smoothing Techniques for Curve Estimation*, Lecture Notes in Mathematics, vol. 757, 23–68. New York: Springer.

Gutierrez, R. G., J. M. Linhart, and J. S. Pitblado. 2003. From the help desk: Local polynomial regression and Stata plugins. *Stata Journal* 3: 412–419.

Nadaraya, E. A. 1964. On estimating regression. *Theory of Probability and Its Application* 9: 141–142.

Sheather, S. J., and M. C. Jones. 1991. A reliable data-based bandwidth selection method for kernel density estimation. *Journal of the Royal Statistical Society, Series B* 53: 683–690.

Watson, G. S. 1964. Smooth regression analysis. *Sankhyā Series A* 26: 359–372.

Also See

[R] **kdensity** — Univariate kernel density estimation

[R] **lowess** — Lowess smoothing

[R] **regress** — Linear regression

[G] **graph twoway lpoly** — Local polynomial smooth plots

[G] **graph twoway lpolyci** — Local polynomial smooth plots with CIs

Title

lrtest — Likelihood-ratio test after estimation

Syntax

`lrtest` *modelspec*$_1$ [*modelspec*$_2$] [, *options*]

where *modelspec* is

name | `.` | (*namelist*)

where *name* is the name under which estimation results were saved using `estimates store`, and "`.`" refers to the last estimation results, whether or not these were already stored.

options	description
`stats`	display statistical information about the two models
`dir`	display descriptive information about the two models
`df(#)`	override the automatic degrees-of-freedom calculation; seldom used
`force`	force testing even when apparently invalid

Description

`lrtest` performs a likelihood-ratio test for the null hypothesis that the parameter vector of a statistical model satisfies some smooth constraint. To conduct the test, both the unrestricted and the restricted models must be fitted using the maximum likelihood method (or some equivalent method), and the results of at least one must be stored using `estimates store`.

modelspec$_1$ and *modelspec*$_2$ specify the restricted and unrestricted model in any order. *modelspec*$_1$ and *modelspec*$_2$ cannot have names in common; for example, `lrtest (A B C) (C D E)` is not allowed since both model specifications include `C`. If *modelspec*$_2$ is not specified, the last estimation result is used; this is equivalent to specifying *modelspec*$_2$ as a period (`.`).

`lrtest` supports composite models specified by a parenthesized list of model names. In a composite model, we assume that the log likelihood and dimension (number of free parameters) of the full model are obtained as the sum of the log-likelihood values and dimensions of the constituting models.

`lrtest` provides an important alternative to `test` for models fitted via maximum likelihood or equivalent methods.

Options

`stats` displays statistical information about the unrestricted and restricted models, including the information indices of Akaike and Schwarz.

`dir` displays descriptive information about the unrestricted and restricted models; see `estimates dir` in [R] **estimates store**.

`df(#)` is seldom specified; it overrides the automatic degrees-of-freedom calculation.

`force` forces the likelihood-ratio test calculations to take place in situations where `lrtest` would normally refuse to do so and issue an error. Such situations arise when one or more assumptions of the test are violated, for example, if the models were fitted with `vce(robust)`, `vce(cluster` *clustvar*`)`, or `pweights`; the dependent variables in the two models differ; the null log likelihoods differ; the samples differ; or the estimation commands differ. If you use the `force` option, there is no guarantee as to the validity or interpretability of the resulting test.

Remarks

The standard way to use `lrtest` is to do the following:

1. Fit either the restricted model or the unrestricted model by using one of Stata's estimation commands and then store the results using `estimates store` *name*.
2. Fit the alternative model (the unrestricted or restricted model) and then type '`lrtest` *name* `.`'. `lrtest` determines for itself which of the two models is the restricted model by comparing the degrees of freedom.

Often you may want to store the alternative model with `estimates store` $name_2$, for instance, if you plan additional tests against models yet to be fitted. The likelihood-ratio test is then obtained as `lrtest` *name* $name_2$.

Remarks are presented under the following headings:

Nested models
Composite models

Nested models

`lrtest` may be used with any estimation command that reports a log likelihood, including `logit`, `poisson`, `streg`, `heckman`, and `stcox`. You must check that one of the model specifications implies a statistical model that is *nested within* the model implied by the other specification. Usually this means that both models are fitted with the same estimation command (e.g., both are fitted by `logit`, with the same dependent variables) and that the set of covariates of one model is a subset of the covariates of the other model. Second, `lrtest` is valid only for models that are fitted by maximum likelihood or by some equivalent method, so it does not apply to models that were fitted with probability weights or clusters. Specifying the `vce(robust)` option similarly would indicate that you are worried about the valid specification of the model, so you would not use `lrtest`. Third, `lrtest` assumes that under the null hypothesis, the test statistic is (approximately) distributed as chi-squared. This assumption is not true for likelihood-ratio tests of "boundary conditions", such as tests for the presence of overdispersion or random effects (Gutierrez, Carter, and Drukker 2001).

▷ Example 1

We have data on infants born with low birthweights along with the characteristics of the mother (Hosmer and Lemeshow 2000; see also [R] **logistic**). We fit the following model:

(*Continued on next page*)

```
. use http://www.stata-press.com/data/r10/lbw2
(Hosmer & Lemeshow data)

. logistic low age lwt race2 race3 smoke ptl ht ui

Logistic regression                               Number of obs   =        189
                                                  LR chi2(8)      =      33.22
                                                  Prob > chi2     =     0.0001
Log likelihood =   -100.724                       Pseudo R2       =     0.1416

------------------------------------------------------------------------------
         low | Odds Ratio   Std. Err.      z    P>|z|     [95% Conf. Interval]
-------------+----------------------------------------------------------------
         age |   .9732636   .0354759    -0.74   0.457     .9061578    1.045339
         lwt |   .9849634   .0068217    -2.19   0.029     .9716834    .9984249
       race2 |   3.534767   1.860737     2.40   0.016     1.259736    9.918406
       race3 |   2.368079   1.039949     1.96   0.050     1.001356    5.600207
       smoke |   2.517698    1.00916     2.30   0.021     1.147676    5.523162
         ptl |   1.719161   .5952579     1.56   0.118     .8721455    3.388787
          ht |   6.249602   4.322408     2.65   0.008     1.611152    24.24199
          ui |     2.1351   .9808153     1.65   0.099     .8677528      5.2534
------------------------------------------------------------------------------
```

We now wish to test the constraint that the coefficients on **age**, **lwt**, **ptl**, and **ht** are all zero or, equivalently in this case, that the odds ratios are all 1. One solution is to type

```
. test age lwt ptl ht

 ( 1)  age = 0
 ( 2)  lwt = 0
 ( 3)  ptl = 0
 ( 4)  ht = 0

           chi2(  4) =   12.38
         Prob > chi2 =    0.0147
```

This test is based on the inverse of the information matrix and is therefore based on a quadratic approximation to the likelihood function; see [R] **test**. A more precise test would be to refit the model, applying the proposed constraints, and then calculate the likelihood-ratio test.

We first save the current model:

```
. estimates store full
```

We then fit the constrained model, which here is the model omitting **age**, **lwt**, **ptl**, and **ht**:

```
. logistic low race2 race3 smoke ui

Logistic regression                               Number of obs   =        189
                                                  LR chi2(4)      =      18.80
                                                  Prob > chi2     =     0.0009
Log likelihood = -107.93404                       Pseudo R2       =     0.0801

------------------------------------------------------------------------------
         low | Odds Ratio   Std. Err.      z    P>|z|     [95% Conf. Interval]
-------------+----------------------------------------------------------------
       race2 |   3.052746   1.498084     2.27   0.023     1.166749    7.987368
       race3 |   2.922593   1.189226     2.64   0.008      1.31646    6.488269
       smoke |   2.945742   1.101835     2.89   0.004      1.41517    6.131701
          ui |   2.419131   1.047358     2.04   0.041      1.03546    5.651783
------------------------------------------------------------------------------
```

That done, **lrtest** compares this model with the model we previously saved:

```
. lrtest full .

Likelihood-ratio test                                 LR chi2(4)  =     14.42
(Assumption: . nested in full)                        Prob > chi2 =    0.0061
```

Let's compare results. `test` reported that `age`, `lwt`, `ptl`, and `ht` were jointly significant at the 1.5% level; `lrtest` reports that they are significant at the 0.6% level. Given the quadratic approximation made by `test`, we could argue that `lrtest`'s results are more accurate.

`lrtest` explicates the assumption that, from a comparison of the degrees of freedom, it has assessed that the last fitted model (`.`) is nested within the model stored as `full`. In other words, `full` is the unconstrained model and `.` is the constrained model.

The names in "`(Assumption: . nested in full)`" are actually links. Click on a name, and the results for that model are replayed.

◁

Aside: the `nestreg` command provides a simple syntax for performing likelihood-ratio tests for nested model specifications; see [R] **nestreg**. In the previous example, we fit a full `logistic` model, used `estimates store` to store the `full` model, fit a constrained `logistic` model, and used `lrtest` to report a likelihood-ratio test between two models. Here is how we could do this with one call to `nestreg`:

```
. nestreg, lr quietly: logistic low (race2 race3 smoke ui) (age lwt ptl ht)
```

❑ Technical Note

`lrtest` determines the degrees of freedom of a model as the rank of the (co)variance matrix `e(V)`. There are two issues here. First, the *numerical* determination of the rank of a matrix is a subtle problem that can, for instance, be affected by the scaling of the variables in the model. The rank of a matrix depends on the number of (independent) linear combinations of coefficients that sum exactly to zero. In the world of numerical mathematics, it is hard to tell whether a very small number is really nonzero or is a real zero that happens to be slightly off because of roundoff error from the finite precision with which computers make floating-point calculations. Whether a small number is being classified as one or the other, typically on the basis of a threshold, affects the determined degrees of freedom. Although Stata generally makes sensible choices, it is bound to make mistakes occasionally. The moral of this story is to make sure that the calculated degrees of freedom are as you expect before interpreting the results.

❑

❑ Technical Note

A second issue involves `regress` and related commands such as `anova`. Mainly for historical reasons, `regress` does not treat the residual variance σ^2 the same way that it treats the regression coefficients. Type `estat vce` after `regress`, and you will see the regression coefficients, not $\widehat{\sigma}^2$. Most estimation commands for models with ancillary parameters (e.g., `streg` and `heckman`) treat all parameters as equals. There is nothing technically wrong with `regress` here; we are usually focused on the regression coefficients, and their estimators are uncorrelated with $\widehat{\sigma}^2$. But, formally, σ^2 adds a degree of freedom to the model, which does not matter if you are comparing two regression models by a likelihood-ratio test. This test depends on the difference in the degrees of freedom, and hence being "off by 1" in each does not matter. But, if you are comparing a regression model with a larger model—e.g., a heteroskedastic regression model fitted by `arch`— the automatic determination of the degrees of freedom is incorrect, and you must specify the `df(#)` option.

❑

▷ Example 2

Returning to the low-birthweight data in the first example, we now wish to test that the coefficient on race2 is equal to that on race3. The base model is still stored under the name full, so we need only fit the constrained model and perform the test. With z as the index of the logit model, the base model is

$$z = \beta_0 + \beta_1 \texttt{age} + \beta_2 \texttt{lwt} + \beta_3 \texttt{race2} + \beta_4 \texttt{race3} + \cdots$$

If $\beta_3 = \beta_4$, this can be written as

$$z = \beta_0 + \beta_1 \texttt{age} + \beta_2 \texttt{lwt} + \beta_3 (\texttt{race2} + \texttt{race3}) + \cdots$$

To fit the constrained model, we create a variable equal to the sum of race2 and race3 and fit the model, which has the sum in place of the two variables.

```
. generate race23 = race2 + race3
. logistic low age lwt race23 smoke ptl ht ui
Logistic regression                               Number of obs   =        189
                                                  LR chi2(7)      =      32.67
                                                  Prob > chi2     =     0.0000
Log likelihood =  -100.9997                       Pseudo R2       =     0.1392

------------------------------------------------------------------------------
         low | Odds Ratio   Std. Err.      z    P>|z|     [95% Conf. Interval]
-------------+----------------------------------------------------------------
         age |   .9716799   .0352638    -0.79   0.429     .9049649    1.043313
         lwt |   .9864971   .0064627    -2.08   0.038     .9739114    .9992453
      race23 |   2.728186   1.080206     2.53   0.011     1.255586    5.927907
       smoke |   2.664498   1.052379     2.48   0.013     1.228633    5.778414
         ptl |   1.709129   .5924775     1.55   0.122     .8663666    3.371691
          ht |   6.116391   4.215585     2.63   0.009      1.58425    23.61385
          ui |    2.09936   .9699702     1.61   0.108     .8487997    5.192407
------------------------------------------------------------------------------
```

Comparing this model with our original model, we obtain

```
. lrtest full .
Likelihood-ratio test                                 LR chi2(1)  =      0.55
(Assumption: . nested in full)                        Prob > chi2 =    0.4577
```

By comparison, typing test race2=race3 after fitting our base model results in a significance level of .4572. Alternatively, we can first store the restricted model, here using the name equal. Next, lrtest is invoked specifying the names of the restricted and unrestricted models (we don't care about the order). This time, we also add the option stats requesting a table of model statistics, including the model selection indices AIC and BIC.

```
. estimates store equal
. lrtest equal full, stats
Likelihood-ratio test                                 LR chi2(1)  =      0.55
(Assumption: equal nested in full)                    Prob > chi2 =    0.4577

-----------------------------------------------------------------------------
       Model |    Obs    ll(null)   ll(model)     df          AIC         BIC
-------------+---------------------------------------------------------------
       equal |    189    -117.336   -100.9997      8     217.9994    243.9334
        full |    189    -117.336    -100.724      9      219.448    248.6237
-----------------------------------------------------------------------------
               Note:  N=Obs used in calculating BIC; see [R] BIC note
```

◁

Composite models

`lrtest` supports composite models; that is, models that can be fitted by fitting a series of simpler models or by fitting models on subsets of the data. Theoretically, a composite model is one in which the likelihood function $L(\theta)$ of the parameter vector θ can be written as the product

$$L(\theta) = L_1(\theta_1) \times L_2(\theta_2) \times \cdots \times L_k(\theta_k)$$

of likelihood terms with $\theta = (\theta_1, \ldots, \theta_k)$ a partitioning of the full parameter vector. In such a case, the full-model likelihood $L(\theta)$ is maximized by maximizing the likelihood terms $L_j(\theta_j)$ in turn. Obviously, $\log L(\widehat{\theta}) = \sum_{j=1}^{k} \log L_j(\widehat{\theta}_j)$. The degrees of freedom for the composite model is obtained as the sum of the degrees of freedom of the constituting models.

▷ Example 3

As an example of the application of composite models, we consider a test of the hypothesis that the coefficients of a statistical model do not differ between different portions ("regimes") of the covariate space. Economists call a test for such a hypothesis a *Chow test*.

We continue the analysis of the data on children of low birthweight by using logistic regression modeling and study whether the regression coefficients are the same among the three races: white, black, and other. A likelihood-ratio Chow test can be obtained by fitting the logistic regression model for each of the races and then comparing the combined results with those of the model previously stored as `full`. Since the full model included dummies for the three races, this version of the Chow test allows the intercept of the logistic regression model to vary between the regimes (races).

```
. logistic low age lwt smoke ptl ht ui if race==1, nolog
Logistic regression                               Number of obs   =         96
                                                  LR chi2(6)      =      13.86
                                                  Prob > chi2     =     0.0312
Log likelihood = -45.927061                       Pseudo R2       =     0.1311

------------------------------------------------------------------------------
         low | Odds Ratio   Std. Err.      z    P>|z|     [95% Conf. Interval]
-------------+----------------------------------------------------------------
         age |   .9869674    .0527756    -0.25   0.806     .8887649    1.096021
         lwt |   .9900874    .0106101    -0.93   0.353     .9695089    1.011103
       smoke |   4.208697    2.680132     2.26   0.024      1.20808    14.66222
         ptl |   1.592145    .7474264     0.99   0.322     .6344379    3.995544
          ht |   2.900166    3.193536     0.97   0.334     .3350554    25.10319
          ui |   1.229523    .9474768     0.27   0.789     .2715165    5.567715
------------------------------------------------------------------------------

. estimates store white
. logistic low age lwt smoke ptl ht ui if race==2, nolog
Logistic regression                               Number of obs   =         26
                                                  LR chi2(6)      =      10.12
                                                  Prob > chi2     =     0.1198
Log likelihood = -12.654157                       Pseudo R2       =     0.2856

------------------------------------------------------------------------------
         low | Odds Ratio   Std. Err.      z    P>|z|     [95% Conf. Interval]
-------------+----------------------------------------------------------------
         age |   .8735313    .1377809    -0.86   0.391     .6412385    1.189974
         lwt |   .9747736    .0166888    -1.49   0.136     .9426068    1.008038
       smoke |   16.50373    24.36988     1.90   0.058     .9134256    298.1884
         ptl |   4.866916    9.331296     0.83   0.409     .1135671    208.5715
          ht |   85.05606    214.6317     1.76   0.078     .6050219    11957.47
          ui |   67.61338    133.3291     2.14   0.033     1.417488     3225.12
------------------------------------------------------------------------------
```

```
. estimates store black

. logistic low age lwt smoke ptl ht ui if race==3, nolog

Logistic regression                             Number of obs   =         67
                                                LR chi2(6)      =      14.06
                                                Prob > chi2     =     0.0289
Log likelihood = -37.228444                     Pseudo R2       =     0.1589

------------------------------------------------------------------------------
         low | Odds Ratio   Std. Err.      z    P>|z|     [95% Conf. Interval]
-------------+----------------------------------------------------------------
         age |   .9263905   .0665385    -1.06   0.287     .8047408    1.066429
         lwt |   .9724499    .015762    -1.72   0.085     .9420424    1.003839
       smoke |   .7979034   .6340581    -0.28   0.776     .1680887    3.787582
         ptl |   2.845675   1.777942     1.67   0.094     .8363061    9.682898
          ht |   7.767504   10.00536     1.59   0.112     .6220776    96.98808
          ui |   2.925006   2.046473     1.53   0.125      .742311    11.52571
------------------------------------------------------------------------------

. estimates store other
```

We are now ready to perform the likelihood-ratio Chow test:

```
. lrtest (full) (white black other), stats

Likelihood-ratio test                                 LR chi2(12) =       9.83
                                                      Prob > chi2 =     0.6310

Assumption: (full) nested in (white, black, other)

------------------------------------------------------------------------------
       Model |    Obs    ll(null)   ll(model)     df          AIC         BIC
-------------+----------------------------------------------------------------
        full |    189    -117.336    -100.724      9      219.448    248.6237
       white |     96   -52.85752   -45.92706      7     105.8541    123.8046
       black |     26   -17.71291   -12.65416      7     39.30831    48.11499
       other |     67   -44.26039   -37.22844      7     88.45689    103.8897
------------------------------------------------------------------------------
               Note:  N=Obs used in calculating BIC; see [R] BIC note
```

We cannot reject the hypothesis that the logistic regression model applies to each of the races at any reasonable significance level. By specifying the option `stats`, we can verify the degrees of freedom of the test: $12 = 7 + 7 + 7 - 9$. We can obtain the same test by fitting an expanded model with interactions between all covariates and the variables `race` by using the `xi` prefix.

```
. xi: logistic low i.race*age i.race*lwt i.race*smoke i.race*ptl i.race*ht i.race*ui
i.race           _Irace_1-3          (naturally coded; _Irace_1 omitted)
i.race*age       _IracXage_#         (coded as above)
i.race*lwt       _IracXlwt_#         (coded as above)
i.race*smoke     _IracXsmoke_#       (coded as above)
i.race*ptl       _IracXptl_#         (coded as above)
i.race*ht        _IracXht_#          (coded as above)
i.race*ui        _IracXui_#          (coded as above)
note: _Irace_2 dropped because of collinearity
note: _Irace_3 dropped because of collinearity
  (output omitted )
note: _Iracc_3 dropped because of collinearity
Logistic regression                               Number of obs   =        189
                                                  LR chi2(20)     =      43.05
                                                  Prob > chi2     =     0.0020
Log likelihood = -95.809661                       Pseudo R2       =     0.1835

------------------------------------------------------------------------------
         low | Odds Ratio   Std. Err.      z    P>|z|     [95% Conf. Interval]
-------------+----------------------------------------------------------------
         age |   .9869674   .0527756    -0.25   0.806      .888765     1.09602
 _IracXage_2 |    .885066   .1474075    -0.73   0.464     .6385697    1.226713
 _IracXage_3 |   .9386232   .0840486    -0.71   0.479     .7875367    1.118695
    _Irace_3 |   100.3769   309.5859     1.49   0.135     .2378648    42358.23
         lwt |   .9900874   .0106101    -0.93   0.353     .9695089    1.011103
 _IracXlwt_2 |   .9845329   .0198857    -0.77   0.440     .9463191     1.02429
 _IracXlwt_3 |   .9821859   .0190847    -0.93   0.355     .9454839    1.020313
    _Irace_2 |   99.62138   402.0819     1.14   0.254     .0365441    271573.6
       smoke |   4.208697   2.680129     2.26   0.024     1.208082     14.6622
_IracXsmok~2 |   3.921338   6.305976     0.85   0.395     .1677265    91.67841
_IracXsmok~3 |   .1895844     .19306    -1.63   0.102      .025763    1.395111
         ptl |   1.592145   .7474262     0.99   0.322     .6344381    3.995543
 _IracXptl_2 |    3.05683   6.034072     0.57   0.571     .0638308    146.3903
 _IracXptl_3 |   1.787322   1.396789     0.74   0.457     .3863583    8.268282
          ht |   2.900166   3.193535     0.97   0.334     .3350557    25.10318
  _IracXht_2 |     29.328   80.74795     1.23   0.220     .1329515    6469.514
  _IracXht_3 |   2.678297   4.538714     0.58   0.561     .0966918    74.18701
          ui |   1.229523   .9474763     0.27   0.789     .2715167    5.567711
  _IracXui_2 |   54.99156   116.4272     1.89   0.058     .8672537    3486.951
  _IracXui_3 |   2.378977   2.476123     0.83   0.405     .3093352    18.29578
------------------------------------------------------------------------------

. lrtest full .
Likelihood-ratio test                                 LR chi2(12)  =       9.83
(Assumption: full nested in .)                        Prob > chi2  =     0.6310
```

Applying `lrtest` for the full model against the model with all interactions yields the same test statistic and p-value as for the full model against the composite model for the three regimes. Here the specification of the model with interactions was convenient, and `logistic` had no problem computing the estimates for the expanded model. In models with more complicated likelihoods, such as Heckman's selection model (see [R] **heckman**) or complicated survival-time models (see [ST] **streg**), fitting the models with all interactions may be numerically demanding and may be much more time consuming than fitting a series of models separately for each regime.

Given the model with all interactions, we could also test the hypothesis of no differences among the regions (races) by a Wald version of the Chow test by using the `testparm` command.

```
. testparm _IracX*
 ( 1)  _IracXage_2 = 0
 ( 2)  _IracXage_3 = 0
 ( 3)  _IracXlwt_2 = 0
 ( 4)  _IracXlwt_3 = 0
 ( 5)  _IracXsmoke_2 = 0
 ( 6)  _IracXsmoke_3 = 0
 ( 7)  _IracXptl_2 = 0
 ( 8)  _IracXptl_3 = 0
 ( 9)  _IracXht_2 = 0
 (10)  _IracXht_3 = 0
 (11)  _IracXui_2 = 0
 (12)  _IracXui_3 = 0
           chi2( 12) =    8.24
         Prob > chi2 =    0.7663
```

We conclude that, here, the Wald version of the Chow test is similar to the likelihood-ratio version of the Chow test.

◁

Saved Results

lrtest saves the following in r():

Scalars

r(p)	level of significance	r(chi2)	LR test statistic
r(df)	degrees of freedom		

Programmers wishing their estimation commands to be compatible with lrtest should note that lrtest requires that the following results be returned:

e(cmd)	name of estimation command
e(ll)	log-likelihood value
e(V)	the (co)variance matrix
e(N)	number of observations

lrtest also verifies that e(N), e(ll_0), and e(depvar) are consistent between two noncomposite models.

Methods and Formulas

lrtest is implemented as an ado-file.

Let L_0 and L_1 be the log-likelihood values associated with the full and constrained models, respectively. The test statistic of the likelihood-ratio test is $\mathrm{LR} = -2(L_1 - L_0)$. If the constrained model is true, LR is approximately χ^2 distributed with $d_0 - d_1$ degrees of freedom, where d_0 and d_1 are the model degrees of freedom associated with the full and constrained models, respectively (Judge et al. 1985, 216–217).

lrtest determines the degrees of freedom of a model as the rank of e(V), computed as the number of nonzero diagonal elements of invsym(e(V)).

References

Gutierrez, R. G., S. L. Carter, and D. M. Drukker. 2001. On boundary-value likelihood-ratio tests. *Stata Technical Bulletin* 60: 15–18. Reprinted in *Stata Technical Bulletin Reprints*, vol. 10, pp. 269–273.

Hosmer, D. W., Jr., and S. Lemeshow. 2000. *Applied Logistic Regression*. 2nd ed. New York: Wiley.

Judge, G. G., W. E. Griffiths, R. C. Hill, H. Lütkepohl, and T.-C. Lee. 1985. *The Theory and Practice of Econometrics*. 2nd ed. New York: Wiley.

Kleinbaum, D. G., and M. Klein. 2002. *Logistic Regression: A Self-Learning Text*. 2nd ed. New York: Springer.

Pérez-Hoyos, S., and A. Tobias. 1999. sg111: A modified likelihood-ratio test command. *Stata Technical Bulletin* 49: 24–25. Reprinted in *Stata Technical Bulletin Reprints*, vol. 9, pp. 171–173.

Wang, Z. 2000. sg133: Sequential and drop one term likelihood-ratio tests. *Stata Technical Bulletin* 54: 46–47. Reprinted in *Stata Technical Bulletin Reprints*, vol. 9, pp. 332–334.

Also See

[R] **test** — Test linear hypotheses after estimation

[R] **testnl** — Test nonlinear hypotheses after estimation

[R] **nestreg** — Nested model statistics

Title

lv — Letter-value displays

Syntax

`lv` [*varlist*] [*if*] [*in*] [, `generate` `tail(#)`]

by is allowed; see [D] **by**.

Description

`lv` shows a letter-value display (Tukey 1977, 44–49; Hoaglin 1983) for each variable in *varlist*. If no variables are specified, letter-value displays are shown for each numeric variable in the data.

Options

Main

`generate` adds four new variables to the data: `_mid`, containing the midsummaries; `_spread`, containing the spreads; `_psigma`, containing the pseudosigmas; and `_z2`, containing the squared values from a standard normal distribution corresponding to the particular letter value. If the variables `_mid`, `_spread`, `_psigma`, and `_z2` already exist, their contents are replaced. At most, only the first 11 observations of each variable are used; the remaining observations contain missing. If *varlist* specifies more than one variable, the newly created variables contain results for the last variable specified. The `generate` option may not be used with the `by` prefix.

`tail(#)` indicates the inverse of the tail density through which letter values are to be displayed: 2 corresponds to the median (meaning half in each tail), 4 to the fourths (roughly the 25th and 75th percentiles), 8 to the eighths, and so on. # may be specified as 4, 8, 16, 32, 64, 128, 256, 512, or 1,024 and defaults to a value of # that has corresponding depth just greater than 1. The default is taken as 1,024 if the calculation results in a number larger than 1,024. Given the intelligent default, this option is rarely specified.

Remarks

Letter-value displays are a collection of observations drawn systematically from the data, focusing especially on the tails rather than the middle of the distribution. The displays are called letter-value displays because letters have been (almost arbitrarily) assigned to tail densities:

Letter	Tail area	Letter	Tail area
M	1/2	B	1/64
F	1/4	A	1/128
E	1/8	Z	1/256
D	1/16	Y	1/512
C	1/32	X	1/1024

▷ Example 1

We have data on the mileage ratings of 74 automobiles. To obtain a letter-value display, we type

```
. use http://www.stata-press.com/data/r10/auto
(1978 Automobile Data)
. lv mpg
#       74              Mileage (mpg)
                ___________________________________
M      37.5  |                 20                  |    spread  pseudosigma
F      19    |        18       21.5         25     |         7     5.216359
E      10    |        15       21.5         28     |        13     5.771728
D       5.5  |        14       22.25        30.5   |      16.5     5.576303
C       3    |        14       24.5         35     |        21     5.831039
B       2    |        12       23.5         35     |        23     5.732448
A       1.5  |        12       25           38     |        26     6.040635
        1    |        12       26.5         41     |        29      6.16562
             |                                     |
             |                                     |   # below      # above
inner fence  |        7.5                 35.5     |         0            1
outer fence  |        -3                    46     |         0            0
```

The decimal points can be made to line up and thus the output made more readable by specifying a display format for the variable; see [U] **12.5 Formats: controlling how data are displayed**.

```
. format mpg %9.2f
. lv mpg
#       74              Mileage (mpg)
                ___________________________________
M      37.5  |                20.00                |    spread  pseudosigma
F      19    |      18.00     21.50       25.00    |      7.00         5.22
E      10    |      15.00     21.50       28.00    |     13.00         5.77
D       5.5  |      14.00     22.25       30.50    |     16.50         5.58
C       3    |      14.00     24.50       35.00    |     21.00         5.83
B       2    |      12.00     23.50       35.00    |     23.00         5.73
A       1.5  |      12.00     25.00       38.00    |     26.00         6.04
        1    |      12.00     26.50       41.00    |     29.00         6.17
             |                                     |
             |                                     |   # below      # above
inner fence  |       7.50                 35.50    |         0            1
outer fence  |      -3.00                 46.00    |         0            0
```

At the top, the number of observations is indicated as 74. The first line shows the statistics associated with M, the letter value that puts half the density in each tail, or the median. The median has *depth* 37.5 (that is, in the ordered data, M is 37.5 observations in from the extremes) and has value 20. The next line shows the statistics associated with F or the fourths. The fourths have depth 19 (that is, in the ordered data, the lower fourth is observation 19, and the upper fourth is observation $74 - 19 + 1$), and the values of the lower and upper fourths are 18 and 25. The number in the middle is the point halfway between the fourths—called a midsummary. If the distribution were perfectly symmetric, the midsummary would equal the median. The spread is the difference between the lower and upper summaries ($25 - 18 = 7$). For fourths, half of the data lie within a 7-mpg band. The pseudosigma is a calculation of the standard deviation using only the lower and upper summaries and assuming that the variable is normally distributed. If the data really were normally distributed, all the pseudosigmas would be roughly equal.

After the letter values, the line labeled with depth 1 reports the minimum and maximum values. Here the halfway point between the extremes is 26.5, which is greater than the median, indicating that 41 is more extreme than 12, at least relative to the median. And with each letter value, the midsummaries are increasing—our data are skewed. The pseudosigmas are also increasing, indicating that the data are spreading out relative to a normal distribution, although, given the evident skewness, this elongation may be an artifact of the skewness.

At the end is an attempt to identify outliers, although the points so identified are merely outside some predetermined cutoff. Points outside the inner fence are called *outside values* or *mild outliers*. Points outside the outer fence are called *severe outliers*. The inner fence is defined as $(3/2)$IQR and the outer fence as 3IQR above and below the F summaries, where the IQR is the spread of the fourths.
◁

❑ Technical Note

The form of the letter-value display has varied slightly with different authors. `lv` displays appear as described by Hoaglin (1983) but as modified by Emerson and Stoto (1983), where they included the midpoint of each of the spreads. This format was later adopted by Hoaglin (1985). If the distribution is symmetric, the midpoints will all be roughly equal. On the other hand, if the midpoints vary systematically, the distribution is skewed.

The pseudosigmas are obtained from the lower and upper summaries for each letter value. For each letter value, they are the standard deviation a normal distribution would have if its spread for the given letter value were to equal the observed spread. If the pseudosigmas are all roughly equal, the data are said to have *neutral elongation*. If the pseudosigmas increase systematically, the data are said to be more elongated than a normal, i.e., have thicker tails. If the pseudosigmas decrease systematically, the data are said to be less elongated than a normal, i.e., have thinner tails.

Interpretation of the number of mild and severe outliers is more problematic. The following discussion is drawn from Hamilton (1991):

Obviously, the presence of any such outliers does not rule out that the data have been drawn from a normal distribution; in large datasets, there will most certainly be observations outside $(3/2)$IQR and 3IQR. Severe outliers, however, make up about two per million (.0002%) of a normal population. In samples, they lie far enough out to have substantial effects on means, standard deviations, and other classical statistics. The .0002%, however, should be interpreted carefully; outliers appear more often in small samples than one might expect from population proportions because of sampling variation in estimated quartiles. Monte Carlo simulation by Hoaglin, Iglewicz, and Tukey (1986) obtained these results on the percentages and numbers of outliers in random samples from a normal population:

	percentage		number	
n	any outliers	severe	any outliers	severe
10	2.83	.362	.283	.0362
20	1.66	.074	.332	.0148
50	1.15	.011	.575	.0055
100	.95	.002	.95	.002
200	.79	.001	1.58	.002
300	.75	.001	2.25	.003
∞	.70	.0002	∞	∞

Thus the presence of any severe outliers in samples of less than 300 is sufficient to reject normality. Hoaglin, Iglewicz, and Tukey (1981) suggested the approximation $.00698 + .4/n$ for the fraction of mild outliers in a sample of size n or, equivalently, $.00698n + .4$ for the number of outliers.
❑

▷ Example 2

The `generate` option adds the variables `_mid`, `_spread`, `_psigma`, and `_z2` to our data, making possible many of the diagnostic graphs suggested by Hoaglin (1985).

```
. lv mpg, generate
  (output omitted )
. list _mid _spread _psigma _z2 in 1/12
```

	_mid	_spread	_psigma	_z2
1.	20	.	.	.
2.	21.5	7	5.216359	.4501955
3.	21.5	13	5.771728	1.26828
4.	22.25	16.5	5.576303	2.188846
5.	24.5	21	5.831039	3.24255
6.	23.5	23	5.732448	4.024532
7.	25	26	6.040635	4.631499
8.	.	.	.	.
9.	.	.	.	.
10.	.	.	.	.
11.	26.5	29	6.16562	5.53073
12.	.	.	.	.

Observations 12 through the end are missing for these new variables. The definition of the observations is always the same. The first observation contains the M summary, the second the F, the third the E, and so on. Observation 11 always contains the summary for depth 1. Observations 8–10—corresponding to letter values Z, Y, and X—contain missing because these statistics were not calculated. We have only 74 observations, and their depth would be 1.

Hoaglin (1985) suggests graphing the midsummary against z^2. If the distribution is not skewed, the points in the resulting graph will be along a horizontal line:

```
. scatter _mid _z2
```

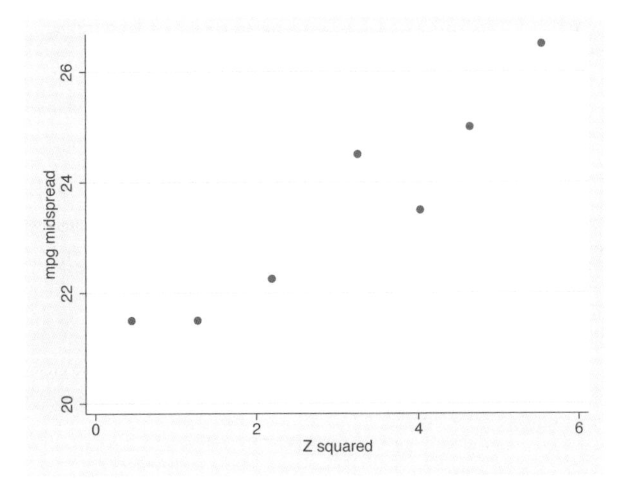

The graph clearly indicates the skewness of the distribution. We might also graph `_psigma` against `_z2` to examine elongation. ◁

Saved Results

`lv` saves the following in `r()`:

Scalars

`r(N)`	number of observations	`r(u_C)`	upper 32nd
`r(min)`	minimum	`r(l_B)`	lower 64th
`r(max)`	maximum	`r(u_B)`	upper 64th
`r(median)`	median	`r(l_A)`	lower 128th
`r(l_F)`	lower 4th	`r(u_A)`	upper 128th
`r(u_F)`	upper 4th	`r(l_Z)`	lower 256th
`r(l_E)`	lower 8th	`r(u_Z)`	upper 256th
`r(u_E)`	upper 8th	`r(l_Y)`	lower 512th
`r(l_D)`	lower 16th	`r(u_Y)`	upper 512th
`r(u_D)`	upper 16th	`r(l_X)`	lower 1024th
`r(l_C)`	lower 32nd	`r(u_X)`	upper 1024th

The lower/upper 8ths, 16ths, ..., 1024ths will be defined only if there are sufficient data.

Methods and Formulas

`lv` is implemented as an ado-file.

Let N be the number of (nonmissing) observations on x, and let $x_{(i)}$ refer to the ordered data when i is an integer. Define $x_{(i+.5)} = (x_{(i)} + x_{(i+1)})/2$; the median is defined as $x_{((N+1)/2)}$.

Define $x_{[d]}$ as the pair of numbers $x_{(d)}$ and $x_{(N+1-d)}$, where d is called the *depth*. Thus $x_{[1]}$ refers to the minimum and maximum of the data. Define $m = (N+1)/2$ as the depth of the median, $f = (\lfloor m \rfloor + 1)/2$ as the depth of the fourths, $e = (\lfloor f \rfloor + 1)/2$ as the depth of the eighths, and so on. Depths are reported on the far left of the letter-value display. The corresponding fourths of the data are $x_{[f]}$, the eighths $x_{[e]}$, and so on. These values are reported inside the display. The middle value is defined as the corresponding midpoint of $x_{[\cdot]}$. The spreads are defined as the difference in $x_{[\cdot]}$.

The corresponding point z_i on a standard normal distribution is obtained as (Hoaglin 1985, 456–457)

$$z_i = \begin{cases} F^{-1}\{(d_i - 1/3)/(N + 1/3)\} & \text{if } d_i > 1 \\ F^{-1}\{0.695/(N + 0.390)\} & \text{otherwise} \end{cases}$$

where d_i is the depth of the letter value. The corresponding pseudosigma is obtained as the ratio of the spread to $-2z_i$ (Hoaglin 1985, 431).

Define $(F_l, F_u) = x_{[f]}$. The inner fence has cutoffs $F_l - \frac{3}{2}(F_u - F_l)$ and $F_u + \frac{3}{2}(F_u - F_l)$. The outer fence has cutoffs $F_l - 3(F_u - F_l)$ and $F_u + 3(F_u - F_l)$.

The inner-fence values reported by `lv` are almost equal to those used by `graph, box` to identify outside points. The only difference is that `graph` uses a slightly different definition of fourths, namely, the 25th and 75th percentiles as defined by `summarize`.

References

Emerson, J. D., and M. A. Stoto. 1983. Transforming data. In *Understanding Robust and Exploratory Data Analysis*, ed. D. C. Hoaglin, F. Mosteller, and J. W. Tukey, 97–128. New York: Wiley.

Fox, J. 1990. Describing univariate distributions. In *Modern Methods of Data Analysis*, ed. J. Fox and J. S. Long, 58–125. Newbury Park, CA: Sage.

Hamilton, L. C. 1991. sed4: Resistant normality check and outlier identification. *Stata Technical Bulletin* 3: 15–18. Reprinted in *Stata Technical Bulletin Reprints*, vol. 1, pp. 86–90.

Hoaglin, D. C. 1983. Letter values: A set of selected order statistics. In *Understanding Robust and Exploratory Data Analysis*, ed. D. C. Hoaglin, F. Mosteller, and J. W. Tukey, 33–57. New York: Wiley.

——. 1985. Using quantiles to study shape. In *Exploring Data Tables, Trends, and Shapes*, ed. D. C. Hoaglin, F. Mosteller, and J. W. Tukey, 417–460. New York: Wiley.

Hoaglin, D. C., B. Iglewicz, and J. W. Tukey. 1981. Small-sample performance of a resistant rule for outlier detection. *1980 Proceedings of the Statistical Computing Section*, 144–152. Washington, DC: American Statistical Association.

——. 1986. Performance of some resistant rules for outlier labeling. *Journal of the American Statistical Association* 81: 991–999.

Tukey, J. W. 1977. *Exploratory Data Analysis*. Reading, MA: Addison–Wesley.

Also See

[R] **diagnostic plots** — Distributional diagnostic plots

[R] **stem** — Stem-and-leaf displays

[R] **summarize** — Summary statistics

Title

matsize — Set the maximum number of variables in a model

Syntax

```
set matsize # [, permanently ]
```

where $10 \le \# \le 11{,}000$ for Stata/MP and Stata/SE and where $10 \le \# \le 800$ for Stata/IC.

Description

`set matsize` sets the maximum number of variables that can be included in any of Stata's estimation commands.

For Stata/MP and Stata/SE, the default value is 400, but it may be changed upward or downward. The upper limit is 11,000.

For Stata/IC, the initial value is 200, but it may be changed upward or downward. The upper limit is 800.

This command may not be used with Small Stata; `matsize` is permanently frozen at 40.

Changing `matsize` has no effect on Mata.

Option

`permanently` specifies that, in addition to making the change right now, the `matsize` setting be remembered and become the default setting when you invoke Stata.

Remarks

`set matsize` controls the internal size of matrices that Stata uses. The default of 200 for Stata/IC, for instance, means that linear regression models are limited to 198 independent variables—198 because the constant uses one position and the dependent variable another, making a total of 200.

You may change `matsize` with data in memory, but increasing `matsize` increases the amount of memory consumed by Stata, increasing the probability of page faults and thus of making Stata run more slowly.

▷ Example 1

We wish to fit a model of `y` on the variables `x1` through `x200`. Without thinking, we type

```
. regress y x1-x200
matsize too small
    You have attempted to create a matrix with more than 200 rows or columns
    or to fit a model with more than 200 variables plus ancillary parameters.
    You need to increase matsize by using the set matsize command; see help
    matsize.
r(908);
```

We realize that we need to increase matsize, so we type

```
. set matsize 250
. regress y x1-x200
  (output omitted)
```

◁

Programmers should note that the current setting of matsize is stored as the c-class value c(matsize); see [P] **creturn**.

Also See

[R] **query** — Display system parameters

[D] **memory** — Memory size considerations

[U] **6 Setting the size of memory**

Title

maximize — Details of iterative maximization

Syntax

Maximum likelihood optimization

mle_cmd ... [, *options*]

Set default maximum iterations

set maxiter # [, permanently]

options	description
[no]log	display an iteration log of the log likelihood; typically, the default
trace	display current parameter vector in iteration log
gradient	display current gradient vector in iteration log
hessian	display current negative Hessian matrix in iteration log
showstep	report steps within an iteration in iteration log
shownrtolerance	report the current value of $\mathbf{g}\mathbf{H}^{-1}\mathbf{g}'$ in iteration log
technique(*algorithm_spec*)	maximization technique
iterate(#)	perform maximum of # iterations; default is iterate(16000)
tolerance(#)	tolerance for the coefficient vector; see *Options* for the defaults
ltolerance(#)	tolerance for the log likelihood; see *Options* for the defaults
gtolerance(#)	optional tolerance for the gradient relative to the coefficients
nrtolerance(#)	tolerance for the scaled gradient; see *Options* for the defaults
nonrtolerance	ignore the nrtolerance() option
difficult	use a different stepping algorithm in nonconcave regions
from(*init_specs*)	initial values for the coefficients

where *algorithm_spec* is

algorithm [# [*algorithm* [#]] ...]

algorithm is { nr | bhhh | dfp | bfgs }

and *init_specs* is one of

matname [, skip copy]

{ [*eqname*:]*name* = # | /*eqname* = # } [...]

[# ...], copy

Description

Stata has two maximum likelihood optimizers: one is used by internally coded commands, and the other is the `ml` command used by estimators implemented as ado-files. Both optimizers use the Newton–Raphson method with step halving (to avoid downhill steps) and special fixups when they encounter nonconcave regions of the likelihood. The two optimizers are similar but differ in their implementation. For information about programming maximum likelihood estimators in ado-files, see [R] **ml** and *Maximum Likelihood Estimation with Stata*, 3rd edition (Gould, Pitblado, and Sribney 2006).

`set maxiter` specifies the default maximum number of iterations for estimation commands that iterate. The initial value is `16000`, and # can be `0` to `16000`. To change the maximum number of iterations performed by a particular estimation command, you need not reset `maxiter`; you can specify the `iterate(#)` option. When `iterate(#)` is not specified, the `maxiter` value is used.

Maximization options

`log` and `nolog` specify whether an iteration log showing the progress of the log likelihood is to be displayed. For most commands, the log is displayed by default, and `nolog` suppresses it. For a few commands (such as the `svy` maximum likelihood estimators), you must specify `log` to see the log.

`trace` adds to the iteration log a display of the current parameter vector.

`gradient` (`ml`-programmed estimators only) adds to the iteration log a display of the current gradient vector.

`hessian` (`ml`-programmed estimators only) adds to the iteration log a display of the current negative Hessian matrix.

`showstep` (`ml`-programmed estimators only) adds to the iteration log a report on the steps within an iteration. This option was added so that developers at StataCorp could view the stepping when they were improving the `ml` optimizer code. At this point, it mainly provides entertainment.

`shownrtolerance` (`ml`-programmed estimators only) adds to the iteration log the current value of the Hessian-scaled gradient, $\mathbf{g}\mathbf{H}^{-1}\mathbf{g}'$, which is compared with the value of `nrtolerance()` to test for convergence. This value is computed and reported only when all other necessary stopping criteria have been met.

`technique(`*algorithm_spec*`)` (`ml`-programmed estimators only) specifies how the likelihood function is to be maximized. The following algorithms are currently implemented in `ml`. For details, see Gould, Pitblado, and Sribney (2006).

`technique(nr)` specifies Stata's modified Newton–Raphson (NR) algorithm.

`technique(bhhh)` specifies the Berndt–Hall–Hall–Hausman (BHHH) algorithm.

`technique(dfp)` specifies the Davidon–Fletcher–Powell (DFP) algorithm.

`technique(bfgs)` specifies the Broyden–Fletcher–Goldfarb–Shanno (BFGS) algorithm.

The default is `technique(nr)`.

You can switch between algorithms by specifying more than one in the `technique()` option. By default, `ml` will use an algorithm for five iterations before switching to the next algorithm. To specify a different number of iterations, include the number after the technique in the option. For example, specifying `technique(bhhh 10 nr 1000)` requests that `ml` perform 10 iterations with the BHHH algorithm followed by 1000 iterations with the NR algorithm, and then switch back to BHHH for 10 iterations, and so on. The process continues until convergence or until the maximum number of iterations is reached.

`iterate(#)` specifies the maximum number of iterations. When the number of iterations equals `iterate()`, the optimizer stops and presents the current results. If convergence is declared before this threshold is reached, it will stop when convergence is declared. Specifying `iterate(0)` is useful for viewing results evaluated at the initial value of the coefficient vector. Specifying `iterate(0)` and `from()` together allows you to view results evaluated at a specified coefficient vector; however, not all commands allow the `from()` option. The default value of `iterate(#)` for both estimators programmed internally and estimators programmed with `ml` is the current value of `set maxiter`, which is `iterate(16000)` by default.

Below we describe the four different types of convergence tolerances used by Stata estimators, and we describe the `nonrtolerance` option. After these descriptions, we explain how the various tolerances are used to determine whether the maximization algorithm has converged.

`tolerance(#)` specifies the tolerance for the coefficient vector. When the relative change in the coefficient vector from one iteration to the next is less than or equal to `tolerance()`, the `tolerance()` convergence criterion is satisfied.

`tolerance(1e-4)` is the default for estimators programmed internally in Stata.

`tolerance(1e-6)` is the default for estimators programmed with `ml`.

`ltolerance(#)` specifies the tolerance for the log likelihood. When the relative change in the log likelihood from one iteration to the next is less than or equal to `ltolerance()`, the `ltolerance()` convergence is satisfied.

`ltolerance(0)` is the default for estimators programmed internally in Stata.

`ltolerance(1e-7)` is the default for estimators programmed with `ml`.

`gtolerance(#)` (`ml`-programmed estimators only) specifies the tolerance for the gradient relative to the coefficients. When $|g_i\, b_i| \leq$ `gtolerance()` for all parameters b_i and the corresponding elements of the gradient g_i, the gradient tolerance criterion is met. By default, this criterion is not checked and so there is no default value for the gradient tolerance.

`nrtolerance(#)` (`ml`-programmed estimators only) specifies the tolerance for the scaled gradient. Convergence is declared when $\mathbf{g}\mathbf{H}^{-1}\mathbf{g}' <$ `nrtolerance()`. `nrtolerance()` differs from `gtolerance()` in that the gradient is scaled by $\mathbf{H}$. The default is `nrtolerance(1e-5)`.

`nonrtolerance` (`ml`-programmed estimators only) specifies that the default `nrtolerance` criterion be turned off.

For internally programmed Stata estimators, convergence is declared when either the `tolerance()` or `ltolerance()` criterion has first been met. No other criteria are checked.

For ml-programmed estimators, by default convergence is declared when the `nrtolerance()` criterion *and* either of the `tolerance()` or `ltolerance()` criterion has been met. If `nonrtolerance` is specified, then convergence is declared when either of the `tolerance()` or `ltolerance()` criterion has been met.

If `gtolerance()` is specified, then the `gtolerance()` criterion must be met *in addition* to any other required criteria for convergence to be declared.

`difficult` (`ml`-programmed estimators only) specifies that the likelihood function is likely to be difficult to maximize because of nonconcave regions. When the message "not concave" appears repeatedly, `ml`'s standard stepping algorithm may not be working well. `difficult` specifies that a different stepping algorithm be used in nonconcave regions. There is no guarantee that `difficult` will work better than the default; sometimes it is better and sometimes it is worse. You should use the `difficult` option only when the default stepper declares convergence and the last iteration is "not concave" or when the default stepper is repeatedly issuing "not concave" messages and producing only tiny improvements in the log likelihood.

`from()` specifies initial values for the coefficients. Not all estimators in Stata support this option. You can specify the initial values in one of three ways: by specifying the name of a vector containing the initial values (e.g., `from(b0)`, where `b0` is a properly labeled vector); by specifying coefficient names with the values (e.g., `from(age=2.1 /sigma=7.4)`); or by specifying a list of values (e.g., `from(2.1 7.4, copy)`). `from()` is intended for use when doing bootstraps (see [R] **bootstrap**) and in other special situations (e.g., with `iterate(0)`). Even when the values specified in `from()` are close to the values that maximize the likelihood, only a few iterations may be saved. Poor values in `from()` may lead to convergence problems.

`skip` specifies that any parameters found in the specified initialization vector that are not also found in the model be ignored. The default action is to issue an error message.

`copy` specifies that the list of values or the initialization vector be copied into the initial-value vector by position rather than by name.

Option for set maxiter

`permanently` specifies that, in addition to making the change right now, the `maxiter` setting be remembered and become the default setting when you invoke Stata.

Remarks

Only in rare circumstances would you ever need to specify any of these options, except `nolog`. The `nolog` option is useful for reducing the amount of output appearing in log files.

The following is an example of an iteration log:

```
Iteration 0:   log likelihood = -3791.0251
Iteration 1:   log likelihood =  -3761.738
Iteration 2:   log likelihood = -3758.0632  (not concave)
Iteration 3:   log likelihood = -3758.0447
Iteration 4:   log likelihood = -3757.5861
Iteration 5:   log likelihood =  -3757.474
Iteration 6:   log likelihood = -3757.4613
Iteration 7:   log likelihood = -3757.4606
Iteration 8:   log likelihood = -3757.4606
```
(table of results omitted)

At iteration 8, the model converged. The message "not concave" at the second iteration is notable. This example was produced using the `heckman` command; its likelihood is not globally concave, so it is not surprising that this message sometimes appears. The other message that is occasionally seen is "backed up". Neither of these messages should be of any concern unless they appear at the final iteration.

If a "not concave" message appears at the last step, there are two possibilities. One is that the result is valid, but there is collinearity in the model that the command did not otherwise catch. Stata checks for obvious collinearity among the independent variables before performing the maximization, but strange collinearities or near collinearities can sometimes arise between coefficients and ancillary parameters. The second, more likely cause for a "not concave" message at the final step is that the optimizer entered a flat region of the likelihood and prematurely declared convergence.

If a "backed up" message appears at the last step, there are also two possibilities. One is that Stata found a perfect maximum and could not step to a better point; if this is the case, all is fine, but this is a highly unlikely occurrence. The second is that the optimizer worked itself into a bad concave spot where the computed gradient and Hessian gave a bad direction for stepping.

If either of these messages appears at the last step, perform the maximization again with the `gradient` option. If the gradient goes to zero, the optimizer has found a maximum that may not be unique but is a maximum. From the standpoint of maximum likelihood estimation, this is a valid result. If the gradient is not zero, it is not a valid result, and you should try tightening up the convergence criterion, or try `ltol(0) tol(1e-7)` or `gtol(0.1)` (with the default `ltol() tol()`) to see if the optimizer can work its way out of the bad region.

If you get repeated "not concave" steps with little progress being made at each step, try specifying the `difficult` option. Sometimes `difficult` works wonderfully, reducing the number of iterations and producing convergence at a good (i.e., concave) point. Other times, `difficult` works poorly, taking much longer to converge than the default stepper.

Saved Results

Maximum likelihood estimators save the following in e():

Scalars

e(N)	number of observations	always saved
e(k)	number of parameters	always saved
e(k_eq)	number of equations	usually saved
e(k_eq_model)	number of equations to include in a model Wald test	usually saved
e(k_dv)	number of dependent variables	usually saved
e(df_m)	model degrees of freedom	always saved
e(r2_p)	pseudo-R-squared	sometimes saved
e(ll)	log likelihood	always saved
e(ll_0)	log likelihood, constant-only model	saved when constant-only model is fitted
e(N_clust)	number of clusters	saved when vce(cluster *clustvar*) is specified; see [U] **20.15 Obtaining robust variance estimates**
e(chi2)	χ^2	usually saved
e(p)	significance of model of test	usually saved
e(ic)	number of iterations	usually saved
e(rank)	rank of e(V)	always saved
e(rank0)	rank of e(V) for constant-only model	saved when constant-only model is fitted
e(rc)	return code	usually saved
e(converged)	1 if converged, 0 otherwise	usually saved

Macros

e(cmd)	name of command	always saved
e(cmdline)	command as typed	always saved
e(depvar)	names of dependent variables	always saved
e(wtype)	weight type	saved when weights are specified or implied
e(wexp)	weight expression	saved when weights are specified or implied
e(title)	title in estimation output	usually saved by commands using ml
e(clustvar)	name of cluster variable	saved when vce(cluster *clustvar*) is specified; see [U] **20.15 Obtaining robust variance estimates**
e(chi2type)	Wald or LR; type of model χ^2 test	usually saved
e(vce)	*vcetype* specified in vce()	saved when command allows (vce())
e(vcetype)	title used to label Std. Err.	sometimes saved
e(opt)	type of optimization	always saved
e(ml_method)	type of ml method	always saved by commands using ml
e(user)	name of likelihood-evaluator program	always saved
e(technique)	from technique() option	sometimes saved
e(crittype)	optimization criterion	always saved
e(properties)	estimator properties	always saved
e(predict)	program used to implement predict	usually saved

Matrices

e(b)	coefficient vector	always saved
e(V)	variance–covariance matrix of the estimators	always saved
e(gradient)	gradient vector	usually saved
e(ilog)	iteration log (up to 20 iterations)	usually saved

Functions

e(sample)	marks estimation sample	always saved

`ml` saves the constraint matrix in matrix `Cns`, which can be obtained by typing

`matrix` *name* `= get(Cns)`.

See *Saved Results* in the manual entry for any maximum likelihood estimator for a list of returned results.

Methods and Formulas

Let L_1 be the log likelihood of the full model (i.e., the log-likelihood value shown on the output), and let L_0 be the log likelihood of the "constant-only" model. The likelihood-ratio χ^2 model test is defined as $2(L_1 - L_0)$. The pseudo-R^2 (McFadden 1974) is defined as $1 - L_1/L_0$. This is simply the log likelihood on a scale where 0 corresponds to the "constant-only" model and 1 corresponds to perfect prediction for a discrete model (in which case the overall log likelihood is 0).

Some maximum likelihood routines can report coefficients in an exponentiated form, e.g., odds ratios in `logistic`. Let b be the unexponentiated coefficient, s its standard error, and b_0 and b_1 the reported confidence interval for b. In exponentiated form, the point estimate is e^b, the standard error $e^b s$, and the confidence interval e^{b_0} and e^{b_1}. The displayed Z (or t) statistics and p-values are the same as those for the unexponentiated results. This is justified since $e^b = 1$ and $b = 0$ are equivalent hypotheses, and normality is more likely to hold in the b metric.

References

Gould, W. W., J. S. Pitblado, and W. M. Sribney. 2006. *Maximum Likelihood Estimation with Stata*. 3rd ed. College Station, TX: Stata Press.

McFadden, D. 1974. Conditional logit analysis of qualitative choice behavior. In *Frontiers of Econometrics*, ed. P. Zarembka, 104–142. New York: Academic Press.

Also See

[R] **ml** — Maximum likelihood estimation

[SVY] **ml for svy** — Maximum pseudolikelihood estimation for survey data

Title

mean — Estimate means

Syntax

`mean` *varlist* [*if*] [*in*] [*weight*] [, *options*]

options	description
Model	
`stdize(`*varname*`)`	variable identifying strata for standardization
`stdweight(`*varname*`)`	weight variable for standardization
`nostdrescale`	do not rescale the standard weight variable
if/in/over	
`over(`*varlist*[, `nolabel`]`)`	group over subpopulations defined by *varlist*; optionally, suppress group labels
SE/Cluster	
`vce(`*vcetype*`)`	*vcetype* may be `analytic`, `cluster` *clustvar*, `bootstrap`, or `jackknife`
Reporting	
`level(`*#*`)`	set confidence level; default is `level(95)`
`noheader`	suppress table header
`nolegend`	suppress table legend

`bootstrap`, `jackknife`, `rolling`, `statsby`, and `svy` are allowed; see [U] **11.1.10 Prefix commands**.
Weights are not allowed with the `bootstrap` prefix.
`aweight`s are not allowed with the `jackknife` prefix.
`vce()` and weights are not allowed with the `svy` prefix.
`fweight`s, `aweight`s, `iweight`s, and `pweight`s are allowed; see [U] **11.1.6 weight**.
See [U] **20 Estimation and postestimation commands** for more capabilities of estimation commands.

Description

`mean` produces estimates of means, along with standard errors.

Options

Model

`stdize(`*varname*`)` specifies that the point estimates be adjusted by direct standardization across the strata identified by *varname*. This option requires the `stdweight()` option.

`stdweight(`*varname*`)` specifies the weight variable associated with the standard strata identified in the `stdize()` option. The standardization weights must be constant within the standard strata.

`nostdrescale` prevents the standardization weights from being rescaled within the `over()` groups. This option requires `stdize()` but is ignored if the `over()` option is not specified.

if/in/over

over(*varlist* [, nolabel]) specifies that estimates be computed for multiple subpopulations, which are identified by the different values of the variables in *varlist*.

When this option is supplied with one variable name, such as over(*varname*), the value labels of *varname* are used to identify the subpopulations. If *varname* does not have labeled values (or there are unlabeled values), the values themselves are used, provided that they are nonnegative integers. Noninteger values, negative values, and labels that are not valid Stata names are substituted with a default identifier.

When over() is supplied with multiple variable names, each subpopulation is assigned a unique default identifier.

nolabel requests that value labels attached to the variables identifying the subpopulations be ignored.

SE/Cluster

vce(*vcetype*) specifies the type of standard error reported, which includes types that are derived from asymptotic theory, that allow for intragroup correlation, and that use bootstrap or jackknife methods; see [R] ***vce_option***.

vce(analytic), the default, uses the analytically derived variance estimator associated with the sample mean.

Reporting

level(*#*); see [R] **estimation options**.

noheader prevents the table header from being displayed. This option implies nolegend.

nolegend prevents the table legend identifying the subpopulations from being displayed.

Remarks

▷ Example 1

Using the fuel data from [R] **ttest**, we estimate the average mileage of the cars without the fuel treatment (mpg1) and those with the fuel treatment (mpg2).

```
. use http://www.stata-press.com/data/r10/fuel
. mean mpg1 mpg2
Mean estimation                    Number of obs    =      12

--------------------------------------------------------------
             |       Mean   Std. Err.     [95% Conf. Interval]
-------------+------------------------------------------------
        mpg1 |         21   .7881701      19.26525    22.73475
        mpg2 |      22.75   .9384465      20.68449    24.81551
--------------------------------------------------------------
```

Using these results, we can test the equality of the mileage between the two groups of cars.

```
. test mpg1 = mpg2
 ( 1)  mpg1 - mpg2 = 0
       F(  1,    11) =    5.04
            Prob > F =    0.0463
```

◁

▷ Example 2

In example 1, the joint observations of mpg1 and mpg2 were used to estimate a covariance between their means.

```
. matrix list e(V)

symmetric e(V)[2,2]
            mpg1        mpg2
mpg1  .62121212
mpg2   .4469697   .88068182
```

If the data were organized this way out of convenience but the two variables represent independent samples of cars (coincidentally of the same sample size), we should reshape the data and use the over() option to ensure that the covariance between the means is zero.

```
. use http://www.stata-press.com/data/r10/fuel

. stack mpg1 mpg2, into(mpg) clear

. mean mpg, over(_stack)

Mean estimation                     Number of obs    =      24

            1: _stack = 1
            2: _stack = 2

--------------------------------------------------------------
        Over |       Mean   Std. Err.     [95% Conf. Interval]
-------------+------------------------------------------------
mpg          |
           1 |         21   .7881701      19.36955    22.63045
           2 |      22.75   .9384465      20.80868    24.69132
--------------------------------------------------------------

. matrix list e(V)

symmetric e(V)[2,2]
            mpg:        mpg:
              1           2
mpg:1  .62121212
mpg:2          0   .88068182
```

Now we can test the equality of the mileage between the two independent groups of cars.

```
. test [mpg]1 = [mpg]2

 ( 1)  [mpg]1 - [mpg]2 = 0

       F(  1,    23) =    2.04
            Prob > F =    0.1667
```

◁

▷ Example 3: standardized means

Suppose that we collected the blood pressure data from [R] **dstdize**, and we wish to obtain standardized high blood pressure rates for each city in 1990 and 1992, using, as the standard, the age, sex, and race distribution of the four cities and two years combined. Our rate is really the mean of a variable that indicates whether a sampled individual has high blood pressure. First, we generate the strata and weight variables from our standard distribution, and then use mean to compute the rates.

```
. use http://www.stata-press.com/data/r10/hbp, clear

. egen strata = group(age race sex) if inlist(year, 1990, 1992)
(675 missing values generated)

. by strata, sort: gen stdw = _N
```

```
. mean hbp, over(city year) stdize(strata) stdweight(stdw)
Mean estimation
N. of std strata =      24          Number of obs    =     455
         Over: city year
    _subpop_1: 1 1990
    _subpop_2: 1 1992
    _subpop_3: 2 1990
    _subpop_4: 2 1992
    _subpop_5: 3 1990
    _subpop_6: 3 1992
    _subpop_7: 5 1990
    _subpop_8: 5 1992
```

Over	Mean	Std. Err.	[95% Conf.	Interval]
hbp				
_subpop_1	.058642	.0296273	.0004182	.1168657
_subpop_2	.0117647	.0113187	-.0104789	.0340083
_subpop_3	.0488722	.0238958	.0019121	.0958322
_subpop_4	.014574	.007342	.0001455	.0290025
_subpop_5	.1011211	.0268566	.0483425	.1538998
_subpop_6	.0810577	.0227021	.0364435	.1256719
_subpop_7	.0277778	.0155121	-.0027066	.0582622
_subpop_8	.0548926	.	.	.

The standard error of the high blood pressure rate estimate is missing for city 5 in 1992 because there was only one individual with high blood pressure; that individual was the only person observed in the stratum of white males 30–35 years old.

By default, mean rescales the standard weights within the over() groups. In the following, we use the nostdrescale option to prevent this, thus reproducing the results in [R] **dstdize**.

```
. mean hbp, over(city year) nolegend stdize(strata) stdweight(stdw)
> nostdrescale
Mean estimation
N. of std strata =      24          Number of obs    =     455
```

Over	Mean	Std. Err.	[95% Conf.	Interval]
hbp				
_subpop_1	.0073302	.0037034	.0000523	.0146082
_subpop_2	.0015432	.0014847	-.0013745	.004461
_subpop_3	.0078814	.0038536	.0003084	.0154544
_subpop_4	.0025077	.0012633	.000025	.0049904
_subpop_5	.0155271	.0041238	.007423	.0236312
_subpop_6	.0081308	.0022772	.0036556	.012606
_subpop_7	.0039223	.0021904	-.0003822	.0082268
_subpop_8	.0088735	0	.	.

◁

Saved Results

`mean` saves the following in `e()`:

Scalars

`e(N)`	number of observations	`e(N_clust)`	number of clusters
`e(N_over)`	number of subpopulations	`e(k_eq)`	number of equations in `e(b)`
`e(N_stdize)`	number of standard strata	`e(df_r)`	sample degrees of freedom

Macros

`e(cmd)`	`mean`	`e(cluster)`	name of cluster variable
`e(cmdline)`	command as typed	`e(over)`	*varlist* from `over()`
`e(varlist)`	*varlist*	`e(over_labels)`	labels from `over()` variables
`e(stdize)`	*varname* from `stdize()`	`e(over_namelist)`	names from `e(over_labels)`
`e(stdweight)`	*varname* from `stdweight()`	`e(vce)`	*vcetype* specified in `vce()`
`e(wtype)`	weight type	`e(vcetype)`	title used to label Std. Err.
`e(wexp)`	weight expression	`e(estat_cmd)`	program used to implement `estat`
`e(title)`	title in estimation output	`e(properties)`	`b V`

Matrices

`e(b)`	vector of mean estimates
`e(V)`	(co)variance estimates
`e(_N)`	vector of numbers of nonmissing observations
`e(_N_stdsum)`	number of nonmissing observations within the standard strata
`e(_p_stdize)`	standardizing proportions
`e(error)`	error code corresponding to `e(b)`

Functions

`e(sample)`	marks estimation sample

Methods and Formulas

`mean` is implemented as an ado-file.

The mean estimator

Let y be the variable on which we want to calculate the mean and y_j an individual observation on y, where $j = 1, \dots, n$ and n is the sample size. Let w_j be the weight, and if no weight is specified, define $w_j = 1$ for all j. For `aweights`, the w_j are normalized to sum to n. See the next section for `pweighted` data.

Let W be the sum of the weights

$$W = \sum_{j=1}^{n} w_j$$

The mean is defined as

$$\overline{y} = \frac{1}{W} \sum_{j=1}^{n} w_j y_j$$

The default variance estimator for the mean is

$$\widehat{V}(\overline{y}) = \frac{1}{W(W-1)} \sum_{j=1}^{n} w_j (y_j - \overline{y})^2$$

The standard error of the mean is the square root of the variance.

If x, x_j, and $\overline{x}$ are similarly defined for another variable (observed jointly with y), the covariance estimator between $\overline{x}$ and $\overline{y}$ is

$$\widehat{\mathrm{Cov}}(\overline{x}, \overline{y}) = \frac{1}{W(W-1)} \sum_{j=1}^{n} w_j (x_j - \overline{x})(y_j - \overline{y})$$

Survey data

See [SVY] **variance estimation**, [SVY] **direct standardization**, and [SVY] **poststratification** for discussions that provide background information for the following formulas. The following formulas are derived from the fact that the mean is a special case of the ratio estimator where the denominator variable is one, $x_j = 1$; see [R] **ratio**.

The survey mean estimator

Let Y_j be a survey item for the jth individual in the population, where $j = 1, \dots, M$ and M is the size of the population. The associated population mean for the item of interest is $\overline{Y} = Y/M$ where

$$Y = \sum_{j=1}^{M} Y_j$$

Let y_j be the survey item for the jth sampled individual from the population, where $j = 1, \dots, m$ and m is the number of observations in the sample.

The estimator for the mean is $\overline{y} = \widehat{Y}/\widehat{M}$, where

$$\widehat{Y} = \sum_{j=1}^{m} w_j y_j \qquad \text{and} \qquad \widehat{M} = \sum_{j=1}^{m} w_j$$

and w_j is a sampling weight. The score variable for the mean estimator is

$$z_j(\overline{y}) = \frac{y_j - \overline{y}}{\widehat{M}} = \frac{\widehat{M} y_j - \widehat{Y}}{\widehat{M}^2}$$

The standardized mean estimator

Let D_g denote the set of sampled observations that belong to the gth standard stratum and define $I_{D_g}(j)$ to indicate if the jth observation is a member of the gth standard stratum; where $g = 1, \dots, L_D$ and L_D is the number of standard strata. Also let π_g denote the fraction of the population that belongs to the gth standard stratum, thus $\pi_1 + \cdots + \pi_{L_D} = 1$. π_g is derived from the `stdweight()` option.

The estimator for the standardized mean is

$$\overline{y}^D = \sum_{g=1}^{L_D} \pi_g \frac{\widehat{Y}_g}{\widehat{M}_g}$$

where

$$\widehat{Y}_g = \sum_{j=1}^{m} I_{D_g}(j)\, w_j y_j \qquad \text{and} \qquad \widehat{M}_g = \sum_{j=1}^{m} I_{D_g}(j)\, w_j$$

The score variable for the standardized mean is

$$z_j(\overline{y}^D) = \sum_{g=1}^{L_D} \pi_g I_{D_g}(j) \frac{\widehat{M}_g y_j - \widehat{Y}_g}{\widehat{M}_g^2}$$

The poststratified mean estimator

Let P_k denote the set of sampled observations that belong to poststratum k and define $I_{P_k}(j)$ to indicate if the jth observation is a member of poststratum k; where $k = 1, \ldots, L_P$ and L_P is the number of poststrata. Also let M_k denote the population size for poststratum k. P_k and M_k are identified by specifying the `poststrata()` and `postweight()` options on `svyset`.

The estimator for the poststratified mean is

$$\overline{y}^P = \frac{\widehat{Y}^P}{\widehat{M}^P} = \frac{\widehat{Y}^P}{M}$$

where

$$\widehat{Y}^P = \sum_{k=1}^{L_P} \frac{M_k}{\widehat{M}_k} \widehat{Y}_k = \sum_{k=1}^{L_P} \frac{M_k}{\widehat{M}_k} \sum_{j=1}^{m} I_{P_k}(j)\, w_j y_j$$

and

$$\widehat{M}^P = \sum_{k=1}^{L_P} \frac{M_k}{\widehat{M}_k} \widehat{M}_k = \sum_{k=1}^{L_P} M_k = M$$

The score variable for the poststratified mean is

$$z_j(\overline{y}^P) = \frac{z_j(\widehat{Y}^P)}{M} = \frac{1}{M} \sum_{k=1}^{L_P} I_{P_k}(j) \frac{M_k}{\widehat{M}_k} \left(y_j - \frac{\widehat{Y}_k}{\widehat{M}_k} \right)$$

The standardized poststratified mean estimator

The estimator for the standardized poststratified mean is

$$\overline{y}^{DP} = \sum_{g=1}^{L_D} \pi_g \frac{\widehat{Y}_g^P}{\widehat{M}_g^P}$$

where

$$\widehat{Y}_g^P = \sum_{k=1}^{L_P} \frac{M_k}{\widehat{M}_k} \widehat{Y}_{g,k} = \sum_{k=1}^{L_P} \frac{M_k}{\widehat{M}_k} \sum_{j=1}^{m} I_{D_g}(j) I_{P_k}(j)\, w_j y_j$$

and

$$\widehat{M}_g^P = \sum_{k=1}^{L_P} \frac{M_k}{\widehat{M}_k} \widehat{M}_{g,k} = \sum_{k=1}^{L_P} \frac{M_k}{\widehat{M}_k} \sum_{j=1}^{m} I_{D_g}(j) I_{P_k}(j)\, w_j$$

The score variable for the standardized poststratified mean is

$$z_j(\overline{y}^{DP}) = \sum_{g=1}^{L_D} \pi_g \frac{\widehat{M}_g^P z_j(\widehat{Y}_g^P) - \widehat{Y}_g^P z_j(\widehat{M}_g^P)}{(\widehat{M}_g^P)^2}$$

where

$$z_j(\widehat{Y}_g^P) = \sum_{k=1}^{L_P} I_{P_k}(j) \frac{M_k}{\widehat{M}_k} \left\{ I_{D_g}(j) y_j - \frac{\widehat{Y}_{g,k}}{\widehat{M}_k} \right\}$$

and

$$z_j(\widehat{M}_g^P) = \sum_{k=1}^{L_P} I_{P_k}(j) \frac{M_k}{\widehat{M}_k} \left\{ I_{D_g}(j) - \frac{\widehat{M}_{g,k}}{\widehat{M}_k} \right\}$$

Subpopulation estimation

Let S denote the set of sampled observations that belong to the subpopulation of interest, and define $I_S(j)$ to indicate if the jth observation falls within the subpopulation.

The estimator for the subpopulation mean is $\overline{y}^S = \widehat{Y}^S / \widehat{M}^S$, where

$$\widehat{Y}^S = \sum_{j=1}^{m} I_S(j)\, w_j y_j \qquad \text{and} \qquad \widehat{M}^S = \sum_{j=1}^{m} I_S(j)\, w_j$$

Its score variable is

$$z_j(\overline{y}^S) = I_S(j) \frac{y_j - \overline{y}^S}{\widehat{M}^S} = I_S(j) \frac{\widehat{M}^S y_j - \widehat{Y}^S}{(\widehat{M}^S)^2}$$

The estimator for the standardized subpopulation mean is

$$\overline{y}^{DS} = \sum_{g=1}^{L_D} \pi_g \frac{\widehat{Y}_g^S}{\widehat{M}_g^S}$$

where

$$\widehat{Y}_g^S = \sum_{j=1}^{m} I_{D_g}(j) I_S(j)\, w_j y_j \qquad \text{and} \qquad \widehat{M}_g^S = \sum_{j=1}^{m} I_{D_g}(j) I_S(j)\, w_j$$

Its score variable is

$$z_j(\overline{y}^{DS}) = \sum_{g=1}^{L_D} \pi_g I_{D_g}(j) I_S(j) \frac{\widehat{M}_g^S y_j - \widehat{Y}_g^S}{(\widehat{M}_g^S)^2}$$

The estimator for the poststratified subpopulation mean is

$$\overline{y}^{PS} = \frac{\widehat{Y}^{PS}}{\widehat{M}^{PS}}$$

where

$$\widehat{Y}^{PS} = \sum_{k=1}^{L_P} \frac{M_k}{\widehat{M}_k} \widehat{Y}_k^S = \sum_{k=1}^{L_P} \frac{M_k}{\widehat{M}_k} \sum_{j=1}^{m} I_{P_k}(j) I_S(j)\, w_j y_j$$

and

$$\widehat{M}^{PS} = \sum_{k=1}^{L_P} \frac{M_k}{\widehat{M}_k} \widehat{M}_k^S = \sum_{k=1}^{L_P} \frac{M_k}{\widehat{M}_k} \sum_{j=1}^{m} I_{P_k}(j) I_S(j)\, w_j$$

Its score variable is

$$z_j(\overline{y}^{PS}) = \frac{\widehat{M}^{PS} z_j(\widehat{Y}^{PS}) - \widehat{Y}^{PS} z_j(\widehat{M}^{PS})}{(\widehat{M}^{PS})^2}$$

where

$$z_j(\widehat{Y}^{PS}) = \sum_{k=1}^{L_P} I_{P_k}(j) \frac{M_k}{\widehat{M}_k} \left\{ I_S(j)\, y_j - \frac{\widehat{Y}_k^S}{\widehat{M}_k} \right\}$$

and

$$z_j(\widehat{M}^{PS}) = \sum_{k=1}^{L_P} I_{P_k}(j) \frac{M_k}{\widehat{M}_k} \left\{ I_S(j) - \frac{\widehat{M}_k^S}{\widehat{M}_k} \right\}$$

The estimator for the standardized poststratified subpopulation mean is

$$\overline{y}^{DPS} = \sum_{g=1}^{L_D} \pi_g \frac{\widehat{Y}_g^{PS}}{\widehat{M}_g^{PS}}$$

where

$$\widehat{Y}_g^{PS} = \sum_{k=1}^{L_p} \frac{M_k}{\widehat{M}_k} \widehat{Y}_{g,k}^S = \sum_{k=1}^{L_p} \frac{M_k}{\widehat{M}_k} \sum_{j=1}^{m} I_{D_g}(j) I_{P_k}(j) I_S(j)\, w_j y_j$$

and

$$\widehat{M}_g^{PS} = \sum_{k=1}^{L_p} \frac{M_k}{\widehat{M}_k} \widehat{M}_{g,k}^S = \sum_{k=1}^{L_p} \frac{M_k}{\widehat{M}_k} \sum_{j=1}^{m} I_{D_g}(j) I_{P_k}(j) I_S(j)\, w_j$$

Its score variable is

$$z_j(\overline{y}^{DPS}) = \sum_{g=1}^{L_D} \pi_g \frac{\widehat{M}_g^{PS} z_j(\widehat{Y}_g^{PS}) - \widehat{Y}_g^{PS} z_j(\widehat{M}_g^{PS})}{(\widehat{M}_g^{PS})^2}$$

where

$$z_j(\widehat{Y}_g^{PS}) = \sum_{k=1}^{L_P} I_{P_k}(j) \frac{M_k}{\widehat{M}_k} \left\{ I_{D_g}(j) I_S(j)\, y_j - \frac{\widehat{Y}_{g,k}^S}{\widehat{M}_k} \right\}$$

and

$$z_j(\widehat{M}_g^{PS}) = \sum_{k=1}^{L_P} I_{P_k}(j) \frac{M_k}{\widehat{M}_k} \left\{ I_{D_g}(j) I_S(j) - \frac{\widehat{M}_{g,k}^S}{\widehat{M}_k} \right\}$$

References

Bakker, A. 2003. The early history of average values and implications for education. *Journal of Statistics Education* 11: 1.

Cochran, W. G. 1977. *Sampling Techniques*. 3rd ed. New York: Wiley.

Stuart, A., and J. K. Ord. 1994. *Kendall's Advanced Theory of Statistics: Distribution Theory, Vol. I*. 6th ed. London: Arnold.

Also See

[R] **mean postestimation** — Postestimation tools for mean

[R] **ameans** — Arithmetic, geometric, and harmonic means

[R] **proportion** — Estimate proportions

[R] **ratio** — Estimate ratios

[R] **summarize** — Summary statistics

[R] **total** — Estimate totals

[SVY] **direct standardization** — Direct standardization of means, proportions, and ratios

[SVY] **poststratification** — Poststratification for survey data

[SVY] **subpopulation estimation** — Subpopulation estimation for survey data

[SVY] **svy estimation** — Estimation commands for survey data

[SVY] **variance estimation** — Variance estimation for survey data

[U] **20 Estimation and postestimation commands**

Title

mean postestimation — Postestimation tools for mean

Description

The following postestimation commands are available for mean:

command	description
estat	VCE
estat (svy)	postestimation statistics for survey data
estimates	cataloging estimation results
lincom	point estimates, standard errors, testing, and inference for linear combinations of coefficients
nlcom	point estimates, standard errors, testing, and inference for nonlinear combinations of coefficients
test	Wald tests for simple and composite linear hypotheses
testnl	Wald tests of nonlinear hypotheses

See the corresponding entries in the *Stata Base Reference Manual* for details, but see [SVY] **estat** for details about estat (svy).

Remarks

▷ Example 1

We have a dataset with monthly rates of returns on the Dow and Nasdaq stock indices. We can use mean to compute the average quarterly rates of return for the two indices separately;

```
. use http://www.stata-press.com/data/r10/rates
. mean dow nasdaq
Mean estimation                     Number of obs    =      357
```

	Mean	Std. Err.	[95% Conf.	Interval]
dow	.2489137	6.524386	-12.58227	13.0801
nasdaq	10.78477	4.160821	2.601887	18.96765

If you chose just one of the indices for your portfolio, you either did rather well or rather poorly, depending on which one you picked. However, as we now show with the postestimation command lincom, if you diversified your portfolio, you would have earned a respectable 5.5% rate of return without having to guess which index would be the better performer.

```
. lincom .5*dow + .5*nasdaq
 ( 1)  .5 dow + .5 nasdaq = 0
```

	Coef.	Std. Err.	t	P>|t|	[95% Conf.	Interval]
(1)	5.51684	4.262673	1.29	0.196	-2.866347	13.90003

◁

Methods and Formulas

All postestimation commands listed above are implemented as ado-files.

Also See

[R] **mean** — Estimate means

[SVY] **svy postestimation** — Postestimation tools for svy

Title

meta — Meta-analysis

Remarks

Stata does not have a meta-analysis command. Stata users, however, have developed an excellent suite of commands for performing meta-analysis, many of which have been published in the *Stata Journal* (SJ) or the *Stata Technical Bulletin* (STB).

For information about meta-analysis, in addition to the articles that have appeared in the *Stata Journal* and the *Stata Technical Bulletin*, see Abramson and Abramson (2001, chap. F); Dohoo, Martin, and Stryhn (2003, chap. 24); Edgger, Smith, and Altman (2001); and Sutton et al. (2000).

Also see the meta-analysis frequently asked question on the Stata web site at

http://www.stata.com/support/faqs/stat/meta.html

Issue	insert	author(s)	command	description
STB-38	sbe16	S. Sharp, J. Sterne	`meta`	meta-analysis for an outcome of two exposures or two treatment regimens
STB-42	sbe16.1	S. Sharp, J. Sterne	`meta`	update of sbe16
STB-43	sbe16.2	S. Sharp, J. Sterne	`meta`	update; *install this version*
STB-41	sbe19	T. J. Steichen	`metabias`	performs the Begg and Mazumdar (1994) adjusted rank correlation test for publication bias and the Egger et al. (1997) regression asymmetry test for publication bias
STB-44	sbe19.1	T. J. Steichen, M. Egger, J. Sterne	`metabias`	update of sbe19
STB-57	sbe19.2	T. J. Steichen	`metabias`	update of sbe19
STB-58	sbe19.3	T. J. Steichen	`metabias`	update of sbe19
STB-61	sbe19.4	T. J. Steichen	`metabias`	update of sbe19
SJ-3-4	sbe19.4	T. J. Steichen	`metabias`	update; *install this version*
STB-41	sbe20	A. Tobias	`galbr`	performs the Galbraith plot (1988), which is useful for investigating heterogeneity in meta-analysis
STB-56	sbe20.1	A. Tobias	`galbr`	update; *install this version*
STB-42	sbe22	J. Sterne	`metacum`	performs cumulative meta-analysis, using fixed- or random-effects models, and graphs the result
STB-42	sbe23	S. Sharp	`metareg`	extends a random-effects meta-analysis to estimate the extent to which one or more covariates, with values defined for each study in the analysis, explains heterogeneity in the treatment effects

(*Continued on next page*)

STB-44	sbe24	M. J. Bradburn, J. J. Deeks, D. G. Altman	`metan`, `funnel`, `labbe`	meta-analysis of studies with two groups funnel plot of precision versus treatment effect L'Abbé plot
STB-45	sbe24.1	M. J. Bradburn, J. J. Deeks, D. G. Altman	`funnel`	update; *install this version*
STB-47	sbe26	A. Tobias	`metainf`, `meta`	graphical technique to look for influential studies in the meta-analysis estimate
STB-56	sbe26.1	A. Tobias	`metainf`	update; *install this version*
STB-49	sbe28	A. Tobias	`metap`	combines p-values by using either Fisher's method or Edgington's method
STB-56	sbe28.1	A. Tobias	`metap`	update; *install this version*
STB-57	sbe39	T. J. Steichen	`metatrim`	performs the Duval and Tweedie (2000) nonparametric "trim and fill" method of accounting for publication bias in meta-analysis
STB-58	sbe39.1	T. J. Steichen	`metatrim`	update of sbe39
STB-61	sbe39.2	T. J. Steichen	`metatrim`	update; *install this version*
SJ-4-2	st0061	J. Sterne, R.Harbord R. Harbord	`metafunnel`	funnel plots
SJ-4-2	pr0012	T. J. Steichen		submenu and dialogs for meta-analysis commands
SJ-6-1	st0096	N. Orsini, R. Bellocco, S. Greenland	`glst`	GLS for trend estimation of summarized dose-response data

More commands may be available; enter Stata and type `search meta analysis`.

To download and install from the Internet the Sharp and Stern `meta` command, for instance, in Stata you could do the following:

1. Select **Help > SJ and User-written Programs**.
2. Click on *STB*.
3. Click on *stb43*.
4. Click on *sbe16_2*.
5. Click on *click here to install*.

Or you could do the following:

1. Navigate to the appropriate STB issue:
   a. Type `net from http://www.stata.com`
      Type `net cd stb`
      Type `net cd stb43`

   or

   b. Type `net from http://www.stata.com/stb/stb43`
2. Type `net describe sbe16_2`
3. Type `net install sbe16_2`

References

Abramson, J. H., and Z. H. Abramson. 2001. *Making Sense of Data: A Self-Instruction Manual on the Interpretation of Epidemiological Data*. 3rd ed. New York: Oxford University Press.

Begg, C. B., and M. Mazumdar. 1994. Operating characteristics of a rank correlation test for publication bias. *Biometrics* 50: 1088–1101.

Bradburn, M. J., J. J. Deeks, and D. G. Altman. 1998a. sbe24: metan—an alternative meta-analysis command. *Stata Technical Bulletin* 44: 4–15. Reprinted in *Stata Technical Bulletin Reprints*, vol. 8, pp. 86–100.

——. 1998b. sbe24.1: Correction to funnel plot. *Stata Technical Bulletin* 45: 21. Reprinted in *Stata Technical Bulletin Reprints*, vol. 8, p. 100.

Dohoo, I., W. Martin, and H. Stryhn. 2003. *Veterinary Epidemiologic Research*. Charlottetown, Prince Edward Island: AVC.

Duval, S. J., and R. L. Tweedie. 2000. A non-parametric "Trim and Fill" method of accounting for publication bias in meta-analysis. *Journal of American Statistical Association* 95: 89–98.

Egger, M., G. D. Smith, and D. G. Altman, ed. 2001. *Systematic Reviews in Health Care: Meta-analysis in Context*. London: BMJ Books.

Egger, M., G. D. Smith, M. Schneider, and C. Minder. 1997. Bias in meta-analysis detected by a simple, graphical test. *British Medical Journal* 315: 629–634.

Galbraith, R. F. 1988. A note on graphical display of estimated odds ratios from several clinical trials. *Statistics in Medicine* 7: 889–894.

L'Abbé, K. A., A. S. Detsky, and K. O'Rourke. 1987. Meta-analysis in clinical research. *Annals of Internal Medicine* 107: 224–233.

Orsini, N., R. Bellocco, and S. Greenland. 2006. GLS for trend estimation of summarized dose-response data. *Stata Journal* 6: 40–57.

Sharp, S. 1998. sbe23: Meta-analysis regression. *Stata Technical Bulletin* 42: 16–22. Reprinted in *Stata Technical Bulletin Reprints*, vol. 7, pp. 148–155.

Sharp, S., and J. Sterne. 1997. sbe16: Meta-analysis. *Stata Technical Bulletin* 38: 9–14. Reprinted in *Stata Technical Bulletin Reprints*, vol. 7, pp. 100–106.

——. 1998a. sbe16.1: New syntax and output for the meta-analysis command. *Stata Technical Bulletin* 42: 6–8. Reprinted in *Stata Technical Bulletin Reprints*, vol. 7, pp. 106–108.

——. 1998b. sbe16.2: Corrections to the meta-analysis command. *Stata Technical Bulletin* 43: 15. Reprinted in *Stata Technical Bulletin Reprints*, vol. 8, p. 84.

Steichen, T. J. 1998. sbe19: Tests for publication bias in meta-analysis. *Stata Technical Bulletin* 41: 9–15. Reprinted in *Stata Technical Bulletin Reprints*, vol. 7, pp. 125–133.

——. 2000a. sbe19.2: Update of tests for publication bias in meta-analysis. *Stata Technical Bulletin* 57: 4. Reprinted in *Stata Technical Bulletin Reprints*, vol. 10, p. 70.

——. 2000b. sbe39: Nonparametric trim and fill analysis of publication bias in meta-analysis. *Stata Technical Bulletin* 57: 8–14. Reprinted in *Stata Technical Bulletin Reprints*, vol. 10, pp. 108–117.

——. 2000c. sbe19.3: Tests for publication bias in meta-analysis: Erratum. *Stata Technical Bulletin* 58: 8. Reprinted in *Stata Technical Bulletin Reprints*, vol. 10, p. 71.

——. 2000d. sbe39.1: Nonparametric trim and fill analysis of publication bias in meta-analysis: Erratum. *Stata Technical Bulletin* 58: 8–9. Reprinted in *Stata Technical Bulletin Reprints*, vol. 10, pp. 117–118.

——. 2001a. sbe19.4: Update to metabias to work under version 7. *Stata Technical Bulletin* 61: 11. Reprinted in *Stata Technical Bulletin Reprints*, vol. 10, pp. 71–72.

——. 2001b. sbe39.2: Update of metatrim to work under version 7. *Stata Technical Bulletin* 61: 11. Reprinted in *Stata Technical Bulletin Reprints*, vol. 10, p. 118.

——. 2004. Submenu and dialogs for meta-analysis commands. *Stata Journal* 4: 124–126.

Steichen, T. J., M. Egger, and J. Sterne. 1998. sbe19.1: Tests for publication bias in meta-analysis. *Stata Technical Bulletin* 44: 3–4. Reprinted in *Stata Technical Bulletin Reprints*, vol. 8, pp. 84–85.

Sterne, J. 1998. sbe22: Cumulative meta-analysis. *Stata Technical Bulletin* 42: 13–16. Reprinted in *Stata Technical Bulletin Reprints*, vol. 7, pp. 143–147.

Sterne, J. A. C., and R. M. Harbord. 2004. Funnel plots in meta-analysis. *Stata Journal* 4: 127–141.

Sterne, J. A. C., R. Harris, and R. M. Harbord. 2007. FAQ: What meta-analysis features are available in Stata? http://www.stata.com/support/faqs/stat/meta.html.

Sutton, A. J., K. R. Abrams, D. R. Jones, T. A. Sheldon, and F. Song. 2000. *Methods for Meta-Analysis in Medical Research*. New York: Wiley.

Tobias, A. 1998. sbe20: Assessing heterogeneity in meta-analysis: The Galbraith plot. *Stata Technical Bulletin* 41: 15–17. Reprinted in *Stata Technical Bulletin Reprints*, vol. 7, pp. 133–136.

——. 1999a. sbe26: Assessing the influence of a single study in the meta-analysis estimate. *Stata Technical Bulletin* 47: 15–17. Reprinted in *Stata Technical Bulletin Reprints*, vol. 8, p. 108–110.

——. 1999b. sbe28: Meta-analysis of p-values. *Stata Technical Bulletin* 49: 15–17. Reprinted in *Stata Technical Bulletin Reprints*, vol. 9, pp. 138–140.

——. 2000a. sbe20.1: Update of galbr. *Stata Technical Bulletin* 56: 14. Reprinted in *Stata Technical Bulletin Reprints*, vol. 10, p. 72.

——. 2000b. sbe26.1: Update of metainf. *Stata Technical Bulletin* 56: 15. Reprinted in *Stata Technical Bulletin Reprints*, vol. 10, p. 72.

——. 2000c. sbe28.1: Update of metap. *Stata Technical Bulletin* 56: 15. Reprinted in *Stata Technical Bulletin Reprints*, vol. 10, p. 73.

Title

mfp — Multivariable fractional polynomial models

Syntax

`mfp` *regression_cmd yvar xvarlist* [*if*] [*in*] [*weight*] [, *options*]

options	description
Model 2	
`sequential`	use the Royston and Altman model-selection algorithm; default uses closed-test algorithm
`cycles(`*#*`)`	maximum number of iteration cycles; default is `cycles(5)`
`dfdefault(`*#*`)`	default maximum degrees of freedom; default is `dfdefault(4)`
`adjust(`*adj_list*`)`	adjustment for each predictor
`alpha(`*alpha_list*`)`	*p*-values for testing between FP models; default is `alpha(0.05)`
`df(`*df_list*`)`	degrees of freedom for each predictor
`powers(`*numlist*`)`	list of fractional polynomial powers to use; default is `powers(-2 -1(.5)1 2 3)`
Adv. model	
`xorder(+`|`-`|`n)`	order of entry into model-selection algorithm; default is `xorder(+)`
`select(`*select_list*`)`	nominal *p*-values for selection on each predictor
`xpowers(`*xp_list*`)`	fractional polynomial powers for each predictor
`zero(`*varlist*`)`	treat nonpositive values of specified predictors as zero when FP transformed
`catzero(`*varlist*`)`	add indicator variable for specified predictors
regression_cmd_options	other options accepted by chosen regression commands
Reporting	
`level(`*#*`)`	set confidence level; default is `level(95)`
`all`	include out-of-sample observations in generated variables

All weight types supported by *regression_cmd* are allowed; see [U] **11.1.6 weight**.
See [U] **20 Estimation and postestimation commands** for more capabilities of estimation commands.
`fracgen` may be used to create new variables containing fractional polynomial powers. See [R] **fracpoly**.

where

regression_cmd may be `clogit`, `cnreg`, `glm`, `logistic`, `logit`, `mlogit`, `nbreg`, `ologit`, `oprobit`, `poisson`, `probit`, `qreg`, `regress`, `stcox`, `streg`, or `xtgee`.

yvar is not allowed for `streg` and `stcox`. For these commands, you must first `stset` your data.

xvarlist has elements of type *varlist* and/or `(`*varlist*`)`, e.g., `x1 x2 (x3 x4 x5)`
Elements enclosed in parentheses are tested jointly for inclusion in the model and are not eligible for fractional polynomial transformation.

Description

mfp selects the fractional polynomial (FP) model that best predicts the outcome variable from the right-hand-side variables in *xvarlist*.

Options

Model 2

sequential chooses the sequential FP selection algorithm (see *Methods of FP model selection*).

cycles(*#*) sets the maximum number of iteration cycles permitted. cycles(5) is the default.

dfdefault(*#*) determines the default maximum degrees of freedom (df) for a predictor. The default is dfdefault(4) (second degree FP).

adjust(*adj_list*) defines the adjustment for the covariates *xvar1*, *xvar2*, ... of *xvarlist*. The default is adjust(mean), except for binary covariates, where it is adjust(*#*), with # being the lower of the two distinct values of the covariate. A typical item in *adj_list* is *varlist*:{mean | # | no}. Items are separated by commas. The first item is special in that *varlist* is optional, and, if it is omitted, the default is reset to the specified value (mean, #, or no). For example, adjust(no, age:mean) sets the default to no (i.e., no adjustment) and the adjustment for age to mean.

alpha(*alpha_list*) sets the significance levels for testing between FP models of different degree. The rules for *alpha_list* are the same as those for *df_list* in the df() option (see below). The default nominal p-value (significance level, selection level) is 0.05 for all variables.

Example: alpha(0.01) specifies that all variables have an FP selection level of 1%.

Example: alpha(0.05, weight:0.1) specifies that all variables except weight have an FP selection level of 5%; weight has a level of 10%.

df(*df_list*) sets the degrees of freedom (df) for each predictor. The df (not counting the regression constant, _cons) are twice the degree of the FP, so, for example, an *xvar* fitted as a second-degree FP (FP2) has 4 df. The first item in *df_list* may be either # or *varlist*:#. Subsequent items must be *varlist*:#. Items are separated by commas, and *varlist* is specified in the usual way for variables. With the first type of item, the df for all predictors are taken to be #. With the second type of item, all members of *varlist* (which must be a subset of *xvarlist*) have # df.

The default number of degrees of freedom for a predictor of type *varlist* specified in *xvarlist* but not in *df_list* is assigned according to the number of distinct (unique) values of the predictor, as follows:

# of distinct values	default df
1	(invalid predictor)
2–3	1
4–5	min(2, dfdefault())
≥ 6	dfdefault()

Example: df(4)
All variables have 4 df.

Example: df(2, weight displ:4)
weight and displ have 4 df; all other variables have 2 df.

Example: `df(weight displ:4, mpg:2)`
`weight` and `displ` have 4 df, `mpg` has 2 df; all other variables have default df.

`powers(`*numlist*`)` is the set of fractional polynomial powers to be used. The default set is −2, −1, −0.5, 0, 0.5, 1, 2, 3 (0 means log).

Adv. model

`xorder(+|-|n)` determines the order of entry of the covariates into the model-selection algorithm. The default is `xorder(+)`, which enters them in decreasing order of significance in a multiple linear regression (most significant first). `xorder(-)` places them in reverse significance order, whereas `xorder(n)` respects the original order in *xvarlist*.

`select(`*select_list*`)` sets the nominal p-values (significance levels) for variable selection by backward elimination. A variable is dropped if its removal causes a nonsignificant increase in deviance. The rules for *select_list* are the same as those for *df_list* in the `df()` option (see above). Using the default selection level of 1 for all variables forces them all into the model. Setting the nominal p-value to be 1 for a given variable forces it into the model, leaving others to be selected or not. The nominal p-value for elements of *xvarlist* bound by parentheses is specified by including `(`*varlist*`)` in *select_list*.

Example: `select(0.05)`
All variables have a nominal p-value of 5%.

Example: `select(0.05, weight:1)`
All variables except `weight` have a nominal p-value of 5%; `weight` is forced into the model.

Example: `select(a (b c):0.05)`
All variables except `a`, `b`, and `c` are forced into the model. `b` and `c` are tested jointly with 2 df at the 5% level, and `a` is tested singly at the 5% level.

`xpowers(`*xp_list*`)` sets the permitted fractional polynomial powers for covariates individually. The rules for *xp_list* are the same as for *df_list* in the `df()` option. The default selection is the same as those for the `powers()` option.

Example: `xpowers(-1 0 1)`
All variables have powers −1, 0, 1.

Example: `xpowers(x5:-1 0 1)`
All variables except `x5` have default powers; `x5` has powers −1, 0, 1.

`zero(`*varlist*`)` treats negative and zero values of members of *varlist* as zero when FP transformations are applied. By default, such variables are subjected to a preliminary linear transformation to avoid negative and zero values (see [R] **fracpoly**). *varlist* must be part of *xvarlist*.

`catzero(`*varlist*`)` is a variation on `zero()`; see *Zeros and zero categories* below. *varlist* must be part of *xvarlist*.

regression_cmd_options may be any of the options appropriate to *regression_cmd*.

Reporting

`level(`*#*`)` specifies the confidence level, as a percentage, for confidence intervals. The default is `level(95)` or as set by `set level`; see [U] **20.7 Specifying the width of confidence intervals**.

`all` includes out-of-sample observations when generating the FP variables. By default, the generated FP variables contain missing values outside the estimation sample.

Remarks

Remarks are presented under the following headings:

For elements in *xvarlist* not enclosed in parentheses, `mfp` leaves variables in the data named `I`*xvar*`__1`, `I`*xvar*`__2`, ..., where *xvar* represents the first four letters of the name of *xvar1*, and so on, for *xvar2*, *xvar3*, etc. The new variables contain the best-fitting fractional polynomial powers of *xvar1*, *xvar2*, ...

Iteration report

By default, for each continuous predictor, x, `mfp` compares null, linear, and FP1 models for x with an FP2 model. The deviance for each of these nested submodels is given in the column headed "Deviance". The line labeled "Final" gives the deviance for the selected model and its powers. All the other predictors currently selected are included, with their transformations (if any). For models specified as having 1 df, the only choice is whether the variable enters the model.

Estimation algorithm

The estimation algorithm in `mfp` processes the *xvars* in turn. Initially, `mfp` silently arranges *xvarlist* in order of increasing p-value (i.e., of decreasing statistical significance) for omitting each predictor from the model comprising *xvarlist*, with each term linear. The aim is to model relatively important variables before unimportant ones. This approach may help to reduce potential model-fitting difficulties caused by collinearity or, more generally, "concurvity" among the predictors. See the `xorder()` option above for details on how to change the ordering.

At the initial cycle, the best-fitting FP function for *xvar1* (the first of *xvarlist*) is determined, with all the other variables assumed to be linear. Either the default or the alternative procedure is used (see *Methods of FP model selection* below). The functional form (but not the estimated regression coefficients) for *xvar1* is kept, and the process is repeated for *xvar2*, *xvar3*, etc. The first iteration concludes when all the variables have been processed in this way. The next cycle is similar, except that the functional forms from the initial cycle are retained for all variables except the one currently being processed.

A variable whose functional form is prespecified to be linear (i.e., to have 1 df) is tested for exclusion within the above procedure when its nominal p-value (selection level) according to `select()` is less than 1; otherwise, it is included.

Updating of FP functions and candidate variables continues until the functions and variables included in the overall model do not change (convergence). Convergence is usually achieved within 1–4 cycles.

Methods of FP model selection

`mfp` includes two algorithms for FP model selection, both of which are types of backward elimination. They start from a most-complex permitted FP model and attempt to simplify the model by reducing the degree. The default algorithm resembles a closed-test procedure, a sequence of tests maintaining the overall type I error rate at a prespecified nominal level, such as 5%. All significance tests are approximate; therefore, the algorithm is not precisely a closed-test procedure.

The closed-test algorithm for choosing an FP model with maximum permitted degree $m = 2$ (i.e., an FP2 model with 4 df) for one continuous predictor, x, is as follows:

1. Inclusion: test FP2 against the null model for x on 4 df at the significance level determined by `select()`. If x is significant, continue; otherwise, drop x from the model.
2. Nonlinearity: test FP2 against a straight line in x on 3 df at the significance level determined by `alpha()`. If significant, continue; otherwise, stop, with the chosen model for x being a straight line.
3. Simplification: test FP2 against FP1 on 2 df at the significance level determined by `alpha()`. If significant, the final model is FP2; otherwise, it is FP1.

The first step is omitted if x is to be retained in the model, that is, if its nominal p-value, according to the `select()` option, is 1.

An alternative algorithm is available with the `sequential` option, as originally suggested by Royston and Altman (1994):

1. Test FP2 against FP1 on 2 df at the `alpha()` significance level. If significant, the final model is FP2; otherwise, continue.
2. Test FP1 against a straight line on 1 df at the `alpha()` level. If significant, the final model is FP1; otherwise, continue.
3. Test a straight line against omitting x on 1 df at the `select()` level. If significant, the final model is a straight line; otherwise, drop x.

The final step is omitted if x is to be retained in the model, that is, if its nominal p-value, according to the `select()` option, is 1.

If x is uninfluential, the overall type I error rate of this procedure is about double that of the closed-test procedure, for which the rate is close to the nominal value. This inflated type I error rate confers increased apparent power to detect nonlinear relationships.

Zeros and zero categories

The `zero()` option permits fitting an FP model to the positive values of a covariate, taking nonpositive values as zero. An application is the assessment of the effect of cigarette smoking as a risk factor in an epidemiological study. Nonsmokers may be qualitatively different from smokers, so the effect of smoking (regarded as a continuous variable) may not be continuous between one and zero cigarettes. To allow for this, the risk may be modeled as constant for the nonsmokers and as an FP function of the number of cigarettes for the smokers:

```
. generate byte nonsmokr = cond(n_cigs==0, 1, 0) if n_cigs != .
. mfp logit case n_cigs nonsmokr age, zero(n_cigs) df(4, nonsmokr:1)
```

Omission of `zero(n_cigs)` would cause `n_cigs` to be transformed before analysis by the addition of a suitable constant, probably 1.

A closely related approach involves the `catzero()` option. The command

```
. mfp logit case n_cigs age, catzero(n_cigs)
```

would achieve a similar result to the previous command but with important differences. First, `mfp` would create the equivalent of the binary variable `nonsmokr` automatically and include it in the model. Second, the two smoking variables would be treated as one predictor in the model. With the `select()` option active, the two variables would be tested jointly for inclusion in the model.

▷ Example 1

We illustrate two of the analyses performed by Sauerbrei and Royston (1999). We use `brcancer.dta`, which contains prognostic factors data from the German Breast Cancer Study Group of patients with node-positive breast cancer. The response variable is recurrence-free survival time (`rectime`), and the censoring variable is `censrec`. There are 686 patients with 299 events. We use Cox regression to predict the log hazard of recurrence from prognostic factors, of which five are continuous (`x1`, `x3`, `x5`, `x6`, `x7`) and three are binary (`x2`, `x4a`, `x4b`). Hormonal therapy (`hormon`) is known to reduce recurrence rates and is forced into the model. We use `mfp` to build a model from the initial set of eight predictors by using the backfitting model-selection algorithm. We set the nominal *p*-value for variable and FP selection to 0.05 for all variables except `hormon`, for which it is set to 1:

```
. use http://www.stata-press.com/data/r10/brcancer
(German breast cancer data)
. stset rectime, fail(censrec)
  (output omitted)
. mfp stcox x1 x2 x3 x4a x4b x5 x6 x7 hormon, nohr alpha(.05)
> select(.05, hormon:1)
Deviance for model with all terms untransformed =  3471.637, 686 observations
Variable      Model (vs.)    Deviance  Dev diff.   P      Powers    (vs.)
-----------------------------------------------------------------------
x5            null    FP2    3503.610    61.366  0.000*   .          .5 3
              lin.           3471.637    29.393  0.000+   1
              FP1            3449.203     6.959  0.031+   0
              Final          3442.244                     .5 3
x6            null    FP2    3464.113    29.917  0.000*   .          -2 .5
              lin.           3442.244     8.048  0.045+   1
              FP1            3435.550     1.354  0.508    .5
              Final          3435.550                     .5
[hormon included with 1 df in model]
x4a           null    lin.   3440.749     5.199  0.023*   .          1
              Final          3435.550                     1
x3            null    FP2    3436.832     3.560  0.469    .          -2 3
              Final          3436.832                     .
x2            null    lin.   3437.589     0.756  0.384    .          1
              Final          3437.589                     .
x4b           null    lin.   3437.848     0.259  0.611    .          1
              Final          3437.848                     .
x1            null    FP2    3437.893    18.085  0.001*   .          -2 -.5
              lin.           3437.848    18.040  0.000+   1
              FP1            3433.628    13.820  0.001+   -2
              Final          3419.808                     -2 -.5
x7            null    FP2    3420.805     3.715  0.446    .          -.5 3
              Final          3420.805                     .
-----------------------------------------------------------------------
End of Cycle 1: deviance =     3420.805
-----------------------------------------------------------------------
x5            null    FP2    3494.867    74.143  0.000*   .          -2 -1
              lin.           3451.795    31.071  0.000+   1
              FP1            3428.023     7.299  0.026+   0
              Final          3420.724                     -2 -1
x6            null    FP2    3452.093    32.704  0.000*   .          0 0
              lin.           3427.703     8.313  0.040+   1
              FP1            3420.724     1.334  0.513    .5
              Final          3420.724                     .5
[hormon included with 1 df in model]
```

```
x4a          null   lin.   3425.310      4.586  0.032*    .         1
             Final         3420.724                      1

x3           null   FP2    3420.724      5.305  0.257     .         -.5 0
             Final         3420.724                       .

x2           null   lin.   3420.724      0.214  0.644     .         1
             Final         3420.724                       .

x4b          null   lin.   3420.724      0.145  0.703     .         1
             Final         3420.724                       .

x1           null   FP2    3440.057     19.333  0.001*    .         -2 -.5
             lin.          3440.038     19.314  0.000+   1
             FP1           3436.949     16.225  0.000+   -2
             Final         3420.724                      -2 -.5

x7           null   FP2    3420.724      2.152  0.708     .         -1 3
             Final         3420.724                       .

Fractional polynomial fitting algorithm converged after 2 cycles.

Transformations of covariates:

-> gen double Ix1__1 = X^-2-.0355294635 if e(sample)
-> gen double Ix1__2 = X^-.5-.4341573547 if e(sample)
   (where: X = x1/10)
-> gen double Ix5__1 = X^-2-3.983723313 if e(sample)
-> gen double Ix5__2 = X^-1-1.99592668 if e(sample)
   (where: X = x5/10)
-> gen double Ix6__1 = X^.5-.3331600619 if e(sample)
   (where: X = (x6+1)/1000)

Final multivariable fractional polynomial model for _t
```

Variable	Initial df	Initial Select	Initial Alpha	Final Status	Final df	Final Powers
x1	4	0.0500	0.0500	in	4	-2 -.5
x2	1	0.0500	0.0500	out	0	
x3	4	0.0500	0.0500	out	0	
x4a	1	0.0500	0.0500	in	1	1
x4b	1	0.0500	0.0500	out	0	
x5	4	0.0500	0.0500	in	4	-2 -1
x6	4	0.0500	0.0500	in	2	.5
x7	4	0.0500	0.0500	out	0	
hormon	1	1.0000	0.0500	in	1	1

```
Cox regression -- Breslow method for ties
Entry time _t0                                  Number of obs   =        686
                                                LR chi2(7)      =     155.62
                                                Prob > chi2     =     0.0000
Log likelihood = -1710.3619                     Pseudo R2       =     0.0435
```

_t	Coef.	Std. Err.	z	P>|z|	[95% Conf.	Interval]
Ix1__1	44.73377	8.256682	5.42	0.000	28.55097	60.91657
Ix1__2	-17.92302	3.909611	-4.58	0.000	-25.58571	-10.26032
x4a	.5006982	.2496324	2.01	0.045	.0114276	.9899687
Ix5__1	.0387904	.0076972	5.04	0.000	.0237041	.0538767
Ix5__2	-.5490645	.0864255	-6.35	0.000	-.7184554	-.3796736
Ix6__1	-1.806966	.3506314	-5.15	0.000	-2.494191	-1.119741
hormon	-.4024169	.1280843	-3.14	0.002	-.6534575	-.1513763

```
Deviance: 3420.724.
```

Some explanation of the output from the model-selection algorithm is desirable. Consider the first few lines of output in the iteration log:

```
1. Deviance for model with all terms untransformed =  3471.637, 686 observations
   Variable       Model (vs.)    Deviance  Dev diff.   P       Powers   (vs.)
   ----------------------------------------------------------------------------
2. x5             null    FP2    3503.610    61.366  0.000*     .          .5 3
3.                lin.           3471.637    29.393  0.000+    1
4.                FP1            3449.203     6.959  0.031+    0
5.                Final          3442.244                      .5 3
```

Line 1 gives the deviance ($-2 \times \log$ partial likelihood) for the Cox model with all terms linear, the place where the algorithm starts. The model is modified variable by variable in subsequent steps. The most significant linear term turns out to be `x5`, which is therefore processed first. Line 2 compares the best-fitting FP2 for `x5` with a model omitting `x5`. The FP has powers (0.5, 3), and the test for inclusion of `x5` is highly significant. The reported deviance of 3503.610 is for the null model, not for the FP2 model. The deviance for the FP2 model may be calculated by subtracting the deviance difference (`Dev diff.`) from the reported deviance, giving $3{,}503.610 - 61.366 = 3{,}442.244$. Line 3 shows that the FP2 model is also a significantly better fit than a straight line (`lin.`) and line 4 that FP2 is also somewhat better than FP1 ($p = 0.031$). Thus at this stage in the model-selection procedure, the final model for `x5` (line 5) is FP2 with powers (0.5, 3). The overall model with an FP2 for `x5` and all other terms linear has a deviance of 3,442.244.

After all the variables have been processed (cycle 1) and reprocessed (cycle 2) in this way, convergence is achieved since the functional forms (FP powers and variables included) after cycle 2 are the same as they were after cycle 1. The model finally chosen is Model II as given in tables 3 and 4 of Sauerbrei and Royston (1999). Because of scaling of variables, the regression coefficients reported there are different, but the model and its deviance are identical. The model includes `x1` with powers $(-2, -0.5)$, `x4a`, `x5` with powers $(-2, -1)$, and `x6` with power 0.5. There is strong evidence of nonlinearity for `x1` and for `x5`, the deviance differences for comparison with a straight-line model (`FP2 vs lin.`) being, respectively, 19.3 and 31.1 at convergence (cycle 2). Predictors `x2`, `x3`, `x4b` and `x7` are dropped, as may be seen from their status `out` in the table `Final multivariable fractional polynomial model for _t` (the assumed *depvar* when using `stcox`).

All predictors except `x4a` and `hormon`, which are binary, have been adjusted to the mean of the original variable. For example, the mean of `x1` (age) is 53.05 years. The first FP-transformed variable for `x1` is `x1^-2` and is created by the expression `gen double Ix1__1 = X^-2-.0355 if e(sample)`. The value .0355 is obtained from $(53.05/10)^{-2}$. The division by 10 is applied automatically to improve the scaling of the regression coefficient for `Ix1__1`.

According to Sauerbrei and Royston (1999), medical knowledge dictates that the estimated risk function for `x5` (number of positive nodes), which was based on the above FP with powers $(-2, -1)$, should be monotonic, but it was not. They improved Model II by estimating a preliminary exponential transformation, `x5e` $= \exp(-0.12 \cdot$ `x5`$)$, for `x5` and fitting a degree 1 FP for `x5e`, thus obtaining a monotonic risk function. The value of -0.12 was estimated univariately using nonlinear Cox regression with the ado-file `boxtid` (Royston and Ambler 1999a,b). To ensure a negative exponent, Sauerbrei and Royston (1999) restricted the powers for `x5e` to be positive. Their Model III may be fitted by using the following command:

```
. mfp stcox x1 x2 x3 x4a x4b x5e x6 x7 hormon, alpha(.05) select(.05, hormon:1)
> df(x5e:2) xpowers(x5e:0.5 1 2 3)
```

Other than the customization for `x5e`, the command is the same as it was before. The resulting model is as reported in table 4 of Sauerbrei and Royston (1999):

```
. use http://www.stata-press.com/data/r10/brcancer, clear
(German breast cancer data)
```

```
. stset rectime, fail(censrec)
```

(output omitted)

```
. mfp stcox x1 x2 x3 x4a x4b x5e x6 x7 hormon, nohr alpha(.05)
> select(.05, hormon:1) df(x5e:2) xpowers(x5e:0.5 1 2 3)
```

(output omitted)

```
Final multivariable fractional polynomial model for _t

Variable |  -----Initial-----           -----Final-----
         |  df   Select    Alpha     Status   df    Powers
---------+-----------------------------------------------
      x1 |   4   0.0500    0.0500      in      4    -2 -.5
      x2 |   1   0.0500    0.0500      out     0
      x3 |   4   0.0500    0.0500      out     0
     x4a |   1   0.0500    0.0500      in      1    1
     x4b |   1   0.0500    0.0500      out     0
     x5e |   2   0.0500    0.0500      in      1    1
      x6 |   4   0.0500    0.0500      in      2    .5
      x7 |   4   0.0500    0.0500      out     0
  hormon |   1   1.0000    0.0500      in      1    1

Cox regression -- Breslow method for ties
Entry time _t0                                  Number of obs   =        686
                                                LR chi2(6)      =     153.11
                                                Prob > chi2     =     0.0000
Log likelihood =  1711.6186                     Pseudo R2       =     0.0428

------------------------------------------------------------------------------
          _t |      Coef.   Std. Err.      z    P>|z|     [95% Conf. Interval]
-------------+----------------------------------------------------------------
     Ix1__1  |   43.55382   8.253433     5.28   0.000     27.37738    59.73025
     Ix1__2  |  -17.48136   3.911882    -4.47   0.000    -25.14851   -9.814212
        x4a  |   .5174351   .2493739     2.07   0.038     .0286713    1.006199
    Ix5e__1  |  -1.981213   .2268903    -8.73   0.000    -2.425909   -1.536516
     Ix6__1  |   -1.84008   .3508432    -5.24   0.000     -2.52772    -1.15244
     hormon  |  -.3944998    .128097    -3.08   0.002    -.6455654   -.1434342
------------------------------------------------------------------------------
Deviance: 3423.237.
```

◁

(Continued on next page)

Saved Results

In addition to what *regression_cmd* saves, `mfp` saves the following in `e()`:

Scalars

`e(fp_nx)`	number of predictors in *xvarlist*
`e(fp_dev)`	deviance of final model fitted
`e(Fp_id#)`	initial degrees of freedom for the #th element of *xvarlist*
`e(Fp_fd#)`	final degrees of freedom for the #th element of *xvarlist*
`e(Fp_al#)`	FP selection level for the #th element of *xvarlist*
`e(Fp_se#)`	backward elimination selection level for the #th element of *xvarlist*

Macros

`e(fp_cmd)`	`fracpoly`
`e(fp_cmd2)`	`mfp`
`e(cmdline)`	command as typed
`e(fp_fvl)`	variables in final model
`e(fp_depv)`	*yvar*
`e(fp_opts)`	estimation command options
`e(fp_x1)`	first variable in *xvarlist*
`e(fp_x2)`	second variable in *xvarlist*
...	
`e(fp_x`*N*`)`	last variable in *xvarlist*, *N*=`e(fp_nx)`
`e(fp_k1)`	power for first variable in *xvarlist* (*)
`e(fp_k2)`	power for second variable in *xvarlist* (*)
...	
`e(fp_k`*N*`)`	power for last var. in *xvarlist* (*), *N*=`e(fp_nx)`

Note: (*) contains '`.`' if variable is not selected in final model.

Methods and Formulas

`mfp` is implemented as an ado-file.

Acknowledgments

`mfp` is an update of `mfracpol` by Royston and Ambler (1998).

References

Ambler, G., and P. Royston. 2001. Fractional polynomial model selection procedures: Investigation of Type I error rate. *Journal of Statistical Simulation and Computation* 69: 89–108.

Royston, P., and D. G. Altman. 1994. Regression using fractional polynomials of continuous covariates: Parsimonious parametric modelling (with discussion). *Applied Statistics* 43: 429–467.

Royston, P., and G. Ambler. 1998. sg81: Multivariable fractional polynomials. *Stata Technical Bulletin* 43: 24–32. Reprinted in *Stata Technical Bulletin Reprints*, vol. 8, pp. 123–132.

——. 1999a. sg81.1: Multivariable fractional polynomials: Update. *Stata Technical Bulletin* 49: 17–23. Reprinted in *Stata Technical Bulletin Reprints*, vol. 9, pp. 161–168.

——. 1999b. sg81.2: Multivariable fractional polynomials: Update. *Stata Technical Bulletin* 50: 25. Reprinted in *Stata Technical Bulletin Reprints*, vol. 9, p. 168.

——. 1999c. sg112: Nonlinear regression models involving power of exponential functions of covariates. *Stata Technical Bulletin* 49: 25–30. Reprinted in *Stata Technical Bulletin Reprints*, vol. 9, pp. 173–179.

——. 1999d. sg112.1: Nonlinear regression models involving power of exponential functions: Update. *Stata Technical Bulletin* 50: 26. Reprinted in *Stata Technical Bulletin Reprints*, vol. 9, p. 180.

Royston, P., and W. Sauerbrei. 2007. Multivariable modeling with cubic regression splines: A principled approach. *Stata Journal* 7: 45–70.

Sauerbrei, W., and P. Royston. 1999. Building multivariable prognostic and diagnostic models: Transformation of the predictors by using fractional polynomials. *Journal of the Royal Statistical Society, Series A* 162: 71–94.

——. 2002. Corrigendum: Building multivariable prognostic and diagnostic models: Transformation of the predictors by using fractional polynomials. *Journal of the Royal Statistical Society, Series A* 165: 299–300.

Also See

[R] **mfp postestimation** — Postestimation tools for mfp

[R] **fracpoly** — Fractional polynomial regression

[U] **20 Estimation and postestimation commands**

Title

mfp postestimation — Postestimation tools for mfp

Description

The following postestimation commands are of special interest after `mfp`:

command	description
`fracplot`	plot data and fit from most recently fitted fractional polynomial model
`fracpred`	create variable containing prediction, deviance residuals, or SEs of fitted values

For `fracplot` and `fracpred`, see [R] **fracpoly postestimation**.

The following standard postestimation commands are also available if available after *regression_cmd*:

command	description
`adjust`	adjusted predictions of $\mathbf{x}\beta$, probabilities, or $\exp(\mathbf{x}\beta)$
`estat`	AIC, BIC, VCE, and estimation sample summary
`estimates`	cataloging estimation results
`lincom`	point estimates, standard errors, testing, and inference for linear combinations of coefficients
`linktest`	link test for model specification
`lrtest`	likelihood-ratio test
`mfx`	marginal effects or elasticities
`nlcom`	point estimates, standard errors, testing, and inference for nonlinear combinations of coefficients
`test`	Wald tests for simple and composite linear hypotheses
`testnl`	Wald tests of nonlinear hypotheses

See the corresponding entries in the *Stata Base Reference Manual* for details.

Methods and Formulas

All postestimation commands listed above are implemented as ado-files.

Also See

[R] **mfp** — Multivariable fractional polynomial models

[R] **fracpoly postestimation** — Postestimation tools for fracpoly

[U] **20 Estimation and postestimation commands**

Title

mfx — Obtain marginal effects or elasticities after estimation

Syntax

`mfx` [`compute`] [*if*] [*in*] [, *options*]

`mfx replay` [, `level(#)`]

options	description
Model	
`predict(`*predict_option*`)`	calculate marginal effects (elasticities) for *predict_option*
`varlist(`*varlist*`)`	calculate marginal effects (elasticities) for *varlist*
`dydx`	calculate marginal effects; the default
`eyex`	calculate elasticities in the form of $\partial \log y / \partial \log x$
`dyex`	calculate elasticities in the form of $\partial y / \partial \log x$
`eydx`	calculate elasticities in the form of $\partial \log y / \partial x$
`nodiscrete`	treat dummy (indicator) variables as continuous
`nose`	do not calculate standard errors
Model 2	
`at(`*atlist*`)`	calculate marginal effects (elasticities) at these values
`noesample`	do not restrict calculation of means and medians to the estimation sample
`nowght`	ignore weights when calculating means and medians
Adv. model	
`nonlinear`	do not use the linear method
`force`	calculate marginal effects and standard errors when it would otherwise refuse to do so
Reporting	
`level(#)`	set confidence level; default is `level(95)`
`diagnostics(beta)`	report suitability of marginal-effect calculation
`diagnostics(vce)`	report suitability of standard-error calculation
`diagnostics(all)`	report all diagnostic information
`tracelvl(#)`	report increasing levels of detail during calculations

where *atlist* is *numlist* or *matname* or

[`mean` | `median` | `zero`] [*varname* = # [, *varname* = #] [...]]

where `mean` is the default.

Description

`mfx` numerically calculates the marginal effects or the elasticities and their standard errors after estimation. Exactly what `mfx` can calculate is determined by the previous estimation command and the `predict(`*predict_option*`)` option. The values at which the marginal effects or elasticities are to be evaluated is determined by the `at(`*atlist*`)` option. By default, `mfx` calculates the marginal effects or elasticities at the means of the independent variables by using the default prediction option associated with the previous estimation command.

Some disciplines use the term *partial effects*, rather than marginal effects, for what is computed by `mfx`.

`mfx replay` replays the results of the previous `mfx` computation.

Options

Model

`predict(`*predict_option*`)` specifies the function (that is, the form of y) for which to calculate the marginal effects or elasticities. The default is to use the default `predict` option of the preceding estimation command. To see which `predict` options are available, see `help` for that estimation command.

`varlist(`*varlist*`)` specifies the variables for which to calculate marginal effects (elasticities). The default is all variables.

`dydx` specifies that marginal effects be calculated. This is the default.

`eyex` specifies that elasticities be calculated in the form of $\partial \log y/\partial \log x$.

`dyex` specifies that elasticities be calculated in the form of $\partial y/\partial \log x$.

`eydx` specifies that elasticities be calculated in the form of $\partial \log y/\partial x$.

`nodiscrete` treats dummy variables as continuous. A dummy variable is one that takes on the value 0 or 1 in the estimation sample. If `nodiscrete` is not specified, the marginal effect of a dummy variable is calculated as the discrete change in y as the dummy variable changes from 0 to 1. This option is irrelevant to the computation of the elasticities because all dummy variables are treated as continuous when computing elasticities.

`nose` specifies that standard errors of the marginal effects (elasticities) not be computed.

Model 2

`at(`*atlist*`)` specifies the values at which the marginal effects (elasticities) are to be calculated. The default is to calculate at the means of the independent variables.

`at(`*numlist*`)` specifies that the marginal effects (elasticities) be calculated at *numlist*. For instance,

```
. use http://www.stata-press.com/data/r10/auto
. probit foreign mpg weight price
. mfx, predict(xb eq(#2)) at(200 3000 0.5)
```

computes the marginal effects for the second equation, setting `disp` = 200, `weight` = 3000, and `foreign` = 0.5.

The order of the values in the *numlist* is the same as the variables in the preceding estimation command, from left to right, without repetition. For instance,

```
. sureg (price disp weight) (mpg foreign disp)
. mfx, predict(xb) at(200 3000 0.5)
```

`at(`*matname*`)` specifies the points in a matrix format. The ordering of the variables is the same as that of *numlist*. For instance,

```
. probit foreign mpg weight price
. matrix A = (21, 3000, 6000)
. mfx, at(A)
```

`at(`[`mean` | `median` | `zero`] [*varname* = # [, *varname* = #] [...]]`)` specifies that the marginal effects (elasticities) be calculated at means, at medians of the independent variables, or at zeros. It also allows users to specify particular values for one or more independent variables, assuming that the rest are means, medians, or zeros.

```
. probit foreign mpg weight price
. mfx, at(mean mpg=30)
```

`at(`*varname* = # [, *varname* = #] [...]`)` specifies that the marginal effects or the elasticities be calculated at particular values for one or more independent variables, assuming that the rest are means.

```
. probit foreign mpg weight price
. mfx, at(mpg=30)
```

`noesample` affects `at(`*atlist*`)`, any offsets used in the preceding estimation, and the determination of dummy variables. It specifies that the whole dataset be considered instead of only those marked in the `e(sample)` defined by the previous estimation command.

`nowght` affects only `at(`*atlist*`)` and offsets. It specifies that weights be ignored when calculating the means or medians for the *atlist* and when calculating the means for any offsets.

Adv. model

`nonlinear` specifies that y, the function to be calculated for the marginal effects or the elasticities, does not meet the linear-form restriction. By default, `mfx` assumes that y meets the linear-form restriction, unless one or more dependent variables are shared by multiple equations or the previous estimation command was `nl` (see *Using mfx after nl* below). For instance, predictions after

```
. heckman mpg price, sel(foreign=rep78)
```

meet the linear-form restriction, but those after

```
. heckman mpg price, sel(foreign=rep78 price)
```

do not. If y meets the linear-form restriction, specifying `nonlinear` should produce the same results as not specifying it. However, the nonlinear method is generally more time consuming. Most likely, you do not need to specify `nonlinear` after an official Stata command. For user-written commands, if you are not sure whether y is of linear form, specifying `nonlinear` is a safe choice.

`force` specifies that marginal effects and their standard errors be calculated when it would otherwise refuse to do so. Such cases arise, for instance, when the marginal effect is a function of a random quantity other than the coefficients of the model (e.g., a residual). If you specify this option, there is no guarantee that the resulting marginal effects and standard errors are correct.

Reporting

`level(`*#*`)` specifies the confidence level, as a percentage, for confidence intervals. The default is `level(95)` or as set by `set level`; see [U] **23.5 Specifying the width of confidence intervals**.

`diagnostics(`*diaglist*`)` asks `mfx` to display various diagnostic information.

`diagnostics(beta)` shows the information used to determine whether the prediction option is suitable for computing marginal effects.

`diagnostics(vce)` shows the information used to determine whether the prediction option is suitable for computing the standard errors of the marginal effects.

`diagnostics(all)` shows all the above diagnostic information.

`tracelvl(`*#*`)` shows increasing levels of detail during calculations. # may be 1, 2, 3, or 4. Level 1 shows the marginal effects and standard errors as they are computed, and which method, either linear or nonlinear, was used. Level 2 shows, in addition, the components of the matrix of partial derivatives needed for each standard error as they are computed. Level 3 shows counts of iterations in obtaining a suitable finite difference for each numerical derivative. Level 4 shows the values of these finite differences.

Remarks

Remarks are presented under the following headings:

Obtaining marginal effects after single-equation (SE) estimation
Obtaining marginal effects after multiple-equation (ME) estimation
Specifying the evaluation points
Obtaining three forms of elasticities
Using mfx after nl

Obtaining marginal effects after single-equation (SE) estimation

Before running `mfx`, type `help` *estimation_cmd* to see what can be predicted after the estimation and to see the default prediction.

▷ Example 1

We fit a logit model with the auto dataset:

```
. use http://www.stata-press.com/data/r10/auto
(1978 Automobile Data)

. logit foreign mpg price, nolog

Logistic regression                               Number of obs   =         74
                                                  LR chi2(2)      =      17.14
                                                  Prob > chi2     =     0.0002
Log likelihood = -36.462189                       Pseudo R2       =     0.1903

------------------------------------------------------------------------------
     foreign |      Coef.   Std. Err.      z    P>|z|     [95% Conf. Interval]
-------------+----------------------------------------------------------------
         mpg |   .2338353   .0671449     3.48   0.000     .1022338    .3654368
       price |    .000266   .0001166     2.28   0.022     .0000375    .0004945
       _cons |  -7.648111   2.043673    -3.74   0.000    -11.65364   -3.642586
------------------------------------------------------------------------------
```

To determine the marginal effects of `mpg` and `price` for the probability of a positive outcome at their mean values, we can issue the `mfx` command without the `predict` option because the default prediction after `logit` is the probability of a positive outcome. The calculation is requested at the mean values by default.

```
. mfx

Marginal effects after logit
      y  = Pr(foreign) (predict)
         = .26347633
```

variable	dy/dx	Std. Err.	z	P>|z|	[95% C.I.	]	X
mpg	.0453773	.0131	3.46	0.001	.019702	.071053	21.2973
price	.0000516	.00002	2.31	0.021	7.8e-06	.000095	6165.26

The first line of the output indicates that the marginal effects were calculated after a `logit` estimation. The second line of the output describes the form of y and the `predict` command that we would type to calculate y separately. The third line of the output gives the value of y given the values of X, which are displayed in the last column of the table.

To calculate the marginal effects at particular data points, say, `mpg` = 20, `price` = 6000, specify the `at()` option:

```
. mfx, at(mpg=20, price=6000)

Marginal effects after logit
      y  = Pr(foreign) (predict)
         =  .20176601
```

variable	dy/dx	Std. Err.	z	P>|z|	[95% C.I.	]	X
mpg	.0376607	.00961	3.92	0.000	.018834	.056488	20
price	.0000428	.00002	2.47	0.014	8.8e-06	.000077	6000

To calculate the marginal effects for the linear prediction (`xb`) instead of the probability, specify `predict(xb)`. The marginal effects for the linear prediction are the coefficients themselves.

```
. mfx, predict(xb)

Marginal effects after logit
      y  = Linear prediction (predict, xb)
         =-1.0279779
```

variable	dy/dx	Std. Err.	z	P>|z|	[95% C.I.	]	X
mpg	.2338353	.06714	3.48	0.000	.102234	.365437	21.2973
price	.000266	.00012	2.28	0.022	.000038	.000495	6165.26

If there is a dummy variable as an independent variable, `mfx` calculates the discrete change as the dummy variable changes from 0 to 1.

```
. generate goodrep = 0

. replace goodrep = 1 if rep > 3
(34 real changes made)
```

(Continued on next page)

```
. logit foreign mpg goodrep, nolog
Logistic regression                               Number of obs   =         74
                                                  LR chi2(2)      =      26.27
                                                  Prob > chi2     =     0.0000
Log likelihood = -31.898321                       Pseudo R2       =     0.2917

------------------------------------------------------------------------------
     foreign |      Coef.   Std. Err.      z    P>|z|     [95% Conf. Interval]
-------------+----------------------------------------------------------------
         mpg |   .1079219   .0565077     1.91   0.056    -.0028311    .2186749
     goodrep |   2.435068   .7128444     3.42   0.001     1.037918    3.832217
       _cons |  -4.689347   1.326547    -3.54   0.000     -7.28933   -2.089363
------------------------------------------------------------------------------

. mfx
Marginal effects after logit
      y  = Pr(foreign) (predict)
         =  .21890034
------------------------------------------------------------------------------
variable |      dy/dx    Std. Err.     z    P>|z|  [    95% C.I.   ]      X
---------+--------------------------------------------------------------------
     mpg |   .0184528      .01017    1.81   0.070  -.001475  .038381   21.2973
goodrep* |   .4271707      .10432    4.09   0.000   .222712   .63163   .459459
------------------------------------------------------------------------------
(*) dy/dx is for discrete change of dummy variable from 0 to 1
```

If `nodiscrete` is specified, `mfx` treats the dummy variable as continuous.

```
. mfx, nodiscrete
Marginal effects after logit
      y  = Pr(foreign) (predict)
         =  .21890034
------------------------------------------------------------------------------
variable |      dy/dx    Std. Err.     z    P>|z|  [    95% C.I.   ]      X
---------+--------------------------------------------------------------------
     mpg |   .0184528      .01017    1.81   0.070  -.001475  .038381   21.2973
 goodrep |   .4163552      .10733    3.88   0.000   .205994  .626716   .459459
------------------------------------------------------------------------------
```

◁

❑ Technical Note

By default, `mfx` uses the estimation sample to determine which independent variables are dummies. A variable is declared a dummy if its only values in the estimation sample are zero or one. This determination may be affected by the option `noesample`. For example,

```
. replace rep78=rep78-3
(69 real changes made)
. logit foreign mpg rep78 if rep78==0|rep78==1, nolog
Logistic regression                               Number of obs   =         48
                                                  LR chi2(2)      =      21.88
                                                  Prob > chi2     =     0.0000
Log likelihood = -16.050909                       Pseudo R2       =     0.4053

------------------------------------------------------------------------------
     foreign |      Coef.   Std. Err.      z    P>|z|     [95% Conf. Interval]
-------------+----------------------------------------------------------------
         mpg |    .347422    .120328     2.89   0.004     .1115834    .5832605
       rep78 |   2.105463   .9194647     2.29   0.022     .3033449     3.90758
       _cons |  -9.683393   2.876752    -3.37   0.001    -15.32172   -4.045063
------------------------------------------------------------------------------
```

```
. mfx

Marginal effects after logit
      y  = Pr(foreign) (predict)
         =   .1357189
```

variable	dy/dx	Std. Err.	z	P>|z|	[95% C.I.]		X
mpg	.0407523	.01531	2.66	0.008	.010755	.07075	20.2708
rep78*	.3027042	.14694	2.06	0.039	.01471	.590699	.375

(*) dy/dx is for discrete change of dummy variable from 0 to 1

When noesample is specified, the value of rep78 is considered for all observations in the dataset. Since observations with rep78 not equal to zero or one do exist, mfx will conclude that it is not a dummy variable.

```
.table rep78
```

Repair Record 1978	Freq.
-2	2
-1	8
0	30
1	18
2	11

```
. mfx, noesample

Marginal effects after logit
      y  = Pr(foreign) (predict)
         =  .19312144
```

variable	dy/dx	Std. Err.	z	P>|z|	[95% C.I.]		X
mpg	.0541372	.02043	2.65	0.008	.014089	.094185	21.2973
rep78	.3280849	.14321	2.29	0.022	.047401	.608769	.405797

The qualifiers if and in have no effect on the determination of dummy variables.

❑

Obtaining marginal effects after multiple-equation (ME) estimation

If you have not read the discussion above on using mfx after SE estimations, please do so. As a general introduction to the ME models, the following examples will demonstrate mfx after heckman and mlogit.

(Continued on next page)

▷ Example 2

```
. use http://www.stata-press.com/data/r10/auto, clear
(1978 Automobile Data)

. heckman mpg weight length, sel(foreign = displacement) nolog
Heckman selection model                         Number of obs      =        74
(regression model with sample selection)        Censored obs       =        52
                                                Uncensored obs     =        22

                                                Wald chi2(2)       =      7.27
Log likelihood = -87.58426                      Prob > chi2        =    0.0264

------------------------------------------------------------------------------
             |      Coef.   Std. Err.      z    P>|z|     [95% Conf. Interval]
-------------+----------------------------------------------------------------
mpg          |
      weight |  -.0039923   .0071948    -0.55   0.579    -.0180939    .0101092
      length |  -.1202545   .2093074    -0.57   0.566    -.5304895    .2899805
       _cons |   56.72567   21.68463     2.62   0.009     14.22458    99.22676
-------------+----------------------------------------------------------------
foreign      |
displacement |  -.0250297   .0067241    -3.72   0.000    -.0382088   -.0118506
       _cons |   3.223625   .8757406     3.68   0.000     1.507205    4.940045
-------------+----------------------------------------------------------------
     /athrho |  -.9840858   .8112212    -1.21   0.225     -2.57405    .6058785
    /lnsigma |   1.724306   .2794524     6.17   0.000     1.176589    2.272022
-------------+----------------------------------------------------------------
         rho |  -.7548292    .349014                     -.9884463    .5412193
       sigma |   5.608626   1.567344                      3.243293    9.698997
      lambda |  -4.233555   3.022645                     -10.15783    1.690721
------------------------------------------------------------------------------
LR test of indep. eqns. (rho = 0):   chi2(1) =     1.37   Prob > chi2 = 0.2413
------------------------------------------------------------------------------
```

heckman estimated two equations, mpg and foreign; see [R] **heckman**. Two of the prediction options after heckman are the expected value of the dependent variable and the probability of being observed. To obtain the marginal effects of all the independent variables for the expected value of the dependent variable, we specify predict(yexpected) with mfx.

```
. mfx, predict(yexpected)

Marginal effects after heckman
      y  = E(mpg*|Pr(foreign)) (predict, yexpected)
         = .56522778
------------------------------------------------------------------------------
variable |      dy/dx    Std. Err.     z    P>|z|  [    95% C.I.   ]      X
---------+--------------------------------------------------------------------
  weight |  -.0001725      .00041   -0.42   0.675  -.000979  .000634   3019.46
  length |  -.0051953      .01002   -0.52   0.604   -.02483   .01444   187.932
displa~t |  -.0340055      .02793   -1.22   0.223  -.088739  .020728   197.297
------------------------------------------------------------------------------
```

To calculate the marginal effects for the probability of being observed, we specify predict(psel) with mfx. Since only the independent variables in equation foreign affect the probability of being observed, some of the marginal effects will be zero. Using the option varlist(*varlist*) with mfx will restrict the calculation of marginal effects to the independent variables in the *varlist*.

```
. mfx, predict(psel)

Marginal effects after heckman
      y  = Pr(foreign) (predict, psel)
         =  .04320292

------------------------------------------------------------------------------
variable |      dy/dx    Std. Err.     z    P>|z|  [    95% C.I.   ]      X
---------+--------------------------------------------------------------------
  weight |          0          0       .        .       0         0   3019.46
  length |          0          0       .        .       0         0   187.932
displa~t |  -.0022958     .00153   -1.50   0.133  -.005287  .000696   197.297
------------------------------------------------------------------------------

. mfx, predict(psel) varlist(displacement)

Marginal effects after heckman
      y  = Pr(foreign) (predict, psel)
         =  .04320292

------------------------------------------------------------------------------
variable |      dy/dx    Std. Err.     z    P>|z|  [    95% C.I.   ]      X
---------+--------------------------------------------------------------------
displa~t |  -.0022958     .00153   -1.50   0.133  -.005287  .000696   197.297
------------------------------------------------------------------------------
```

◁

▷ Example 3

`predict` after `mlogit`, unlike most other estimation commands, can predict multiple new variables by issuing `predict` only once; see [R] **mlogit**. To calculate the marginal effects for the probability of more than one outcome, we run `mfx` separately for each outcome.

```
. mlogit rep78 mpg, nolog

Multinomial logistic regression                   Number of obs   =         69
                                                  LR chi2(4)      =      15.88
                                                  Prob > chi2     =     0.0032
Log likelihood = -85.752375                       Pseudo R2       =     0.0847

------------------------------------------------------------------------------
       rep78 |      Coef.   Std. Err.      z    P>|z|     [95% Conf. Interval]
-------------+----------------------------------------------------------------
1            |
         mpg |   .0708122   .1471461     0.48   0.630    -.2175888    .3592132
       _cons |  -4.137144    3.15707    -1.31   0.190    -10.32489    2.050599
-------------+----------------------------------------------------------------
2            |
         mpg |  -.0164251   .0926724    -0.18   0.859    -.1980597    .1652095
       _cons |  -1.005118   1.822129    -0.55   0.581    -4.576426    2.566189
-------------+----------------------------------------------------------------
4            |
         mpg |   .0958626   .0633329     1.51   0.130    -.0282676    .2199927
       _cons |  -2.474187   1.341131    -1.84   0.065    -5.102756     .154381
-------------+----------------------------------------------------------------
5            |
         mpg |   .2477469   .0764076     3.24   0.001     .0979908     .397503
       _cons |  -6.653164   1.841793    -3.61   0.000    -10.26301   -3.043316
------------------------------------------------------------------------------
(rep78==3 is the base outcome)
```

```
. mfx, predict(outcome(1))

Marginal effects after mlogit
      y  = Pr(rep78==1) (predict, outcome(1))
         =  .03233059
```

variable	dy/dx	Std. Err.	z	P>|z|	[	95% C.I.]	X
mpg	.0004712	.0045	0.10	0.917	-.00835	.009292	21.2899

◁

Specifying the evaluation points

By default, `mfx` evaluates the marginal effects at the means of the independent variables. To evaluate elsewhere, you would specify the option `at()` with `mfx`. This option allows several different syntaxes.

▷ Example 4

Using the *numlist* and *matname* syntax, we must specify the evaluation points in the same order as the variables in the preceding estimation command, that is, from left to right, without repetition.

```
. sureg (price disp weight) (mpg foreign disp)

Seemingly unrelated regression
```

Equation	Obs	Parms	RMSE	"R-sq"	chi2	P
price	74	2	2466.937	0.2909	29.98	0.0000
mpg	74	2	4.061588	0.5004	74.03	0.0000

	Coef.	Std. Err.	z	P>|z|	[95% Conf.	Interval]
price						
displacement	2.660349	7.043538	0.38	0.706	-11.14473	16.46543
weight	1.747666	.8322723	2.10	0.036	.1164423	3.37889
_cons	363.3701	1441.681	0.25	0.801	-2462.273	3189.014
mpg						
foreign	-.6825672	1.30813	-0.52	0.602	-3.246455	1.881321
displacement	-.0465529	.0065556	-7.10	0.000	-.0594017	-.033704
_cons	30.68498	1.632312	18.80	0.000	27.4857	33.88425

```
. mfx, predict(xb) at(200 3000 0.5)

Marginal effects after sureg
      y  = Linear prediction (predict, xb)
         =  6138.4383
```

variable	dy/dx	Std. Err.	z	P>|z|	[	95% C.I.]	X
displa~t	2.660349	7.04354	0.38	0.706	-11.1447	16.4654	200
weight	1.747666	.83227	2.10	0.036	.116442	3.37889	3000
foreign*	0	0	.	.	0	0	.5

(*) dy/dx is for discrete change of dummy variable from 0 to 1

◁

❑ Technical Note

When using the *numlist* or *matname* syntax together with `varlist()`, you must specify values for all the independent variables, not just those for which marginal effects will be calculated. These values are used in the estimation of the marginal effects, and although they don't display in the output of `mfx`, they are included in the saved results.

```
. probit foreign mpg price, nolog
Probit regression                               Number of obs   =         74
                                                LR chi2(2)      =      17.53
                                                Prob > chi2     =     0.0002
Log likelihood = -36.266068                     Pseudo R2       =     0.1947

------------------------------------------------------------------------------
     foreign |      Coef.   Std. Err.      z    P>|z|     [95% Conf. Interval]
-------------+----------------------------------------------------------------
         mpg |   .1404876   .0373595     3.76   0.000     .0672644    .2137108
       price |   .0001571   .0000641     2.45   0.014     .0000315    .0002827
       _cons |  -4.592058   1.115907    -4.12   0.000    -6.779195   -2.404921
------------------------------------------------------------------------------

. capture noisily mfx, at(6000) varlist(price)
numlist too short in at()

. mfx, at(20 6000) varlist(price)
Marginal effects after probit
      y  = Pr(foreign) (predict)
         =  .2005512
------------------------------------------------------------------------------
variable |      dy/dx    Std. Err.     z    P>|z|  [    95% C.I.   ]      X
---------+--------------------------------------------------------------------
   price |   .0000441      .00002    2.62   0.009   .000011  .000077     6000
------------------------------------------------------------------------------

. matrix list e(Xmfx_X)
e(Xmfx_X)[1,2]
      mpg  price
r1     20   6000
```

❑

Obtaining three forms of elasticities

`mfx` can also be used to obtain all three forms of elasticities.

option	elasticity
`eyex`	$\partial \log y/\partial \log x$
`dyex`	$\partial y/\partial \log x$
`eydx`	$\partial \log y/\partial x$

▷ Example 5

We fit a regression model with the auto dataset. The marginal effects for the predicted value y after a `regress` are the same as the coefficients. To obtain the elasticities of form $\partial \log y/\partial \log x$, we specify the `eyex` option:

```
. regress mpg weight length

      Source |       SS       df       MS              Number of obs =      74
-------------+------------------------------           F(  2,    71) =   69.34
       Model |  1616.08062     2  808.040312           Prob > F      =  0.0000
    Residual |  827.378835    71   11.653223           R-squared     =  0.6614
-------------+------------------------------           Adj R-squared =  0.6519
       Total |  2443.45946    73  33.4720474           Root MSE      =  3.4137

------------------------------------------------------------------------------
         mpg |      Coef.   Std. Err.      t    P>|t|     [95% Conf. Interval]
-------------+----------------------------------------------------------------
      weight |  -.0038515    .001586    -2.43   0.018    -.0070138   -.0006891
      length |  -.0795935   .0553577    -1.44   0.155    -.1899736    .0307867
       _cons |   47.88487    6.08787     7.87   0.000       35.746    60.02374
------------------------------------------------------------------------------

. mfx, eyex

Elasticities after regress
      y  = Fitted values (predict)
         = 21.297297
------------------------------------------------------------------------------
variable |      ey/ex    Std. Err.     z    P>|z|  [    95% C.I.   ]      X
---------+--------------------------------------------------------------------
  weight |  -.5460497      .22509   -2.43   0.015  -.987208 -.104891   3019.46
  length |  -.7023518      .48867   -1.44   0.151  -1.66012  .255414   187.932
------------------------------------------------------------------------------
```

The first line of the output indicates that the elasticities were calculated after a `regress` estimation. The title of the second column of the table gives the form of the elasticities, $\partial \log y/\partial \log x$, the percent change in y for a 1% change in x.

If the independent variables have been log-transformed already, we will want the elasticities of the form $\partial \log y/\partial x$ instead.

```
. generate lnweight = ln(weight)

. generate lnlength = ln(length)

. regress mpg lnweight lnlength
      Source |       SS       df       MS              Number of obs =      74
-------------+------------------------------           F(  2,    71) =   74.00
       Model |  1651.28916     2  825.644581           Prob > F      =  0.0000
    Residual |  792.170298    71  11.1573281           R-squared     =  0.6758
-------------+------------------------------           Adj R-squared =  0.6667
       Total |  2443.45946    73  33.4720474           Root MSE      =  3.3403

------------------------------------------------------------------------------
         mpg |      Coef.   Std. Err.      t    P>|t|     [95% Conf. Interval]
-------------+----------------------------------------------------------------
    lnweight |   -13.5974   4.692504    -2.90   0.005    -22.95398   -4.240811
    lnlength |  -9.816726   10.40316    -0.94   0.349    -30.56005    10.92659
       _cons |   181.1196   22.18429     8.16   0.000     136.8853    225.3538
------------------------------------------------------------------------------

. mfx, eydx

Elasticities after regress
      y  = Fitted values (predict)
         = 21.297297
------------------------------------------------------------------------------
variable |      ey/dx    Std. Err.     z    P>|z|  [    95% C.I.   ]      X
---------+--------------------------------------------------------------------
lnweight |  -.6384565      .22064   -2.89   0.004   -1.0709 -.206009   7.97875
lnlength |  -.4609376      .48855   -0.94   0.345  -1.41847  .496594   5.22904
------------------------------------------------------------------------------
```

Although the interpretation is the same, the results for `eyex` and `eydx` differ since we are fitting different models.

If the dependent variable were log-transformed, we would specify `dyex` instead. ◁

Using mfx after nl

You must specify the independent variables by using the `variables()` option when using the interactive version of `nl` to obtain marginal effects. Otherwise, `mfx` has no way of distinguishing the independent variables from the parameters of your model and will therefore exit with an error message.

Instead of typing

```
. nl (mpg = {b0} + {b1}*gear^{b2=1})
```

type

```
. nl (mpg = {b0} + {b1}*gear^{b2=1}), variables(gear)
```

If you use the programmed substitutable expression or function evaluator program versions of `nl`, you do not need to use the `variables()` option.

Saved Results

In addition to the `e()` results from the preceding estimation, `mfx` saves the following in `e()`:

Scalars

`e(Xmfx_y)`	value of y given X
`e(Xmfx_off)`	value of mean of the offset variable or log of the exposure variable
`e(Xmfx_off#)`	value of mean of the offset variable for equation #

Macros

`e(Xmfx_type)`	`dydx`, `eyex`, `eydx` or `dyex`
`e(Xmfx_discrete)`	`discrete` or `nodiscrete`
`e(Xmfx_cmd)`	`mfx`
`e(Xmfx_label_p)`	label for prediction in output
`e(Xmfx_predict)`	*predict_option* specified in `predict()`
`e(Xmfx_dummy)`	corresponding to independent variables; `1` means dummy, `0` means continuous
`e(Xmfx_variables)`	corresponding to independent variables; `1` means marginal effect calculated, `0` otherwise
`e(Xmfx_method)`	`linear` or `nonlinear`

Matrices

`e(Xmfx_dydx)`	marginal effects
`e(Xmfx_se_dydx)`	standard errors of the marginal effects
`e(Xmfx_eyex)`	elasticities of form `eyex`
`e(Xmfx_se_eyex)`	standard errors of elasticities of form `eyex`
`e(Xmfx_eydx)`	elasticities of form `eydx`
`e(Xmfx_se_eydx)`	standard errors of elasticities of form `eydx`
`e(Xmfx_dyex)`	elasticities of form `dyex`
`e(Xmfx_se_dyex)`	standard errors of elasticities of form `dyex`
`e(Xmfx_X)`	values around which marginal effects (elasticities) were estimated

Methods and Formulas

`mfx` is implemented as an ado-file.

After an estimation, `mfx` calculates marginal effects (elasticities) and their standard errors. A marginal effect of a continuous independent variable x is the partial derivative, with respect to x, of the prediction function f specified in `mfx`'s `predict` option; see Greene (2003, 668) for more information about marginal effects. If no prediction function is specified, the default prediction for the preceding estimation command is used. This derivative is evaluated at the values of the independent variables specified in the `at()` option of `mfx` or, if none is specified, at the default values that are the means of the independent variables. If there were any offsets in the preceding estimation, the derivative is evaluated at the means of the offset variables.

For a dummy variable—one that takes only the values zero or one in the estimation sample—a difference rather than a derivative is computed. The difference is the value of the prediction function at one, minus its value at zero.

For a continuous variable, the derivative is calculated numerically, which means that it approximates the derivative by using the following formula with an appropriately small h,

$$\frac{\partial y}{\partial x_j} = \lim_{h \to 0} \frac{f(x_1, \ldots, x_j + h, \ldots, x_p, \beta_0, \ldots, \beta_p) - f(x_1, \ldots, x_j, \ldots, x_p, \beta_0, \ldots, \beta_p)}{h} \quad (1)$$

The delta method is used to estimate the variance of the marginal effect (Greene 2003, 70). The marginal effect is a function of only the coefficients of the model since all other variables are held constant at the values at which the marginal effect is sought.

$$\mathrm{Var}\left(\frac{\partial y}{\partial x_j}\right) = \mathbf{D}_j' \mathbf{V} \mathbf{D}_j$$

where $\mathbf{V}$ is the variance–covariance matrix from the estimation and $\mathbf{D}_j$ is the column vector whose kth entry is the partial derivative of the marginal effect of x_j, with respect to the coefficient of the kth independent variable:

$$(\mathbf{D}_j)_k = \frac{\partial}{\partial \beta_k} \frac{\partial y}{\partial x_j}$$

Thus to compute one standard error, the derivative of the marginal effect is computed with respect to each coefficient in the model.

Computing the derivative of a function f with respect to a variable x can be time consuming because an iterative algorithm must be used to find an appropriately small change in x for use in (1). The command `mfx` avoids this type of iteration as much as possible. If the independent variables and coefficients appear in the formula for the prediction function f only in the sum

$$\mathbf{x}\boldsymbol{\beta} = \beta_0 + \mathbf{x}\boldsymbol{\beta}_x = \beta_0 + \sum_{j=1}^{p} x_j \beta_j$$

a marked simplification in the computation of the marginal effects and their standard errors can be made.

An example of a prediction that satisfies this condition is the predicted probability of success following `logistic`:

$$f(x_1, \ldots x_p, \beta_0, \ldots, \beta_p) = \frac{\exp(\mathbf{x}\boldsymbol{\beta})}{1 + \exp(\mathbf{x}\boldsymbol{\beta})}$$

An example when this condition is not satisfied is the predicted hazard ratio following `streg` without the option `noconstant`:

$$f(x_1, \dots x_p, \beta_0, \dots, \beta_p) = \exp(\mathbf{x}\boldsymbol{\beta}_x)$$

The constant β_0 is missing from the sum.

For this condition to be satisfied after multiple-equation estimation, say, an estimation with two equations, the variables and coefficients can only appear as part of the two sums: that of the first equation $\mathbf{x}\boldsymbol{\beta}$ and that of the second equation $\mathbf{z}\boldsymbol{\gamma}$. If the same variable appears in both equations, even the linear predictor for the first equation does not satisfy this condition: if $x_{j_0} = z_{k_0}$ for some j_0 and k_0, then

$$\mathbf{x}\boldsymbol{\beta} = \sum_{j=0}^{p} x_j \beta_j = z_{k_0}\beta_{j_0} + \sum_{j=0}^{j_0-1} x_j\beta_j + \sum_{j=j_0+1}^{p} x_j\beta_j$$

so z_{k_0} is appearing, but not as part of the sum $\mathbf{z}\boldsymbol{\gamma} = \sum_{k=0}^{q} z_k\gamma_k$.

If this condition is satisfied, the linear-form restriction has been met, and the linear method, to be described below, is used to estimate the marginal effects and their standard errors. If not, the usual method, described above, is used and is called the nonlinear method. Following a multiple-equation estimation with any independent variables common to more than one equation, the nonlinear method will always be used.

We begin our description of the linear method with the easiest case, a single-equation estimation. Using the chain rule, we can write the marginal effect as

$$\frac{\partial y}{\partial x_j} = \frac{dy}{d(\mathbf{x}\boldsymbol{\beta})}\frac{\partial(\mathbf{x}\boldsymbol{\beta})}{\partial x_j} = \frac{dy}{d(\mathbf{x}\boldsymbol{\beta})}\beta_j$$

The same derivative, $dy/d(\mathbf{x}\boldsymbol{\beta})$, is used for every x_j. To calculate it, we use the same formula in reverse. Since it doesn't matter which variable is used, we use the first one, x_1:

$$\frac{dy}{d(\mathbf{x}\boldsymbol{\beta})} = \frac{1}{\beta_1}\frac{\partial y}{\partial x_1}$$

Therefore, only one derivative $\partial y/\partial x_1$ needs to be calculated by the usual nonlinear method, and all marginal effects are then obtained by multiplying the derivative by the appropriate coefficient. Thus the linear method is generally much faster than the nonlinear method.

To compute the standard errors, we need the second derivatives. If j is not equal to k, using the chain rule we have

$$\frac{\partial}{\partial\beta_k}\frac{\partial y}{\partial x_j} = \beta_j\, x_k \frac{d}{d(\mathbf{x}\boldsymbol{\beta})}\frac{dy}{d(\mathbf{x}\boldsymbol{\beta})} \tag{2}$$

If j is equal to k, using the product rule we have

$$\frac{\partial}{\partial\beta_j}\frac{\partial y}{\partial x_j} = \beta_j\, x_j \frac{d}{d(\mathbf{x}\boldsymbol{\beta})}\frac{dy}{d(\mathbf{x}\boldsymbol{\beta})} + \frac{dy}{d(\mathbf{x}\boldsymbol{\beta})} \tag{3}$$

We obtain the second derivative, again using the chain rule,

$$\frac{d}{d(\mathbf{x}\boldsymbol{\beta})}\frac{dy}{d(\mathbf{x}\boldsymbol{\beta})} = \frac{1}{\beta_1^2}\frac{\partial^2 y}{\partial x_1^2}$$

Now we turn to multiple equations. The linear method will be used only when there are no variables in common between the equations. For marginal effects, the formulas developed above apply in each equation separately. For standard errors, we will consider the case of two equations. Suppose that x_j and x_k are both in the first equation and j is not equal to k. Then $\partial/\partial\beta_k(\partial y/\partial x_j)$ is calculated as in (2) above, since x_j does not appear in the second equation, making $\partial(\mathbf{z}\boldsymbol{\gamma})/\partial x_j = 0$, and β_k is in equation one, making $\partial(\mathbf{z}\boldsymbol{\gamma})/\partial\beta_k = 0$. If x_j and x_k are both in the first equation and j is equal to k, we use the product rule and obtain the same as (3) above. Now suppose that x_j is in equation one and z_k is in equation two. Then

$$\frac{\partial}{\partial\gamma_k}\frac{\partial y}{\partial x_j} = \beta_j\ z_k\ \frac{d}{d(\mathbf{z}\boldsymbol{\gamma})}\frac{dy}{d(\mathbf{x}\boldsymbol{\beta})}$$

The second derivative is calculated by $d/d(\mathbf{z}\boldsymbol{\gamma})\{dy/d(\mathbf{x}\boldsymbol{\beta})\} = 1/(\gamma_1\beta_1)\{\partial^2 y/(\partial z_1 \partial x_1)\}$.

For multiple equations, it is possible to have an equation that is a constant only, such as an ancillary parameter. Then it is not possible to obtain $d^2y/d(\mathbf{z}\boldsymbol{\gamma})d(\mathbf{x}\boldsymbol{\beta})$ by converting to derivatives with respect to the independent variables, so it is evaluated directly. For example, if (2) had only a constant term so that $\mathbf{z}\boldsymbol{\gamma} = \gamma_0$, then $d^2y/d(\mathbf{z}\boldsymbol{\gamma})d(\mathbf{x}\boldsymbol{\beta}) = 1/\beta_1\ \partial/\partial x_1(dy/d\gamma_0)$.

References

Bartus, T. 2005. Estimation of marginal effects using margeff. *Stata Journal* 5: 309–329.

Baum, C. F. 2006. *An Introduction to Modern Econometrics Using Stata.* College Station, TX: Stata Press.

Greene, W. H. 2003. *Econometric Analysis.* 5th ed. Upper Saddle River, NJ: Prentice Hall.

Also See

[U] **20 Estimation and postestimation commands**

[R] **predict** — Obtain predictions, residuals, etc., after estimation

Title

mkspline — Linear and restricted cubic spline construction

Syntax

Linear spline with knots at specified points

mkspline *newvar*$_1$ #$_1$ [*newvar*$_2$ #$_2$ [...]] *newvar*$_k$ = *oldvar* [*if*] [*in*] [, marginal displayknots]

Linear spline with knots equally spaced or at percentiles of data

mkspline *stubname* # = *oldvar* [*if*] [*in*] [*weight*] [, marginal pctile displayknots]

Restricted cubic spline

mkspline *stubname* = *oldvar* [*if*] [*in*] [*weight*], cubic [nknots(#) knots(*numlist*) displayknots]

fweights are allowed with the second and third syntax; see [U] **11.1.6 weight**.

Description

mkspline creates variables containing a linear spline or a restricted cubic spline of *oldvar*.

In the first syntax, mkspline creates *newvar*$_1$, ..., *newvar*$_k$ containing a linear spline of *oldvar* with knots at the specified #$_1$, ..., #$_{k-1}$.

In the second syntax, mkspline creates # variables named *stubname*1, ..., *stubname*# containing a linear spline of *oldvar*. The knots are equally spaced over the range of *oldvar* or are placed at the percentiles of *oldvar*.

In the third syntax, mkspline creates variables containing a restricted cubic spline of *oldvar*. This is also known as a natural spline. The location and spacing of the knots is determined by the specification of the nknots() and knots() options.

Options

Options

marginal is allowed with the first or second syntax. It specifies that the new variables be constructed so that, when used in estimation, the coefficients represent the change in the slope from the preceding interval. The default is to construct the variables so that, when used in estimation, the coefficients measure the slopes for the interval.

`displayknots` displays the values of the knots that were used in creating the linear or restricted cubic spline.

`pctile` is allowed only with the second syntax. It specifies that the knots be placed at percentiles of the data rather than being equally spaced based on the range.

`nknots(#)` is allowed only with the third syntax. It specifies the number of knots that are to be used for a restricted cubic spline. This number must be between 3 and 7 unless the knot locations are specified using `knots()`. The default number of knots is 5.

`knots(`*numlist*`)` is allowed only with the third syntax. It specifies the exact location of the knots to be used for a restricted cubic spline. The values of these knots must be given in increasing order. When this option is omitted, the default knot values are based on Harrell's recommended percentiles with the additional restriction that the smallest knot may not be less than the fifth-smallest value of *oldvar* and the largest knot may not be greater than the fifth-largest value of *oldvar*. If both `nknots()` and `knots()` are given, they must specify the same number of knots.

Remarks

Remarks are presented under the following headings:

Linear splines
Restricted cubic splines

Linear splines

Linear splines allow estimating the relationship between y and x as a piecewise linear function, which is a function composed of linear segments—straight lines. One linear segment represents the function for values of x below x_0, another linear segment handles values between x_0 and x_1, and so on. The linear segments are arranged so that they join at x_0, x_1, ..., which are called the knots. An example of a piecewise linear function is shown below.

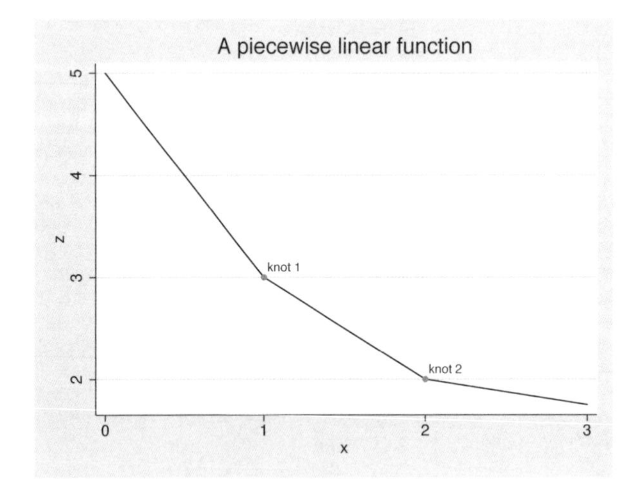

▷ Example 1

We wish to fit a model of log income on education and age by using a piecewise linear function for age:

$$\texttt{lninc} = b_0 + b_1\,\texttt{educ} + f(\texttt{age}) + u$$

The knots are to be placed at 10-year intervals: 20, 30, 40, 50, and 60.

```
. use http://www.stata-press.com/data/r10/mksp1
. mkspline age1 20 age2 30 age3 40 age4 50 age5 60 age6 = age, marginal
. regress lninc educ age1-age6
  (output omitted)
```

Since we specified the `marginal` option, we could test whether the age effect is the same in the 30–40 and 40–50 intervals by asking whether the `age4` coefficient is zero. With the `marginal` option, coefficients measure the change in slope from the preceding group. Specifying `marginal` changes only the interpretation of the coefficients; the same model is fitted in either case. Without the `marginal` option, the interpretation of the coefficients would have been

$$\frac{dy}{d\texttt{age}} = \begin{cases} a_1 & \text{if } \texttt{age} < 20 \\ a_2 & \text{if } 20 \le \texttt{age} < 30 \\ a_3 & \text{if } 30 \le \texttt{age} < 40 \\ a_4 & \text{if } 40 \le \texttt{age} < 50 \\ a_5 & \text{if } 50 \le \texttt{age} < 60 \\ a_6 & \text{otherwise.} \end{cases}$$

With the `marginal` option, the interpretation is

$$\frac{dy}{d\texttt{age}} = \begin{cases} a_1 & \text{if } \texttt{age} < 20 \\ a_1 + a_2 & \text{if } 20 \le \texttt{age} < 30 \\ a_1 + a_2 + a_3 & \text{if } 30 \le \texttt{age} < 40 \\ a_1 + a_2 + a_3 + a_4 & \text{if } 40 \le \texttt{age} < 50 \\ a_1 + a_2 + a_3 + a_4 + a_5 & \text{if } 50 \le \texttt{age} < 60 \\ a_1 + a_2 + a_3 + a_4 + a_5 + a_6 & \text{otherwise.} \end{cases}$$

◁

▷ Example 2

Say that we have a binary outcome variable called `outcome`. We are beginning an analysis and wish to parameterize the effect of dosage on outcome. We wish to divide the data into five equal-width groups of dosage for the piecewise linear function.

```
. use http://www.stata-press.com/data/r10/mksp2
. mkspline dose 5 = dosage, displayknots

             |     knot1      knot2      knot3      knot4
-------------+--------------------------------------------
      dosage |        20         40         60         80

. logistic outcome dose1-dose5
  (output omitted)
```

mkspline dose 5 = dosage creates five variables—dose1, dose2, ..., dose5—equally spacing the knots over the range of dosage. Since dosage varied between 0 and 100, the mkspline command above has the same effect as typing

```
. mkspline dose1 20 dose2 40 dose3 60 dose4 80 dose5 = dosage
```

The pctile option sets the knots to divide the data into five equal sample-size groups rather than five equal-width ranges. Typing

```
. mkspline pctdose 5 = dosage, pctile displayknots
```

	knot1	knot2	knot3	knot4
dosage	16	36.4	55.6	82

places the knots at the 20th, 40th, 60th, and 80th percentiles of the data.

◁

Restricted cubic splines

A linear spline can be used to fit many functions well. However, a restricted cubic spline may be a better choice than a linear spline when working with a very curved function. When using a restricted cubic spline, one obtains a continuous smooth function that is linear before the first knot, a piecewise cubic polynomial between adjacent knots, and linear again after the last knot.

▷ Example 3

Returning to the data from example 1, we may feel that a curved function is a better fit. First, we will use the knots() option to specify the five knots that we used previously.

```
. use http://www.stata-press.com/data/r10/mksp1, clear
. mkspline agesp = age, cubic knots(20 30 40 50 60)
. regress lninc educ agesp*
(output omitted)
```

Harrell (2001, 23) recommends placing knots at equally spaced percentiles of the original variable's marginal distribution. If we do not specify the knots() option, variables will be created containing a restricted cubic spline with five knots determined by Harrell's default percentiles.

```
. use http://www.stata-press.com/data/r10/mksp1, clear
. mkspline agesp = age, cubic displayknots
. regress lninc educ agesp*
(output omitted)
```

◁

Methods and Formulas

mkspline is implemented as an ado-file.

Linear splines

Let V_i, $i = 1, \ldots, n$, be the variables to be created; k_i, $i = 1, \ldots, n-1$, be the corresponding knots; and $\mathcal{V}$ be the original variable (the command is `mkspline` V_1 k_1 V_2 k_2 ... V_n `=` $\mathcal{V}$). Then

$$V_1 = \min(\mathcal{V}, k_1)$$
$$V_i = \max\Big\{\min(\mathcal{V}, k_i), k_{i-1}\Big\} - k_{i-1} \quad i = 2, \ldots, n$$

If the `marginal` option is specified, the definitions are

$$V_1 = \mathcal{V}$$
$$V_i = \max(0, \mathcal{V} - k_{i-1}) \quad i = 2, \ldots, n$$

In the second syntax, `mkspline` *stubname # =* $\mathcal{V}$, so let m and M be the minimum and maximum of $\mathcal{V}$. Without the `pctile` option, knots are set at $m + (M - m)(i/n)$ for $i = 1, \ldots, n-1$. If `pctile` is specified, knots are set at the $100(i/n)$ percentiles, for $i = 1, \ldots, n-1$. Percentiles are calculated by `centile`; see [R] **centile**.

Restricted cubic splines

Let k_i, $i = 1, \ldots, n$, be the knot values; V_i, $i = 1, \ldots, n-1$, be the variables to be created; and $\mathcal{V}$ be the original variable. Then

$$V_1 = \mathcal{V}$$
$$V_{i+1} = \frac{(\mathcal{V} - k_i)^3_+ - (k_n - k_{n-1})^{-1}\{(\mathcal{V} - k_{n-1})^3_+ (k_n - k_i) + (\mathcal{V} - k_n)^3_+ (k_{n-1} - k_i)\}}{(k_n - k_1)^2}$$
$$i = 1, \ldots, n-2$$

where

$$(u)_+ = \begin{cases} u, \text{ if } & u > 0 \\ 0, \text{ if } & u \le 0 \end{cases}$$

Without the `knots()` option, the locations of the knots are determined by the percentiles recommended in Harrell (2001, 23). These percentiles are based on the chosen number of knots as follows:

No. of knots	Percentiles						
3	10	50	90				
4	5	35	65	95			
5	5	27.5	50	72.5	95		
6	5	23	41	59	77	95	
7	2.5	18.33	34.17	50	65.83	81.67	97.5

Harrell provides default percentiles when the number of knots is between 3 and 7. When using a number of knots outside this range, the location of the knots must be specified in `knots()`.

Acknowledgment

The restricted cubic spline portion of `mkspline` is based on the `rc_spline` command by William Dupont.

References

Dupont, W. D., and F. E. Harrell, Jr. 2007. Personal communication.

Gould, W. W. 1993. sg19: Linear splines and piecewise linear functions. *Stata Technical Bulletin* 15: 13–17. Reprinted in *Stata Technical Bulletin Reprints*, vol. 3, pp. 98–104.

Greene, W. H. 2003. *Econometric Analysis*. 5th ed. Upper Saddle River, NJ: Prentice Hall.

Harrell, F. E., Jr. 2001. *Regression Modeling Strategies: With Applications to Linear Models, Logistic Regression, and Survival Analysis*. New York: Springer.

Newson, R. 2000. sg151: B-splines parameterized by their values at reference points on the x-axis. *Stata Technical Bulletin* 57: 20–27. Reprinted in *Stata Technical Bulletin Reprints*, vol. 10, pp. 221–230.

Panis, C. 1994. sg24: The piecewise linear spline transformation. *Stata Technical Bulletin* 18: 27–29. Reprinted in *Stata Technical Bulletin Reprints*, vol. 3, pp. 146–149.

Also See

[R] **fracpoly** — Fractional polynomial regression

Title

ml — Maximum likelihood estimation

Syntax

ml model in interactive mode

```
ml model    method progname eq [eq ...] [if] [in] [weight]
            [, model_options svy diparm_options]
```

ml model in noninteractive mode

```
ml model    method progname eq [eq ...] [if] [in] [weight], maximize
            [model_options svy diparm_options noninteractive_options]
```

Noninteractive mode is invoked by specifying the `maximize` option. Use `maximize` when `ml` will be used as a subroutine of another ado-file or program and you want to carry forth the problem, from definition to posting of results, in one command.

```
ml clear

ml query

ml check

ml search     [ [/]eqname[:] #lb #ub ] [...] [, search_options]

ml plot       [eqname:]name [# [# [#]]] [, saving(filename[, replace])]

ml init       { [eqname:]name=# | /eqname=# } [...]
ml init       # [# ...], copy
ml init       matname [, copy skip]

ml report

ml trace      { on | off }

ml count      [clear | on | off]

ml maximize   [, ml_maximize_options display_options eform_option]

ml graph      [#] [, saving(filename[, replace])]

ml display    [, display_options eform_option]

ml footnote
```

ml score *newvar* [*if*] [*in*] [, equation(*eqname*) forcescores missing]

ml score *newvarlist* [*if*] [*in*] [, missing]

ml score [*type*] *stub** [*if*] [*in*] [, missing]

where *method* is { lf | d0 | d1 | d1debug | d2 | d2debug },

eq is the equation to be estimated, enclosed in parentheses, and optionally with a name to be given to the equation, preceded by a colon,

([*eqname*:] [$varlist_y$ =] [$varlist_x$] [, *eq_options*])

or *eq* is the name of a parameter, such as sigma, with a slash in front

/*eqname* which is equivalent to (*eqname*:)

and *diparm_options* is one or more diparm(*diparm_args*) options where *diparm_args* is either __sep__ or anything accepted by the _diparm command; see help _diparm.

eq_options	description
noconstant	do not include an intercept in the equation
offset($varname_o$)	include $varname_o$ in model with coefficient constrained to 1
exposure($varname_e$)	include ln($varname_e$) in model with coefficient constrained to 1

model_options	description
vce(*vcetype*)	*vcetype* may be robust, cluster *clustvar*, oim, opg, bootstrap, or jackknife
constraints(*numlist*)	constraints by number to be applied
constraints(*matname*)	matrix that contains the constraints to be applied
nocnsnotes	do not display notes when constraints are dropped
title(*string*)	place a title on the estimation output
nopreserve	do not preserve the estimation subsample in memory
collinear	keep collinear variables within equations
missing	keep observations containing variables with missing values
lf0($\#_k$ $\#_{ll}$)	number of parameters and log-likelihood value of the constant-only model
continue	specifies that a model has been fitted and sets the initial values $\mathbf{b}_0$ for the model to be fitted based on those results
waldtest(#)	perform a Wald test; see *Options for use with ml model in interactive or noninteractive mode* below
obs(#)	number of observations
noscvars	do not create and pass score variables to likelihood-evaluator program; seldom used
crittype(*string*)	describe the criterion optimized by ml
subpop(*varname*)	compute estimates for the single subpopulation
nosvyadjust	carry out Wald test as $W/k \sim F(k,d)$

technique(nr)	Stata's modified Newton–Raphson (NR) algorithm
technique(bhhh)	Berndt–Hall–Hall–Hausman (BHHH) algorithm
technique(dfp)	Davidon–Fletcher–Powell (DFP) algorithm
technique(bfgs)	Broyden–Fletcher–Goldfarb–Shanno (BFGS) algorithm

noninteractive_options	description
init(*ml_init_args*)	set the initial values $\mathbf{b}_0$
search(on)	equivalent to ml search, repeat(0); the default
search(norescale)	equivalent to ml search, repeat(0) norescale
search(quietly)	same as search(on), except that output is suppressed
search(off)	prevents calling ml search
repeat(*#*)	ml search's repeat() option; see below
bounds(*ml_search_bounds*)	specify bounds for ml search
nowarning	suppress "convergence not achieved" message of iterate(0)
novce	substitute the zero matrix for the variance matrix
score(*newvars*)	new variables containing the contribution to the score
maximize_options	control the maximization process; seldom used

search_options	description
repeat(*#*)	number of random attempts to find better initial-value vector; default is repeat(10) in interactive mode and repeat(0) in noninteractive mode
restart	use random actions to find starting values; not recommended
norescale	do not rescale to improve parameter vector; not recommended
maximize_options	control the maximization process; seldom used

ml_maximize_options	description
nowarning	suppress "convergence not achieved" message of iterate(0)
novce	substitute the zero matrix for the variance matrix
score(*newvars* | *stub**)	new variables containing the contribution to the score
nooutput	suppress display of results
noclear	do not clear ml problem definition after model has converged
maximize_options	control the maximization process; seldom used

(*Continued on next page*)

display_options	description
noheader	suppress header display above the coefficient table
nofootnote	suppress footnote display below the coefficient table
level(#)	set confidence level; default is level(95)
first	display coefficient table reporting results for first equation only
neq(#)	display coefficient table reporting first # equations
showeqns	display equation names in the coefficient table
plus	display coefficient table ending in dashes–plus-sign–dashes

eform_option	description
eform(*string*)	display exponentiated coefficients; column title is "*string*"
eform	display exponentiated coefficients; column title is "exp(b)"
hr	report hazard ratios
irr	report incidence-rate ratios
or	report odds ratios
rrr	report relative-risk ratios

fweights, aweights, iweights, and pweights are allowed; see [U] **11.1.6 weight**. With all but method lf, you must write your likelihood-evaluation program carefully if pweights are to be specified, and pweights may not be specified with method d0. See Gould, Pitblado, and Sribney (2006, chap. 4) for details.

See [U] **20 Estimation and postestimation commands** for more capabilities of estimation commands.

To redisplay results, type ml display.

Syntax of subroutines for use by method d0, d1, and d2 evaluators

mleval *newvar* = *vecname* [, eq(#)]

mleval *scalarname* = *vecname* , scalar [eq(#)]

mlsum *scalarname*$_{\text{lnf}}$ = *exp* [*if*] [, noweight]

mlvecsum *scalarname*$_{\text{lnf}}$ *rowvecname* = *exp* [*if*] [, eq(#)]

mlmatsum *scalarname*$_{\text{lnf}}$ *matrixname* = *exp* [*if*] [, eq(#[,#])]

mlmatbysum *scalarname*$_{\text{lnf}}$ *matrixname* *varname*$_a$ *varname*$_b$ [*varname*$_c$] [*if*] ,
by(*varname*) [eq(#[,#])]

Syntax of user-written evaluator

Summary of notation

The log-likelihood function is $\ln L(\theta_{1j}, \theta_{2j}, \dots, \theta_{Ej})$, where $\theta_{ij} = \mathbf{x}_{ij}\mathbf{b}_i$, $j = 1, \dots, N$ indexes observations, and $i = 1, \dots, E$ indexes the linear equations defined by ml model. If the likelihood satisfies the linear-form restrictions, it can be decomposed as $\ln L = \sum_{j=1}^{N} \ln \ell(\theta_{1j}, \theta_{2j}, \dots, \theta_{Ej})$.

Method lf evaluators:

```
program progname
        version 10
        args lnf theta1 [theta2 ... ]
        /* if you need to create any intermediate results: */
        tempvar tmp1 tmp2 ...
        quietly gen double `tmp1' = ...
        ...
        quietly replace `lnf' = ...
end
```

where

`lnf' variable to be filled in with observation-by-observation values of $\ln\ell_j$
`theta1' variable containing evaluation of first equation $\theta_{1j}=\mathbf{x}_{1j}\mathbf{b}_1$
`theta2' variable containing evaluation of second equation $\theta_{2j}=\mathbf{x}_{2j}\mathbf{b}_2$

Method d0 evaluators:

```
program progname
        version 10
        args todo b lnf
        tempvar theta1 theta2 ...
        mleval `theta1' = `b', eq(1)
        mleval `theta2' = `b', eq(2) // if there is a θ2
        ...
        // if you need to create any intermediate results:
        tempvar tmp1 tmp2 ...
        gen double `tmp1' = ...
        ...
        mlsum `lnf' = ...
end
```

where

`todo' always contains 0 (may be ignored)
`b' full parameter row vector $\mathbf{b}=(\mathbf{b}_1,\mathbf{b}_2,\ldots,\mathbf{b}_E)$
`lnf' scalar to be filled in with overall $\ln L$

Method d1 evaluators:

```
program progname
        version 10
        args todo b lnf g [negH g1 [g2 ... ] ]
        tempvar theta1 theta2 ...
        mleval `theta1' = `b', eq(1)
        mleval `theta2' = `b', eq(2) // if there is a θ2
        ...
        // if you need to create any intermediate results:
        tempvar tmp1 tmp2 ...
        gen double `tmp1' = ...
        ...
        mlsum `lnf' = ...
        if (`todo'==0 | `lnf'>=.) exit
        tempname d1 d2 ...
        mlvecsum `lnf' `d1' = formula for ∂lnℓj/∂θ1j, eq(1)
        mlvecsum `lnf' `d2' = formula for ∂lnℓj/∂θ2j, eq(2)
        ...
        matrix `g' = (`d1',`d2', ... )
end
```

where

'todo'	contains 0 or 1 0⇒'lnf' to be filled in; 1⇒'lnf' and 'g' to be filled in
'b'	full parameter row vector $\mathbf{b}$=($\mathbf{b}_1$,$\mathbf{b}_2$,...,$\mathbf{b}_E$)
'lnf'	scalar to be filled in with overall $\ln L$
'g'	row vector to be filled in with overall $\mathbf{g}=\partial \ln L/\partial \mathbf{b}$
'negH'	argument to be ignored
'g1'	variable optionally to be filled in with $\partial \ln\ell_j/\partial \mathbf{b}_1$
'g2'	variable optionally to be filled in with $\partial \ln\ell_j/\partial \mathbf{b}_2$
...	

Method d2 evaluators:

```
program progname
        version 10
        args todo b lnf g negH [g1 [g2 ... ] ]
        tempvar theta1 theta2 ...
        mleval 'theta1' = 'b', eq(1)
        mleval 'theta2' = 'b', eq(2) // if there is a θ2
        ...
        // if you need to create any intermediate results:
        tempvar tmp1 tmp2 ...
        gen double 'tmp1' = ...
        ...
        mlsum 'lnf' = ...
        if ('todo'==0 | 'lnf'>=.) exit
        tempname d1 d2 ...
        mlvecsum 'lnf' 'd1' = formula for ∂lnℓj/∂θ1j, eq(1)
        mlvecsum 'lnf' 'd2' = formula for ∂lnℓj/∂θ2j, eq(2)
        ...
        matrix 'g' = ('d1','d2', ... )
        if ('todo'==1 | 'lnf'>=.) exit
        tempname d11 d12 d22 ...
        mlmatsum 'lnf' 'd11' = formula for −∂²lnℓj/∂θ²1j, eq(1)
        mlmatsum 'lnf' 'd12' = formula for −∂²lnℓj/∂θ1j∂θ2j, eq(1,2)
        mlmatsum 'lnf' 'd22' = formula for −∂²lnℓj/∂θ²2j, eq(2)
        ...
        matrix 'negH' = ('d11','d12', ... \ 'd12','d22', ... )
end
```

where

'todo'	contains 0, 1, or 2 0⇒'lnf' to be filled in; 1⇒'lnf' and 'g' to be filled in; 2⇒'lnf', 'g', and 'negH' to be filled in
'b'	full parameter row vector $\mathbf{b}$=($\mathbf{b}_1$,$\mathbf{b}_2$,...,$\mathbf{b}_E$)
'lnf'	scalar to be filled in with overall $\ln L$
'g'	row vector to be filled in with overall $\mathbf{g}=\partial \ln L/\partial \mathbf{b}$
'negH'	matrix to be filled in with overall negative Hessian $-\mathbf{H}=-\partial^2 \ln L/\partial\mathbf{b}\partial\mathbf{b}'$
'g1'	variable optionally to be filled in with $\partial \ln\ell_j/\partial \mathbf{b}_1$
'g2'	variable optionally to be filled in with $\partial \ln\ell_j/\partial \mathbf{b}_2$
...	

Global macros for use by all evaluators

$ML_y1	name of first dependent variable
$ML_y2	name of second dependent variable, if any
...	
$ML_samp	variable containing 1 if observation to be used; 0 otherwise
$ML_w	variable containing weight associated with observation or 1 if no weights specified

Method lf evaluators can ignore `$ML_samp`, but restricting calculations to the `$ML_samp==1` subsample will speed execution. Method lf evaluators must ignore `$ML_w`; application of weights is handled by the method itself.

Methods d0, d1, and d2 can ignore `$ML_samp` as long as `ml model`'s `nopreserve` option is not specified. Methods d0, d1, and d2 will run more quickly if `nopreserve` is specified. Method d0, d1, and d2 evaluators can ignore `$ML_w` only if they use `mlsum`, `mlvecsum`, `mlmatsum`, and `mlmatbysum` to produce all final results.

Description

`ml model` defines the current problem.

`ml clear` clears the current problem definition. This command is rarely used because, when you type `ml model`, any previous problem is automatically cleared.

`ml query` displays a description of the current problem.

`ml check` verifies that the log-likelihood evaluator you have written works. We strongly recommend using this command.

`ml search` searches for (better) initial values. We recommend using this command.

`ml plot` provides a graphical way of searching for (better) initial values.

`ml init` provides a way to specify initial values.

`ml report` reports $\ln L$'s values, gradient, and negative Hessian at the initial values or current parameter estimates $\mathbf{b}_0$.

`ml trace` traces the execution of the user-defined log-likelihood evaluation program.

`ml count` counts the number of times the user-defined log-likelihood evaluation program is called; this command is seldom used. `ml count clear` clears the counter. `ml count on` turns on the counter. `ml count` without arguments reports the current values of the counter. `ml count off` stops counting calls.

`ml maximize` maximizes the likelihood function and reports results. Once `ml maximize` has successfully completed, the previously mentioned `ml` commands may no longer be used unless `noclear` is specified. `ml graph` and `ml display` may be used whether or not `noclear` is specified.

`ml graph` graphs the log-likelihood values against the iteration number.

`ml display` redisplays results.

`ml footnote` displays a warning message when the model did not converge within the specified number of iterations.

`ml score` creates new variables containing the equation-level scores. The variables generated by `ml score` are equivalent to those generated by specifying the `score()` option of `ml maximize` (and `ml model ..., ...maximize`).

progname is the name of a program you write to evaluate the log-likelihood function. In this documentation, it is referred to as the user-written evaluator, the likelihood evaluator, or sometimes simply as the evaluator. The program you write is written in the style required by the method you choose. The methods are lf, d0, d1, and d2. Thus if you choose to use method lf, your program is called a method lf evaluator. Method lf evaluators are required to evaluate the observation-by-observation log likelihood $\ln \ell_j$, $j = 1, \ldots, N$. Method d0 evaluators are required to evaluate the overall log likelihood $\ln L$. Method d1 evaluators are required to evaluate the overall log likelihood and its gradient vector $\mathbf{g} = \partial \ln L/\partial \mathbf{b}$. Method d2 evaluators are required to evaluate the overall log likelihood, its gradient, and its negative Hessian matrix $-H = -\partial^2 \ln L/\partial \mathbf{b} \partial \mathbf{b}'$.

`mleval` is a subroutine used by method d0, d1, and d2 evaluators to evaluate the coefficient vector $\mathbf{b}$ that they are passed.

`mlsum` is a subroutine used by method d0, d1, and d2 evaluators to define the value $\ln L$ that is to be returned.

`mlvecsum` is a subroutine used by method d1 and d2 evaluators to define the gradient vector $\mathbf{g}$ that is to be returned. It is suitable for use only when the likelihood function meets the linear-form restrictions.

`mlmatsum` is a subroutine used by method d2 evaluators to define the negative Hessian matrix, $-\mathbf{H}$, that is to be returned. It is suitable for use only when the likelihood function meets the linear-form restrictions.

`mlmatbysum` is a subroutine used by method d2 evaluators to help define the negative Hessian matrix, $-\mathbf{H}$, that is to be returned. It is suitable for use when the likelihood function contains terms made up of grouped sums, such as in panel-data models. For such models, use `mlmatsum` to compute the observation-level outer products and `mlmatbysum` to compute the group-level outer products. `mlmatbysum` requires that the data be sorted by the variable identified in the `by()` option.

Options for use with ml model in interactive or noninteractive mode

`vce(`*vcetype*`)` specifies the type of standard error reported, which includes types that are robust to some kinds of misspecification, that allow for intragroup correlation, that are derived from asymptotic theory, and that use bootstrap or jackknife methods; see [R] ***vce_option***.

`vce(robust)`, `vce(cluster` *clustvar*`)`, `pweight`, and `svy` will work with a method lf evaluator; all you need do is specify them.

These options will not work with a method d0 evaluator, and specifying these options will produce an error message.

With method d1 or d2 evaluators in which the likelihood function satisfies the linear-form restrictions, these options will work only if you fill in the equation scores; otherwise, specifying these options will produce an error message.

`constraints(`*numlist* | *matname*`)` specifies the linear constraints to be applied during estimation. `constraints(`*numlist*`)` specifies the constraints by number. Constraints are defined by using the `constraint` command; see [R] **constraint**. `constraint(`*matname*`)` specifies a matrix that contains the constraints.

`nocnsnotes` prevents notes from being displayed when constraints are dropped. A constraint will be dropped if it is inconsistent, contradicts other constraints, or causes some other error when the constraint matrix is being built. Constraints are checked in the order in which they are specified.

`title(`*string*`)` specifies the title for the estimation output when results are complete.

`nopreserve` specifies that `ml` need not ensure that only the estimation subsample is in memory when the user-written likelihood evaluator is called. `nopreserve` is irrelevant when you use method lf.

For the other methods, if `nopreserve` is not specified, `ml` saves the data in a file (preserves the original dataset) and drops the irrelevant observations before calling the user-written evaluator. This way, even if the evaluator does not restrict its attentions to the `$ML_samp==1` subsample, results will still be correct. Later, `ml` automatically restores the original dataset.

ml need not go through these machinations for method lf because the user-written evaluator calculates observation-by-observation values, and ml itself sums the components.

ml goes through these machinations if and only if the estimation sample is a subsample of the data in memory. If the estimation sample includes every observation in memory, ml does not preserve the original dataset. Thus programmers must not alter the original dataset unless they preserve the data themselves.

We recommend that interactive users of ml not specify nopreserve; the speed gain is not worth the possibility of getting incorrect results.

We recommend that programmers specify nopreserve, but only after verifying that their evaluator really does restrict its attentions solely to the $ML_samp==1 subsample.

collinear specifies that ml not remove the collinear variables within equations. There is no reason to leave collinear variables in place, but this option is of interest to programmers who, in their code, have already removed collinear variables and do not want ml to waste computer time checking again.

missing specifies that observations containing variables with missing values not be eliminated from the estimation sample. There are two reasons you might want to specify missing:

Programmers may wish to specify missing because, in other parts of their code, they have already eliminated observations with missing values and do not want ml to waste computer time looking again.

You may wish to specify missing if your model explicitly deals with missing values. Stata's heckman command is a good example of this. In such cases, there will be observations where missing values are allowed and other observations where they are not—where their presence should cause the observation to be eliminated. If you specify missing, it is your responsibility to specify an if *exp* that eliminates the irrelevant observations.

lf0($\#_k$ $\#_{ll}$) is typically used by programmers. It specifies the number of parameters and log-likelihood value of the constant-only model so that ml can report a likelihood-ratio test rather than a Wald test. These values may have been analytically determined, or they may have been determined by a previous fitting of the constant-only model on the estimation sample.

Also see the continue option directly below.

If you specify lf0(), it must be safe for you to specify the missing option, too, else how did you calculate the log likelihood for the constant-only model on the same sample? You must have identified the estimation sample, and done so correctly, so there is no reason for ml to waste time rechecking your results. All of which is to say, do not specify lf0() unless you are certain your code identifies the estimation sample correctly.

lf0(), even if specified, is ignored if vce(robust), vce(cluster *clustvar*), pweight, or svy is specified because, in that case, a likelihood-ratio test would be inappropriate.

continue is typically specified by programmers and does two things:

First, it specifies that a model has just been fitted by either ml or some other estimation command, such as logit, and that the likelihood value stored in e(ll) and the number of parameters stored in e(b) as of that instant are the relevant values of the constant-only model. The current value of the log likelihood is used to present a likelihood-ratio test unless vce(robust), vce(cluster *clustvar*), pweight, svy, or constraints() is specified. A likelihood-ratio test is inappropriate when vce(robust), vce(cluster *clustvar*), pweight, or svy is specified. We suggest using lrtest when constraints() is specified.

Second, `continue` sets the initial values, $\mathbf{b}_0$, for the model about to be fitted according to the `e(b)` currently stored.

The comments made about specifying `missing` with `lf0()` apply equally well here.

`waldtest(#)` is typically specified by programmers. By default, `ml` presents a Wald test, but that is overridden if option `lf0()` or `continue` is specified. A Wald test is performed if `vce(robust)`, `vce(cluster` *clustvar*`)`, or `pweight` is specified.

`waldtest(0)` prevents even the Wald test from being reported.

`waldtest(-1)` is the default. It specifies that a Wald test be performed by constraining all coefficients except the intercept to 0 in the first equation. Remaining equations are to be unconstrained. A Wald test is performed if neither `lf0()` nor `continue` was specified, and a Wald test is forced if `vce(robust)`, `vce(cluster` *clustvar*`)`, or `pweight` was specified.

`waldtest(`k`)` for $k \leq -1$ specifies that a Wald test be performed by constraining all coefficients except intercepts to 0 in the first $|k|$ equations; remaining equations are to be unconstrained. A Wald test is performed if neither `lf0()` nor `continue` was specified, and a Wald test is forced if `vce(robust)`, `vce(cluster` *clustvar*`)`, or `pweight` was specified.

`waldtest(`k`)` for $k \geq 1$ works like the options above, except that it forces a Wald test to be reported even if the information to perform the likelihood-ratio test is available and even if none of `vce(robust)`, `vce(cluster` *clustvar*`)`, or `pweight` was specified. `waldtest(`k`)`, $k \geq 1$, may not be specified with `lf0()`.

`obs(#)` is used mostly by programmers. It specifies that the number of observations reported and ultimately stored in `e(N)` be #. Ordinarily, `ml` works that out for itself. Programmers may want to specify this option when, for the likelihood evaluator to work for N observations, they first had to modify the dataset so that it contained a different number of observations.

`noscvars` is used mostly by programmers. It specifies that method d0, d1, or d2 is being used but that the likelihood-evaluation program does not calculate or use arguments `` `g1' ``, `` `g2' ``, etc., which are the score vectors. Thus `ml` can save a little time by not generating and passing those arguments.

`crittype(`*string*`)` is used mostly by programmers. It allows programmers to supply a string (up to 32 characters long) that describes the criterion that is being optimized by `ml`. The default is `"log likelihood"` for nonrobust and `"log pseudolikelihood"` for robust estimation.

`svy` indicates that `ml` is to pick up the `svy` settings set by `svyset` and use the robust variance estimator. This option requires the data to be `svyset`. `svy` may not be specified with `vce()` or *weights*.

`subpop(`*varname*`)` specifies that estimates be computed for the single subpopulation defined by the observations for which *varname* $\neq 0$. Typically, *varname* $= 1$ defines the subpopulation, and *varname* $= 0$ indicates observations not belonging to the subpopulation. For observations whose subpopulation status is uncertain, *varname* should be set to missing (‘`.`’). This option requires the `svy` option.

`nosvyadjust` specifies that the model Wald test be carried out as $W/k \sim F(k,d)$, where W is the Wald test statistic, k is the number of terms in the model excluding the constant term, d is the total number of sampled PSUs minus the total number of strata, and $F(k,d)$ is an F distribution with k numerator degrees of freedom and d denominator degrees of freedom. By default, an adjusted Wald test is conducted: $(d-k+1)W/(kd) \sim F(k,d-k+1)$. See Korn and Graubard (1990) for a discussion of the Wald test and the adjustments thereof. This option requires the `svy` option.

`technique(`*algorithm_spec*`)` specifies how the likelihood function is to be maximized. The following algorithms are currently implemented in `ml`. For details, see Gould, Pitblado, and Sribney (2006).

`technique(nr)` specifies Stata's modified Newton–Raphson (NR) algorithm.

`technique(bhhh)` specifies the Berndt–Hall–Hall–Hausman (BHHH) algorithm.

`technique(dfp)` specifies the Davidon–Fletcher–Powell (DFP) algorithm.

`technique(bfgs)` specifies the Broyden–Fletcher–Goldfarb–Shanno (BFGS) algorithm.

The default is `technique(nr)`.

You can switch between algorithms by specifying more than one in the `technique()` option. By default, `ml` will use an algorithm for five iterations before switching to the next algorithm. To specify a different number of iterations, include the number after the technique in the option. For example, `technique(bhhh 10 nr 1000)` requests that `ml` perform 10 iterations using the BHHH algorithm, followed by 1,000 iterations using the NR algorithm, and then switch back to BHHH for 10 iterations, and so on. The process continues until convergence or until reaching the maximum number of iterations.

Options for use with ml model in noninteractive mode

The following extra options are for use with `ml model` in noninteractive mode. Noninteractive mode is for programmers who use `ml` as a subroutine and want to issue one command that will carry forth the estimation from start to finish.

`maximize` is required. It specifies noninteractive mode.

`init(`*ml_init_args*`)` sets the initial values $\mathbf{b}_0$. *ml_init_args* are whatever you would type after the `ml init` command.

`search(on | norescale | quietly | off)` specifies whether `ml search` is to be used to improve the initial values. `search(on)` is the default and is equivalent to separately running `ml search, repeat(0)`. `search(norescale)` is equivalent to separately running `ml search, repeat(0) norescale`. `search(quietly)` is equivalent to `search(on)`, except that it suppresses `ml search`'s output. `search(off)` prevents calling `ml search`.

`repeat(`*#*`)` is `ml search`'s `repeat()` option. `repeat(0)` is the default.

`bounds(`*ml_search_bounds*`)` specifies the search bounds. The command `ml model` issues is `ml search` *ml_search_bounds*`, repeat(`*#*`)`. Specifying search bounds is optional.

`nowarning`, `novce`, and `score()` are `ml maximize`'s equivalent options.

maximize_options: `difficult`, `technique(`*algorithm_spec*`)`, `iterate(`*#*`)`, `[no]log`, `trace`, `gradient`, `showstep`, `hessian`, `shownrtolerance`, `tolerance(`*#*`)`, `ltolerance(`*#*`)`, `gtolerance(`*#*`)`, `nrtolerance(`*#*`)`, `nonrtolerance`, `from(`*init_specs*`)`; see [R] **maximize**. These options are seldom used.

Options for use when specifying equations

`noconstant` specifies that the equation not include an intercept.

`offset(`$varname_o$`)` specifies that the equation be $\mathbf{xb} + varname_o$—that it include $varname_o$ with coefficient constrained to be 1.

`exposure(`$varname_e$`)` is an alternative to `offset(`$varname_o$`)`; it specifies that the equation be $\mathbf{xb} + \ln(varname_e)$. The equation is to include $\ln(varname_e)$ with coefficient constrained to be 1.

Options for use with ml search

`repeat(#)` specifies the number of random attempts that are to be made to find a better initial-value vector. The default is `repeat(10)`.

`repeat(0)` specifies that no random attempts be made. More precisely, `repeat(0)` specifies that no random attempts be made if the first initial-value vector is a feasible starting point. If it is not, `ml search` will make random attempts, even if you specify `repeat(0)`, because it has no alternative. The `repeat()` option refers to the number of random attempts to be made to improve the initial values. When the initial starting value vector is not feasible, `ml search` will make up to 1,000 random attempts to find starting values. It stops when it finds one set of values that works and then moves into its improve-initial-values logic.

`repeat(`k`)`, $k > 0$, specifies the number of random attempts to be made to improve the initial values.

`restart` specifies that random actions be taken to obtain starting values and that the resulting starting values not be a deterministic function of the current values. Generally, you should not specify this option because, with `restart`, `ml search` intentionally does not produce as good a set of starting values as it could. `restart` is included for use by the optimizer when it gets into serious trouble. The random actions ensure that the optimizer and `ml search`, working together, do not cause an endless loop.

`restart` implies `norescale`, which is why we recommend that you do not specify `restart`. In testing, sometimes `rescale` worked so well that, even after randomization, the rescaler would bring the starting values right back to where they had been the first time and thus defeat the intended randomization.

`norescale` specifies that `ml search` not engage in its rescaling actions to improve the parameter vector. We do not recommend specifying this option because rescaling tends to work so well.

maximize_options: [`no`]`log`, `trace`; see [R] **maximize**. These options are seldom used.

Option for use with ml plot

`saving(`*filename*[`, replace`]`)` specifies that the graph be saved in *filename*`.gph`. See [G] ***saving_option***.

Options for use with ml init

`copy` specifies that the list of numbers or the initialization vector be copied into the initial-value vector by position rather than by name.

`skip` specifies that any parameters found in the specified initialization vector that are not also found in the model be ignored. The default action is to issue an error message.

Options for use with ml maximize

`nowarning` is allowed only with `iterate(0)`. `nowarning` suppresses the "convergence not achieved" message. Programmers might specify `iterate(0) nowarning` when they have a vector **b** already containing the final estimates and want `ml` to calculate the variance matrix and postestimation results. In that case, specify `init(b) search(off) iterate(0) nowarning nolog`.

`novce` is allowed only with `iterate(0)`. `novce` substitutes the zero matrix for the variance matrix, which in effect posts estimation results as fixed constants.

`score(`*newvars* | *stub*`*)` creates new variables containing the contributions to the score for each equation and ancillary parameter in the model; see [U] **20.16 Obtaining scores**.

If `score(`*newvars*`)` is specified, the *newvars* must contain k new variables, one for each equation in the model. If `score(`*stub*`*)` is specified, variables named *stub*`1`, *stub*`2`, ..., *stub*`k` are created.

The first variable contains $\partial \ln L_j / \partial(\mathbf{x}_{1j}\mathbf{b}_1)$, the second variable contains $\partial \ln L_j / \partial(\mathbf{x}_{2j}\mathbf{b}_2)$, and so on.

`nooutput` suppresses display of results. This option is different from prefixing `ml maximize` with `quietly` in that the iteration log is still displayed (assuming that `nolog` is not specified).

`noclear` specifies that the ml problem definition not be cleared after the model has converged. Perhaps you are having convergence problems and intend to run the model to convergence. If so, use `ml search` to see if those values can be improved, and then restart the estimation.

maximize_options: `difficult`, `iterate(`*#*`)`, [`no`]`log`, `trace`, `gradient`, `showstep`, `hessian`, `shownrtolerance`, `tolerance(`*#*`)`, `ltolerance(`*#*`)`, `gtolerance(`*#*`)`, `nrtolerance(`*#*`)`, `nonrtolerance`; see [R] **maximize**. These options are seldom used.

display_options; see *Options for use with ml display* below.

eform_option; see *Options for use with ml display* below.

Option for use with ml graph

`saving(`*filename*[`, replace`]`)` specifies that the graph be saved in *filename*`.gph`. See [G] ***saving_option***.

Options for use with ml display

`noheader` suppresses the header display above the coefficient table that displays the final log-likelihood value, the number of observations, and the model significance test.

`nofootnote` suppresses the footnote display below the coefficient table, which displays a warning if the model fitted did not converge within the specified number of iterations. Use `ml footnote` to display the warning if (1) you add to the coefficient table using the `plus` option or (2) you have your own footnotes and want the warning to be last.

`level(`*#*`)` is the standard confidence-level option. It specifies the confidence level, as a percentage, for confidence intervals of the coefficients. The default is `level(95)` or as set by `set level`; see [U] **20.7 Specifying the width of confidence intervals**.

`first` displays a coefficient table reporting results for the first equation only, and the report makes it appear that the first equation is the only equation. This option is used by programmers who estimate ancillary parameters in the second and subsequent equations and who wish to report the values of such parameters themselves.

`neq(`*#*`)` is an alternative to `first`. `neq(`*#*`)` displays a coefficient table reporting results for the first *#* equations. This option is used by programmers who estimate ancillary parameters in the $\# + 1$ and subsequent equations and who wish to report the values of such parameters themselves.

`showeqns` is a seldom-used option that displays the equation names in the coefficient table. `ml display` uses the numbers stored in `e(k_eq)` and `e(k_aux)` to determine how to display the coefficient table. `e(k_eq)` identifies the number of equations, and `e(k_aux)` identifies how many of these are for ancillary parameters. The `first` option is implied when `showeqns` is not specified and all but the first equation are for ancillary parameters.

`plus` displays the coefficient table, but rather than ending the table in a line of dashes, ends it in dashes–plus-sign–dashes. This is so that programmers can write additional display code to add more results to the table and make it appear as if the combined result is one table. Programmers typically specify `plus` with options `first` or `neq()`. This option implies `nofootnote`.

eform_option: `eform(`*string*`)`, `eform`, `hr`, `irr`, `or`, and `rrr` display the coefficient table in exponentiated form: for each coefficient, $\exp(b)$ rather than b is displayed, and standard errors and confidence intervals are transformed. Display of the intercept, if any, is suppressed. *string* is the table header that will be displayed above the transformed coefficients and must be 11 characters or shorter in length—for example, `eform("Odds ratio")`. The options `eform`, `hr`, `irr`, `or`, and `rrr` provide a default *string* equivalent to "`exp(b)`", "`Haz. Ratio`", "`IRR`", "`Odds Ratio`", and "`RRR`", respectively. These options may not be combined.

`ml display` looks at `e(k_eform)` to determine how many equations are affected by an *eform_option*; by default, only the first equation is affected.

Options for use with mleval

`eq(`*#*`)` specifies the equation number i for which $\theta_{ij} = \mathbf{x}_{ij}\mathbf{b}_i$ is to be evaluated. `eq(1)` is assumed if `eq()` is not specified.

`scalar` asserts that the ith equation is known to evaluate to a constant, meaning that the equation was specified as `()`, `(`*name*`:)`, or `/`*name* on the `ml model` statement. If you specify this option, the new variable created is created as a scalar. If the ith equation does not evaluate to a scalar, an error message is issued.

Option for use with mlsum

`noweight` specifies that weights (`$ML_w`) be ignored when summing the likelihood function.

Option for use with mlvecsum

`eq(`*#*`)` specifies the equation for which a gradient vector $\partial \ln L/\partial \mathbf{b}_i$ is to be constructed. The default is `eq(1)`.

Option for use with mlmatsum

`eq(`*#*[`,`*#*]`)` specifies the equations for which the negative Hessian matrix is to be constructed. The default is `eq(1)`, which is the same as `eq(1,1)`, which means $-\partial^2 \ln L/\partial \mathbf{b}_1 \partial \mathbf{b}_1'$. Specifying `eq(`$i$`,`$j$`)` results in $-\partial^2 \ln L/\partial \mathbf{b}_i \partial \mathbf{b}_j'$.

Options for use with mlmatbysum

`by(`*varname*`)` is required and specifies the group variable.

`eq(`*#*[`,`*#*]`)` specifies the equations for which the negative Hessian matrix is to be constructed. The default is `eq(1)`, which is the same as `eq(1,1)`, which means $-\partial^2 \ln L/\partial \mathbf{b}_1 \partial \mathbf{b}_1'$. Specifying `eq(`*i*`,`*j*`)` results in $-\partial^2 \ln L/\partial \mathbf{b}_i \partial \mathbf{b}_j'$.

Options for use with ml score

`equation(`*eqname*`)` identifies which equation the observation scores are to come from. This option may be used only when generating one variable.

`forcescores` causes `ml score` to generate scores from results generated from methods d1 and d2 even though the estimation command does not have the `ml_score` property.

`missing` specifies that observations containing variables with missing values not be eliminated from the estimation sample.

Remarks

For a thorough discussion of `ml`, see *Maximum Likelihood Estimation with Stata*, 3rd edition (Gould, Pitblado, and Sribney 2006). The book provides a tutorial introduction to `ml`, notes on advanced programming issues, and a discourse on maximum likelihood estimation from both theoretical and practical standpoints. See *Survey options and ml* at the end of *Remarks* for examples of the new `svy` options. For more information about survey estimation, see [SVY] **survey**, [SVY] **svy estimation**, and [SVY] **variance estimation**.

`ml` requires that you write a program that evaluates the log-likelihood function and, possibly, its first and second derivatives. The style of the program you write depends upon the method you choose: methods lf and d0 require that your program evaluate the log likelihood only, method d1 requires that your program evaluate the log likelihood and gradient, and method d2 requires that your program evaluate the log likelihood, gradient, and negative Hessian. Methods lf and d0 differ from each other in that, with method lf, your program is required to produce observation-by-observation log-likelihood values $\ln \ell_j$ and it is assumed that $\ln L = \sum_j \ln \ell_j$; with method d0, your program is only required to produce the overall value $\ln L$.

Once you have written the program—called an evaluator—you define a model to be fitted using `ml model` and obtain estimates using `ml maximize`. You might type

```
. ml model ...
. ml maximize
```

but we recommend that you type

```
. ml model ...
. ml check
. ml search
. ml maximize
```

`ml check` verifies your evaluator has no obvious errors, and `ml search` finds better initial values.

You fill in the `ml model` statement with (1) the method you are using, (2) the name of your program, and (3) the "equations". You write your evaluator in terms of θ_1, θ_2, ..., each of which has a linear equation associated with it. That linear equation might be as simple as $\theta_i = b_0$, it might be $\theta_i = b_1$`mpg` $+ b_2$`weight` $+ b_3$, or it might omit the intercept b_3. The equations are specified in parentheses on the `ml model` line.

Suppose that you are using method lf and the name of your evaluator program is myprog. The statement

```
. ml model lf myprog (mpg weight)
```

would specify one equation with $\theta_i = b_1\texttt{mpg} + b_2\texttt{weight} + b_3$. If you wanted to omit b_3, you would type

```
. ml model lf myprog (mpg weight, nocons)
```

and if all you wanted was $\theta_i = b_0$, you would type

```
. ml model lf myprog ()
```

With multiple equations, you list the equations one after the other; so, if you typed

```
. ml model lf myprog (mpg weight) ()
```

you would be specifying $\theta_1 = b_1\texttt{mpg} + b_2\texttt{weight} + b_3$ and $\theta_2 = b_4$. You would write your likelihood in terms of θ_1 and θ_2. If the model was linear regression, θ_1 might be the $\mathbf{xb}$ part and θ_2 the variance of the residuals.

When you specify the equations, you also specify any dependent variables. If you typed

```
. ml model lf myprog (price = mpg weight) ()
```

price would be the one and only dependent variable, and that would be passed to your program in $ML_y1. If your model had two dependent variables, you could type

```
. ml model lf myprog (price displ = mpg weight) ()
```

Then $ML_y1 would be price, and $ML_y2 would be displ. You can specify however many dependent variables are necessary and specify them on any equation. It does not matter on which equation you specify them; the first one specified is placed in $ML_y1, the second in $ML_y2, and so on.

▷ Example 1: Method lf

Using method lf, we want to produce observation-by-observation values of the log likelihood. The probit log-likelihood function is

$$\ln \ell_j = \begin{cases} \ln \Phi(\theta_{1j}) & \text{if } y_j = 1 \\ \ln \Phi(-\theta_{1j}) & \text{if } y_j = 0 \end{cases}$$

$$\theta_{1j} = \mathbf{x}_j \mathbf{b}_1$$

The following is the method lf evaluator for this likelihood function:

```
program myprobit
        version 10
        args lnf theta1
        quietly replace `lnf' = ln(normal(`theta1')) if $ML_y1==1
        quietly replace `lnf' = ln(normal(-`theta1')) if $ML_y1==0
end
```

If we wanted to fit a model of foreign on mpg and weight, we would type

```
. use http://www.stata-press.com/data/r10/auto
(1978 Automobile Data)
. ml model lf myprobit (foreign = mpg weight)
. ml maximize
```

The '`foreign =`' part specifies that y is `foreign`. The '`mpg weight`' part specifies that $\theta_{1j} = b_1\texttt{mpg}_j + b_2\texttt{weight}_j + b_3$. The result of running this is

```
. ml model lf myprobit (foreign = mpg weight)
. ml maximize
initial:       log likelihood = -51.292891
alternative:   log likelihood = -45.055272
rescale:       log likelihood = -45.055272
Iteration 0:   log likelihood = -45.055272
Iteration 1:   log likelihood = -27.904114
Iteration 2:   log likelihood = -26.858048
Iteration 3:   log likelihood = -26.844198
Iteration 4:   log likelihood = -26.844189
Iteration 5:   log likelihood = -26.844189
                                                  Number of obs   =         74
                                                  Wald chi2(2)    =      20.75
Log likelihood = -26.844189                       Prob > chi2     =     0.0000
```

foreign	Coef.	Std. Err.	z	P>|z|	[95% Conf.	Interval]
mpg	-.1039503	.0515689	-2.02	0.044	-.2050235	-.0028772
weight	-.0023355	.0005661	-4.13	0.000	-.003445	-.0012261
_cons	8.275464	2.554142	3.24	0.001	3.269438	13.28149

◁

▷ Example 2: Method lf for two-equation, two-dependent-variable model

A two-equation, two-dependent-variable model is a little different. Rather than receiving one θ, our program will receive two. Rather than there being one dependent variable in `$ML_y1`, there will be dependent variables in `$ML_y1` and `$ML_y2`. For instance, the Weibull regression log-likelihood function is

$$\ln \ell_j = -(t_j e^{-\theta_{1j}})^{\exp(\theta_{2j})} + d_j\{\theta_{2j} - \theta_{1j} + (e^{\theta_{2j}} - 1)(\ln t_j - \theta_{1j})\}$$
$$\theta_{1j} = \mathbf{x}_j \mathbf{b}_1$$
$$\theta_{2j} = s$$

where t_j is the time of failure or censoring and $d_j = 1$ if failure and 0 if censored. We can make the log likelihood a little easier to program by introducing some extra variables:

$$p_j = \exp(\theta_{2j})$$
$$M_j = \{t_j \exp(-\theta_{1j})\}^{p_j}$$
$$R_j = \ln t_j - \theta_{1j}$$
$$\ln \ell_j = -M_j + d_j\{\theta_{2j} - \theta_{1j} + (p_j - 1)R_j\}$$

The method lf evaluator for this is

```
program myweib
        version 10
        args lnf theta1 theta2
        tempvar p M R
        quietly gen double `p' = exp(`theta2')
        quietly gen double `M' = ($ML_y1*exp(-`theta1'))^`p'
        quietly gen double `R' = ln($ML_y1)-`theta1'
        quietly replace `lnf' = -`M' + $ML_y2*(`theta2'-`theta1' + (`p'-1)*`R')
end
```

We can fit a model by typing

```
. ml model lf myweib (studytime died = drug2 drug3 age) ()
. ml maximize
```

Note that we specified '()' for the second equation. The second equation corresponds to the Weibull shape parameter s, and the linear combination we want for s contains just an intercept. Alternatively, we could type

```
. ml model lf myweib (studytime died = drug2 drug3 age) /s
```

Typing /s means the same thing as typing (s:), and both really mean the same thing as (). The s, either after a slash or in parentheses before a colon, labels the equation. It makes the output look prettier, and that is all:

```
. use http://www.stata-press.com/data/r10/cancer, clear
(Patient Survival in Drug Trial)
. gen drug3 = 1 if drug==3
(34 missing values generated)
. gen drug2 = 1 if drug==2
(34 missing values generated)
. gen drug1 = 1 if drug==1
(28 missing values generated)
. replace drug3 = 0 if drug3==.
(34 real changes made)
. replace drug2 = 0 if drug2==.
(34 real changes made)
. replace drug1 = 0 if drug1==.
(28 real changes made)
. ml model lf myweib (studytime died = drug2 drug3 age) /s
. ml maximize
initial:       log likelihood =      -744
alternative:   log likelihood = -356.14276
rescale:       log likelihood = -200.80201
rescale eq:    log likelihood = -136.69232
Iteration 0:   log likelihood = -136.69232  (not concave)
Iteration 1:   log likelihood = -124.11726
Iteration 2:   log likelihood = -113.90443
Iteration 3:   log likelihood = -110.30539
Iteration 4:   log likelihood = -110.26747
Iteration 5:   log likelihood = -110.26736
Iteration 6:   log likelihood = -110.26736

                                                  Number of obs   =         48
                                                  Wald chi2(3)    =      35.25
Log likelihood = -110.26736                       Prob > chi2     =     0.0000

------------------------------------------------------------------------------
             |      Coef.   Std. Err.      z    P>|z|     [95% Conf. Interval]
-------------+----------------------------------------------------------------
eq1          |
       drug2 |   1.012966   .2903917     3.49   0.000     .4438086    1.582123
       drug3 |    1.45917   .2821195     5.17   0.000     .9062261    2.012114
         age |  -.0671728   .0205688    -3.27   0.001    -.1074868   -.0268587
       _cons |   6.060723   1.152845     5.26   0.000     3.801188    8.320259
-------------+----------------------------------------------------------------
s            |
       _cons |   .5573333   .1402154     3.97   0.000     .2825162    .8321504
------------------------------------------------------------------------------
```

◁

▷ Example 3: Method d0

Method d0 evaluators receive $\mathbf{b} = (\mathbf{b}_1, \mathbf{b}_2, \ldots, \mathbf{b}_E)$, the coefficient vector, rather than the already evaluated $\theta_1, \theta_2, \ldots, \theta_E$, and they are required to evaluate the overall log-likelihood $\ln L$ rather than $\ln \ell_j$, $j = 1, \ldots, N$.

Use `mleval` to produce the thetas from the coefficient vector.

Use `mlsum` to sum the components that enter into $\ln L$.

In the case of Weibull, $\ln L = \sum \ln \ell_j$, and our method d0 evaluator is

```
program weib0
        version 10
        args todo b lnf
        tempvar theta1 theta2
        mleval `theta1' = `b', eq(1)
        mleval `theta2' = `b', eq(2)
        local t "$ML_y1"            // this is just for readability
        local d "$ML_y2"
        tempvar p M R
        quietly gen double `p' = exp(`theta2')
        quietly gen double `M' = (`t'*exp(-`theta1'))^`p'
        quietly gen double `R' = ln(`t')-`theta1'
        mlsum `lnf' = -`M' + `d'*(`theta2'-`theta1' + (`p'-1)*`R')
end
```

To fit our model using this evaluator, we would type

```
. ml model d0 weib0 (studytime died = drug2 drug3 age) /s
. ml maximize
```

◁

❑ Technical Note

Method d0 does not require $\ln L = \sum_j \ln \ell_j$, $j = 1, \ldots, N$, as method lf does. Your likelihood function might have independent components only for groups of observations. Panel-data estimators have a log-likelihood value $\ln L = \sum_i \ln L_i$, where i indexes the panels, each of which contains multiple observations. Conditional logistic regression has $\ln L = \sum_k \ln L_k$, where k indexes the risk pools. Cox regression has $\ln L = \sum_{(t)} \ln L_{(t)}$, where (t) denotes the ordered failure times.

To evaluate such likelihood functions, first calculate the within-group log-likelihood contributions. This usually involves `generate` and `replace` statements prefixed with `by`, as in

```
tempvar sumd
by group: gen double `sumd' = sum($ML_y1)
```

Structure your code so that the log-likelihood contributions are recorded in the last observation of each group. Say that a variable is named `` `cont' ``. To sum the contributions, code

```
tempvar last
quietly by group: gen byte `last' = (_n==_N)
mlsum `lnf' = `cont' if `last'
```

You must inform `mlsum` which observations contain log-likelihood values to be summed. First, you do not want to include intermediate results in the sum. Second, `mlsum` does not skip missing values. Rather, if `mlsum` sees a missing value among the contributions, it sets the overall result, `` `lnf' ``, to missing. That is how `ml maximize` is informed that the likelihood function could not be evaluated at the particular value of $\mathbf{b}$. `ml maximize` will then take action to escape from what it thinks is an infeasible area of the likelihood function.

When the likelihood function violates the linear-form restriction $\ln L = \sum_j \ln \ell_j$, $j = 1, \ldots, N$, with $\ln \ell_j$ being a function solely of values within the jth observation, use method d0. In the following examples, we will demonstrate methods d1 and d2 with likelihood functions that meet this linear-form restriction. The d1 and d2 methods themselves do not require the linear-form restriction, but the utility routines `mlvecsum` and `mlmatsum` do. Using method d1 or d2 when the restriction is violated is a difficult; however, `mlmatbysum` may be of some help for method d2 evaluators. ❑

▷ Example 4: Method d1

Method d1 evaluators are required to produce the gradient vector $\mathbf{g} = \partial \ln L/\partial \mathbf{b}$, as well as the overall log-likelihood value. Using `mlvecsum`, we can obtain $\partial \ln L/\partial \mathbf{b}$ from $\partial \ln L/\partial \theta_i$, $i = 1, \ldots, E$. The derivatives of the Weibull log-likelihood function are

$$\frac{\partial \ln \ell_j}{\partial \theta_{1j}} = p_j(M_j - d_j)$$

$$\frac{\partial \ln \ell_j}{\partial \theta_{2j}} = d_j - R_j p_j(M_j - d_j)$$

The method d1 evaluator for this is

```
program weib1
        version 10
        args todo b lnf g                                    // g is new
        tempvar t1 t2
        mleval `t1' = `b', eq(1)
        mleval `t2' = `b', eq(2)
        local t "$ML_y1"
        local d "$ML_y2"
        tempvar p M R
        quietly gen double `p' = exp(`t2')
        quietly gen double `M' = (`t'*exp(-`t1'))^`p'
        quietly gen double `R' = ln(`t')-`t1'
        mlsum `lnf' = -`M' + `d'*(`t2'-`t1' + (`p'-1)*`R')
        if (`todo'==0 | `lnf'>=.) exit                       /* <-- new */
        tempname d1 d2                                       /* <-- new */
        mlvecsum `lnf' `d1' = `p'*(`M'-`d'), eq(1)           /* <-- new */
        mlvecsum `lnf' `d2' = `d' - `R'*`p'*(`M'-`d'), eq(2) /* <-- new */
        matrix `g' = (`d1',`d2')                             /* <-- new */
end
```

We obtained this code by starting with our method d0 evaluator and then adding the extra lines method d1 requires. To fit our model using this evaluator, we could type

```
. ml model d1 weib1 (studytime died = drug2 drug3 age) /s
. ml maximize
```

but we recommend substituting method d1debug for method d1 and typing

```
. ml model d1debug weib1 (studytime died = drug2 drug3 age) /s
. ml maximize
```

Method d1debug will compare the derivatives we calculate with numerical derivatives and thus verify that our program is correct. Once we are certain the program is correct, then we would switch from method d1debug to method d1. ◁

▷ Example 5: Method d2

Method d2 evaluators are required to produce $-\mathbf{H} = -\partial^2 \ln L/\partial\mathbf{b}\partial\mathbf{b}'$, the negative Hessian matrix, as well as the gradient and log-likelihood value. mlmatsum will help calculate $\partial^2 \ln L/\partial\mathbf{b}\partial\mathbf{b}'$ from the negative second derivatives with respect to θ. For the Weibull model, these negative second derivatives are

$$-\frac{\partial^2 \ln \ell_j}{\partial\theta_{1j}^2} = p_j^2 M_j$$

$$-\frac{\partial^2 \ln \ell_j}{\partial\theta_{1j}\partial\theta_{2j}} = -p_j(M_j - d_j + R_j p_j M_j)$$

$$-\frac{\partial^2 \ln \ell_j}{\partial\theta_{2j}^2} = p_j R_j(R_j p_j M_j + M_j - d_j)$$

The method d2 evaluator is

```
program weib2
        version 10
        args todo b lnf g negH                              // negH added

        tempvar t1 t2
        mleval `t1' = `b', eq(1)
        mleval `t2' = `b', eq(2)

        local t "$ML_y1"
        local d "$ML_y2"

        tempvar p M R
        quietly gen double `p' = exp(`t2')
        quietly gen double `M' = (`t'*exp(-`t1'))^`p'
        quietly gen double `R' = ln(`t')-`t1'

        mlsum `lnf' = -`M' + `d'*(`t2'-`t1' + (`p'-1)*`R')
        if (`todo'==0 | `lnf'>=.) exit

        tempname d1 d2
        mlvecsum `lnf' `d1' = `p'*(`M'-`d'), eq(1)
        mlvecsum `lnf' `d2' = `d' - `R'*`p'*(`M'-`d'), eq(2)
        matrix `g' = (`d1',`d2')
        if (`todo'==1 | `lnf'>=.) exit                      // new from here down

        tempname d11 d12 d22
        mlmatsum `lnf' `d11' = `p'^2 * `M', eq(1)
        mlmatsum `lnf' `d12' = -`p'*(`M'-`d' + `R'*`p'*`M'), eq(1,2)
        mlmatsum `lnf' `d22' = `p'*`R'*(`R'*`p'*`M' + `M' - `d') , eq(2)
        matrix `negH' = (`d11',`d12' \ `d12'',`d22')
end
```

We started with our previous method d1 evaluator and added the lines that method d2 requires. We could now fit a model by typing

```
. ml model d2 weib2 (studytime died = drug2 drug3 age) /s
. ml maximize
```

but we would recommend substituting method d2debug for method d2 and typing

```
. ml model d2debug weib2 (studytime died = drug2 drug3 age) /s
. ml maximize
```

Method d2debug will compare the first and second derivatives we calculate with numerical derivatives and thus verify that our program is correct. Once we are certain the program is correct, then we would switch from method d2debug to method d2. ◁

As we stated earlier, to produce the robust variance estimator with method lf, there is nothing to do except specify `vce(robust)`, `vce(cluster` *clustvar*`)`, or `pweight`. For method d0, these options do not work. For methods d1 and d2, these options will work if your likelihood function meets the linear-form restrictions and you fill in the equation scores. The equation scores are defined as

$$\frac{\partial \ln \ell_j}{\partial \theta_{1j}}, \quad \frac{\partial \ln \ell_j}{\partial \theta_{2j}}, \quad \ldots$$

Your evaluator will be passed variables, one for each equation, which you fill in with the equation scores. For *both* method d1 and d2, these variables are passed in the sixth and subsequent positions of the argument list. That is, you must process the arguments as

```
args todo b lnf g negH g1 g2 ...
```

Note that for method d1, the `‘negH’` argument is not used; it is merely a placeholder.

▷ Example 6: Robust variance estimates

If you have used `mlvecsum` in your method d1 or d2 evaluator, it is easy to turn it into a program that allows the computation of the robust variance estimator. The expression that you specified on the right-hand side of `mlvecsum` is the equation score.

Here we turn the program that we gave earlier in the method d1 example into one that allows `vce(robust)`, `vce(cluster` *clustvar*`)`, or `pweight`.

```
program weib1
        version 10
        args todo b lnf g negH g1 g2         // negH, g1, and g2 are new
        tempvar t1 t2
        mleval ‘t1’ = ‘b’, eq(1)
        mleval ‘t2’ = ‘b’, eq(2)
        local t "$ML_y1"
        local d "$ML_y2"
        tempvar p M R
        quietly gen double ‘p’ = exp(‘t2’)
        quietly gen double ‘M’ = (‘t’*exp(-‘t1’))^‘p’
        quietly gen double ‘R’ = ln(‘t’)-‘t1’
        mlsum ‘lnf’ = -‘M’ + ‘d’*(‘t2’-‘t1’ + (‘p’-1)*‘R’)
        if (‘todo’==0 | ‘lnf’>=.) exit
        tempname d1 d2
        quietly replace ‘g1’ = ‘p’*(‘M’-‘d’)               /* <-- new     */
        quietly replace ‘g2’ = ‘d’ - ‘R’*‘p’*(‘M’-‘d’)     /* <-- new     */
        mlvecsum ‘lnf’ ‘d1’  = ‘g1’, eq(1)                 /* <-- changed */
        mlvecsum ‘lnf’ ‘d2’  = ‘g2’, eq(2)                 /* <-- changed */
        matrix ‘g’ = (‘d1’,‘d2’)
end
```

To fit our model and get the robust variance estimates, we type

```
. ml model d1 weib1 (studytime died = drug2 drug3 age) /s, vce(robust)
. ml maximize
```

◁

Survey options and ml

ml can handle stratification, poststratification, multiple stages of clustering, and finite population corrections. Specifying the svy option implies that the data come from a survey design and also implies that the survey linearized variance estimator is to be used; see [SVY] **variance estimation**.

▷ Example 7

Suppose that we are interested in a probit analysis of data from a survey in which q1 is the answer to a yes/no question and d1, d2, d3 are demographic responses. The following is a d2 evaluator for the probit model that meets the requirements for vce(robust) (linear form and computes the scores).

```
program myd2probit
        version 10
        args todo b lnf g negH g1
        tempvar z Fz lnfj
        mleval `z' = `b'
        quietly gen double `Fz'   = normal( `z')  if $ML_y1 == 1
        quietly replace    `Fz'   = normal(-`z')  if $ML_y1 == 0
        quietly gen double `lnfj' = log(`Fz')
        mlsum `lnf' = `lnfj'
        if (`todo'==0 | `lnf' >= .) exit
        quietly replace `g1' =  normalden(`z')/`Fz'  if $ML_y1 == 1
        quietly replace `g1' = -normalden(`z')/`Fz'  if $ML_y1 == 0
        mlvecsum `lnf' `g' = `g1', eq(1)
        if (`todo'==1 | `lnf' >= .) exit
        mlmatsum `lnf' `negH' = `g1'*(`g1'+`z'), eq(1,1)
end
```

To fit a model, we svyset the data, then use svy with ml.

```
. svyset psuid [pw=w], strata(strid)
. ml model d2 myd2probit (q1 = d1 d2 d3), svy
. ml maximize
```

We could also use the subpop() option to make inferences about the subpopulation identified by the variable sub:

```
. svyset psuid [pw=w], strata(strid)
. ml model d2 myd2probit (q1 = d1 d2 d3), svy subpop(sub)
. ml maximize
```

◁

Saved Results

For results saved by ml without the svy option, see [R] **maximize**.

For results saved by ml with the svy option, see [SVY] **svy**.

References

Gould, W. W., J. S. Pitblado, and W. M. Sribney. 2006. *Maximum Likelihood Estimation with Stata*. 3rd ed. College Station, TX: Stata Press.

Korn, E. L., and B. I. Graubard. 1990. Simultaneous testing of regression coefficients with complex survey data: Use of Bonferroni t statistics. *American Statistician* 44: 270–276.

Also See

[R] **maximize** — Details of iterative maximization

[R] **nl** — Nonlinear least-squares estimation

[M-5] **optimize( )** — Function optimization

Title

mlogit — Multinomial (polytomous) logistic regression

Syntax

`mlogit` *depvar* [*indepvars*] [*if*] [*in*] [*weight*] [, *options*]

options	description
Model	
`noconstant`	suppress constant term
`baseoutcome(#)`	value of *depvar* that will be the base outcome
`constraints(clist)`	apply specified linear constraints
`collinear`	keep collinear variables
SE/Robust	
`vce(vcetype)`	*vcetype* may be `oim`, `robust`, `cluster` *clustvar*, `bootstrap`, or `jackknife`
Reporting	
`level(#)`	set confidence level; default is `level(95)`
`rrr`	report relative-risk ratios
Max options	
maximize_options	control the maximization process; seldom used

where *clist* has the form #[-#] [, #[-#] ...]

`bootstrap`, `by`, `jackknife`, `rolling`, `statsby`, `svy`, and `xi` are allowed; see [U] **11.1.10 Prefix commands**.
Weights are not allowed with the `bootstrap` prefix.
`vce()` and weights are not allowed with the `svy` prefix.
`fweight`s, `iweight`s, and `pweight`s are allowed; see [U] **11.1.6 weight**.
See [U] **20 Estimation and postestimation commands** for more capabilities of estimation commands.

Description

`mlogit` fits maximum-likelihood multinomial logit models, also known as polytomous logistic regression. You can define constraints to perform constrained estimation. Some people refer to conditional logistic regression as multinomial logit. If you are one of them, see [R] **clogit**.

See [R] **logistic** for a list of related estimation commands.

The model can have a maximum of 50 outcomes with Stata/MP, Stata/SE, or Stata/IC and 20 outcomes with Small Stata.

Options

Model

`noconstant`; see [R] **estimation options**.

`baseoutcome(#)` specifies the value of *depvar* to be treated as the base outcome. The default is to choose the most frequent outcome.

`constraints(clist)`, `collinear`; see [R] **estimation options**.

SE/Robust

`vce(vcetype)` specifies the type of standard error reported, which includes types that are derived from asymptotic theory, that are robust to some kinds of misspecification, that allow for intragroup correlation, and that use bootstrap or jackknife methods; see [R] ***vce_option***.

If specifying `vce(bootstrap)` or `vce(jackknife)`, you must also specify `baseoutcome()`.

Reporting

`level(#)`; see [R] **estimation options**.

`rrr` reports the estimated coefficients transformed to relative-risk ratios, i.e., e^b rather than b; see *Description of the model* below for an explanation of this concept. Standard errors and confidence intervals are similarly transformed. This option affects how results are displayed, not how they are estimated. `rrr` may be specified at estimation or when replaying previously estimated results.

Max options

maximize_options: `iterate(#)`, `[no]log`, `trace`, `tolerance(#)`, `ltolerance(#)`; see [R] **maximize**. These options are seldom used.

Remarks

Remarks are presented under the following headings:

Description of the model
Fitting unconstrained models
Fitting constrained models

`mlogit` fits maximum likelihood models with discrete dependent (left-hand-side) variables when the dependent variable takes on more than two outcomes and the outcomes have no natural ordering. If the dependent variable takes on only two outcomes, estimates are identical to those produced by `logistic` or `logit`; see [R] **logistic** or [R] **logit**. If the outcomes are ordered, see [R] **ologit**.

Description of the model

For an introduction to multinomial logit models, see Aldrich and Nelson (1984, 73–77), Greene (2003, chap. 21), Hosmer and Lemeshow (2000, 260–287), Long (1997, chap. 6), and Long and Freese (2006, chap. 6 and 7). For a description emphasizing the difference in assumptions and data requirements for conditional and multinomial logit, see Judge et al. (1985, 768–772).

Consider the outcomes 1, 2, 3, ..., m recorded in y, and the explanatory variables X. Assume that there are $m = 3$ outcomes: "buy an American car", "buy a Japanese car", and "buy a European car". The values of y are then said to be "unordered". Even though the outcomes are coded 1, 2, and 3, the numerical values are arbitrary because $1 < 2 < 3$ does not imply that outcome 1 (buy American) is less than outcome 2 (buy Japanese) is less than outcome 3 (buy European). This unordered categorical property of y distinguishes the use of `mlogit` from `regress` (which is appropriate for a continuous dependent variable), from `ologit` (which is appropriate for ordered categorical data), and from `logit` (which is appropriate for two outcomes, which can be thought of as ordered).

In the multinomial logit model, you estimate a set of coefficients, $\beta^{(1)}$, $\beta^{(2)}$, and $\beta^{(3)}$, corresponding to each outcome:

$$\Pr(y=1) = \frac{e^{X\beta^{(1)}}}{e^{X\beta^{(1)}} + e^{X\beta^{(2)}} + e^{X\beta^{(3)}}}$$

$$\Pr(y=2) = \frac{e^{X\beta^{(2)}}}{e^{X\beta^{(1)}} + e^{X\beta^{(2)}} + e^{X\beta^{(3)}}}$$

$$\Pr(y=3) = \frac{e^{X\beta^{(3)}}}{e^{X\beta^{(1)}} + e^{X\beta^{(2)}} + e^{X\beta^{(3)}}}$$

The model, however, is unidentified in the sense that there is more than one solution to $\beta^{(1)}$, $\beta^{(2)}$, and $\beta^{(3)}$ that leads to the same probabilities for $y = 1$, $y = 2$, and $y = 3$. To identify the model, you arbitrarily set one of $\beta^{(1)}$, $\beta^{(2)}$, or $\beta^{(3)}$ to 0—it does not matter which. That is, if you arbitrarily set $\beta^{(1)} = 0$, the remaining coefficients $\beta^{(2)}$ and $\beta^{(3)}$ will measure the change relative to the $y = 1$ group. If you instead set $\beta^{(2)} = 0$, the remaining coefficients $\beta^{(1)}$ and $\beta^{(3)}$ will measure the change relative to the $y = 2$ group. The coefficients will differ because they have different interpretations, but the predicted probabilities for $y = 1$, 2, and 3 will still be the same. Thus either parameterization will be a solution to the same underlying model.

Setting $\beta^{(1)} = 0$, the equations become

$$\Pr(y=1) = \frac{1}{1 + e^{X\beta^{(2)}} + e^{X\beta^{(3)}}}$$

$$\Pr(y=2) = \frac{e^{X\beta^{(2)}}}{1 + e^{X\beta^{(2)}} + e^{X\beta^{(3)}}}$$

$$\Pr(y=3) = \frac{e^{X\beta^{(3)}}}{1 + e^{X\beta^{(2)}} + e^{X\beta^{(3)}}}$$

The relative probability of $y = 2$ to the base outcome is

$$\frac{\Pr(y=2)}{\Pr(y=1)} = e^{X\beta^{(2)}}$$

Let us call this ratio the relative risk, and let us further assume that X and $\beta_k^{(2)}$ are vectors equal to $(x_1, x_2, \ldots, x_k)$ and $(\beta_1^{(2)}, \beta_2^{(2)}, \ldots, \beta_k^{(2)})'$, respectively. The ratio of the relative risk for a one-unit change in x_i is then

$$\frac{e^{\beta_1^{(2)}x_1 + \cdots + \beta_i^{(2)}(x_i+1) + \cdots + \beta_k^{(2)}x_k}}{e^{\beta_1^{(2)}x_1 + \cdots + \beta_i^{(2)}x_i + \cdots + \beta_k^{(2)}x_k}} = e^{\beta_i^{(2)}}$$

Thus the exponentiated value of a coefficient is the relative-risk ratio for a one-unit change in the corresponding variable (risk is measured as the risk of the outcome relative to the base outcome).

Fitting unconstrained models

▷ Example 1

We have data on the type of health insurance available to 616 psychologically depressed subjects in the United States (Tarlov et al. 1989; Wells et al. 1989). The insurance is categorized as either an indemnity plan (i.e., regular fee-for-service insurance, which may have a deductible or coinsurance rate) or a prepaid plan (a fixed up-front payment allowing subsequent unlimited use as provided, for instance, by an HMO). The third possibility is that the subject has no insurance whatsoever. We wish to explore the demographic factors associated with each subject's insurance choice. One of the demographic factors in our data is the race of the participant, coded as white or nonwhite:

```
. use http://www.stata-press.com/data/r10/sysdsn3
(Health insurance data)

. tabulate insure nonwhite, chi2 col

+-------------------+
| Key               |
|-------------------|
|     frequency     |
| column percentage |
+-------------------+

           |       nonwhite
    insure |         0          1 |     Total
-----------+----------------------+----------
 Indemnity |       251         43 |       294 
           |     50.71      35.54 |     47.73 
-----------+----------------------+----------
   Prepaid |       208         69 |       277 
           |     42.02      57.02 |     44.97 
-----------+----------------------+----------
  Uninsure |        36          9 |        45 
           |      7.27       7.44 |      7.31 
-----------+----------------------+----------
     Total |       495        121 |       616 
           |    100.00     100.00 |    100.00 

          Pearson chi2(2) =   9.5599   Pr = 0.008
```

Although `insure` appears to take on the values `Indemnity`, `Prepaid`, and `Uninsure`, it actually takes on the values 1, 2, and 3. The words appear because we have associated a value label with the numeric variable `insure`; see [U] **12.6.3 Value labels**.

When we fit a multinomial logit model, we can tell `mlogit` which outcome to use as the base outcome, or we can let `mlogit` choose. To fit a model of `insure` on `nonwhite`, letting `mlogit` choose the base outcome, we type

```
. mlogit insure nonwhite

Iteration 0:   log likelihood = -556.59502
Iteration 1:   log likelihood = -551.78935
Iteration 2:   log likelihood = -551.78348
Iteration 3:   log likelihood = -551.78348

Multinomial logistic regression                   Number of obs   =        616
                                                  LR chi2(2)      =       9.62
                                                  Prob > chi2     =     0.0081
Log likelihood = -551.78348                       Pseudo R2       =     0.0086

------------------------------------------------------------------------------
      insure |      Coef.   Std. Err.      z    P>|z|     [95% Conf. Interval]
-------------+----------------------------------------------------------------
Prepaid      |
    nonwhite |   .6608212   .2157321     3.06   0.002     .2379942    1.083648
       _cons |  -.1879149   .0937644    -2.00   0.045    -.3716896   -.0041401
-------------+----------------------------------------------------------------
Uninsure     |
    nonwhite |   .3779585    .407589     0.93   0.354    -.4209012    1.176818
       _cons |  -1.941934   .1782185   -10.90   0.000    -2.291236   -1.592632
------------------------------------------------------------------------------
(insure==Indemnity is the base outcome)
```

mlogit chose the indemnity outcome as the base outcome and presented coefficients for the outcomes prepaid and uninsured. According to the model, the probability of prepaid for whites (nonwhite = 0) is

$$\Pr(\texttt{insure} = \texttt{Prepaid}) = \frac{e^{-.188}}{1 + e^{-.188} + e^{-1.942}} = 0.420$$

Similarly, for nonwhites, the probability of prepaid is

$$\Pr(\texttt{insure} = \texttt{Prepaid}) = \frac{e^{-.188+.661}}{1 + e^{-.188+.661} + e^{-1.942+.378}} = 0.570$$

These results agree with the column percentages presented by tabulate since the mlogit model is fully saturated. That is, there are enough terms in the model to fully explain the column percentage in each cell. The model chi-squared and the tabulate chi-squared are in almost perfect agreement; both test that the column percentages of insure are the same for both values of nonwhite.

◁

▷ Example 2

By specifying the baseoutcome() option, we can control which outcome of the dependent variable is treated as the base. Left to its own, mlogit chose to make outcome 1, indemnity, the base outcome. To make outcome 2, prepaid, the base, we would type

(Continued on next page)

```
. mlogit insure nonwhite, base(2)
Iteration 0:   log likelihood = -556.59502
Iteration 1:   log likelihood = -551.78935
Iteration 2:   log likelihood = -551.78348
Iteration 3:   log likelihood = -551.78348
Multinomial logistic regression                   Number of obs   =        616
                                                  LR chi2(2)      =       9.62
                                                  Prob > chi2     =     0.0081
Log likelihood = -551.78348                       Pseudo R2       =     0.0086

------------------------------------------------------------------------------
      insure |      Coef.   Std. Err.      z    P>|z|     [95% Conf. Interval]
-------------+----------------------------------------------------------------
Indemnity    |
    nonwhite |  -.6608212   .2157321    -3.06   0.002    -1.083648   -.2379942
       _cons |   .1879149   .0937644     2.00   0.045     .0041401    .3716896
-------------+----------------------------------------------------------------
Uninsure     |
    nonwhite |  -.2828628   .3977302    -0.71   0.477      -1.0624    .4966741
       _cons |  -1.754019   .1805145    -9.72   0.000    -2.107821   -1.400217
------------------------------------------------------------------------------
(insure==Prepaid is the base outcome)
```

The `baseoutcome()` option requires that we specify the numeric value of the outcome, so we could not type `base(Prepaid)`.

Although the coefficients now appear to be different, the summary statistics reported at the top are identical. With this parameterization, the probability of prepaid insurance for whites is

$$\Pr(\texttt{insure} = \texttt{Prepaid}) = \frac{1}{1 + e^{.188} + e^{-1.754}} = 0.420$$

This is the same answer we obtained previously. ◁

▷ Example 3

By specifying `rrr`, which we can do at estimation time or when we redisplay results, we see the model in terms of relative-risk ratios:

```
. mlogit, rrr
Multinomial logistic regression                   Number of obs   =        616
                                                  LR chi2(2)      =       9.62
                                                  Prob > chi2     =     0.0081
Log likelihood = -551.78348                       Pseudo R2       =     0.0086

------------------------------------------------------------------------------
      insure |        RRR   Std. Err.      z    P>|z|     [95% Conf. Interval]
-------------+----------------------------------------------------------------
Indemnity    |
    nonwhite |    .516427   .1114099    -3.06   0.002     .3383588    .7882073
-------------+----------------------------------------------------------------
Uninsure     |
    nonwhite |   .7536232   .2997387    -0.71   0.477     .3456254    1.643247
------------------------------------------------------------------------------
(insure==Prepaid is the base outcome)
```

Looked at this way, the relative risk of choosing an indemnity over a prepaid plan is 0.516 for nonwhites relative to whites.

To illustrate, from the output and discussions of examples 1 and 2 we find that

$$\Pr(\texttt{insure} = \texttt{Indemnity} \mid \text{white}) = \frac{1}{1 + e^{-.188} + e^{-1.942}} = 0.507$$

and thus the relative risk of choosing indemnity over prepaid (for whites) is

$$\frac{\Pr(\texttt{insure} = \texttt{Indemnity} \mid \text{white})}{\Pr(\texttt{insure} = \texttt{Prepaid} \mid \text{white})} = \frac{0.507}{0.420} = 1.207$$

For nonwhites,

$$\Pr(\texttt{insure} = \texttt{Indemnity} \mid \text{not white}) = \frac{1}{1 + e^{-.188+.661} + e^{-1.942+.378}} = 0.355$$

and thus the relative risk of choosing indemnity over prepaid (for nonwhites) is

$$\frac{\Pr(\texttt{insure} = \texttt{Indemnity} \mid \text{not white})}{\Pr(\texttt{insure} = \texttt{Prepaid} \mid \text{not white})} = \frac{0.355}{0.570} = 0.623$$

The ratio of these two relative risks, hence the name "relative-risk ratio", is $0.623/1.207 = 0.516$, as given in the output under the heading "RRR".

◁

❑ Technical Note

In models where only two categories are considered, the `mlogit` model reduces to standard `logit`. Consequently the exponentiated regression coefficients, labeled as `RRR` within `mlogit`, are equal to the odds ratios as given when option `or` is specified under `logit`.

As such, always referring to `mlogit`'s exponentiated coefficients as odds ratios may be tempting. However, the discussion in example 3 demonstrates that doing so would be incorrect. In general `mlogit` models, the exponentiated coefficients are ratios of relative risks, not ratios of odds.

❑

▷ Example 4

One of the advantages of `mlogit` over `tabulate` is that we can include continuous variables and multiple categorical variables in the model. In examining the data on insurance choice, we decide that we want to control for age, gender, and site of study (the study was conducted in three sites):

(Continued on next page)

```
. mlogit insure age male nonwhite site2 site3
Iteration 0:   log likelihood = -555.85446
Iteration 1:   log likelihood = -534.72983
Iteration 2:   log likelihood = -534.36536
Iteration 3:   log likelihood = -534.36165
Iteration 4:   log likelihood = -534.36165
Multinomial logistic regression                   Number of obs   =        615
                                                  LR chi2(10)     =      42.99
                                                  Prob > chi2     =     0.0000
Log likelihood = -534.36165                       Pseudo R2       =     0.0387

------------------------------------------------------------------------------
      insure |      Coef.   Std. Err.      z    P>|z|     [95% Conf. Interval]
-------------+----------------------------------------------------------------
Prepaid      |
         age |  -.011745   .0061946    -1.90   0.058    -.0238862    .0003962
        male |  .5616934   .2027465     2.77   0.006     .1643175    .9590693
    nonwhite |  .9747768   .2363213     4.12   0.000     .5115955    1.437958
       site2 |  .1130359   .2101903     0.54   0.591    -.2989296    .5250013
       site3 | -.5879879   .2279351    -2.58   0.010    -1.034733   -.1412433
       _cons |  .2697127   .3284422     0.82   0.412    -.3740222    .9134476
-------------+----------------------------------------------------------------
Uninsure     |
         age | -.0077961   .0114418    -0.68   0.496    -.0302217    .0146294
        male |  .4518496   .3674867     1.23   0.219     -.268411     1.17211
    nonwhite |  .2170589   .4256361     0.51   0.610    -.6171725     1.05129
       site2 | -1.211563   .4705127    -2.57   0.010    -2.133751   -.2893747
       site3 | -.2078123   .3662926    -0.57   0.570    -.9257327     .510108
       _cons | -1.286943   .5923219    -2.17   0.030    -2.447872   -.1260135
------------------------------------------------------------------------------
(insure==Indemnity is the base outcome)
```

These results suggest that the inclination of nonwhites to choose prepaid care is even stronger than it was without controlling. We also see that subjects in site 2 are less likely to be uninsured.

◁

Fitting constrained models

mlogit can fit models with subsets of coefficients constrained to be zero, with subsets of coefficients constrained to be equal both within and across equations, and with subsets of coefficients arbitrarily constrained to equal linear combinations of other estimated coefficients.

Before fitting a constrained model, you define the constraints with the `constraint` command; see [R] **constraint**. Once the constraints are defined, you estimate using `mlogit`, specifying the `constraint()` option. Typing `constraint(4)` would use the constraint you previously saved as 4. Typing `constraint(1,4,6)` would use the previously stored constraints 1, 4, and 6. Typing `constraint(1-4,6)` would use the previously stored constraints 1, 2, 3, 4, and 6.

Sometimes you will not be able to specify the constraints without knowing the omitted outcome. In such cases, assume that the omitted outcome is whatever outcome is convenient for you, and include the `baseoutcome()` option when you type the `mlogit` command.

▷ Example 5

We can use constraints to test hypotheses, among other things. In our insurance-choice model, let's test the hypothesis that there is no distinction between having indemnity insurance and being uninsured. Indemnity-style insurance was the omitted outcome, so we type

```
. test [Uninsure]

 ( 1)  [Uninsure]age = 0
 ( 2)  [Uninsure]male = 0
 ( 3)  [Uninsure]nonwhite = 0
 ( 4)  [Uninsure]site2 = 0
 ( 5)  [Uninsure]site3 = 0

           chi2(  5) =    9.31
         Prob > chi2 =    0.0973
```

If indemnity had not been the omitted outcome, we would have typed `test [Uninsure=Indemnity]`.

The results produced by `test` are an approximation based on the estimated covariance matrix of the coefficients. Since the probability of being uninsured is low, the log likelihood may be nonlinear for the uninsured. Conventional statistical wisdom is not to trust the asymptotic answer under these circumstances but to perform a likelihood-ratio test instead.

To use Stata's `lrtest` (likelihood-ratio test) command, we must fit both the unconstrained and constrained models. The unconstrained model is the one we have previously fitted. Following the instruction in [R] **lrtest**, we first save the unconstrained model results:

```
. estimates store unconstrained
```

To fit the constrained model, we must refit our model with all the coefficients except the constant set to 0 in the `Uninsure` equation. We define the constraint and then refit:

```
. constraint 1 [Uninsure]

. mlogit insure age male nonwhite site2 site3, constr(1)

Iteration 0:   log likelihood = -555.85446
Iteration 1:   log likelihood = -539.80523
Iteration 2:   log likelihood = -539.75644
Iteration 3:   log likelihood = -539.75643

Multinomial logistic regression                   Number of obs   =        615
                                                  LR chi2(5)      =      32.20
                                                  Prob > chi2     =     0.0000
Log likelihood = -539.75643                       Pseudo R2       =     0.0290

 ( 1)  [Uninsure]age = 0
 ( 2)  [Uninsure]male = 0
 ( 3)  [Uninsure]nonwhite = 0
 ( 4)  [Uninsure]site2 = 0
 ( 5)  [Uninsure]site3 = 0

------------------------------------------------------------------------------
      insure |      Coef.   Std. Err.      z    P>|z|     [95% Conf. Interval]
-------------+----------------------------------------------------------------
Prepaid      |
         age |  -.0107025   .0060039    -1.78   0.075    -.0224699    .0010649
        male |   .4963616   .1939683     2.56   0.010     .1161908    .8765324
    nonwhite |    .942137   .2252094     4.18   0.000     .5007347    1.383539
       site2 |   .2530912   .2029465     1.25   0.212    -.1446767    .6508591
       site3 |  -.5521774   .2187237    -2.52   0.012    -.9808678   -.1234869
       _cons |   .1792752   .3171372     0.57   0.572    -.4423023    .8008527
-------------+----------------------------------------------------------------
Uninsure     |
         age |  (dropped)
        male |  (dropped)
    nonwhite |  (dropped)
       site2 |  (dropped)
       site3 |  (dropped)
       _cons |   -1.87351   .1601099   -11.70   0.000     -2.18732     -1.5597
------------------------------------------------------------------------------
(insure==Indemnity is the base outcome)
```

We can now perform the likelihood-ratio test:

```
. lrtest unconstrained .
Likelihood-ratio test                                 LR chi2(5)  =     10.79
(Assumption: . nested in unconstrained)               Prob > chi2 =    0.0557
```

The likelihood-ratio chi-squared is 10.79 with 5 degrees of freedom—just slightly greater than the magic $p = .05$ level—so we should not call this difference significant.

◁

❑ Technical Note

In certain circumstances, you should fit a multinomial logit model with conditional logit; see [R] **clogit**. With substantial data manipulation, `clogit` can handle the same class of models with some interesting additions. For example, if we had available the price and deductible of the most competitive insurance plan of each type, `mlogit` could not use this information, but `clogit` could.

❑

Saved Results

`mlogit` saves the following in `e()`:

Scalars

`e(N)`	number of observations	`e(ll_0)`	log likelihood, constant-only model
`e(k_out)`	number of outcomes	`e(N_clust)`	number of clusters
`e(k_eq_model)`	number of equations in model Wald test	`e(chi2)`	χ^2
		`e(ibaseout)`	base outcome number
`e(df_m)`	model degrees of freedom	`e(baseout)`	the value of *depvar* to be treated as the base outcome
`e(r2_p)`	pseudo-R-squared		
`e(ll)`	log likelihood		

Macros

`e(cmd)`	`mlogit`	`e(chi2type)`	`Wald` or `LR`; type of model χ^2 test
`e(cmdline)`	command as typed	`e(vce)`	*vcetype* specified in `vce()`
`e(depvar)`	name of dependent variable	`e(vcetype)`	title used to label Std. Err.
`e(wtype)`	weight type	`e(crittype)`	optimization criterion
`e(wexp)`	weight expression	`e(properties)`	`b V`
`e(title)`	title in estimation output	`e(footnote)`	program used to implement the footnote display
`e(clustvar)`	name of cluster variable		
`e(eqnames)`	names of equations	`e(predict)`	program used to implement `predict`
`e(baselab)`	value label corresponding to base outcome		

Matrices

`e(b)`	coefficient vector	`e(V)`	variance–covariance matrix of the estimators
`e(out)`	outcome values		

Functions

`e(sample)`	marks estimation sample

Methods and Formulas

`mlogit` is implemented as an ado-file.

The multinomial logit model is described in Greene (2003, chap. 21).

Suppose that there are k categorical outcomes and—without loss of generality—let the base outcome be 1. The probability that the response for the jth observation is equal to the ith outcome is

$$p_{ij} = \Pr(y_j = i) = \begin{cases} \dfrac{1}{1+\sum\limits_{m=2}^{k} \exp(\mathbf{x}_j\boldsymbol{\beta}_m)}, & \text{if} \quad i=1 \\ \dfrac{\exp(\mathbf{x}_j\boldsymbol{\beta}_i)}{1+\sum\limits_{m=2}^{k} \exp(\mathbf{x}_j\boldsymbol{\beta}_m)}, & \text{if} \quad i>1 \end{cases}$$

where $\mathbf{x}_j$ is the row vector of observed values of the independent variables for the jth observation and $\boldsymbol{\beta}_m$ is the coefficient vector for outcome m. The log pseudolikelihood is

$$\ln L = \sum_j w_j \sum_{i=1}^{k} I_i(y_j) \ln p_{ik}$$

where w_j is an optional weight and

$$I_i(y_j) = \begin{cases} 1, \text{ if } y_j = i \\ 0, \text{ otherwise} \end{cases}$$

Newton–Raphson maximum likelihood is used; see [R] **maximize**.

For constrained equations, the set of constraints is orthogonalized, and a subset of maximizable parameters is selected. For example, a parameter that is constrained to zero is not a maximizable parameter. If two parameters are constrained to be equal to each other, only one is a maximizable parameter.

Let $\mathbf{r}$ be the vector of maximizable parameters. $\mathbf{r}$ is physically a subset of the solution parameters, $\mathbf{b}$. A matrix, $\mathbf{T}$, and a vector, $\mathbf{m}$, are defined as

$$\mathbf{b} = \mathbf{Tr} + \mathbf{m}$$

so that

$$\frac{\partial f}{\partial \mathbf{b}} = \frac{\partial f}{\partial \mathbf{r}}\mathbf{T}'$$

$$\frac{\partial^2 f}{\partial \mathbf{b}^2} = \mathbf{T}\frac{\partial^2 f}{\partial \mathbf{r}^2}\mathbf{T}'$$

$\mathbf{T}$ consists of a block form in which one part is a permutation of the identity matrix and the other part describes how to calculate the constrained parameters from the maximizable parameters.

References

Aldrich, J. H., and F. D. Nelson. 1984. *Linear Probability, Logit, and Probit Models.* Newbury Park, CA: Sage.

Freese, J., and J. S. Long. 2000. sg155: Tests for the multinomial logit model. *Stata Technical Bulletin* 58: 19–25. Reprinted in *Stata Technical Bulletin Reprints*, vol. 10, pp. 247–255.

Greene, W. H. 2003. *Econometric Analysis.* 5th ed. Upper Saddle River, NJ: Prentice Hall.

Haan, P., and A. Uhlendorff. 2006. Estimation of multinomial logit models with unobserved heterogeneity using maximum simulated likelihood. *Stata Journal* 6: 229–245.

Hamilton, L. C. 1993. sqv8: Interpreting multinomial logistic regression. *Stata Technical Bulletin* 13: 24–28. Reprinted in *Stata Technical Bulletin Reprints*, vol. 3, pp. 176–181.

——. 2006. *Statistics with Stata (Updated for Version 9).* Belmont, CA: Duxbury.

Hendrickx, J. 2000. sbe37: Special restrictions in multinomial logistic regression. *Stata Technical Bulletin* 56: 18–26. Reprinted in *Stata Technical Bulletin Reprints*, vol. 10, pp. 93–103.

Hosmer, D. W., Jr., and S. Lemeshow. 2000. *Applied Logistic Regression.* 2nd ed. New York: Wiley.

Judge, G. G., W. E. Griffiths, R. C. Hill, H. Lütkepohl, and T.-C. Lee. 1985. *The Theory and Practice of Econometrics.* 2nd ed. New York: Wiley.

Kleinbaum, D. G., and M. Klein. 2002. *Logistic Regression: A Self-Learning Text.* 2nd ed. New York: Springer.

Long, J. S. 1997. *Regression Models for Categorical and Limited Dependent Variables.* Thousand Oaks, CA: Sage.

Long, J. S., and J. Freese. 2006. *Regression Models for Categorical Dependent Variables Using Stata.* 2nd ed. College Station, TX: Stata Press.

Tarlov, A. R., J. E. Ware, Jr., S. Greenfield, E. C. Nelson, E. Perrin, and M. Zubkoff. 1989. The medical outcomes study. *Journal of the American Medical Association* 262: 925–930.

Wells, K. E., R. D. Hays, M. A. Burnam, W. H. Rogers, S. Greenfield, and J. E. Ware, Jr. 1989. Detection of depressive disorder for patients receiving prepaid or fee-for-service care. *Journal of the American Medical Association* 262: 3298–3302.

Xu, J., and J. S. Long. 2005. Confidence intervals for predicted outcomes in regression models for categorical outcomes. *Stata Journal* 5: 537–559.

Also See

[R] **mlogit postestimation** — Postestimation tools for mlogit

[R] **clogit** — Conditional (fixed-effects) logistic regression

[R] **logistic** — Logistic regression, reporting odds ratios

[R] **logit** — Logistic regression, reporting coefficients

[R] **mlogit** — Multinomial (polytomous) logistic regression

[R] **nlogit** — Nested logit regression

[R] **ologit** — Ordered logistic regression

[R] **rologit** — Rank-ordered logistic regression

[R] **slogit** — Stereotype logistic regression

[SVY] **svy estimation** — Estimation commands for survey data

[U] **20 Estimation and postestimation commands**

Title

mlogit postestimation — Postestimation tools for mlogit

Description

The following postestimation commands are available for `mlogit`:

command	description
`adjust`	adjusted predictions of $\mathbf{x}\beta$ or $\exp(\mathbf{x}\beta)$
`estat`	AIC, BIC, VCE, and estimation sample summary
`estat` (svy)	postestimation statistics for survey data
`estimates`	cataloging estimation results
`hausman`	Hausman's specification test
`lincom`	point estimates, standard errors, testing, and inference for linear combinations of coefficients
`lrtest`[1]	likelihood-ratio test
`mfx`	marginal effects or elasticities
`nlcom`	point estimates, standard errors, testing, and inference for nonlinear combinations of coefficients
`predict`	predictions, residuals, influence statistics, and other diagnostic measures
`predictnl`	point estimates, standard errors, testing, and inference for generalized predictions
`suest`	seemingly unrelated estimation
`test`	Wald tests for simple and composite linear hypotheses
`testnl`	Wald tests of nonlinear hypotheses

[1] `lrtest` is not appropriate with `svy` estimation results.

See the corresponding entries in the *Stata Base Reference Manual* for details, but see [SVY] **estat** for details about `estat` (svy).

Syntax for predict

`predict` [*type*] { *stub** | *newvar* | *newvarlist* } [*if*] [*in*] [, *statistic* `outcome(`*outcome*`)`]

`predict` [*type*] { *stub** | *newvarlist* } [*if*] [*in*], `scores`

statistic	description
Main	
`pr`	probability of a positive outcome; the default
`xb`	linear prediction
`stdp`	standard error of the linear prediction
`stddp`	standard error of the difference in two linear predictions

You specify one new variable with `pr` and you specify `outcome()`, or you specify *stub** or k new variables with `pr`, where k is the number of outcomes.

You specify one new variable with `xb`, `stdp`, and `stddp` and you specify `outcome()`.

These statistics are available both in and out of sample; type `predict ... if e(sample) ...` if wanted only for the estimation sample.

Options for predict

Main

`pr`, the default, calculates the probability of each of the categories of the dependent variable or the probability of the level specified in `outcome(`*outcome*`)`. If you specify the `outcome(`*outcome*`)` option, you need to specify only one new variable; otherwise, you must specify a new variable for each category of the dependent variable.

`xb` calculates the linear prediction. You must also specify the `outcome(`*outcome*`)` option.

`stdp` calculates the standard error of the linear prediction. You must also specify the `outcome(`*outcome*`)` option.

`stddp` calculates the standard error of the difference in two linear predictions. You must specify option `outcome(`*outcome*`)`, and here you specify the two particular outcomes of interest inside the parentheses, for example, `predict sed, stddp outcome(1,3)`.

`outcome(`*outcome*`)` specifies the outcome for which the statistic is to be calculated. `equation()` is a synonym for `outcome()`: it does not matter which you use. `outcome()` or `equation()` can be specified using

`#1`, `#2`, ..., where `#1` means the first category of the dependent variable, `#2` the second category, etc.;

the values of the dependent variable; or

the value labels of the dependent variable if they exist.

`scores` calculates equation-level score variables. The number of score variables created will be one less than the number of outcomes in the model. If the number of outcomes in the model were k, then

the first new variable will contain $\partial \ln L/\partial(\mathbf{x}_j\boldsymbol{\beta}_1)$;

the second new variable will contain $\partial \ln L/\partial(\mathbf{x}_j\boldsymbol{\beta}_2)$;

...

the $(k-1)$st new variable will contain $\partial \ln L/\partial(\mathbf{x}_j\boldsymbol{\beta}_{k-1})$.

Remarks

Remarks are presented under the following headings:

Obtaining predicted values
Calculating marginal effects
Testing hypotheses about coefficients

Obtaining predicted values

▷ Example 1

After estimation, we can use `predict` to obtain predicted probabilities, index values, and standard errors of the index, or differences in the index. For instance, in example 4 of [R] **mlogit**, we fitted a model of insurance choice on various characteristics. We can obtain the predicted probabilities for outcome 1 by typing

```
. use http://www.stata-press.com/data/r10/sysdsn3
(Health insurance data)
. mlogit insure age male nonwhite site2 site3
  (output omitted)
. predict p1 if e(sample), outcome(1)
(option pr assumed; predicted probability)
(29 missing values generated)
. summarize p1
```

Variable	Obs	Mean	Std. Dev.	Min	Max
p1	615	.4764228	.1032279	.1698142	.71939

We included `if e(sample)` to restrict the calculation to the estimation sample. In example 4 of [R] **mlogit**, the multinomial logit model was fitted on 615 observations, so there must be missing values in our dataset.

Although we typed `outcome(1)`, specifying `1` for the indemnity outcome, we could have typed `outcome(Indemnity)`. For instance, to obtain the probabilities for prepaid, we could type

```
. predict p2 if e(sample), outcome(prepaid)
(option pr assumed; predicted probability)
equation prepaid not found
r(303);
. predict p2 if e(sample), outcome(Prepaid)
(option pr assumed; predicted probability)
(29 missing values generated)
. summarize p2
```

Variable	Obs	Mean	Std. Dev.	Min	Max
p2	615	.4504065	.1125962	.1964103	.7885724

We must specify the label exactly as it appears in the underlying value label (or how it appears in the `mlogit` output), including capitalization.

Here we have used `predict` to obtain probabilities for the same sample on which we estimated. That is not necessary. We could use another dataset that had the independent variables defined (in our example, `age`, `male`, `nonwhite`, `site2`, and `site3`) and use `predict` to obtain predicted probabilities; here, we would not specify `if e(sample)`.

◁

▷ Example 2

`predict` can also be used to obtain the index values—the $\sum x_i \widehat{\beta}_i^{(k)}$—as well as the probabilities:

```
. predict idx1, outcome(Indemnity) xb
(1 missing value generated)
. summarize idx1
```

Variable	Obs	Mean	Std. Dev.	Min	Max
idx1	643	0	0	0	0

The indemnity outcome was our base outcome—the outcome for which all the coefficients were set to 0—so the index is always 0. For the prepaid and uninsured outcomes, we type

```
. predict idx2, outcome(Prepaid) xb
(1 missing value generated)
. predict idx3, outcome(Uninsure) xb
(1 missing value generated)
. summarize idx2 idx3
    Variable |       Obs        Mean    Std. Dev.       Min        Max
-------------+--------------------------------------------------------
        idx2 |       643   -.0566113    .4962973  -1.298198   1.700719
        idx3 |       643   -1.980747    .6018139  -3.112741  -.8258458
```

We can obtain the standard error of the index by specifying the `stdp` option:

```
. predict se2, outcome(Prepaid) stdp
(1 missing value generated)
. list p2 idx2 se2 in 1/5

     +----------------------------------+
     |       p2        idx2         se2 |
     |----------------------------------|
  1. | .3709022   -.4831167    .2437772 |
  2. | .4977667     .055111    .1694686 |
  3. | .4113073   -.1712106    .1793498 |
  4. | .5424927    .3788345    .2513701 |
  5. |        .   -.0925817    .1452616 |
     +----------------------------------+
```

We obtained the probability `p2` in the previous example.

Finally, `predict` can calculate the standard error of the difference in the index values between two outcomes with the `stddp` option:

```
. predict se_2_3, outcome(Prepaid,Uninsure) stddp
(1 missing value generated)
. list idx2 idx3 se_2_3 in 1/5

     +-----------------------------------+
     |      idx2        idx3      se_2_3 |
     |-----------------------------------|
  1. | -.4831167   -3.073253    .5469354 |
  2. |   .055111   -2.715986    .4331917 |
  3. | -.1712106   -1.579621    .3053815 |
  4. |  .3788345   -1.462007    .4492552 |
  5. | -.0925817   -2.814022    .4024784 |
     +-----------------------------------+
```

In the first observation, the difference in the indexes is $-.483-(-3.073)=2.59$. The standard error of that difference is .547.

◁

▷ Example 3

It is more difficult to interpret the results from `mlogit` than those from `clogit` or `logit` since there are multiple equations. For example, suppose that one of the independent variables in our model takes on the values 0 and 1 and we are attempting to understand the effect of this variable. Assume that the coefficient on this variable for the second outcome, $\beta^{(2)}$, is positive. We might then be tempted to reason that the probability of the second outcome is higher if the variable is 1 rather than 0. Most of the time, that will be true, but occasionally we will be surprised. The probability of some other outcome could increase even more (say, $\beta^{(3)} > \beta^{(2)}$), and thus the probability of outcome 2 would actually fall relative to that outcome. We can use `predict` to help interpret such results.

Continuing with our previously fitted insurance-choice model, we wish to describe the model's predictions by race. For this purpose, we can use the method of recycled predictions, in which we vary characteristics of interest across the whole dataset and average the predictions. That is, we have data on both whites and nonwhites, and our individuals have other characteristics as well. We will first pretend that all the people in our data are white but hold their other characteristics constant. We then calculate the probabilities of each outcome. Next we will pretend that all the people in our data are nonwhite, still holding their other characteristics constant. Again we calculate the probabilities of each outcome. The difference in those two sets of calculated probabilities, then, is the difference due to race, holding other characteristics constant.

```
. gen byte nonwhold = nonwhite                      // save real race
. replace nonwhite = 0                              // make everyone white
(126 real changes made)
. predict wpind, outcome(Indemnity)                 // predict probabilities
(option pr assumed; predicted probability)
(1 missing value generated)
. predict wpp, outcome(Prepaid)
(option pr assumed; predicted probability)
(1 missing value generated)
. predict wpnoi, outcome(Uninsure)
(option pr assumed; predicted probability)
(1 missing value generated)
. replace nonwhite=1                                // make everyone nonwhite
(644 real changes made)
. predict nwpind, outcome(Indemnity)
(option pr assumed; predicted probability)
(1 missing value generated)
. predict nwpp, outcome(Prepaid)
(option pr assumed; predicted probability)
(1 missing value generated)
. predict nwpnoi, outcome(Uninsure)
(option pr assumed; predicted probability)
(1 missing value generated)
. replace nonwhite=nonwhold                         // restore real race
(518 real changes made)
. summarize wp* nwp*, sep(3)

    Variable |       Obs        Mean    Std. Dev.       Min        Max
-------------+--------------------------------------------------------
       wpind |       643    .5141673    .0872679   .3092903     .71939
         wpp |       643    .4082052    .0993286   .1964103   .6502247
       wpnoi |       643    .0776275    .0360283   .0273596   .1302816
-------------+--------------------------------------------------------
      nwpind |       643    .3112809    .0817693   .1511329    .535021
        nwpp |       643     .630078    .0979976   .3871782   .8278881
      nwpnoi |       643    .0586411    .0287185   .0209648   .0933874
```

In [R] **mlogit**, we presented a cross-tabulation of insurance type and race. Those values were unadjusted. The means reported above are the values adjusted for age, sex, and site. Combining the results gives

	Unadjusted		Adjusted	
	white	nonwhite	white	nonwhite
Indemnity	.51	.36	.52	.31
Prepaid	.42	.57	.41	.63
Uninsured	.07	.07	.08	.06

We find, for instance, after adjusting for age, sex, and site, that although 57% of nonwhites in our data had prepaid plans, 63% of nonwhites chose prepaid plans.

◁

❑ Technical Note

You can use `predict` to classify predicted values and compare them with the observed outcomes to interpret a multinomial logit model. This is a variation on the notions of sensitivity and specificity for logistic regression. Here we will classify indemnity and prepaid as definitely predicting indemnity, definitely predicting prepaid, and ambiguous.

```
. predict indem, outcome(Indemnity) index              // obtain indexes
(1 missing value generated)
. predict prepaid, outcome(Prepaid) index
(1 missing value generated)
. gen diff = prepaid-indem                             // obtain difference
(1 missing value generated)
. predict sediff, outcome(Indemnity,Prepaid) stddp    // & its standard error
(1 missing value generated)
. gen type = 1 if diff/sediff < -1.96                  // definitely indemnity
(504 missing values generated)
. replace type = 3 if diff/sediff > 1.96               // definitely prepaid
(100 real changes made)
. replace type = 2 if type>=. & diff/sediff < .        // ambiguous
(404 real changes made)
. label def type 1 "Def Ind" 2 "Ambiguous" 3 "Def Prep"
. label values type type                               // label results
. tabulate insure type
```

insure	type Def Ind	Ambiguous	Def Prep	Total
Indemnity	78	183	33	294
Prepaid	44	177	56	277
Uninsure	12	28	5	45
Total	134	388	94	616

We can see that the predictive power of this model is modest. There are many misclassifications in both directions, though there are more correctly classified observations than misclassified observations.

Also the uninsured look overwhelmingly as though they might have come from the indemnity system rather than from the prepaid system.

❑

Calculating marginal effects

You can also compute marginal effects to interpret the results from multinomial logit models effectively. The marginal effects show how the probabilities of each outcome change with respect to changes in the regressors. See example 3 in [R] **mfx** for an example of computing marginal effects after `mlogit`.

Testing hypotheses about coefficients

▷ Example 4

test tests hypotheses about the coefficients just as after any estimation command; see [R] **test**. Note, however, test's syntax for dealing with multiple equation models. Because test bases its results on the estimated covariance matrix, we might prefer a likelihood-ratio test; see example 5 in [R] **mlogit** for an example of lrtest.

If we simply list variables after the test command, we are testing that the corresponding coefficients are zero across all equations:

```
. test site2 site3
 ( 1)  [Prepaid]site2 = 0
 ( 2)  [Uninsure]site2 = 0
 ( 3)  [Prepaid]site3 = 0
 ( 4)  [Uninsure]site3 = 0
           chi2(  4) =   19.74
         Prob > chi2 =    0.0006
```

We can test that all the coefficients (except the constant) in an equation are zero by simply typing the outcome in square brackets:

```
. test [Uninsure]
 ( 1)  [Uninsure]age = 0
 ( 2)  [Uninsure]male = 0
 ( 3)  [Uninsure]nonwhite = 0
 ( 4)  [Uninsure]site2 = 0
 ( 5)  [Uninsure]site3 = 0
           chi2(  5) =    9.31
         Prob > chi2 =    0.0973
```

We specify the outcome just as we do with predict; we can specify the label if the outcome variable is labeled, or we can specify the numeric value of the outcome. We would have obtained the same test as above if we had typed test [3] since 3 is the value of insure for the outcome uninsured.

We can combine the two syntaxes. To test that the coefficients on the site variables are 0 in the equation corresponding to the outcome prepaid, we can type

```
. test [Prepaid]: site2 site3
 ( 1)  [Prepaid]site2 = 0
 ( 2)  [Prepaid]site3 = 0
           chi2(  2) =   10.78
         Prob > chi2 =    0.0046
```

We specified the outcome and then followed that with a colon and the variables we wanted to test.

We can also test that coefficients are equal across equations. To test that all coefficients except the constant are equal for the prepaid and uninsured outcomes, we can type

```
. test [Prepaid=Uninsure]
 ( 1)  [Prepaid]age - [Uninsure]age = 0
 ( 2)  [Prepaid]male - [Uninsure]male = 0
 ( 3)  [Prepaid]nonwhite - [Uninsure]nonwhite = 0
 ( 4)  [Prepaid]site2 - [Uninsure]site2 = 0
 ( 5)  [Prepaid]site3 - [Uninsure]site3 = 0
           chi2(  5) =   13.80
         Prob > chi2 =    0.0169
```

To test that only the site variables are equal, we can type

```
. test [Prepaid=Uninsure]: site2 site3

 ( 1)  [Prepaid]site2 - [Uninsure]site2 = 0
 ( 2)  [Prepaid]site3 - [Uninsure]site3 = 0

           chi2(  2) =   12.68
         Prob > chi2 =    0.0018
```

Finally we can test any arbitrary constraint by simply entering the equation and specifying the coefficients as described in [U] **13.5 Accessing coefficients and standard errors**. The following hypothesis is senseless but illustrates the point:

```
. test (([Prepaid]age+[Uninsure]site2)/2 = 2-[Uninsure]nonwhite)

 ( 1)  .5 [Prepaid]age + [Uninsure]nonwhite + .5 [Uninsure]site2 = 2

           chi2(  1) =   22.45
         Prob > chi2 =    0.0000
```

See [R] **test** for more information about `test`. The information there about combining hypotheses across `test` commands (the `accum` option) also applies after `mlogit`.

◁

Methods and Formulas

All postestimation commands listed above are implemented as ado-files.

Also See

[R] **mlogit** — Multinomial (polytomous) logistic regression

[U] **20 Estimation and postestimation commands**

Title

more — The —more— message

Syntax

Tell Stata to pause or not pause for –more– messages

```
set more { on | off } [ , permanently ]
```

Set number of lines between –more– messages

```
set pagesize #
```

Description

set more on, which is the default, tells Stata to wait until you press a key before continuing when a —more— message is displayed.

set more off tells Stata not to pause or display the —more— message.

set pagesize # sets the number of lines between —more— messages. The permanently option is not allowed with set pagesize.

Option

permanently specifies that, in addition to making the change right now, the more setting be remembered and become the default setting when you invoke Stata.

Remarks

When you see —more— at the bottom of the screen,

Press ...	and Stata ...
letter *l* or *Enter*	displays the next line
letter *q*	acts as if you pressed *Break*
space bar or any other key	displays the next screen

You can also click the **More** button or click on —more— to display the next screen.

—more— is Stata's way of telling you that it has something more to show you but that showing it to you will cause the information on the screen to scroll off.

If you type set more off, —more— conditions will never arise, and Stata's output will scroll by at full speed.

If you type set more on, —more— conditions will be restored at the appropriate places.

Programmers should see [P] **more** for information on the more programming command.

Also See

[R] **query** — Display system parameters

[P] **creturn** — Return c-class values

[P] **more** — Pause until key is pressed

[U] **7 —more— conditions**

Title

mprobit — Multinomial probit regression

Syntax

`mprobit` *depvar* [*indepvars*] [*if*] [*in*] [*weight*] [, *options*]

options	description
Model	
`noconstant`	suppress constant terms
`baseoutcome(`*#* | *lbl*`)`	outcome used to normalize location
`probitparam`	use the probit variance parameterization
`constraints(`*constraints*`)`	apply specified linear constraints
`collinear`	keep collinear variables
SE/Robust	
`vce(`*vcetype*`)`	*vcetype* may be `oim`, `robust`, `cluster` *clustvar*, `opg`, `bootstrap`, or `jackknife`
Reporting	
`level(`*#*`)`	set confidence level; default is `level(95)`
Int options	
`intpoints(`*#*`)`	number of quadrature points
Max options	
maximize_options	control the maximization process; seldom used

`bootstrap`, `by`, `jackknife`, `rolling`, `statsby`, `svy`, and `xi` are allowed; see [U] **11.1.10 Prefix commands**.
Weights are not allowed with the `bootstrap` prefix.
`vce()` and weights are not allowed with the `svy` prefix.
`fweights`, `iweights`, and `pweights` are allowed; see [U] **11.1.6 weight**.
See [U] **20 Estimation and postestimation commands** for more capabilities of estimation commands.

Description

`mprobit` fits multinomial probit (MNP) models via maximum likelihood. *depvar* contains the outcome for each observation, and *indepvars* are the associated covariates. The error terms are assumed to be independent, standard normal, random variables. See [R] **asmprobit** for the case where the latent-variable errors are correlated or heteroskedastic and you have alternative-specific variables.

Options

Model

`noconstant` suppresses the $J - 1$ constant terms.

`baseoutcome(`*#* | *lbl*`)` specifies the outcome used to normalize the location of the latent variable. The base outcome may be specified as a number or a label. The default is to use the most frequent outcome. The coefficients associated with the base outcome are zero.

`probitparam` specifies to use the probit variance parameterization by fixing the variance of the differenced latent errors between the scale and the base alternatives to be one. The default is to make the variance of the base and scale latent errors one, thereby making the variance of the difference to be two.

`constraints(`*constraints*`)`, `collinear`; see [R] **estimation options**.

SE/Robust

`vce(`*vcetype*`)` specifies the type of standard error reported, which includes types that are derived from asymptotic theory, that are robust to some kinds of misspecification, that allow for intragroup correlation, and that use bootstrap or jackknife methods; see [R] ***vce_option***.

If specifying `vce(bootstrap)` or `vce(jackknife)`, you must also specify `baseoutcome()`.

Reporting

`level(`*#*`)`; see [R] **estimation options**.

Int options

`intpoints(`*#*`)` specifies the number of Gaussian quadrature points to use in approximating the likelihood. The default is 15.

Max options

maximize_options: `difficult`, `technique(`*algorithm_spec*`)`, `iterate(`*#*`)`, `[no]log`, `trace`, `gradient`, `showstep`, `hessian`, `shownrtolerance`, `tolerance(`*#*`)`, `ltolerance(`*#*`)`, `gtolerance(`*#*`)`, `nrtolerance(`*#*`)`, `nonrtolerance`, `from(`*init_specs*`)`; see [R] **maximize**. These options are seldom used.

Setting the optimization type to `technique(bhhh)` resets the default *vcetype* to `vce(opg)`.

Remarks

The MNP model is used with discrete dependent variables that take on more than two outcomes that do not have a natural ordering. The stochastic error terms for this implementation of the model are assumed to have independent, standard normal distributions. To use `mprobit`, you must have one observation for each decision maker in the sample. See [R] **asmprobit** for another implementation of the MNP model that permits correlated and heteroskedastic errors and is suitable when you have data for each alternative that a decision maker faced.

The MNP model is frequently motivated using a latent-variable framework. The latent variable for the jth alternative, $j = 1, \dots, J$, is

$$\eta_{ij} = \mathbf{z}_i \boldsymbol{\alpha}_j + \xi_{ij}$$

where the $1 \times q$ row vector $\mathbf{z}_i$ contains the observed independent variables for the ith decision maker. Associated with $\mathbf{z}_i$ are the J vectors of regression coefficients $\boldsymbol{\alpha}_j$. The $\xi_{i,1}, \dots, \xi_{i,J}$ are distributed independently and identically standard normal. The decision maker chooses the alternative k such that $\eta_{ik} \geq \eta_{im}$ for $m \neq k$.

Suppose that case i chooses alternative k, and take the difference between latent variable η_{ik} and the $J - 1$ others:

$$\begin{aligned} v_{ijk} &= \eta_{ij} - \eta_{ik} \\ &= \mathbf{z}_i(\boldsymbol{\alpha}_j - \boldsymbol{\alpha}_k) + \xi_{ij} - \xi_{ik} \qquad (1) \\ &= \mathbf{z}_i \boldsymbol{\gamma}_{j'} + \epsilon_{ij'} \end{aligned}$$

where $j' = j$ if $j < k$ and $j' = j-1$ if $j > k$ so that $j' = 1, \ldots, J-1$. $\mathrm{Var}(\epsilon_{ij'}) = \mathrm{Var}(\xi_{ij} - \xi_{ik}) = 2$ and $\mathrm{Cov}(\epsilon_{ij'}, \epsilon_{il'}) = 1$ for $j' \neq l'$. The probability that alternative k is chosen is

$$\begin{aligned}\Pr(i \text{ chooses } k) &= \Pr(v_{i1k} \leq 0, \ldots, v_{i,J-1,k} \leq 0)\\ &= \Pr(\epsilon_{i1} \leq -\mathbf{z}_i\gamma_1, \ldots, \epsilon_{i,J-1} \leq -\mathbf{z}_i\gamma_{J-1})\end{aligned}$$

Hence, evaluating the likelihood function involves computing probabilities from the multivariate normal distribution. That all the covariances are equal simplifies the problem somewhat; see *Methods and Formulas* for details.

In (1), not all J of the $\boldsymbol{\alpha}_j$ are identifiable. To remove the indeterminacy, $\boldsymbol{\alpha}_l$ is set to the zero vector, where l is the base outcome as specified in the `baseoutcome()` option. That fixes the lth latent variable to zero so that the remaining variables measure the attractiveness of the other alternatives relative to the base.

▷ Example 1

As discussed in [R] **mlogit**, we have data on the type of health insurance available to 616 psychologically depressed subjects in the United States (Tarlov et al. 1989; Wells et al. 1989). Patients may have either an indemnity (fee-for-service) plan or a prepaid plan such as an HMO, or the patient may be uninsured. Demographic variables include age, gender, race, and site. Indemnity insurance is the most popular alternative, so `mprobit` will choose it as the base outcome by default.

```
. use http://www.stata-press.com/data/r10/sysdsn3
(Health insurance data)

. mprobit insure age male nonwhite site2 site3

Iteration 0:   log likelihood = -535.89424
Iteration 1:   log likelihood = -534.56173
Iteration 2:   log likelihood = -534.52835
Iteration 3:   log likelihood = -534.52833

Multinomial probit regression                     Number of obs   =        615
                                                  Wald chi2(10)   =      40.18
Log likelihood = -534.52833                       Prob > chi2     =     0.0000
```

insure	Coef.	Std. Err.	z	P>|z|	[95% Conf.	Interval]
Prepaid						
age	-.0098536	.0052688	-1.87	0.061	-.0201802	.000473
male	.4774678	.1718316	2.78	0.005	.1406841	.8142515
nonwhite	.8245003	.1977582	4.17	0.000	.4369013	1.212099
site2	.0973956	.1794546	0.54	0.587	-.2543289	.4491201
site3	-.495892	.1904984	-2.60	0.009	-.869262	-.1225221
_cons	.22315	.2792424	0.80	0.424	-.324155	.7704549
Uninsure						
age	-.0050814	.0075327	-0.67	0.500	-.0198452	.0096823
male	.3332637	.2432986	1.37	0.171	-.1435929	.8101203
nonwhite	.2485859	.2767734	0.90	0.369	-.29388	.7910518
site2	-.6899485	.2804497	-2.46	0.014	-1.23962	-.1402771
site3	-.1788447	.2479898	-0.72	0.471	-.6648957	.3072063
_cons	-.9855917	.3891873	-2.53	0.011	-1.748385	-.2227986

```
(insure=Indemnity is the base outcome)
```

◁

The likelihood function for `mprobit` is derived under the assumption that all decision-making units face the same choice set, which is the union of all outcomes observed in the dataset. If that is not true for your model, then an alternative is to use the `asmprobit` command, which does not require this assumption. To do that, you will need to expand the dataset so that each decision maker has k_i observations, where k_i is the number of alternatives in the choice set faced by decision maker i. You will also need to create a binary variable to indicate the choice made by each decision maker. Moreover, you will need to use the `correlation(independent)` and `stddev(homoskedastic)` options with `asmprobit` unless you have alternative-specific variables.

Saved Results

`mprobit` saves the following in `e()`:

Scalars

`e(N)`	number of observations	`e(ll)`	log simulated-likelihood
`e(N_clust)`	number of clusters	`e(chi2)`	χ^2
`e(k_indvars)`	number of independent variables	`e(p)`	significance
`e(k)`	number of parameters	`e(const)`	`0` if `noconstant` is specified; `1` otherwise
`e(k_eq)`	number of equations		
`e(k_eq_model)`	number of equations in model Wald test	`e(rank)`	rank of `e(V)`
		`e(ic)`	number of iterations
`e(k_out)`	number of outcomes	`e(rc)`	return code
`e(k_points)`	number of quadrature points	`e(converged)`	`1` if converged, `0` otherwise
`e(i_base)`	base outcome index	`e(probitparam)`	`1` if `probitparam` is specified, `0` otherwise
`e(df_m)`	model degrees of freedom		

Macros

`e(cmd)`	`mprobit`	`e(out#)`	outcome labels, #=1,...,`e(k_out)`
`e(cmdline)`	command as typed	`e(vce)`	*vcetype* specified in `vce()`
`e(depvar)`	name of dependent variable	`e(vcetype)`	title used to label Std. Err.
`e(indvars)`	independent variables	`e(opt)`	type of optimization
`e(wtype)`	weight type	`e(ml_method)`	type of `ml` method
`e(wexp)`	weight expression	`e(user)`	name of likelihood-evaluator program
`e(title)`	title in estimation output	`e(technique)`	maximization technique
`e(clustvar)`	name of cluster variable	`e(crittype)`	optimization criterion
`e(chi2type)`	`Wald`, type of model χ^2 test	`e(properties)`	`b V`
`e(outeqs)`	outcome equations	`e(predict)`	program used to implement `predict`

Matrices

`e(b)`	coefficient vector	`e(outcomes)`	outcome values
`e(ilog)`	iteration log (up to 20 iterations)	`e(V)`	variance–covariance matrix of the estimators
`e(gradient)`	gradient vector		

Functions

`e(sample)`	marks estimation sample

Methods and Formulas

`mprobit` is implemented as an ado-file.

As discussed in *Remarks*, the latent variables for a J-alternative model are $\eta_{ij} = \mathbf{z}_i\boldsymbol{\alpha}_j + \xi_{ij}$, for $j = 1, \dots, J$, $i = 1, \dots, n$, and $\{\xi_{i,1}, \dots, \xi_{i,J}\} \sim$ i.i.d.$N(0,1)$. The experimenter observes alternative k for the ith observation if $\eta_{ik} > \eta_{il}$ for $l \neq k$. For $j \neq k$, let

$$\begin{aligned} v_{ij'} &= \eta_{ij} - \eta_{ik} \\ &= \mathbf{z}_i(\boldsymbol{\alpha}_j - \boldsymbol{\alpha}_k) + \xi_{ij} - \xi_{ik} \\ &= \mathbf{z}_i\boldsymbol{\gamma}_{j'} + \epsilon_{ij'} \end{aligned}$$

where $j' = j$ if $j < k$ and $j' = j - 1$ if $j > k$ so that $j' = 1, \dots, J-1$. $\epsilon_i = (\epsilon_{i1}, \dots, \epsilon_{i,J-1}) \sim \mathrm{MVN}(\mathbf{0}, \boldsymbol{\Sigma})$, where

$$\boldsymbol{\Sigma} = \begin{pmatrix} 2 & 1 & 1 & \dots & 1 \\ 1 & 2 & 1 & \dots & 1 \\ 1 & 1 & 2 & \dots & 1 \\ \vdots & \vdots & \vdots & \ddots & \vdots \\ 1 & 1 & 1 & \dots & 2 \end{pmatrix}$$

Denote the deterministic part of the model as $\lambda_{ij'} = \mathbf{z}_i\boldsymbol{\gamma}_{j'}$; the probability that subject i chooses outcome k is

$$\begin{aligned} \Pr(y_i = k) &= \Pr(v_{i1} \le 0, \dots, v_{i,J-1} \le 0) \\ &= \Pr(\epsilon_{i1} \le -\lambda_{i1}, \dots, \epsilon_{i,J-1} \le -\lambda_{i,J-1}) \\ &= \frac{1}{(2\pi)^{(J-1)/2}\,|\boldsymbol{\Sigma}|^{1/2}} \int_{-\infty}^{-\lambda_{i1}} \cdots \int_{-\infty}^{-\lambda_{i,J-1}} \exp\left(-\tfrac{1}{2}\mathbf{z}'\boldsymbol{\Sigma}^{-1}\mathbf{z}\right) d\mathbf{z} \end{aligned}$$

Because of the exchangeable correlation structure of $\boldsymbol{\Sigma}$ ($\rho_{ij} = 1/2$ for all $i \neq j$), we can use Dunnett's (1989) result to reduce the multidimensional integral to one dimension:

$$\Pr(y_i = k) = \frac{1}{\sqrt{\pi}} \int_0^\infty \left\{ \prod_{j=1}^{J-1} \Phi\left(-z\sqrt{2} - \lambda_{ij}\right) + \prod_{j=1}^{J-1} \Phi\left(z\sqrt{2} - \lambda_{ij}\right) \right\} e^{-z^2} dz$$

Gaussian quadrature is used to approximate this integral, resulting in the K-point quadrature formula

$$\Pr(y_i = k) \approx \frac{1}{2} \sum_{k=1}^{K} w_k \left\{ \prod_{j=1}^{J-1} \Phi\left(-\sqrt{2x_k} - \lambda_{ij}\right) + \prod_{j=1}^{J-1} \Phi\left(\sqrt{2x_k} - \lambda_{ij}\right) \right\}$$

where w_k and x_k are the weights and roots of the Laguerre polynomial of order K. In `mprobit`, K is specified by option `intpoints()`.

References

Dunnett, C. W. 1989. Algorithm AS 251: Multivariate normal probability integrals with product correlation structure. *Journal of the Royal Statistical Society, Series C* 38: 564–579.

Haan, P., and A. Uhlendorff. 2006. Estimation of multinomial logit models with unobserved heterogeneity using maximum simulated likelihood. *Stata Journal* 6: 229–245.

Tarlov, A. R., J. E. Ware, Jr., S. Greenfield, E. C. Nelson, E. Perrin, and M. Zubkoff. 1989. The medical outcomes study. An application of methods for monitoring the results of medical care. *Journal of the American Medical Association* 262: 925–930.

Wells, K. E., R. D. Hays, M. A. Burnam, W. H. Rogers, S. Greenfield, and J. E. Ware, Jr. 1989. Detection of depressive disorder for patients receiving prepaid or fee-for-service care. *Journal of the American Medical Association* 262: 3298–3302.

Also See

[R] **mprobit postestimation** — Postestimation tools for mprobit

[R] **asmprobit** — Alternative-specific multinomial probit regression

[R] **mlogit** — Multinomial (polytomous) logistic regression

[R] **clogit** — Conditional (fixed-effects) logistic regression

[R] **nlogit** — Nested logit regression

[R] **ologit** — Ordered logistic regression

[R] **oprobit** — Ordered probit regression

[SVY] **svy estimation** — Estimation commands for survey data

[U] **20 Estimation and postestimation commands**

Title

mprobit postestimation — Postestimation tools for mprobit

Description

The following postestimation commands are available for `mprobit`:

command	description
`adjust`	adjusted predictions of $\mathbf{x}\beta$
`estat`	AIC, BIC, VCE, and estimation sample summary
`estat` (svy)	postestimation statistics for survey data
`estimates`	cataloging estimation results
`lincom`	point estimates, standard errors, testing, and inference for linear combinations of coefficients
`lrtest`[1]	likelihood-ratio test
`mfx`	marginal effects or elasticities
`nlcom`	point estimates, standard errors, testing, and inference for nonlinear combinations of coefficients
`predict`	predicted probabilities, linear predictions, and standard errors
`predictnl`	point estimates, standard errors, testing, and inference for generalized predictions
`suest`	seemingly unrelated estimation
`test`	Wald tests for simple and composite linear hypotheses
`testnl`	Wald tests of nonlinear hypotheses

[1] `lrtest` is not appropriate with `svy` estimation results.

See the corresponding entries in the *Stata Base Reference Manual* for details, but see [SVY] **estat** for details about `estat` (svy).

Syntax for predict

`predict` [*type*] {*stub** | *newvar* | *newvarlist*} [*if*] [*in*] [, *statistic* `outcome(`*outcome*`)`]

`predict` [*type*] {*stub** | *newvarlist*} [*if*] [*in*], `scores`

statistic	description
Main	
`pr`	probability of a positive outcome; the default
`xb`	linear prediction
`stdp`	standard error of the linear prediction

You specify one new variable with `pr` and you specify `outcome()`, or you specify *stub** or *k* new variables with `pr`, where *k* is the number of outcomes.

You specify one new variable with `xb` and `stdp` and you specify `outcome()`.

These statistics are available both in and out of sample; type `predict ... if e(sample) ...` if wanted only for the estimation sample.

Options for predict

Main

pr, the default, calculates the probability of each of the categories of the dependent variable or the probability of the level specified in outcome(*outcome*). If you specify the outcome(*outcome*) option, you need to specify only one new variable; otherwise, you must specify a new variable for each category of the dependent variable.

xb calculates the linear prediction $\mathbf{x}_i\boldsymbol{\alpha}_j$ for alternative j and individual i. The index j corresponds to the outcome specified in outcome().

stdp calculates the standard error of the linear prediction.

outcome(*outcome*) specifies the outcome for which the statistic is to be calculated. equation() is a synonym for outcome(): it does not matter which you use. outcome() or equation() can be specified using

#1, #2, ..., where #1 means the first category of the dependent variable, #2 the second category, etc.;

the values of the dependent variable; or

the value labels of the dependent variable if they exist.

scores calculates the equation-level score variables. The jth new variable will contain the scores for the jth fitted equation.

Remarks

Once you have fitted a multinomial probit model, you can use predict to obtain probabilities that an individual will choose each of the alternatives for the estimation sample, as well as other samples; see [U] **20 Estimation and postestimation commands** and [R] **predict**.

▷ Example 1

In [R] **mprobit** we fitted the multinomial probit model to a dataset containing the type of health insurance available to 616 psychologically depressed subjects in the United States (Tarlov et al. 1989; Wells et al. 1989). We can obtain the predicted probabilities by typing

```
. use http://www.stata-press.com/data/r10/sysdsn3
(Health insurance data)
. mprobit insure age male nonwhite site2 site3
  (output omitted)
. predict p1-p3
(option pr assumed; predicted probabilities)
```

```
. list p1-p3 insure in 1/10
```

	p1	p2	p3	insure
1.	.5961306	.3741824	.029687	Indemnity
2.	.4719296	.4972289	.0308415	Prepaid
3.	.4896086	.4121961	.0981953	Indemnity
4.	.3730529	.5416623	.0852848	Prepaid
5.	.5063069	.4629773	.0307158	.
6.	.4768125	.4923548	.0308327	Prepaid
7.	.5035672	.4657016	.0307312	Prepaid
8.	.3326361	.5580404	.1093235	.
9.	.4758165	.4384811	.0857024	Uninsure
10.	.5734057	.3316601	.0949342	Prepaid

`insure` contains a missing value for observations 5 and 8. Because of that, those two observations were not used in the estimation. However, because none of the independent variables is missing, `predict` can still calculate the probabilities. Had we typed

```
. predict p1-p3 if e(sample)
```

`predict` would have filled in missing values for `p1`, `p2`, and `p3` for those observations since they were not used in the estimation.

◁

Methods and Formulas

All postestimation commands listed above are implemented as ado-files.

References

Tarlov, A. R., J. E. Ware, Jr., S. Greenfield, E. C. Nelson, E. Perrin, and M. Zubkoff. 1989. The medical outcomes study. An application of methods for monitoring the results of medical care. *Journal of the American Medical Association* 262: 925–930.

Wells, K. E., R. D. Hays, M. A. Burnam, W. H. Rogers, S. Greenfield, and J. E. Ware, Jr. 1989. Detection of depressive disorder for patients receiving prepaid or fee-for-service care. *Journal of the American Medical Association* 262: 3298–3302.

Also See

[R] **mprobit** — Multinomial probit regression

[U] **20 Estimation and postestimation commands**

Title

mvreg — Multivariate regression

Syntax

`mvreg` *depvars* = *indepvars* [*if*] [*in*] [*weight*] [, *options*]

options	description
Model	
noconstant	suppress constant term
Reporting	
level(#)	set confidence level; default is level(95)
corr	report correlation matrix
†noheader	suppress header table from above coefficient table
†notable	suppress coefficient table

† noheader and notable are not shown in the dialog box.

depvars and *indepvars* may contain time-series operators; see [U] **11.4.3 Time-series varlists**.

bootstrap, by, jackknife, rolling, statsby, and xi are allowed; see [U] **11.1.10 Prefix commands**.

Weights are not allowed with the bootstrap prefix.

aweights are not allowed with the jackknife prefix.

aweights and fweights are allowed; see [U] **11.1.6 weight**.

See [U] **20 Estimation and postestimation commands** for more capabilities of estimation commands.

Description

`mvreg` fits multivariate regression models.

Options

Model

`noconstant` suppresses the constant term (intercept) in the model.

Reporting

`level(#)` specifies the confidence level, as a percentage, for confidence intervals. The default is `level(95)` or as set by `set level`; see [U] **20.7 Specifying the width of confidence intervals**.

`corr` displays the correlation matrix of the residuals between the equations.

The following options are available with `mvreg` but are not shown in the dialog box:

`noheader` suppresses display of the table reporting F statistics, R-squared, and root mean squared error above the coefficient table.

`notable` suppresses display of the coefficient table.

Remarks

Multivariate regression differs from multiple regression in that *several* dependent variables are jointly regressed on the same independent variables. Multivariate regression is related to Zellner's seemingly unrelated regression (see [R] **sureg**), but since the same set of independent variables is used for each dependent variable, the syntax is simpler, and the calculations are faster.

The individual coefficients and standard errors produced by `mvreg` are identical to those that would be produced by `regress` estimating each equation separately. The difference is that `mvreg`, being a joint estimator, also estimates the between-equation covariances, so you can test coefficients across equations and, in fact, the `test` syntax makes such tests more convenient.

▷ Example 1

Using the automobile data, we fit a multivariate regression for space variables (`headroom`, `trunk`, and `turn`) in terms of a set of other variables, including three performance variables (`displacement`, `gear_ratio`, and `mpg`):

```
. use http://www.stata-press.com/data/r10/auto
(1978 Automobile Data)

. mvreg headroom trunk turn = price mpg displ gear_ratio length weight

Equation          Obs  Parms        RMSE    "R-sq"          F        P
----------------------------------------------------------------------
headroom           74      7    .7390205    0.2996   4.777213   0.0004
trunk              74      7    3.052314    0.5326    12.7265   0.0000
turn               74      7    2.132377    0.7844   40.62042   0.0000

------------------------------------------------------------------------------
             |      Coef.   Std. Err.      t    P>|t|     [95% Conf. Interval]
-------------+----------------------------------------------------------------
headroom     |
       price |  -.0000528    .000038    -1.39   0.168    -.0001286    .0000229
         mpg |  -.0093774   .0260463    -0.36   0.720     -.061366    .0426112
displacement |   .0031025   .0024999     1.24   0.219    -.0018873    .0080922
  gear_ratio |   .2108071   .3539588     0.60   0.553    -.4956976    .9173118
      length |    .015886    .012944     1.23   0.224    -.0099504    .0417223
      weight |  -.0000868   .0004724    -0.18   0.855    -.0010296    .0008561
       _cons |  -.4525117   2.170073    -0.21   0.835    -4.783995    3.878972
-------------+----------------------------------------------------------------
trunk        |
       price |   .0000445   .0001567     0.28   0.778    -.0002684    .0003573
         mpg |  -.0220919   .1075767    -0.21   0.838    -.2368159    .1926322
displacement |   .0032118   .0103251     0.31   0.757    -.0173971    .0238207
  gear_ratio |  -.2271321   1.461926    -0.16   0.877    -3.145149    2.690885
      length |    .170811   .0534615     3.20   0.002     .0641014    .2775206
      weight |  -.0015944    .001951    -0.82   0.417    -.0054885    .0022997
       _cons |  -13.28253   8.962868    -1.48   0.143    -31.17249    4.607429
-------------+----------------------------------------------------------------
turn         |
       price |  -.0002647   .0001095    -2.42   0.018    -.0004833   -.0000462
         mpg |  -.0492948   .0751542    -0.66   0.514    -.1993031    .1007136
displacement |   .0036977   .0072132     0.51   0.610    -.0106999    .0180953
  gear_ratio |  -.1048432   1.021316    -0.10   0.919    -2.143399    1.933712
      length |    .072128   .0373487     1.93   0.058    -.0024204    .1466764
      weight |   .0027059    .001363     1.99   0.051    -.0000145    .0054264
       _cons |   20.19157   6.261549     3.22   0.002     7.693467    32.68967
------------------------------------------------------------------------------
```

We should have specified the `corr` option so that we would also see the correlations between the residuals of the equations. We can correct our omission because `mvreg`—like all estimation commands—typed without arguments redisplays results. The `noheader` and `notable` (read "no-table") options suppress redisplaying the output we have already seen:

```
. mvreg, notable noheader corr
Correlation matrix of residuals:

           headroom     trunk      turn
headroom    1.0000
   trunk    0.4986    1.0000
    turn    0.1090   -0.0628    1.0000

Breusch-Pagan test of independence: chi2(3) =    19.566, Pr = 0.0002
```

The Breusch–Pagan test is significant, so the residuals of these three space variables are not independent of each other.

The three performance variables among our independent variables are `mpg`, `displacement`, and `gear_ratio`. We can jointly test the significance of these three variables in all the equations by typing

```
. test mpg displacement gear_ratio
 ( 1)  [headroom]mpg = 0
 ( 2)  [trunk]mpg = 0
 ( 3)  [turn]mpg = 0
 ( 4)  [headroom]displacement = 0
 ( 5)  [trunk]displacement = 0
 ( 6)  [turn]displacement = 0
 ( 7)  [headroom]gear_ratio = 0
 ( 8)  [trunk]gear_ratio = 0
 ( 9)  [turn]gear_ratio = 0

       F(  9,    67) =    0.33
            Prob > F =    0.9622
```

These three variables are not, as a group, significant. We might have suspected this from their individual significance in the individual regressions, but this multivariate test provides an overall assessment with one p-value.

We can also perform a test for the joint significance of all three equations:

```
. test [headroom]
  (output omitted )
. test [trunk], accum
  (output omitted )
. test [turn], accum
 ( 1)  [headroom]price = 0
 ( 2)  [headroom]mpg = 0
 ( 3)  [headroom]displacement = 0
 ( 4)  [headroom]gear_ratio = 0
 ( 5)  [headroom]length = 0
 ( 6)  [headroom]weight = 0
 ( 7)  [trunk]price = 0
 ( 8)  [trunk]mpg = 0
 ( 9)  [trunk]displacement = 0
 (10)  [trunk]gear_ratio = 0
 (11)  [trunk]length = 0
 (12)  [trunk]weight = 0
 (13)  [turn]price = 0
 (14)  [turn]mpg = 0
 (15)  [turn]displacement = 0
```

```
 (16)  [turn]gear_ratio = 0
 (17)  [turn]length = 0
 (18)  [turn]weight = 0

       F( 18,    67) =   19.34
            Prob > F =    0.0000
```

The set of variables as a whole is strongly significant. We might have suspected this, too, from the individual equations.

◁

❑ Technical Note

The `mvreg` command provides a good way to deal with multiple comparisons. If we wanted to assess the effect of `length`, we might be dissuaded from interpreting any of its coefficients except that in the `trunk` equation. `[trunk]length`—the coefficient on `length` in the `trunk` equation—has a p-value of .002, but in the other two equations, it has p-values of only .224 and .058.

A conservative statistician might argue that there are 18 tests of significance in `mvreg`'s output (not counting those for the intercept), so p-values more than $.05/18 = .0028$ should be declared insignificant at the 5% level. A more aggressive but, in our opinion, reasonable approach would be to first note that the three equations are jointly significant, so we are justified in making some interpretation. Then we would work through the individual variables using `test`, possibly using $.05/6 = .0083$ (6 because there are six independent variables) for the 5% significance level. For instance, examining `length`:

```
. test length

 ( 1)  [headroom]length = 0
 ( 2)  [trunk]length = 0
 ( 3)  [turn]length = 0

       F(  3,    67) =    4.94
            Prob > F =    0.0037
```

The reported significance level of .0037 is less than .0083, so we will declare this variable significant. `[trunk]length` is certainly significant with its p-value of .002, but what about in the remaining two equations with p-values .224 and .058? We perform a joint test:

```
. test [headroom]length [turn]length

 ( 1)  [headroom]length = 0
 ( 2)  [turn]length = 0

       F(  2,    67) =    2.91
            Prob > F =    0.0613
```

At this point, reasonable statisticians could disagree. The .06 significance value suggests no interpretation, but these were the two least-significant values out of three, so we would expect the p-value to be a little high. Perhaps an equivocal statement is warranted: there seems to be an effect, but chance cannot be excluded.

❑

(Continued on next page)

Saved Results

mvreg saves the following in e():

Scalars

e(N)	number of observations
e(k)	number of parameters (including constant)
e(k_eq)	number of equations
e(df_r)	residual degrees of freedom
e(chi2)	Breusch–Pagan χ^2 (corr only)
e(df_chi2)	degrees of freedom for Breusch–Pagan χ^2 (corr only)

Macros

e(cmd)	mvreg
e(cmdline)	command as typed
e(depvar)	names of dependent variables
e(eqnames)	names of equations
e(wtype)	weight type
e(wexp)	weight expression
e(r2)	R-squared for each equation
e(rmse)	RMSE for each equation
e(F)	F statistic for each equation
e(p_F)	significance of F for each equation
e(properties)	b V
e(estat_cmd)	program used to implement estat
e(predict)	program used to implement predict

Matrices

e(b)	coefficient vector
e(V)	variance–covariance matrix of the estimators
e(Sigma)	$\widehat{\boldsymbol{\Sigma}}$ matrix

Functions

e(sample)	marks estimation sample

Methods and Formulas

mvreg is implemented as an ado-file.

Given q equations and p independent variables (including the constant), the parameter estimates are given by the $p \times q$ matrix

$$\mathbf{B} = (\mathbf{X}'\mathbf{W}\mathbf{X})^{-1}\mathbf{X}'\mathbf{W}\mathbf{Y}$$

where $\mathbf{Y}$ is an $n \times q$ matrix of dependent variables and $\mathbf{X}$ is a $n \times p$ matrix of independent variables. $\mathbf{W}$ is a weighting matrix equal to $\mathbf{I}$ if no weights are specified. If weights are specified, let $\mathbf{v}$: $1 \times n$ be the specified weights. If fweight frequency weights are specified, $\mathbf{W} = \text{diag}(\mathbf{v})$. If aweight analytic weights are specified, $\mathbf{W} = \text{diag}\{\mathbf{v}/(\mathbf{1}'\mathbf{v})(\mathbf{1}'\mathbf{1})\}$, meaning that the weights are normalized to sum to the number of observations.

The residual covariance matrix is

$$\mathbf{R} = \{\mathbf{Y}'\mathbf{W}\mathbf{Y} - \mathbf{B}'(\mathbf{X}'\mathbf{W}\mathbf{X})\mathbf{B}\}/(n-p)$$

The estimated covariance matrix of the estimates is $\mathbf{R} \otimes (\mathbf{X}'\mathbf{W}\mathbf{X})^{-1}$. These results are identical to those produced by sureg when the same list of independent variables is specified repeatedly; see [R] **sureg**.

The Breusch and Pagan (1980) χ^2 statistic—a Lagrange multiplier statistic—is given by

$$\lambda = n \sum_{i=1}^{q} \sum_{j=1}^{i-1} r_{ij}^2$$

where r_{ij} is the estimated correlation between the residuals of the equations and n is the number of observations. It is distributed as χ^2 with $q(q-1)/2$ degrees of freedom.

Reference

Breusch, T., and A. Pagan. 1980. The LM test and its applications to model specification in econometrics. *Review of Economic Studies* 47: 239–254.

Also See

[R] **mvreg postestimation** — Postestimation tools for mvreg

[MV] **manova** — Multivariate analysis of variance and covariance

[R] **nlsur** — Estimation of nonlinear systems of equations

[R] **reg3** — Three-stage estimation for systems of simultaneous equations

[R] **regress** — Linear regression

[R] **regress postestimation** — Postestimation tools for regress

[R] **sureg** — Zellner's seemingly unrelated regression

[U] **20 Estimation and postestimation commands**

Title

mvreg postestimation — Postestimation tools for mvreg

Description

The following postestimation commands are available for `mvreg`:

command	description
`adjust`	adjusted predictions of $\mathbf{x}\beta$ or $\exp(\mathbf{x}\beta)$
`estat`	VCE and estimation sample summary
`estimates`	cataloging estimation results
`lincom`	point estimates, standard errors, testing, and inference for linear combinations of coefficients
`nlcom`	point estimates, standard errors, testing, and inference for nonlinear combinations of coefficients
`predict`	predictions, residuals, influence statistics, and other diagnostic measures
`predictnl`	point estimates, standard errors, testing, and inference for generalized predictions
`test`	Wald tests for simple and composite linear hypotheses
`testnl`	Wald tests of nonlinear hypotheses

See the corresponding entries in the *Stata Base Reference Manual* for details.

Syntax for predict

`predict` [*type*] *newvar* [*if*] [*in*] [, `equation(`*eqno*[, *eqno*]`)` *statistic*]

statistic	description
Main	
`xb`	linear prediction; the default
`stdp`	standard error of the linear prediction
`residuals`	residuals
`difference`	difference between the linear predictions of two equations
`stddp`	standard error of the difference in linear predictions

These statistics are available both in and out of sample; type `predict ... if e(sample) ...` if wanted only for the estimation sample.

Options for predict

Main

`equation(`*eqno*[, *eqno*]`)` specifies the equation to which you are referring.

`equation()` is filled in with one *eqno* for options `xb`, `stdp`, and `residuals`. `equation(#1)` would mean the calculation is to be made for the first equation, `equation(#2)` would mean the second, and so on. You could also refer to the equations by their names. `equation(income)` would refer to the equation named income and `equation(hours)` to the equation named hours.

If you do not specify `equation()`, results are the same as if you specified `equation(#1)`.

`difference` and `stddp` refer to between-equation concepts. To use these options, you must specify two equations, e.g., `equation(#1,#2)` or `equation(income,hours)`. When two equations must be specified, `equation()` is required. With `equation(#1,#2)`, `difference` computes the prediction of `equation(#1)` minus the prediction of `equation(#2)`.

`xb`, the default, calculates the fitted values—the prediction of $\mathbf{x}_j\mathbf{b}$ for the specified equation.

`stdp` calculates the standard error of the prediction for the specified equation (the standard error of the predicted expected value or mean for the observation's covariate pattern). The standard error of the prediction is also referred to as the standard error of the fitted value.

`residuals` calculates the residuals.

`difference` calculates the difference between the linear predictions of two equations in the system.

`stddp` is allowed only after you have previously fitted a multiple-equation model. The standard error of the difference in linear predictions $(\mathbf{x}_{1j}\mathbf{b} - \mathbf{x}_{2j}\mathbf{b})$ between equations 1 and 2 is calculated.

For more information on using `predict` after multiple-equation estimation commands, see [R] **predict**.

Methods and Formulas

All postestimation commands listed above are implemented as ado-files.

Also See

[R] **mvreg** — Multivariate regression

[U] **20 Estimation and postestimation commands**

Title

nbreg — Negative binomial regression

Syntax

Negative binomial regression model

nbreg *depvar* [*indepvars*] [*if*] [*in*] [*weight*] [, *nbreg_options*]

Generalized negative binomial model

gnbreg *depvar* [*indepvars*] [*if*] [*in*] [*weight*] [, *gnbreg_options*]

nbreg_options	description
Model	
noconstant	suppress constant term
dispersion(mean)	parameterization of dispersion; dispersion(mean) is the default
dispersion(constant)	constant dispersion for all observations
exposure(*varname*$_e$)	include ln(*varname*$_e$) in model with coefficient constrained to 1
offset(*varname*$_o$)	include *varname*$_o$ in model with coefficient constrained to 1
constraints(*constraints*)	apply specified linear constraints
collinear	keep collinear variables
SE/Robust	
vce(*vcetype*)	*vcetype* may be oim, robust, cluster *clustvar*, opg, bootstrap, or jackknife
Reporting	
level(#)	set confidence level; default is level(95)
nolrtest	suppress likelihood-ratio test
irr	report incidence-rate ratios
Max options	
maximize_options	control the maximization process; seldom used

depvar, *indepvars*, *varname*$_e$, and *varname*$_o$ may contain time-series operators; see [U] **11.4.3 Time-series varlists**.

gnbreg_options	description
Model	
noconstant	suppress constant term
lnalpha(*varlist*)	dispersion model variables
exposure(*varname*$_e$)	include ln(*varname*$_e$) in model with coefficient constrained to 1
offset(*varname*$_o$)	include *varname*$_o$ in model with coefficient constrained to 1
constraints(*constraints*)	apply specified linear constraints
collinear	keep collinear variables
SE/Robust	
vce(*vcetype*)	*vcetype* may be oim, robust, cluster *clustvar*, opg, bootstrap, or jackknife
Reporting	
level(#)	set confidence level; default is level(95)
irr	report incidence-rate ratios
Max options	
maximize_options	control the maximization process; seldom used

bootstrap, by (nbreg only), jackknife, nestreg (nbreg only), rolling, statsby, stepwise, svy, and xi are allowed; see [U] **11.1.10 Prefix commands**.

Weights are not allowed with the bootstrap prefix.

vce() and weights are not allowed with the svy prefix.

fweights, iweights, and pweights are allowed; see [U] **11.1.6 weight**.

See [U] **20 Estimation and postestimation commands** for more capabilities of estimation commands.

Description

nbreg fits a negative binomial regression model of *depvar* on *indepvars*, where *depvar* is a nonnegative count variable. In this model, the count variable is believed to be generated by a Poisson-like process, except that the variation is greater than that of a true Poisson. This extra variation is referred to as overdispersion. See [R] **poisson** before reading this entry.

gnbreg fits a generalization of the negative binomial mean-dispersion model; the shape parameter α may also be parameterized.

If you have panel data, see [XT] **xtnbreg**.

Options for nbreg

Model

noconstant; see [R] **estimation options**.

dispersion(mean | constant) specifies the parameterization of the model. dispersion(mean), the default, yields a model with dispersion equal to $1+\alpha\exp(\mathbf{x}_j\boldsymbol{\beta}+\text{offset}_j)$; that is, the dispersion is a function of the expected mean: $\exp(\mathbf{x}_j\boldsymbol{\beta}+\text{offset}_j)$. dispersion(constant) has dispersion equal to $1+\delta$; that is, it is a constant for all observations.

exposure(*varname*$_e$), offset(*varname*$_o$), constraints(*constraints*), collinear; see [R] **estimation options**.

SE/Robust

vce(*vcetype*) specifies the type of standard error reported, which includes types that are derived from asymptotic theory, that are robust to some kinds of misspecification, that allow for intragroup correlation, and that use bootstrap or jackknife methods; see [R] ***vce_option***.

Reporting

level(*#*); see [R] **estimation options**.

nolrtest suppresses fitting the Poisson model. Without this option, a comparison Poisson model is fitted, and the likelihood is used in a likelihood-ratio test of the null hypothesis that the dispersion parameter is zero.

irr reports estimated coefficients transformed to incidence-rate ratios, that is, e^{β_i} rather than β_i. Standard errors and confidence intervals are similarly transformed. This option affects how results are displayed, not how they are estimated or stored. irr may be specified at estimation or when replaying previously estimated results.

Max options

maximize_options: difficult, technique(*algorithm_spec*), iterate(*#*), [no]log, trace, gradient, showstep, hessian, shownrtolerance, tolerance(*#*), ltolerance(*#*), gtolerance(*#*), nrtolerance(*#*), nonrtolerance, from(*init_specs*); see [R] **maximize**. These options are seldom used.

Setting the optimization type to technique(bhhh) resets the default *vcetype* to vce(opg).

Options for gnbreg

Model

noconstant; see [R] **estimation options**.

lnalpha(*varlist*) allows you to specify a linear equation for $\ln\alpha$. Specifying lnalpha(male old) means that $\ln\alpha = \gamma_0 + \gamma_1$male $+ \gamma_2$old, where γ_0, γ_1, and γ_2 are parameters to be estimated along with the other model coefficients. If this option is not specified, gnbreg and nbreg will produce the same results because the shape parameter will be parameterized as a constant.

exposure(*varname*$_e$), offset(*varname*$_o$), constraints(*constraints*), collinear; see [R] **estimation options**.

SE/Robust

vce(*vcetype*) specifies the type of standard error reported, which includes types that are derived from asymptotic theory, that are robust to some kinds of misspecification, that allow for intragroup correlation, and that use bootstrap or jackknife methods; see [R] ***vce_option***.

Reporting

level(*#*); see [R] **estimation options**.

irr reports estimated coefficients transformed to incidence-rate ratios, that is, e^{β_i} rather than β_i. Standard errors and confidence intervals are similarly transformed. This option affects how results are displayed, not how they are estimated or stored. irr may be specified at estimation or when replaying previously estimated results.

Max options

maximize_options: `difficult`, `technique(`*algorithm_spec*`)`, `iterate(#)`, `[no]log`, `trace`, `gradient`, `showstep`, `hessian`, `shownrtolerance`, `tolerance(#)`, `ltolerance(#)`, `gtolerance(#)`, `nrtolerance(#)`, `nonrtolerance`, `from(`*init_specs*`)`; see [R] **maximize**. These options are seldom used.

Setting the optimization type to `technique(bhhh)` resets the default *vcetype* to `vce(opg)`.

Remarks

Remarks are presented under the following headings:

Introduction to negative binomial regression
nbreg
gnbreg

Introduction to negative binomial regression

Negative binomial regression models the number of occurrences (counts) of an event when the event has extra-Poisson variation, that is, when it has overdispersion. The Poisson regression model is

$$y_j \sim \text{Poisson}(\mu_j)$$

where

$$\mu_j = \exp(\mathbf{x}_j\boldsymbol{\beta} + \text{offset}_j)$$

for observed counts y_j with covariates $\mathbf{x}_j$ for the jth observation. One derivation of the negative binomial mean-dispersion model is that individual units follow a Poisson regression model, but there is an omitted variable ν_j, such that e^{ν_j} follows a gamma distribution with mean 1 and variance α:

$$y_j \sim \text{Poisson}(\mu_j^*)$$

where

$$\mu_j^* = \exp(\mathbf{x}_j\boldsymbol{\beta} + \text{offset}_j + \nu_j)$$

and

$$e^{\nu_j} \sim \text{Gamma}(1/\alpha, \alpha)$$

With this parameterization, a Gamma(a, b) distribution will have expectation ab and variance ab^2.

We refer to α as the overdispersion parameter. The larger α is, the greater the overdispersion. The Poisson model corresponds to $\alpha = 0$. `nbreg` parameterizes α as $\ln\alpha$. `gnbreg` allows $\ln\alpha$ to be modeled as $\ln\alpha_j = \mathbf{z}_j\boldsymbol{\gamma}$, a linear combination of covariates $\mathbf{z}_j$.

`nbreg` will fit two different parameterizations of the negative binomial model. The default, described above and also given by the option `dispersion(mean)`, has dispersion for the jth observation equal to $1 + \alpha\exp(\mathbf{x}_j\boldsymbol{\beta} + \text{offset}_j)$. This is seen by noting that the above implies that

$$\mu_j^* \sim \text{Gamma}(1/\alpha, \alpha\mu_j)$$

and thus

$$\begin{aligned}\text{Var}(y_j) &= E\left\{\text{Var}(y_j|\mu_j^*)\right\} + \text{Var}\left\{E(y_j|\mu_j^*)\right\} \\ &= E(\mu_j^*) + \text{Var}(\mu_j^*) \\ &= \mu_j(1 + \alpha\mu_j)\end{aligned}$$

The alternative parameterization, given by the option `dispersion(constant)`, has dispersion equal to $1 + \delta$; that is, it is constant for all observations. This is so because the constant-dispersion model assumes instead that

$$\mu_j^* \sim \text{Gamma}(\mu_j/\delta, \delta)$$

and thus $\text{Var}(y_j) = \mu_j(1 + \delta)$. The Poisson model corresponds to $\delta = 0$.

For detailed derivations of both models, see Cameron and Trivedi (1998, 70–77). In particular, note that the mean-dispersion model is known as the NB2 model in their terminology, whereas the constant-dispersion model is referred to as the NB1 model.

See Long and Freese (2006) for a discussion of the negative binomial regression model with Stata examples and for a discussion of other regression models for count data.

nbreg

It is not uncommon to posit a Poisson regression model and observe a lack of model fit. The following data appeared in Rodríguez (1993):

```
. use http://www.stata-press.com/data/r10/rod93
. list, sepby(cohort)
```

	cohort	age_mos	deaths	exposure
1.	1	0.5	168	278.4
2.	1	2.0	48	538.8
3.	1	4.5	63	794.4
4.	1	9.0	89	1,550.8
5.	1	18.0	102	3,006.0
6.	1	42.0	81	8,743.5
7.	1	90.0	40	14,270.0
8.	2	0.5	197	403.2
9.	2	2.0	48	786.0
10.	2	4.5	62	1,165.3
11.	2	9.0	81	2,294.8
12.	2	18.0	97	4,500.5
13.	2	42.0	103	13,201.5
14.	2	90.0	39	19,525.0
15.	3	0.5	195	495.3
16.	3	2.0	55	956.7
17.	3	4.5	58	1,381.4
18.	3	9.0	85	2,604.5
19.	3	18.0	87	4,618.5
20.	3	42.0	70	9,814.5
21.	3	90.0	10	5,802.5

```
. generate logexp = ln(exposure)
. quietly tab cohort, gen(coh)
. poisson deaths coh2 coh3, offset(logexp)
Iteration 0:   log likelihood = -2160.0544
Iteration 1:   log likelihood = -2159.5162
Iteration 2:   log likelihood = -2159.5159
Iteration 3:   log likelihood = -2159.5159
Poisson regression                                Number of obs   =         21
                                                  LR chi2(2)      =      49.16
                                                  Prob > chi2     =     0.0000
Log likelihood = -2159.5159                       Pseudo R2       =     0.0113

------------------------------------------------------------------------------
      deaths |      Coef.   Std. Err.      z    P>|z|     [95% Conf. Interval]
-------------+----------------------------------------------------------------
        coh2 |  -.3020405   .0573319    -5.27   0.000    -.4144089   -.1896721
        coh3 |   .0742143   .0589726     1.26   0.208    -.0413698    .1897983
       _cons |  -3.899488   .0411345   -94.80   0.000     -3.98011   -3.818866
      logexp |   (offset)
------------------------------------------------------------------------------

. estat gof
         Goodness-of-fit chi2  =  4190.689
         Prob > chi2(18)       =    0.0000
```

The extreme significance of the goodness-of-fit χ^2 indicates that the Poisson regression model is inappropriate, suggesting to us that we should try a negative binomial model:

```
. nbreg deaths coh2 coh3, offset(logexp) nolog
Negative binomial regression                      Number of obs   =         21
                                                  LR chi2(2)      =       0.40
Dispersion     = mean                             Prob > chi2     =     0.8171
Log likelihood = -131.3799                        Pseudo R2       =     0.0015

------------------------------------------------------------------------------
      deaths |      Coef.   Std. Err.      z    P>|z|     [95% Conf. Interval]
-------------+----------------------------------------------------------------
        coh2 |  -.2676187   .7237203    -0.37   0.712    -1.686084    1.150847
        coh3 |  -.4573957   .7236651    -0.63   0.527    -1.875753    .9609618
       _cons |  -2.086731    .511856    -4.08   0.000     -3.08995   -1.083511
      logexp |   (offset)
-------------+----------------------------------------------------------------
    /lnalpha |   .5939963   .2583615                      .0876171    1.100376
-------------+----------------------------------------------------------------
       alpha |   1.811212   .4679475                       1.09157    3.005295
------------------------------------------------------------------------------
Likelihood-ratio test of alpha=0:  chibar2(01) = 4056.27 Prob>=chibar2 = 0.000
```

Our original Poisson model is a special case of the negative binomial—it corresponds to $\alpha = 0$. **nbreg**, however, estimates α indirectly, estimating instead $\ln\alpha$. In our model, $\ln\alpha = 0.594$, meaning that $\alpha = 1.81$ (**nbreg** undoes the transformation for us at the bottom of the output).

To test $\alpha = 0$ (equivalent to $\ln\alpha = -\infty$), **nbreg** performs a likelihood-ratio test. The staggering χ^2 value of 4,056 asserts that the probability that we would observe these data conditional on $\alpha = 0$ is virtually zero, that is, conditional on the process being Poisson. The data are not Poisson. It is not accidental that this χ^2 value is close to the goodness-of-fit statistic from the Poisson regression itself.

❑ Technical Note

The usual Gaussian test of $\alpha = 0$ is omitted since this test occurs on the boundary, invalidating the usual theory associated with such tests. However, the likelihood-ratio test of $\alpha = 0$ has been modified to be valid on the boundary. In particular, the null distribution of the likelihood-ratio test statistic is not the usual χ_1^2, but rather a 50:50 mixture of a χ_0^2 (point mass at zero) and a χ_1^2, denoted as $\overline{\chi}_{01}^2$. See Gutierrez, Carter, and Drukker (2001) for more details.

❑

❑ Technical Note

The negative binomial model deals with cases in which there is more variation than would be expected if the process were Poisson. The negative binomial model is not helpful if there is less than Poisson variation—if the variance of the count variable is less than its mean. However, underdispersion is uncommon. Poisson models arise because of independently generated events. Overdispersion comes about if some of the parameters (causes) of the Poisson processes are unknown. To obtain underdispersion, the sequence of events somehow would have to be regulated; that is, events would not be independent but controlled based on past occurrences.

❑

gnbreg

`gnbreg` is a generalization of `nbreg, dispersion(mean)`. Whereas in `nbreg` one $\ln\alpha$ is estimated, `gnbreg` allows $\ln\alpha$ to vary, observation by observation, as a linear combination of another set of covariates: $\ln\alpha_j = \mathbf{z}_j\gamma$.

We will assume that the number of deaths is a function of age, whereas the $\ln\alpha$ parameter is a function of cohort. To fit the model, we type

```
. gnbreg deaths age_mos, lnalpha(coh2 coh3) offset(logexp)
Fitting constant-only model:
Iteration 0:   log likelihood =   -187.067  (not concave)
Iteration 1:   log likelihood = -137.43798
Iteration 2:   log likelihood = -132.47158
Iteration 3:   log likelihood = -131.57982
Iteration 4:   log likelihood = -131.57948
Iteration 5:   log likelihood = -131.57948
Fitting full model:
Iteration 0:   log likelihood = -124.34327
Iteration 1:   log likelihood = -117.68002
Iteration 2:   log likelihood = -117.56307
Iteration 3:   log likelihood = -117.56164
Iteration 4:   log likelihood = -117.56164
```

```
Generalized negative binomial regression          Number of obs   =         21
                                                  LR chi2(1)      =      28.04
                                                  Prob > chi2     =     0.0000
Log likelihood = -117.56164                       Pseudo R2       =     0.1065

------------------------------------------------------------------------------
             |      Coef.   Std. Err.      z    P>|z|     [95% Conf. Interval]
-------------+----------------------------------------------------------------
deaths       |
     age_mos |  -.0516657   .0051747    -9.98   0.000     -.061808   -.0415233
       _cons |  -1.867225   .2227944    -8.38   0.000    -2.303894   -1.430556
      logexp |   (offset)
-------------+----------------------------------------------------------------
lnalpha      |
        coh2 |   .0939546   .7187747     0.13   0.896    -1.314818    1.502727
        coh3 |   .0815279   .7365476     0.11   0.912    -1.362079    1.525135
       _cons |  -.4759581   .5156502    -0.92   0.356    -1.486614    .5346978
------------------------------------------------------------------------------
```

We find that age is a significant determinant of the number of deaths. The standard errors for the variables in the $\ln\alpha$ equation suggest that the overdispersion parameter does not vary across cohorts. We can test this assertion by typing

```
. test coh2 coh3

 ( 1)  [lnalpha]coh2 = 0
 ( 2)  [lnalpha]coh3 = 0

           chi2(  2) =    0.02
         Prob > chi2 =    0.9904
```

There is no evidence of variation by cohort in these data.

❑ Technical Note

Note the intentional absence of a likelihood-ratio test for $\alpha = 0$ in gnbreg. The test is affected by the same boundary condition that affects the comparison test in nbreg; however, when α is parameterized by more than a constant term, the null distribution becomes intractable. For this reason, we recommend using nbreg to test for overdispersion and, if you have reason to believe that overdispersion exists, only then modeling the overdispersion using gnbreg.

❑

(*Continued on next page*)

Saved Results

nbreg and gnbreg save the following in e():

Scalars

e(N)	number of observations	e(alpha)	the value of alpha
e(k)	number of parameters	e(N_clust)	number of clusters
e(k_eq)	number of equations	e(chi2)	χ^2
e(k_eq_model)	number of equations in model Wald test	e(chi2_c)	χ^2 for comparison test
		e(p)	significance
e(k_dv)	number of dependent variables	e(rank)	rank of e(V)
e(df_m)	model degrees of freedom	e(rank0)	rank of e(V) for constant-only model
e(r2_p)	pseudo-R-squared		
e(ll)	log likelihood	e(ic)	number of iterations
e(ll_0)	log likelihood, constant-only model	e(rc)	return code
e(ll_c)	log likelihood, comparison model	e(converged)	1 if converged, 0 otherwise

Macros

e(cmd)	nbreg or gnbreg	e(vcetype)	title used to label Std. Err.
e(cmdline)	command as typed	e(diparm#)	display transformed parameter #
e(depvar)	name of dependent variable	e(diparm_opt2)	options for displaying transformed parameters
e(wtype)	weight type		
e(wexp)	weight expression	e(opt)	type of optimization
e(title)	title in estimation output	e(ml_method)	type of ml method
e(clustvar)	name of cluster variable	e(user)	name of likelihood-evaluator program
e(offset)	offset (nbreg)		
e(offset1)	offset (gnbreg)	e(technique)	maximization technique
e(chi2type)	Wald or LR; type of model χ^2 test	e(crittype)	optimization criterion
e(chi2_ct)	Wald or LR; type of model χ^2 test corresponding to e(chi2_c)	e(properties)	b V
		e(predict)	program used to implement predict
e(dispers)	mean or constant		
e(vce)	*vcetype* specified in vce()		

Matrices

e(b)	coefficient vector	e(V)	variance–covariance matrix of the estimators
e(ilog)	iteration log (up to 20 iterations)		
e(gradient)	gradient vector		

Functions

e(sample)	marks estimation sample

Methods and Formulas

nbreg and gnbreg are implemented as ado-files.

See [R] **poisson** and Feller (1968, 156–164) for an introduction to the Poisson distribution.

Mean-dispersion model

A negative binomial distribution can be regarded as a gamma mixture of Poisson random variables. The number of times something occurs, y_j, is distributed as Poisson$(\nu_j \mu_j)$. That is, its conditional likelihood is

$$f(y_j \mid \nu_j) = \frac{(\nu_j \mu_j)^{y_j} e^{-\nu_j \mu_j}}{\Gamma(y_j + 1)}$$

where $\mu_j = \exp(\mathbf{x}_j\boldsymbol{\beta} + \text{offset}_j)$ and ν_j is an unobserved parameter with a Gamma$(1/\alpha, \alpha)$ density:

$$g(\nu) = \frac{\nu^{(1-\alpha)/\alpha} e^{-\nu/\alpha}}{\alpha^{1/\alpha}\Gamma(1/\alpha)}$$

This gamma distribution has mean 1 and variance α, where α is our ancillary parameter.

The unconditional likelihood for the jth observation is therefore

$$f(y_j) = \int_0^\infty f(y_j \mid \nu) g(\nu)\, d\nu = \frac{\Gamma(m + y_j)}{\Gamma(y_j + 1)\Gamma(m)}\, p_j^m (1 - p_j)^{y_j}$$

where $p_j = 1/(1 + \alpha\mu_j)$ and $m = 1/\alpha$. Solutions for α are handled by searching for $\ln\alpha$ since α is required to be greater than zero.

The log likelihood (with weights w_j and offsets) is given by

$$m = 1/\alpha \qquad p_j = 1/(1 + \alpha\mu_j) \qquad \mu_j = \exp(\mathbf{x}_j\boldsymbol{\beta} + \text{offset}_j)$$

$$\begin{aligned}\ln L = \sum_{j=1}^n w_j \Big[& \ln\{\Gamma(m + y_j)\} - \ln\{\Gamma(y_j + 1)\} \\ & - \ln\{\Gamma(m)\} + m\ln(p_j) + y_j \ln(1 - p_j) \Big]\end{aligned}$$

For `gnbreg`, α can vary across the observations according to the parameterization $\ln\alpha_j = \mathbf{z}_j\boldsymbol{\gamma}$.

Constant-dispersion model

The constant-dispersion model assumes that y_j is conditionally distributed as Poisson(μ_j^*), where $\mu_j^* \sim$ Gamma$(\mu_j/\delta, \delta)$ for some dispersion parameter δ (by contrast, the mean-dispersion model assumes that $\mu_j^* \sim$ Gamma$(1/\alpha, \alpha\mu_j)$). The log likelihood is given by

$$m_j = \mu_j/\delta \qquad p = 1/(1 + \delta)$$

$$\begin{aligned}\ln L = \sum_{j=1}^n w_j \Big[& \ln\{\Gamma(m_j + y_j)\} - \ln\{\Gamma(y_j + 1)\} \\ & - \ln\{\Gamma(m_j)\} + m_j\ln(p) + y_j \ln(1 - p) \Big]\end{aligned}$$

with everything else defined as before in the calculations for the mean-dispersion model.

Maximization for `gnbreg` is done via the `lf` linear-form method, and for `nbreg` it is done via the `d2` method described in [R] **ml**.

References

Cameron, A. C., and P. K. Trivedi. 1998. *Regression Analysis of Count Data.* Cambridge: Cambridge University Press.

Deb, P., and P. K. Trivedi. 2006. Maximum simulated likelihood estimation of a negative binomial regression model with multinomial endogenous treatment. *Stata Journal* 6: 246–255.

Feller, W. 1968. *An Introduction to Probability Theory and Its Applications*, vol. 1. 3rd ed. New York: Wiley.

Gutierrez, R. G., S. L. Carter, and D. M. Drukker. 2001. On boundary-value likelihood-ratio tests. *Stata Technical Bulletin* 60: 15–18. Reprinted in *Stata Technical Bulletin Reprints*, vol. 10, pp. 269–273.

Hilbe, J. 1998. sg91: Robust variance estimators for MLE Poisson and negative binomial regression. *Stata Technical Bulletin* 45: 26–28. Reprinted in *Stata Technical Bulletin Reprints*, vol. 8, pp. 177–180.

——. 1999. sg102: Zero-truncated Poisson and negative binomial regression. *Stata Technical Bulletin* 47: 37–40. Reprinted in *Stata Technical Bulletin Reprints*, vol. 8, pp. 233–236.

Long, J. S. 1997. *Regression Models for Categorical and Limited Dependent Variables*. Thousand Oaks, CA: Sage.

Long, J. S., and J. Freese. 2001. Predicted probabilities for count models. *Stata Journal* 1: 51–57.

——. 2006. *Regression Models for Categorical Dependent Variables Using Stata*. 2nd ed. College Station, TX: Stata Press.

Miranda, A., and S. Rabe-Hesketh. 2006. Maximum likelihood estimation of endogenous switching and sample selection models for binary, ordinal, and count variables. *Stata Journal* 6: 285–308.

Rodríguez, G. 1993. sbe10: An improvement to poisson. *Stata Technical Bulletin* 11: 11–14. Reprinted in *Stata Technical Bulletin Reprints*, vol. 2, pp. 94–98.

Rogers, W. H. 1991. sbe1: Poisson regression with rates. *Stata Technical Bulletin* 1: 11–12. Reprinted in *Stata Technical Bulletin Reprints*, vol. 1, pp. 62–64.

——. 1993. sg16.4: Comparison of nbreg and glm for negative binomial. *Stata Technical Bulletin* 16: 7. Reprinted in *Stata Technical Bulletin Reprints*, vol. 3, pp. 82–84.

Also See

[R] **nbreg postestimation** — Postestimation tools for nbreg and gnbreg

[R] **glm** — Generalized linear models

[R] **poisson** — Poisson regression

[R] **zinb** — Zero-inflated negative binomial regression

[R] **ztnb** — Zero-truncated negative binomial regression

[SVY] **svy estimation** — Estimation commands for survey data

[XT] **xtnbreg** — Fixed-effects, random-effects, & population-averaged negative binomial models

[U] **20 Estimation and postestimation commands**

Title

nbreg postestimation — Postestimation tools for nbreg and gnbreg

Description

The following postestimation commands are available for `nbreg` and `gnbreg`:

command	description
`adjust`[1]	adjusted predictions of $\mathbf{x}\beta$ or $\exp(\mathbf{x}\beta)$
`estat`	AIC, BIC, VCE, and estimation sample summary
`estat` (svy)	postestimation statistics for survey data
`estimates`	cataloging estimation results
`lincom`	point estimates, standard errors, testing, and inference for linear combinations of coefficients
`linktest`	link test for model specification
`lrtest`[2]	likelihood-ratio test
`mfx`	marginal effects or elasticities
`nlcom`	point estimates, standard errors, testing, and inference for nonlinear combinations of coefficients
`predict`	predictions, residuals, influence statistics, and other diagnostic measures
`predictnl`	point estimates, standard errors, testing, and inference for generalized predictions
`suest`	seemingly unrelated estimation
`test`	Wald tests for simple and composite linear hypotheses
`testnl`	Wald tests of nonlinear hypotheses

[1] `adjust` is not appropriate with time-series operators.
[2] `lrtest` is not appropriate with `svy` estimation results.

See the corresponding entries in the *Stata Base Reference Manual* for details, but see [SVY] **estat** for details about `estat` (svy).

Syntax for predict

`predict` [*type*] *newvar* [*if*] [*in*] [, *statistic* `nooffset`]

`predict` [*type*] { *stub*∗ | $newvar_{\rm reg}$ $newvar_{\rm disp}$ } [*if*] [*in*] , `scores`

statistic	description
Main	
`n`	number of events; the default
`ir`	incidence rate (equivalent to `predict ..., n nooffset`)
`xb`	linear prediction
`stdp`	standard error of the linear prediction

In addition, relevant only after `gnbreg` are

statistic	description
Main	
`alpha`	predicted values of α_j
`lnalpha`	predicted values of $\ln\alpha_j$
`stdplna`	standard error of predicted $\ln\alpha_j$

These statistics are available both in and out of sample; type `predict ... if e(sample) ...` if wanted only for the estimation sample.

Options for predict

Main

`n`, the default, calculates the predicted number of events, which is $\exp(\mathbf{x}_j\boldsymbol{\beta})$ if neither `offset(`*varname*$_o$`)` nor `exposure(`*varname*$_e$`)` was specified when the model was fitted; $\exp(\mathbf{x}_j\boldsymbol{\beta}+\text{offset}_j)$ if `offset()` was specified; or $\exp(\mathbf{x}_j\boldsymbol{\beta}) \times \text{exposure}_j$ if `exposure()` was specified.

`ir` calculates the incidence rate $\exp(\mathbf{x}_j\boldsymbol{\beta})$, which is the predicted number of events when exposure is 1. This is equivalent to specifying both `n` and `nooffset` options.

`xb` calculates the linear prediction, which is $\mathbf{x}_j\boldsymbol{\beta}$ if neither `offset()` nor `exposure()` was specified; $\mathbf{x}_j\boldsymbol{\beta}+\text{offset}_j$ if `offset()` was specified; or $\mathbf{x}_j\boldsymbol{\beta}+\ln(\text{exposure}_j)$ if `exposure()` was specified; see `nooffset` below.

`stdp` calculates the standard error of the linear prediction.

`alpha`, `lnalpha`, and `stdplna` are relevant after `gnbreg` estimation only; they produce the predicted values of α_j, $\ln\alpha_j$, and the standard error of the predicted $\ln\alpha_j$, respectively.

`nooffset` is relevant only if you specified `offset()` or `exposure()` when you fitted the model. It modifies the calculations made by `predict` so that they ignore the offset or exposure variable; the linear prediction is treated as $\mathbf{x}_j\boldsymbol{\beta}$ rather than as $\mathbf{x}_j\boldsymbol{\beta}+\text{offset}_j$ or $\mathbf{x}_j\boldsymbol{\beta}+\ln(\text{exposure}_j)$. Specifying `predict ..., nooffset` is equivalent to specifying `predict ..., ir`.

`scores` calculates equation-level score variables.

The first new variable will contain $\partial\ln L/\partial(\mathbf{x}_j\boldsymbol{\beta})$.

The second new variable will contain $\partial\ln L/\partial(\ln\alpha_j)$ for `dispersion(mean)` and `gnbreg`.

The second new variable will contain $\partial\ln L/\partial(\ln\delta)$ for `dispersion(constant)`.

Remarks

After `nbreg` and `gnbreg`, `predict` returns the predicted number of events:

```
. nbreg deaths coh2 coh3, nolog
Negative binomial regression                      Number of obs   =         21
                                                  LR chi2(2)      =       0.14
Dispersion     = mean                             Prob > chi2     =     0.9307
Log likelihood = -108.48841                       Pseudo R2       =     0.0007

------------------------------------------------------------------------------
      deaths |      Coef.   Std. Err.      z    P>|z|     [95% Conf. Interval]
-------------+----------------------------------------------------------------
        coh2 |   .0591305   .2978419     0.20   0.843    -.5246289      .64289
        coh3 |  -.0538792   .2981621    -0.18   0.857    -.6382662    .5305077
       _cons |   4.435906   .2107213    21.05   0.000       4.0229    4.848912
-------------+----------------------------------------------------------------
    /lnalpha |  -1.207379   .3108622                     -1.816657   -.5980999
-------------+----------------------------------------------------------------
       alpha |     .29898   .0929416                      .1625683    .5498555
------------------------------------------------------------------------------
Likelihood-ratio test of alpha=0:  chibar2(01) =  434.62 Prob>=chibar2 = 0.000
. predict count
(option n assumed; predicted number of events)
. summarize deaths count
    Variable |       Obs        Mean    Std. Dev.       Min        Max
-------------+--------------------------------------------------------
      deaths |        21    84.66667    48.84192         10        197
       count |        21    84.66667     4.00773         80   89.57143
```

Methods and Formulas

All postestimation commands listed above are implemented as ado-files.

In the following, we use the same notation as in [R] **nbreg**.

Mean-dispersion model

The equation-level scores are given by

$$\begin{aligned}\text{score}(\mathbf{x}\boldsymbol{\beta})_j &= p_j(y_j - \mu_j)\\ \text{score}(\tau)_j &= -m\left\{\frac{\alpha_j(\mu_j - y_j)}{1+\alpha_j\mu_j} - \ln(1+\alpha_j\mu_j) + \psi(y_j+m) - \psi(m)\right\}\end{aligned}$$

where $\tau_j = \ln\alpha_j$, and $\psi(z)$ is the digamma function.

Constant-dispersion model

The equation-level scores are given by

$$\begin{aligned}\text{score}(\mathbf{x}\boldsymbol{\beta})_j &= m_j\left\{\psi(y_j+m_j) - \psi(m_j) + \ln(p)\right\}\\ \text{score}(\tau)_j &= y_j - (y_j+m_j)(1-p) - \text{score}(\mathbf{x}\boldsymbol{\beta})_j\end{aligned}$$

where $\tau_j = \ln\delta_j$.

Also See

[R] **nbreg** — Negative binomial regression

[U] **20 Estimation and postestimation commands**

Title

nestreg — Nested model statistics

Syntax

Standard estimation command syntax

`nestreg` [, *options*] : *command_name depvar* (*varlist*) [(*varlist*) ...]

[*if*] [*in*] [*weight*] [*command_options*]

Survey estimation command syntax

`nestreg` [, *options*] : `svy` [*vcetype*] [, *svy_options*] : *command_name depvar*

(*varlist*) [(*varlist*) ...] [*if*] [*in*] [, *command_options*]

options	description
Reporting	
`level(#)`	set confidence level; default is `level(95)`
`waldtable`	report Wald test results; the default
`lrtable`	report likelihood-ratio test results
`quietly`	suppress any output from *command_name*
`store(stub)`	store nested estimation results to *stub#*

`by` is allowed; see [U] **11.1.10 Prefix commands**.

Weights are allowed if *command_name* allows them; see [U] **11.1.6 weight**.

A *varlist* in parentheses indicates that this list of variables is to be considered as a block. Each variable in a *varlist* not bound in parentheses will be treated as its own block.

All postestimation commands behave as they would after *command_name* without the `nestreg` prefix; see the postestimation manual entry for *command_name*.

Description

`nestreg` fits nested models by sequentially adding blocks of variables and then reports comparison tests between the nested models.

Options

Reporting

`level(#)` specifies the confidence level, as a percentage, for confidence intervals. The default is `level(95)` or as set by `set level`.

`waldtable` specifies that the table of Wald test results be reported. `waldtable` is the default.

`lrtable` specifies that the table of likelihood-ratio tests be reported. This option is not allowed if `pweights`, the `vce(robust)` option, or the `vce(cluster` *clustvar*`)` option is specified. `lrtable` is also not allowed with the `svy` prefix.

`quietly` suppresses the display of any output from *command_name*.

`store(`*stub*`)` specifies that each model fitted by `nestreg` be stored under the name *stub#*, where # is the nesting order from first to last.

Remarks

Remarks are presented under the following headings:

Estimation commands

`nestreg` removes collinear predictors and observations with missing values from the estimation sample before calling *command_name*.

The following Stata commands are supported by `nestreg`:

```
clogit      logit      qreg
cloglog     nbreg      regress
cnreg       ologit     scobit
glm         oprobit    stcox
intreg      poisson    streg
logistic    probit     tobit
```

You do not supply a *depvar* for `streg` or `stcox`; otherwise, *depvar* is required. You must supply two *depvars* for `intreg`.

Wald tests

Use `nestreg` to test the significance of blocks of predictors, building the regression model one block at a time. Using the data from the first example of `test` (see [R] **test**), we wish to test the significance of the following predictors of birth rate: `medage`, `medagesq`, and `region` (already partitioned into four indicator variables, `reg1`, `reg2`, `reg3`, and `reg4`).

```
. use http://www.stata-press.com/data/r10/census4
(birth rate, median age)

. nestreg: regress brate (medage) (medagesq) (reg2-reg4)

Block  1: medage

      Source |       SS       df       MS              Number of obs =      50
-------------+------------------------------           F(  1,    48) =  164.72
       Model |  32675.1044     1  32675.1044           Prob > F      =  0.0000
    Residual |  9521.71561    48  198.369075           R-squared     =  0.7743
-------------+------------------------------           Adj R-squared =  0.7696
       Total |    42196.82    49  861.159592           Root MSE      =  14.084

------------------------------------------------------------------------------
       brate |      Coef.   Std. Err.      t    P>|t|     [95% Conf. Interval]
-------------+----------------------------------------------------------------
      medage |  -15.24893   1.188141   -12.83   0.000    -17.63785   -12.86002
       _cons |   618.3935   35.15416    17.59   0.000     547.7113    689.0756
------------------------------------------------------------------------------
```

```
Block  2: medagesq

      Source |       SS       df       MS              Number of obs =      50
-------------+------------------------------           F(  2,    47) =  158.75
       Model |  36755.8524     2  18377.9262           Prob > F      =  0.0000
    Residual |  5440.96755    47  115.765267           R-squared     =  0.8711
-------------+------------------------------           Adj R-squared =  0.8656
       Total |    42196.82    49  861.159592           Root MSE      =  10.759

------------------------------------------------------------------------------
       brate |      Coef.   Std. Err.      t    P>|t|     [95% Conf. Interval]
-------------+----------------------------------------------------------------
      medage |  -109.8925   15.96663     6.88   0.000    -142.0132    -77.7718
    medagesq |   1.607332   .2707228     5.94   0.000     1.062708    2.151956
       _cons |   2007.071   235.4316     8.53   0.000     1533.444    2480.698
------------------------------------------------------------------------------

Block  3: reg2 reg3 reg4

      Source |       SS       df       MS              Number of obs =      50
-------------+------------------------------           F(  5,    44) =  100.63
       Model |  38803.419      5  7760.68381           Prob > F      =  0.0000
    Residual |  3393.40095    44  77.1227489           R-squared     =  0.9196
-------------+------------------------------           Adj R-squared =  0.9104
       Total |    42196.82    49  861.159592           Root MSE      =   8.782

------------------------------------------------------------------------------
       brate |      Coef.   Std. Err.      t    P>|t|     [95% Conf. Interval]
-------------+----------------------------------------------------------------
      medage |  -109.0957   13.52452    -8.07   0.000    -136.3526    -81.83886
    medagesq |   1.635208   .2290536     7.14   0.000     1.173581    2.096835
        reg2 |   15.00284   4.252068     3.53   0.001     6.433365    23.57233
        reg3 |   7.366435   3.953336     1.86   0.069    -.6009898    15.33386
        reg4 |   21.39679   4.650602     4.60   0.000     12.02412    30.76946
       _cons |    1947.61   199.8405     9.75   0.000     1544.858    2350.362
------------------------------------------------------------------------------

  +-------------------------------------------------------------+
  |       |          Block  Residual                     Change |
  | Block |       F     df        df   Pr > F       R2    in R2 |
  |-------+-----------------------------------------------------|
  |     1 |  164.72      1        48   0.0000   0.7743          |
  |     2 |   35.25      1        47   0.0000   0.8711   0.0967 |
  |     3 |    8.85      3        44   0.0001   0.9196   0.0485 |
  +-------------------------------------------------------------+
```

This single call to `nestreg` ran `regress` three times, adding a block of predictors to the model for each run as in

```
. regress brate medage

      Source |       SS       df       MS              Number of obs =      50
-------------+------------------------------           F(  1,    48) =  164.72
       Model |  32675.1044     1  32675.1044           Prob > F      =  0.0000
    Residual |  9521.71561    48  198.369075           R-squared     =  0.7743
-------------+------------------------------           Adj R-squared =  0.7696
       Total |    42196.82    49  861.159592           Root MSE      =  14.084

------------------------------------------------------------------------------
       brate |      Coef.   Std. Err.      t    P>|t|     [95% Conf. Interval]
-------------+----------------------------------------------------------------
      medage |  -15.24893   1.188141   -12.83   0.000    -17.63785   -12.86002
       _cons |   618.3935   35.15416    17.59   0.000     547.7113    689.0756
------------------------------------------------------------------------------
```

```
. regress brate medage medagesq

      Source |       SS       df       MS              Number of obs =      50
-------------+------------------------------           F(  2,    47) =  158.75
       Model |  36755.8524     2  18377.9262           Prob > F      =  0.0000
    Residual |  5440.96755    47  115.765267           R-squared     =  0.8711
-------------+------------------------------           Adj R-squared =  0.8656
       Total |    42196.82    49  861.159592           Root MSE      =  10.759

------------------------------------------------------------------------------
       brate |      Coef.   Std. Err.      t    P>|t|     [95% Conf. Interval]
-------------+----------------------------------------------------------------
      medage |  -109.8925   15.96663    -6.88   0.000    -142.0132     -77.7718
    medagesq |   1.607332   .2707228     5.94   0.000     1.062708     2.151956
       _cons |   2007.071   235.4316     8.53   0.000     1533.444     2480.698
------------------------------------------------------------------------------

. regress brate medage medagesq reg2-reg4

      Source |       SS       df       MS              Number of obs =      50
-------------+------------------------------           F(  5,    44) =  100.63
       Model |   38803.419     5  7760.68381           Prob > F      =  0.0000
    Residual |  3393.40095    44  77.1227489           R-squared     =  0.9196
-------------+------------------------------           Adj R-squared =  0.9104
       Total |    42196.82    49  861.159592           Root MSE      =   8.782

------------------------------------------------------------------------------
       brate |      Coef.   Std. Err.      t    P>|t|     [95% Conf. Interval]
-------------+----------------------------------------------------------------
      medage |  -109.0957   13.52452    -8.07   0.000    -136.3526    -81.83886
    medagesq |   1.635208   .2290536     7.14   0.000     1.173581     2.096835
        reg2 |   15.00284   4.252068     3.53   0.001     6.433365     23.57233
        reg3 |   7.366435   3.953336     1.86   0.069    -.6009898     15.33386
        reg4 |   21.39679   4.650602     4.60   0.000     12.02412     30.76946
       _cons |    1947.61   199.8405     9.75   0.000     1544.858     2350.362
------------------------------------------------------------------------------
```

`nestreg` collected the F statistic for the corresponding block of predictors and the model R^2 statistic from each model fitted.

The F statistic for the first block, 164.72, is for a test of the joint significance of the first block of variables; it is simply the F statistic from the regression of `brate` on `medage`. The F statistic for the second block, 35.25, is for a test of the joint significance of the second block of variables in a regression of both the first and second blocks of variables. In our example, it is an F test of `medagesq` in the regression of `brate` on `medage` and `medagesq`. Similarly, the third block's F statistic of 8.85 corresponds to a joint test of `reg2`, `reg3`, and `reg4` in the final regression.

Likelihood-ratio tests

The `nestreg` command provides a simple syntax for performing likelihood-ratio tests for nested model specifications; also see `lrtest`. Using the data from the first example of `lrtest` (see [R] **lrtest**), we wish to jointly test the significance of the following predictors of low birthweight: `age`, `lwt`, `ptl`, and `ht`.

```
. use http://www.stata-press.com/data/r10/lbw
(Hosmer & Lemeshow data)

. xi: nestreg, lr: logistic low (i.race smoke ui) (age lwt ptl ht)
i.race            _Irace_1-3          (naturally coded; _Irace_1 omitted)

Block  1: _Irace_2 _Irace_3 smoke ui

Logistic regression                               Number of obs   =        189
                                                  LR chi2(4)      =      18.80
                                                  Prob > chi2     =     0.0009
Log likelihood = -107.93404                       Pseudo R2       =     0.0801

------------------------------------------------------------------------------
         low | Odds Ratio   Std. Err.      z    P>|z|     [95% Conf. Interval]
-------------+----------------------------------------------------------------
    _Irace_2 |   3.052746   1.498084     2.27   0.023     1.166749    7.987368
    _Irace_3 |   2.922593   1.189226     2.64   0.008      1.31646    6.488269
       smoke |   2.945742   1.101835     2.89   0.004      1.41517    6.131701
          ui |   2.419131   1.047358     2.04   0.041      1.03546    5.651783
------------------------------------------------------------------------------

Block  2: age lwt ptl ht

Logistic regression                               Number of obs   =        189
                                                  LR chi2(8)      =      33.22
                                                  Prob > chi2     =     0.0001
Log likelihood =   -100.724                       Pseudo R2       =     0.1416

------------------------------------------------------------------------------
         low | Odds Ratio   Std. Err.      z    P>|z|     [95% Conf. Interval]
-------------+----------------------------------------------------------------
    _Irace_2 |   3.534767   1.860737     2.40   0.016     1.259736    9.918406
    _Irace_3 |   2.368079   1.039949     1.96   0.050     1.001356    5.600207
       smoke |   2.517698    1.00916     2.30   0.021     1.147676    5.523162
          ui |     2.1351   .9808153     1.65   0.099     .8677528      5.2534
         age |   .9732636   .0354759    -0.74   0.457     .9061578    1.045339
         lwt |   .9849634   .0068217    -2.19   0.029     .9716834    .9984249
         ptl |   1.719161   .5952579     1.56   0.118     .8721455    3.388787
          ht |   6.249602   4.322408     2.65   0.008     1.611152    24.24199
------------------------------------------------------------------------------

  +----------------------------------------------------------------+
  | Block |        LL         LR     df  Pr > LR       AIC       BIC |
  |-------+--------------------------------------------------------|
  |     1 |  -107.934      18.80      4   0.0009  225.8681  242.0768 |
  |     2 |  -100.724      14.42      4   0.0061   219.448  248.6237 |
  +----------------------------------------------------------------+
```

The estimation results from the full model are left in `e()`, so we can later use `estat` and other postestimation commands.

```
. estat gof
Logistic model for low, goodness-of-fit test
       number of observations =       189
 number of covariate patterns =       182
            Pearson chi2(173) =       179.24
                  Prob > chi2 =         0.3567
```

Programming for nestreg

If you want your user-written command (*command_name*) to work with `nestreg`, it must follow standard Stata syntax and allow the `if` qualifier. Furthermore, *command_name* must have `sw` or `swml` as a program property; see [P] **program**. If *command_name* has `swml` as a property, *command_name* must save the log-likelihood value in `e(ll)` and the model degrees of freedom in `e(df_m)`.

Saved Results

`nestreg` saves the following in `r()`:

Matrices

`r(wald)`	matrix corresponding to the Wald table
`r(lr)`	matrix corresponding to the likelihood-ratio table

Methods and Formulas

`nestreg` is implemented as an ado-file.

Acknowledgment

We thank Paul H. Bern, Princeton University, for developing the hierarchical regression command that inspired `nestreg`.

Reference

Acock, A. C. 2006. *A Gentle Introduction to Stata.* College Station, TX: Stata Press.

Also See

[P] **program properties** — Properties of user-defined programs

Title

net — Install and manage user-written additions from the Internet

Syntax

Set current location for net

`net from` *directory_or_url*

Change to a different net directory

`net cd` *path_or_url*

Change to a different net site

`net link` *linkname*

Search for installed packages

`net search` (see [R] **net search**)

Report current net location

`net`

Describe a package

`net describe` *pkgname* [`, from(`*directory_or_url*`)`]

Set location where packages will be installed

`net set ado` *dirname*

Set location where ancillary files will be installed

`net set other` *dirname*

Report net 'from', 'ado', and 'other' settings

`net query`

Install ado- and help files from a package

`net install` *pkgname* [`, all replace force from(`*directory_or_url*`)`]

Install ancillary files from a package

`net get` *pkgname* [`, all replace force from(`*directory_or_url*`)`]

Shortcut to access Stata Journal (SJ) net site

net sj *vol-issue* [*insert*]

Shortcut to access Stata Technical Bulletin (STB) net site

net stb *issue* [*insert*]

List installed packages

ado [, find(*string*) from(*dirname*)]

ado dir [*pkgid*] [, find(*string*) from(*dirname*)]

Describe installed packages

ado describe [*pkgid*] [, find(*string*) from(*dirname*)]

Uninstall an installed package

ado uninstall *pkgid* [, from(*dirname*)]

where

pkgname is name of a package
pkgid is name of a package
or a number in square brackets: [#]
dirname is a directory name
or PLUS (default)
or PERSONAL
or SITE

Description

net downloads and installs additions to Stata. The additions can be obtained from the Internet or from media. The additions can be ado-files (new commands), help files, or even datasets. Collections of files are bound together into *packages*. For instance, the package named zz49 might add the xyz command to Stata. At a minimum, such a package would contain xyz.ado, the code to implement the new command, and xyz.sthlp, the online help to describe it. That the package contains two files is a detail: you use net to download the package zz49, regardless of the number of files.

ado manages the packages you have installed by using net. The ado command lets you list and uninstall previously installed packages.

You can also access the net and ado features by selecting **Help > SJ and User-written Programs**; this is the recommended method to find and install additions to Stata.

Options

all is used with net install and net get. Typing it with either one makes the command equivalent to typing net install followed by net get.

`replace` is for use with `net install` and `net get`. It specifies that the downloaded files replace existing files if any of the files already exist.

`force` specifies that the downloaded files replace existing files if any of the files already exist, even if Stata thinks all the files are the same. `force` implies `replace`.

`find(`*string*`)` is for use with `ado`, `ado dir`, and `ado describe`. It specifies that the descriptions of the packages installed on your computer be searched, and that the package descriptions containing *string* be listed.

`from(`*dirname*`)`, when used with `ado`, specifies where the packages are installed. The default is `from(PLUS)`. `PLUS` is a code word that Stata understands to correspond to a particular directory on your computer that was set at installation time. On Windows computers, `PLUS` probably means the directory `c:\ado\plus`, but it might mean something else. You can find out what it means by typing `sysdir`, but doing so is irrelevant if you use the defaults.

`from(`*directory_or_url*`)`, when used with `net`, specifies the directory or URL where installable packages may be found. The directory or URL is the same as the one that would have been specified with `net from`.

Remarks

For an introduction to using `net` and `ado`, see [U] **28 Using the Internet to keep up to date**. The purpose of this documentation is

1. to briefly, but accurately, describe `net` and `ado` and all their features and
2. to provide documentation to those who wish to set up their own sites to distribute additions to Stata.

Remarks are presented under the following headings:

Definition of a package

A *package* is a collection of files—typically `.ado` and `.sthlp` files—that together provide a new feature in Stata. Packages contain additions that you wish had been part of Stata at the outset. We write such additions, and so do other users.

One source of these additions is the *Stata Journal*, a printed and electronic journal with corresponding software. If you want the journal, you must subscribe, but the software is available for free from our web site.

The purpose of the net and ado commands

The `net` command makes it easy to distribute and install packages. The goal is to get you quickly to a package description page that summarizes the addition, for example,

```
. net describe rte_stat, from(http://www.wemakeitupaswego.edu/faculty/sgazer/)
--------------------------------------------------------------------------------
package rte_stat from http://www.wemakeitupaswego.edu/faculty/sgazer/
--------------------------------------------------------------------------------
TITLE
      rte_stat.  The robust-to-everything statistic; update.
DESCRIPTION/AUTHOR(S)
      S. Gazer, Dept. of Applied Theoretical Mathematics, WMIUAWG Univ.
      Aleph-0 100% confidence intervals proved too conservative for some
      applications; Aleph-1 confidence intervals have been substituted.
      The new robust-to-everything supplants the previous robust-to-
      everything-conceivable statistic.  See "Inference in the absence
      of data" (forthcoming).  After installation, see help rte.
INSTALLATION FILES                                  (type net install rte_stat)
      rte.ado
      rte.sthlp
      nullset.ado
      random.ado
--------------------------------------------------------------------------------
```

If you decide that the addition might prove useful, `net` makes the installation easy:

```
. net install rte_stat
checking rte_stat consistency and verifying not already installed...
installing into c:\ado\plus\ ...
installation complete.
```

The `ado` command helps you manage packages installed with `net`. Perhaps you remember that you installed a package that calculates the robust-to-everything statistic, but you cannot remember the command's name. You could use `ado` to search what you have previously installed for the `rte` command,

```
. ado
[1] package sg145 from http://www.stata.com/stb/stb56
      STB-56 sg145. Scalar measures of fit for regression models.
  (output omitted)
[15] package rte_stat from http://www.wemakeitupaswego.edu/faculty/sgazer
      rte_stat.  The robust-to-everything statistic; update.
  (output omitted)
[21] package st0119 from http://www.stata-journal.com/software/sj7-1
      SJ7-1 st0119.  Rasch analysis
```

or you might type

```
. ado, find("robust-to-everything")
[15] package rte_stat from http://www.wemakeitupaswego.edu/faculty/sgazer
      rte_stat.  The robust-to-everything statistic; update.
```

Perhaps you decide that `rte`, despite the author's claims, is not worth the disk space it occupies. You can use `ado` to erase it:

```
. ado uninstall rte_stat
package rte_stat from http://www.wemakeitupaswego.edu/faculty/sgazer
      rte_stat.  The robust-to-everything statistic; update.
(package uninstalled)
```

ado uninstall is easier than erasing the files by hand because ado uninstall erases every file associated with the package, and, moreover, ado knows where on your computer rte_stat is installed; you would have to hunt for these files.

Content pages

There are two types of pages displayed by net: content pages and package description pages. When you type net from, net cd, net link, or net without arguments, Stata goes to the specified place and displays the content page:

```
. net from http://www.stata.com
---------------------------------------------------------------------------
http://www.stata.com/
StataCorp
---------------------------------------------------------------------------

Welcome to StataCorp.

Below we provide links to sites providing additions to Stata, including
the Stata Journal, STB, and Statalist.  These are NOT THE OFFICIAL UPDATES;
you fetch and install the official updates by typing -update-.

PLACES you could -net link- to:
    sj                  The Stata Journal

DIRECTORIES you could -net cd- to:
    stb                 materials published in the Stata Technical Bulletin
    users               materials written by various people, including StataCorp
                        employees
    meetings            software packages from Stata Users Group meetings
    links               links to other locations providing additions to Stata
---------------------------------------------------------------------------
```

A content page tells you about other content pages and package description pages. The example above lists other content pages only. Below we follow one of the links for the *Stata Journal*:

```
. net link sj
---------------------------------------------------------------------------
http://www.stata-journal.com/
The Stata Journal
---------------------------------------------------------------------------

The Stata Journal is a refereed, quarterly journal containing articles
of interest to Stata users.  For more details and subscription information,
visit the Stata Journal web site at http://www.stata-journal.com.

PLACES you could -net link- to:
    stata               StataCorp web site

DIRECTORIES you could -net cd- to:
    production          Files for authors of the Stata Journal
    software            Software associated with Stata Journal articles
---------------------------------------------------------------------------
```

(Continued on next page)

```
. net cd software
```

```
http://www.stata-journal.com/software/
The Stata Journal

PLACES you could -net link- to:
    stata               StataCorp web site
    stb                 Stata Technical Bulletin (STB) software archive
DIRECTORIES you could -net cd- to:
  (output omitted)
    sj7-1               volume 7, issue 1
  (output omitted)
    sj1-1               volume 1, issue 1
```

```
. net cd sj7-1
```

```
http://www.stata-journal.com/software/sj7-1/
Stata Journal volume 7, issue 1

DIRECTORIES you could -net cd- to:
    ..                  Other Stata Journals
PACKAGES you could -net describe-:
    dm0027              File filtering in Stata: handling complex data
                        formats and navigating log files efficiently
    st0119              Rasch analysis
    st0120              Multivariable regression spline models
    st0121              mhbounds - Sensitivity Analysis for Average
                        Treatment Effects
```

dm0027, st0119, ..., st0121 are links to package description pages.

The links for the *Stata Technical Bulletin* (STB) follow steps similar to those in the *Stata Journal* example above.

```
. net link stb
```

```
http://www.stata.com/stb/
The Stata Technical Bulletin

PLACES you could -net link- to:
    stata               StataCorp web site
DIRECTORIES you could -net cd- to:
  (output omitted)
    stb54               STB-54, March     2000
  (output omitted)
```

```
. net cd stb54
```

```
http://www.stata.com/stb/stb54/
STB-54 March 2000

DIRECTORIES you could -net cd- to:
    ..                  Other STBs
PACKAGES you could -net describe-:
  (output omitted)
```

1. When you type `net from`, you follow that with a location to display the location's content page.

   a. The location could be a URL, such as http://www.stata.com. The content page at that location would then be listed.

   b. The location could be `e:` on a Windows computer or a mounted volume on a Macintosh computer. The content page on that source would be listed. That would work if you had special media obtained from StataCorp or special media prepared by another user.

   c. The location could even be a directory on your computer, but that would work only if that directory contained the right kind of files.

2. Once you have specified a location, typing `net cd` will take you into subdirectories of that location, if there are any. Typing

   ```
   . net from http://www.stata-journal.com
   . net cd software
   ```

   is equivalent to typing

   ```
   . net from http://www.stata-journal.com/software
   ```

   Typing `net cd` displays the content page from that location.

3. Typing `net` without arguments redisplays the current content page, which is the content page last displayed.

4. `net link` is similar to `net cd` in that the result is to change the location, but rather than changing to subdirectories of the current location, `net link` jumps to another location:

   ```
   . net from http://www.stata-journal.com
   ------------------------------------------------------------------------
   http://www.stata-journal.com/
   The Stata Journal
   ------------------------------------------------------------------------

   The Stata Journal is a refereed, quarterly journal containing articles
   of interest to Stata users.  For more details and subscription information,
   visit the Stata Journal web site at http://www.stata-journal.com.

   PLACES you could -net link- to:
       stata             StataCorp web site

   DIRECTORIES you could -net cd- to:
       production        Files for authors of the Stata Journal
       software          Software associated with Stata Journal articles
   ------------------------------------------------------------------------
   ```

   Typing `net link stata` would jump to *http://www.stata.com*:

   ```
   . net link stata
   ------------------------------------------------------------------------
   http://www.stata.com/
   StataCorp
   ------------------------------------------------------------------------

   Welcome to StataCorp.
     (output omitted)
   ------------------------------------------------------------------------
   ```

Package-description pages

Package-description pages describe what could be installed:

```
. net from http://www.stata-journal.com/software/sj7-1
```

```
http://www.stata-journal.com/software/sj7-1/
```
(output omitted)

```
. net describe st0119
```

```
package st0119 from http://www.stata-journal.com/software/sj7-1

TITLE
      SJ7-1 st0119.  Rasch analysis
DESCRIPTION/AUTHOR(S)
      Rasch analysis
      by Jean-Benoit Hardouin, University of Nantes, France
      Support:  jean-benoit.hardouin@univ-nantes.fr
      After installation, type help gammasym, gausshermite,
          geekel2d, raschtest, and raschtestv7
INSTALLATION FILES                              (type net install st0119)
      st0119/raschtest.ado
      st0119/raschtest.hlp
      st0119/raschtestv7.ado
      st0119/raschtestv7.hlp
      st0119/gammasym.ado
      st0119/gammasym.hlp
      st0119/gausshermite.ado
      st0119/gausshermite.hlp
      st0119/geekel2d.ado
      st0119/geekel2d.hlp
ANCILLARY FILES                                 (type net get st0119)
      st0119/data.dta
      st0119/outrasch.do
```

A package description page describes the package and tells you how to install the component files. Package-description pages potentially describe two types of files:

1. Installation files: files that you type `net install` to install and that are required to make the addition work.
2. Ancillary files: additional files you might want to install—you type `net get` to install them—but that you can ignore. Ancillary files are typically datasets that are useful for demonstration purposes. Ancillary files are not really installed in the sense of being copied to an official place for use by Stata itself. They are merely copied into the current directory so that you may use them if you wish.

You install the official files by typing `net install` followed by the package name. For example, to install `st0119`, you would type

```
. net install st0119
checking st0119 consistency and verifying not already installed...
installing into c:\ado\plus\ ...
installation complete.
```

You get the ancillary files—if there are any and if you want them—by typing `net get` followed by the package name:

```
. net get st0119
checking st0119 consistency and verifying not already installed...
copying into current directory...
      copying  data.dta
      copying  outrasch.do
ancillary files successfully copied.
```

Most users ignore the ancillary files.

Once you have installed a package—typed `net install`—use `ado` to redisplay the package description page whenever you wish:

```
. ado describe st0119
-------------------------------------------------------------------------------
[1] package st0119 from http://www.stata-journal.com/software/sj7-1
-------------------------------------------------------------------------------
TITLE
      SJ7-1 st0119.  Rasch analysis
DESCRIPTION/AUTHOR(S)
      Rasch analysis
      by Jean-Benoit Hardouin, University of Nantes, France
      Support:  jean-benoit.hardouin@univ-nantes.fr
      After installation, type help gammasym, gausshermite,
         geekel2d, raschtest, and raschtestv7
INSTALLATION FILES
      r/raschtest.ado
      r/raschtest.hlp
      r/raschtestv7.ado
      r/raschtestv7.hlp
      g/gammasym.ado
      g/gammasym.hlp
      g/gausshermite.ado
      g/gausshermite.hlp
      g/geekel2d.ado
      g/geekel2d.hlp
INSTALLED ON
      24 Apr 2007
-------------------------------------------------------------------------------
```

The package description page shown by `ado` includes the location from which we got the package and when we installed it. It does not mention the ancillary files that were originally part of this package because they are not tracked by `ado`.

Where packages are installed

Packages should be installed in `PLUS` or `SITE`, which are code words that Stata understands and that correspond to some real directories on your computer. Typing `sysdir` will tell you where these are, if you care.

```
. sysdir
   STATA:  C:\Program Files\Stata10\
 UPDATES:  C:\Program Files\Stata10\ado\updates\
    BASE:  C:\Program Files\Stata10\ado\base\
    SITE:  C:\Program Files\Stata10\ado\site\
    PLUS:  c:\ado\plus\
PERSONAL:  c:\ado\personal\
OLDPLACE:  c:\ado\
```

If you type `sysdir`, you may obtain different results.

By default, net installs in the PLUS directory, and ado tells you about what is installed there. If you are on a multiple-user system, you may wish to install some packages in the SITE directory. This way, they will be available to other Stata users. To do that, before using net install, type

```
. net set ado SITE
```

and when reviewing what is installed or removing packages, redirect ado to that directory:

```
. ado ..., from(SITE)
```

In both cases, you type SITE because Stata will understand that SITE means the site ado-directory as defined by sysdir. To install into SITE, you must have write access to that directory.

If you reset where net installs and then, in the same session, wish to install into your private ado-directory, type

```
. net set ado PLUS
```

That is how things were originally. If you are confused as to where you are, type net query.

A summary of the net command

The net command displays content pages and package description pages. Such pages are provided over the Internet, and most users get them there. We recommend that you start at *http://www.stata.com* and work out from there. We also recommend using net search to find packages of interest to you; see [R] **net search**.

net from moves you to a location and displays the content page.

net cd and net link change from your current location to other locations. net cd enters subdirectories of the original location. net link jumps from one location to another, depending on the code on the content page.

net describe lists a package description page. Packages are named, and you type net describe *pkgname*.

net install installs a package into your copy of Stata. net get copies any additional files (ancillary files) to your current directory.

net sj and net stb simplify loading files from the *Stata Journal* and its predecessor, the *Stata Technical Bulletin*.

net sj *vol-issue*

is a synonym for typing

net from http://www.stata-journal.com/software/sj*vol-issue*

whereas

net sj *vol-issue insert*

is a synonym for typing

net from http://www.stata-journal.com/software/sj*vol-issue*
net describe *insert*

net set controls where net installs files. By default, net installs in the PLUS directory; see [P] **sysdir**. net set ado SITE would cause subsequent net commands to install in the SITE directory. net set other sets where ancillary files, such as .dta files, are installed. The default is the current directory.

net query displays the current net from, net set ado, and net set other settings.

A summary of the ado command

The ado command lists the package descriptions of previously installed packages.

Typing ado without arguments is the same as typing ado dir. Both list the names and titles of the packages you have installed.

ado describe lists full package description pages.

ado uninstall removes packages from your computer.

Since you can install packages from a variety of sources, the package names may not always be unique. Thus the packages installed on your computer are numbered sequentially, and you may refer to them by name or by number. For instance, say that you wanted to get rid of the robust-to-everything statistic command you installed. Type

```
. ado, find("robust-to-everything")
[15] package rte_stat from http://www.wemakeitupaswego.edu/faculty/sgazer
     rte_stat.  The robust-to-everything statistic; update.
```

You could then type

```
. ado uninstall rte_stat
```

or

```
. ado uninstall [15]
```

Typing ado uninstall rte_stat would work only if the name rte_stat were unique; otherwise, ado would refuse, and you would have to type the number.

The find() option is allowed with ado dir and ado describe. It searches the package description for the word or phrase you specify, ignoring case (alpha matches Alpha). The complete package description is searched, including the author's name and the name of the files. Thus if rte was the name of a command that you wanted to eliminate, but you could not remember the name of the package, you could type

```
. ado, find(rte)
[15] package rte_stat from http://www.wemakeitupaswego.edu/faculty/sgazer
     rte_stat.  The robust-to-everything statistic; update.
```

Relationship of net and ado to the point-and-click interface

Users may instead select **Help > SJ and User-written Programs**. There are advantages and disadvantages:

1. Flipping through content and package description pages is easier; it is much like a browser. See chapter 20 in the *Getting Started* manual.
2. When browsing a product-description page, note that the .sthlp files are highlighted. You may click on .sthlp files to review them before installing the package.
3. You may not redirect from where ado searches for files.

Creating your own site

The rest of this entry concerns how to create your own site to distribute additions to Stata. The idea is that you have written additions for use with Stata—say, `xyz.ado` and `xyz.sthlp`—and you wish to put them out so that coworkers or researchers at other institutions can easily install them. Or, perhaps you just have a dataset that you and others want to share.

In any case, all you need is a homepage. You place the files that you want to distribute on your homepage (or in a subdirectory), and you add two more files—a content file and a package description file—and you are done.

Format of content and package description files

The content file describes the content page. It must be named `stata.toc`:

——— top of `stata.toc` ———

```
OFF                                    (to make site unavailable temporarily)
* lines starting with * are comments; they are ignored
* blank lines are ignored, too
* v indicates version—specify v 3; old-style toc files do not have this
v 3
* d lines display description text
* the first d line is the title, and the remaining ones are text
* blank d lines display a blank line
d title
d text
d text
d
...
* l lines display links
l word-to-show path-or-url [description]
l word-to-show path-or-url [description]
...
* t lines display other directories within the site
t path [description]
t path [description]
...
* p lines display packages
p pkgname [description]
p pkgname [description]
...
```

——— end of `stata.toc` ———

Package files describe packages and are named *pkgname*`.pkg`:

——— top of *pkgname*.pkg ———

```
* lines starting with * are comments; they are ignored
* blank lines are ignored, too
* v indicates version—specify v 3; old-style pkg files do not have this
v 3
* d lines display package description text
* the first d line is the title, and the remaining ones are text
* blank d lines display a blank line
d title
d text
d Distribution-Date: date
d text
d
...
* f identifies the component files
f [path/]filename [description]
f [path/]filename [description]
...
* e line is optional; it means stop reading
e
```

——— end of *pkgname*.pkg ———

Note the Distribution-Date description line. This line is optional, but recommended. Stata can look for updates to user-written programs with the `adoupdate` command if the package files from which those programs were installed contain a Distribution-Date description line.

Example 1

Say that we want the user to see the following:

```
. net from http://www.university.edu/~me
------------------------------------------------------------------------
http://www.university.edu/~me
Chris Farrar, Uni University
------------------------------------------------------------------------
PACKAGES you could -net describe-:
    xyz               interval-truncated survival

. net describe xyz
------------------------------------------------------------------------
package xyz from http://www.university.edu/~me
------------------------------------------------------------------------
TITLE
      xyz.  interval-truncated survival.

DESCRIPTION/AUTHOR(S)
      C. Farrar, Uni University.

INSTALLATION FILES                              (type net install xyz)
      xyz.ado
      xyz.sthlp
ANCILLARY FILES                                 (type net get xyz)
      sample.dta
------------------------------------------------------------------------
```

The files needed to do this would be

——— top of stata.toc ———

```
v 3
d Chris Farrar, Uni University
p xyz interval-truncated survival
```

——— end of stata.toc ———

top of xyz.pkg

```
v 3
d xyz.  interval-truncated survival.
d C. Farrar, Uni University.
f xyz.ado
f xyz.sthlp
f sample.dta
```

end of xyz.pkg

On his homepage, Chris would place the following files:

stata.toc	(shown above)
xyz.pkg	(shown above)
xyz.ado	file to be delivered (for use by net install)
xyz.sthlp	file to be delivered (for use by net install)
sample.dta	file to be delivered (for use by net get)

Chris does nothing to distinguish ancillary files from installation files.

Example 2

S. Gazer wants to create a more complex site:

```
. net from http://www.wemakeitupaswego.edu/faculty/sgazer
```

```
http://www.wemakeitupaswego.edu/faculty/sgazer
Data-free inference materials

S. Gazer, Department of Applied Theoretical Mathematics
Also see my homepage for the preprint of "Irrefutable inference".
PLACES you could -net link- to:
    stata             StataCorp web site
DIRECTORIES you could -net cd- to:
    ir                irrefutable inference programs (work in progress)
PACKAGES you could -net describe-:
    rtec              Robust-to-everything-conceivable statistic
    rte               Robust-to-everything statistic
```

```
. net describe rte
----------------------------------------------------------------------
package rte from http://www.wemakeitupaswego.edu/faculty/sgazer/
----------------------------------------------------------------------

TITLE
      rte.  The robust-to-everything statistic; update.

DESCRIPTION/AUTHOR(S)
      S. Gazer, Dept. of Applied Theoretical Mathematics, WMIUAWG Univ.
      Aleph-0 100% confidence intervals proved too conservative for some
      applications; Aleph-1 confidence intervals have been substituted.
      The new robust-to-everything supplants the previous robust-to-
      everything-conceivable statistic.  See "Inference in the absence
      of data" (forthcoming).  After installation, see help rte.

      Distribution-Date: 20070420

      Support:  email sgazer@wemakeitupaswego.edu

INSTALLATION FILES                            (type net install rte_stat)
      rte.ado
      rte.sthlp
      nullset.ado
      random.ado

ANCILLARY FILES                               (type net get rte_stat)
      empty.dta
----------------------------------------------------------------------
```

The files needed to do this would be

```
---------------------------------------------------- top of stata.toc ----
v 3
d Data-free inference materials
d S. Gazer, Department of Applied Theoretical Mathematics
d
d Also see my homepage for the preprint of "Irrefutable inference".
l stata http://www.stata.com
t ir irrefutable inference programs (work in progress)
p rtec Robust-to-everything-conceivable statistic
p rte  Robust-to-everything statistic
---------------------------------------------------- end of stata.toc ----
```

```
------------------------------------------------------ top of rte.pkg ----
v 3
d rte.  The robust-to-everything statistic; update.
d {bf:S. Gazer, Dept. of Applied Theoretical Mathematics, WMIUAWG Univ.}
d Aleph-0 100% confidence intervals proved too conservative for some
d applications; Aleph-1 confidence intervals have been substituted.
d The new robust-to-everything supplants the previous robust-to-
d everything-conceivable statistic.  See "Inference in the absence
d of data" (forthcoming).  After installation, see help {bf:rte}.
d
d Distribution-Date: 20070420
d
d Support:  email sgazer@wemakeitupaswego.edu
f rte.ado
f rte.sthlp
f nullset.ado
f random.ado
f empty.dta
------------------------------------------------------ end of rte.pkg ----
```

On his homepage, Mr. Gazer would place the following files:

`stata.toc`	(shown above)
`rte.pkg`	(shown above)
`rte.ado`	(file to be delivered)
`rte.sthlp`	(file to be delivered)
`nullset.ado`	(file to be delivered)
`random.ado`	(file to be delivered)
`empty.dta`	(file to be delivered)
`rtec.pkg`	the other package referred to in `stata.toc`
`rtec.ado`	the corresponding files to be delivered
`rtec.sthlp`	
`ir/stata.toc`	the contents file for when the user types `net cd ir`
`ir/...`	whatever other `.pkg` files are referred to
`ir/...`	whatever other files are to be delivered

If Mr. Gazer later updated the `rte` package, he could change the Distribution-Date description line in his package. Then, if someone who had previously installed the `rte` packaged wanted to obtain the latest version, he could use the `adoupdate` command. See [R] **adoupdate**.

For complex sites, a different structure may prove more convenient:

`stata.toc`	(shown above)
`rte.pkg`	(shown above)
`rtec.pkg`	the other package referred to in `stata.toc`
`rte/`	directory containing rte files to be delivered:
`rte/rte.ado`	(file to be delivered)
`rte/rte.sthlp`	(file to be delivered)
`rte/nullset.ado`	(file to be delivered)
`rte/random.ado`	(file to be delivered)
`rte/empty.dta`	(file to be delivered)
`rtec/`	directory containing rtec files to be delivered:
`rtec/...`	(files to be delivered)
`ir/stata.toc`	the contents file for when the user types `net cd ir`
`ir/*.pkg`	whatever other package files are referred to
`ir/*/...`	whatever other files are to be delivered

If you prefer this structure, it is simply a matter of changing the bottom of the `rte.pkg` from

```
f rte.ado
f rte.sthlp
f nullset.ado
f random.ado
f empty.dta
```

to

```
f rte/rte.ado
f rte/rte.sthlp
f rte/nullset.ado
f rte/random.ado
f rte/empty.dta
```

In writing paths and files, the directory separator forward slash (/) is used, regardless of operating system, because this is what the Internet uses.

It does not matter whether the files you put out are in DOS/Windows, Macintosh, or Unix format (how lines end is recorded differently). When Stata reads the files over the Internet, it will figure out the file format on its own and will automatically translate the files to what is appropriate for the receiver.

Additional package directives

F *filename* is similar to f *filename*, except that, when the file is installed, it will always be copied to the system directories (and not the current directory).

With f *filename*, the file is installed into a directory according to the file's suffix. For instance, xyz.ado would be installed in the system directories, whereas xyz.dta would be installed in the current directory.

Coding F xyz.ado would have the same result as coding f xyz.ado.

Coding F xyz.dta, however, would state that xyz.dta is to be installed in the system directories.

g *platformname filename* is also a variation on f *filename*. It specifies that the file be installed only if the user's operating system is of type *platformname*; otherwise, the file is ignored. The platform names are WIN, WIN64A (64-bit x86-64), and WIN64I (64-bit Itanium) for Windows; MAC (32-bit PowerPC), MACINTEL (32-bit Intel), OSX.PPC (32-bit PowerPC), and OSX.X86 (32-bit Intel), for Macintosh; and AIX, HP64I, LINUX (32-bit x86), LINUX64 (64-bit x86-64), LINUX64I (64-bit Itanium), SOL64, and SOLX8664 (64-bit x86-64) for Unix.

G *platformname filename* is a variation on F *filename*. The file, if not ignored, is to be installed in the system directories.

g *platformname filename1 filename2* is a more detailed version of g *platformname filename*. In this case, *filename1* is the name of the file on the server (the file to be copied), and *filename2* is to be the name of the file on the user's system; e.g., you might code

```
g WIN mydll.forwin mydll.plugin
g LINUX mydll.forlinux mydll.plugin
```

When you specify one *filename*, the result is the same as specifying two identical *filenames*.

G *platformname filename1 filename2* is the install-in-system-directories version of g *platformname filename1 filename2*

h *filename* asserts that *filename* must be loaded, or this package is not to be installed; e.g., you might code

```
g WIN mydll.forwin mydll.plugin
g LINUX mydll.forlinux mydll.plugin
h mydll.plugin
```

if you were offering the plugin mydll.plugin for Windows and Linux only.

SMCL in content and package description files

The text listed on the second and subsequent d lines in both stata.toc and *pkgname*.pkg may contain SMCL as long as you include v 3; see [P] **smcl**.

Thus in rte.pkg, S. Gazer coded the third line as

```
d {bf:S. Gazer, Dept. of Applied Theoretical Mathematics, WMIUAWG Univ.}
```

Error-free file delivery

Most people transport files over the Internet and never worry about the file being corrupted in the process because corruption rarely occurs. If, however, the files must be delivered perfectly or not at all, you can include checksum files in the directory.

For instance, say that `big.dta` is included in your package and that it must be sent perfectly. First, use Stata to make the checksum file for `big.dta`

```
. checksum big.dta, save
```

That command creates a small file called `big.sum`; see [D] **checksum**. Then copy both `big.dta` and `big.sum` to your homepage. If `set checksum` is `on` (the default is `off`), whenever Stata reads *filename*`.`*whatever* over the net, it also looks for *filename*`.sum`. If it finds such a file, it uses the information recorded in it to verify that what was copied was error free.

If you do this, be cautious. If you put `big.dta` and `big.sum` on your homepage and then later change `big.dta` without changing `big.sum`, people will think that there are transmission errors when they try to download `big.dta`.

References

Baum, C. F., and N. J. Cox. 1999. ip29: Metadata for user-written contributions to the Stata programming language. *Stata Technical Bulletin* 52: 10–12. Reprinted in *Stata Technical Bulletin Reprints*, vol. 9, pp. 121–124.

Cox, N. J., and C. F. Baum. 2000. ip29.1: Metadata for user-written contributions to the Stata programming language: Extensions. *Stata Technical Bulletin* 54: 21–22. Reprinted in *Stata Technical Bulletin Reprints*, vol. 9, pp. 124–126.

Also See

[R] **adoupdate** — Update user-written ado-files

[R] **net search** — Search the Internet for installable packages

[R] **search** — Search Stata documentation

[R] **sj** — Stata Journal and STB installation instructions

[R] **ssc** — Install and uninstall packages from SSC

[D] **checksum** — Calculate checksum of file

[P] **smcl** — Stata Markup and Control Language

[R] **update** — Update Stata

[GSM] **20 Updating and extending Stata—Internet functionality**

[GSU] **20 Updating and extending Stata—Internet functionality**

[GSW] **20 Updating and extending Stata—Internet functionality**

[U] **28 Using the Internet to keep up to date**

Title

net search — Search the Internet for installable packages

Syntax

`net search` *word* [*word* ...] [, *options*]

options	description
`or`	list packages that contain any of the keywords; default is all
`nosj`	search non-SJ and non-STB sources
`tocpkg`	search both tables of contents and packages; the default
`toc`	search tables of contents only
`pkg`	search packages only
`everywhere`	search packages for match
`filenames`	search filenames associated with package for match
`errnone`	make return code 111 instead of 0 when no matches found

Description

`net search` searches the Internet for user-written additions to Stata, including, but not limited to, user-written additions published in the *Stata Journal* (SJ) and in the *Stata Technical Bulletin* (STB). `net search` lists the available additions that contain the specified keywords.

The user-written materials found are available for immediate download by using the `net` command or by clicking on the link.

In addition to typing `net search`, you may select **Help > Search...** and choose **Search net resources**. This is the recommended way to search for user-written additions to Stata.

Options

`or` is relevant only when multiple keywords are specified. By default, `net search` lists only packages that include all the keywords. `or` changes the command to list packages that contain any of the keywords.

`nosj` specifies that `net search` not list matches that were published in the SJ or in the STB.

`tocpkg`, `toc`, and `pkg` determine what is searched. `tocpkg` is the default, meaning that both tables of contents (tocs) and packages (pkgs) are searched. `toc` restricts the search to tables of contents. `pkg` restricts the search to packages.

`everywhere` and `filenames` determine where in packages `net search` looks for *keywords*. The default is `everywhere`. `filenames` restricts `net search` to search for matches only in the filenames associated with a package. Specifying `everywhere` implies `pkg`.

`errnone` is a programmer's option that causes the return code to be 111 instead of 0 when no matches are found.

Remarks

net search searches the Internet for user-written additions to Stata. If you want to search the Stata documentation for a particular topic, command, or author, see [R] **search**. net search *word* [*word* ...] (without options) is equivalent to typing search *word* [*word* ...], net.

Remarks are presented under the following headings:

Topic searches
Author searches
Command searches
Where does net search look?
How does net search work?

Topic searches

Example: find what is available about random effects

```
. net search random effect
```

Comments:

1. It is best to search using the singular form of a word. net search random effect will find both "random effect" and "random effects".
2. net search random effect will also find "random-effect" because net search performs a string search and not a word search.
3. net search random effect lists all packages containing the words "random" and "effect", not necessarily used together.
4. If you wanted all packages containing the word "random" or the word "effect", you would type net search random effect, or.

Author searches

Example: find what is available by author Jeroen Weesie

```
. net search weesie
```

Comments:

1. You could type net search jeroen weesie, but that might list fewer results because sometimes the last name is used without the first.
2. You could type net search Weesie, but it would not matter. Capitalization is ignored in the search.

Example: find what is available by Jeroen Weesie, excluding SJ and STB materials

```
. net search weesie, nosj
```

1. The SJ and the STB tend to dominate search results because so much has been published in them. If you know that what you are looking for is not in the SJ or in the STB, specifying the nosj option will narrow the search.
2. net search weesie lists everything that net search weesie, nosj lists, and more. If you just type net search weesie, look down the list. SJ and STB materials are listed first, and non-SJ and non-STB materials are listed last.

Command searches

Example: find the user-written command kursus

```
. net search kursus, file
```

1. You could just type `net search kursus`, and that will list everything `net search kursus, file` lists, and more. Since you know `kursus` is a command, however, there must be a `kursus.ado` file associated with the package. Typing `net search kursus, file` narrows the search.
2. You could also type `net search kursus.ado, file` to narrow the search even more.

Where does net search look?

`net search` looks everywhere, not just at http://www.stata.com.

`net search` begins by looking at http://www.stata.com, but then follows every link, which takes it to other places, and then follows every link again, which takes it to even more places, and so on.

Authors: Please let us know if you have a site that we should include in our search by sending an email to webmaster@stata.com. We will then link to your site from ours to ensure that `net search` finds your materials. That is not strictly necessary, however, as long as your site is directly or indirectly linked from some site that is linked to ours.

How does net search work?

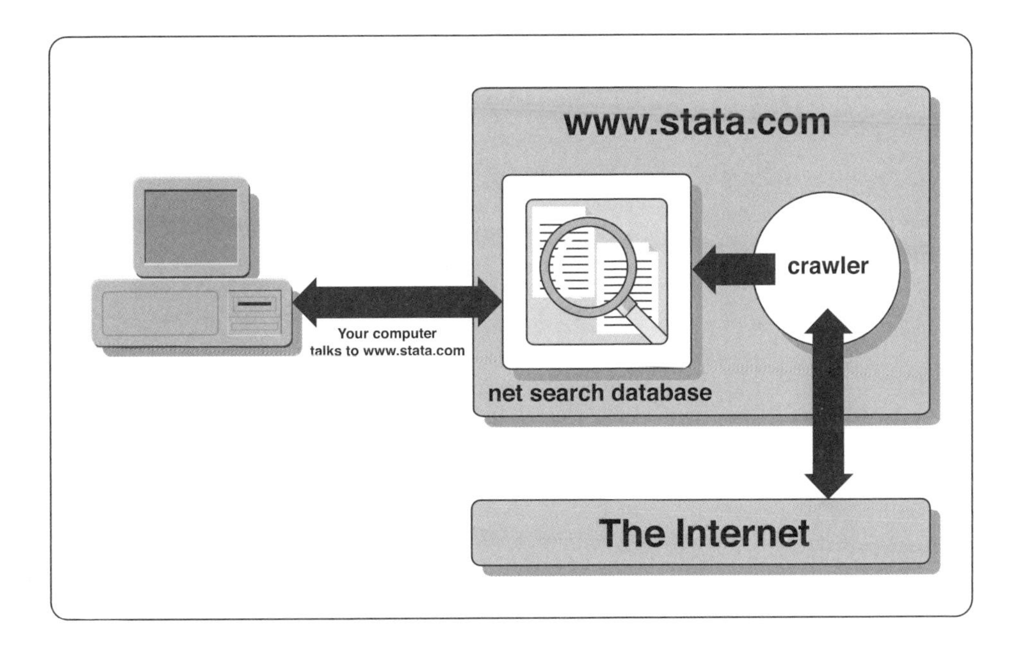

http://www.stata.com maintains a database of Stata resources. When you use `net search`, it contacts http://www.stata.com with your request, http://www.stata.com searches its database, and Stata returns the results to you.

Another part of the system is called the crawler, which searches the web for new Stata resources to add to the `net search` database and verifies that the resources already found are still available. When a new resource becomes available, the crawler takes about 2 days to add it to the database, and, similarly, if a resource disappears, the crawler takes roughly 2 days to remove it from the database.

References

Baum, C. F., and N. J. Cox. 1999. ip29: Metadata for user-written contributions to the Stata programming language. *Stata Technical Bulletin* 52: 10–12. Reprinted in *Stata Technical Bulletin Reprints*, vol. 9, pp. 121–124.

Cox, N. J., and C. F. Baum. 2000. ip29.1: Metadata for user-written contributions to the Stata programming language: Extensions. *Stata Technical Bulletin* 54: 21–22. Reprinted in *Stata Technical Bulletin Reprints*, vol. 9, pp. 124–126.

Gould, W. W., and A. R. Riley. 2000. stata55: Search web for installable packages. *Stata Technical Bulletin* 54: 4–6. Reprinted in *Stata Technical Bulletin Reprints*, vol. 9, pp. 10–13.

Also See

[R] **net** — Install and manage user-written additions from the Internet

[R] **ssc** — Install and uninstall packages from SSC

[R] **sj** — Stata Journal and STB installation instructions

[R] **hsearch** — Search help files

[R] **search** — Search Stata documentation

[R] **adoupdate** — Update user-written ado-files

[R] **update** — Update Stata

Title

netio — Control Internet connections

Syntax

Turn on or off the use of a proxy server

set httpproxy {on | off} [, init]

Set proxy host name

set httpproxyhost ["]*name*["]

Set the proxy port number

set httpproxyport #

Turn on or off proxy authorization

set httpproxyauth {on | off}

Set proxy authorization user ID

set httpproxyuser ["]*name*["]

Set proxy authorization password

set httpproxypw ["]*password*["]

Set time limit for establishing initial connection

set timeout1 *#seconds* [, permanently]

Set time limit for data transfer

set timeout2 *#seconds* [, permanently]

Description

Several commands (e.g., `net`, `news`, and `update`) are designed specifically for use over the Internet. Many other Stata commands that read a file (e.g., `copy`, `type`, and `use`) can also read directly from a URL. All these commands will usually work without your ever needing to concern yourself with the `set` commands discussed here. These `set` commands provide control over network system parameters.

If you experience problems when using Stata's network features, ask your system administrator if your site uses a proxy. A proxy is a server between your computer and the rest of the Internet, and your computer may need to communicate with other computers on the Internet through this proxy. If your site uses a proxy, your system administrator can provide you with its host name and the port your computer can use to communicate with it. If your site's proxy requires you to log in to it before it will respond, your system administrator will provide you with a user ID and password.

set httpproxyhost sets the name of the host to be used as a proxy server. set httpproxyport sets the port number. set httpproxy turns on or off the use of a proxy server, leaving the proxy host name and port intact, even when not in use.

Under the Windows and Macintosh operating systems, when you set httpproxy on, Stata will attempt to obtain the values of httpproxyhost and httpproxyport from the operating system if they have not been previously set. set httpproxy on, init attempts to obtain these values from the operating system, even if they have been previously set.

If the proxy requires authorization (user ID and password), set authorization on via set httpproxyauth on. The proxy user and proxy password must also be set to the appropriate user ID and password by using set httpproxyuser and set httpproxypw.

Stata remembers the various proxy settings between sessions and does not need a permanently option.

set timeout1 changes the time limit in seconds that Stata imposes for establishing the initial connection with a remote host. set timeout2 changes the time limit in seconds that Stata imposes for subsequent data transfer with the host. If these time limits are exceeded, a "connection timed out" message and error code 2 are produced. You should seldom need to change these settings.

Options

init specifies that set httpproxy on attempts to initialize httpproxyhost and httpproxyport from the operating system (Windows and Macintosh only).

permanently specifies that, in addition to making the change right now, the timeout1 and timeout2 settings be remembered and become the default setting when you invoke Stata.

The various httpproxy settings do not have a permanently option because permanently is implied.

Remarks

If you receive an error message, see http://www.stata.com/support/faqs/web/ for the latest information.

1. remote connection failed r(677);

If you see

```
remote connection failed
r(677);
```

then you asked for something to be done over the web, and Stata tried but could not contact the specified host. Stata was able to talk over the network and look up the host but was not able to establish a connection to that host.

Perhaps the host is down; try again later.

If all your web accesses result in this message, then perhaps your network connection is through a proxy server. If it is, then you must tell Stata.

Contact your system administrator. Ask for the name and port of the "http proxy server". Say that you are told

> http proxy server: jupiter.myuni.edu
> port number: 8080

In Stata, type

```
. set httpproxyhost jupiter.myuni.edu
. set httpproxyport 8080
. set httpproxy on
```

Your web accesses should then work.

2. connection timed out r(2);

If you see

```
connection timed out
r(2);
```

then an Internet connection has timed out. This can happen when

a. the connection between you and the host is slow, or

b. the connection between you and the host has disappeared, and so it eventually "timed out".

For (b), wait a while (say, 5 minutes) and try again (sometimes pieces of the Internet can break for up to a day, but that is rare). For (a), you can reset the limits for what constitutes "timed out". There are two numbers to set.

The time to establish the initial connection is **timeout1**. By default, Stata waits 120 seconds before declaring a timeout. You can change the limit:

```
. set timeout1 #seconds
```

You might try doubling the usual limit and specify 240; *#seconds* must be between 1 and 32,000.

The time to retrieve data from an open connection is **timeout2**. By default, Stata waits 300 seconds (5 minutes) before declaring a timeout. To change the limit, type

```
. set timeout2 #seconds
```

You might try doubling the usual limit and specify 600; *#seconds* must be between 1 and 32,000.

Also See

[R] **query** — Display system parameters

[P] **creturn** — Return c-class values

[U] **28 Using the Internet to keep up to date**

Title

news — Report Stata news

Syntax

`news`

Description

`news` displays a brief listing of recent Stata news and information, which it obtains from Stata's web site. `news` requires that your computer be connected to the Internet.

You may also execute `news` by selecting **Help > News**.

Remarks

`news` provides an easy way of displaying a brief list of the latest Stata news:

```
. news

  ___  ____  ____  ____  ____
 /__    /   ____/   /   ____/
___/   /   /___/   /   /___/   News         The latest from http://www.stata.com
---------------------------------------------------------------------------------

8 October 2007.  Official update available for download
    Click here (equivalent to pulling down Help and selecting
    Official Updates) or type update from http://www.stata.com.
28 July 2007.  Stata 10 available
    Stata 10 -- interactive graph editor -- time, date, and date-and-time
    variables -- exact logistic -- exact Poisson -- nested, hierarchical, and
    dynamic panel models -- more survey -- discriminant analysis -- MCA  --
    modern MDS -- power & sample-size for survival studies -- is now available.
    Click here or visit http://www.stata.com for more information.
15 June 2007.  NetCourse schedule updated
    See http://www.stata.com/netcourse/ for more information.
05 February 2007.  New book available from Stata Press
     (output omitted)
<end>
```

Also See

[U] **28 Using the Internet to keep up to date**

Title

nl — Nonlinear least-squares estimation

Syntax

Interactive version

`nl` (*depvar* = <*sexp*>) [*if*] [*in*] [*weight*] [, *options*]

Programmed substitutable expression version

`nl` *sexp_prog* : *depvar* [*varlist*] [*if*] [*in*] [*weight*] [, *options*]

Function evaluator program version

`nl` *func_prog* `@` *depvar* [*varlist*] [*if*] [*in*] [*weight*] ,

{ `parameters(`*namelist*`)` | `nparameters(`*#*`)`} [*options*]

where

depvar is the dependent variable;

<*sexp*> is a substitutable expression;

sexp_prog is a substitutable expression program; and

func_prog is a function evaluator program.

(*Continued on next page*)

options	description
Model	
variables(*varlist*)	variables in model
initial(*initial_values*)	initial values for parameters
* parameters(*namelist*)	parameters in model (function evaluator program version only)
* nparameters(#)	number of parameters in model (function evaluator program version only)
sexp_options	options for substitutable expression program
func_options	options for function evaluator program
Model 2	
lnlsq(#)	use log least-squares where ln(*depvar* − #) is assumed to be normally distributed
noconstant	the model has no constant term; seldom used
hasconstant(*name*)	use *name* as constant term; seldom used
SE/Robust	
vce(*vcetype*)	*vcetype* may be gnr, robust, cluster *clustvar*, bootstrap, jackknife, hac *kernel*, hc2, or hc3
Reporting	
level(#)	set confidence level; default is level(95)
leave	create variables containing derivative of $E(y)$
title(*string*)	display *string* as title above the table of parameter estimates
title2(*string*)	display *string* as subtitle
Opt options	
optimization_options	control the optimization process; seldom used
eps(#)	specify # for convergence criterion; default is eps(1e-5)
delta(#)	specify # for computing derivatives; default is delta(4e-7)

* For function evaluator program version, you must specify parameters(*namelist*) or nparameters(#), or both.

bootstrap, by, jackknife, rolling, statsby, and svy are allowed; see [U] **11.1.10 Prefix commands**.

Weights are not allowed with the bootstrap prefix.

aweights are not allowed with the jackknife prefix.

vce(), leave, and weights are not allowed with the svy prefix.

aweights, fweights, and iweights are allowed; see [U] **11.1.6 weight**.

See [U] **20 Estimation and postestimation commands** for more capabilities of estimation commands.

Description

nl fits an arbitrary nonlinear regression function by least squares. With the interactive version of the command, you enter the function directly on the command line or dialog box by using a *substitutable expression*. If you have a function that you use regularly, you can write a *substitutable expression program* and use the second syntax to avoid having to reenter the function every time. The function evaluator program version gives you the most flexibility in exchange for increased complexity; with this version, your program is given a vector of parameters and a variable list, and your program computes the regression function.

When you write a substitutable expression program or function evaluator program, the first two letters of the name must be `nl`. *sexp_prog* and *func_prog* refer to the name of the program without the first two letters. For example, if you wrote a function evaluator program named `nlregss`, you would type `nl regss @` ... to estimate the parameters.

Options

Model

`variables(`*varlist*`)` specifies the variables in the model. `nl` ignores observations for which any of these variables have missing values. If you do not specify `variables()`, `nl` issues an error message with return code 480 if the estimation sample contains any missing values.

`initial(`*initial_values*`)` specifies the initial values to begin the estimation. You can specify a $1 \times k$ matrix, where k is the number of parameters in the model, or you can specify a parameter name, its initial value, another parameter name, its initial value, and so on. For example, to initialize `alpha` to 1.23 and `delta` to 4.57, you would type

```
nl ... , initial(alpha 1.23 delta 4.57) ...
```

Initial values declared using this option override any that are declared within substitutable expressions. If you specify a parameter that does not appear in your model, `nl` exits with error code 480. If you specify a matrix, the values must be in the same order that the parameters are declared in your model. `nl` ignores the row and column names of the matrix.

`parameters(`*namelist*`)` specifies the names of the parameters in the model. The names of the parameters must adhere to the naming conventions of Stata's variables; see [U] **11.3 Naming conventions**. If you specify both `parameters()` and `nparameters()`, the number of names in the former must match the number specified in the latter; if not, `nl` issues an error message with return code 198.

`nparameters(`*#*`)` specifies the number of parameters in the model. If you do not specify names with the `parameters()` option, `nl` names them `b1`, `b2`, ..., `b`*#*. If you specify both `parameters()` and `nparameters()`, the number of names in the former must match the number specified in the latter; if not, `nl` issues an error message with return code 198.

sexp_options refer to any options allowed by your *sexp_prog*.

func_options refer to any options allowed by your *func_prog*.

Model 2

`lnlsq(`*#*`)` fits the model by using log least-squares, which we define as least squares with shifted lognormal errors. In other words, $\ln(depvar - \#)$ is assumed to be normally distributed. Sums of squares and deviance are adjusted to the same scale as *depvar*.

`noconstant` indicates that the function does not include a constant term. This option is generally not needed, even if there is no constant term in the model, unless the coefficient of variation (over observations) of the partial derivative of the function with respect to a parameter is less than `eps()` and that parameter is not a constant term.

`hasconstant(`*name*`)` indicates that parameter *name* be treated as the constant term in the model and that `nl` should not use its default algorithm to find a constant term. As with `noconstant`, this option is seldom used.

SE/Robust

`vce(`*vcetype*`)` specifies the type of standard error reported, which includes types that are derived from asymptotic theory, that are robust to some kinds of misspecification, that allow for intragroup correlation, and that use bootstrap or jackknife methods; see [R] ***vce_option***.

`vce(gnr)`, the default, uses the conventionally derived variance estimator for nonlinear models fit using Gauss–Newton regression.

`nl` also allows the following:

`vce(hac` *kernel* [#]`)` specifies that a heteroskedasticity- and autocorrelation-consistent (HAC) variance estimate be used. HAC refers to the general form for combining weighted matrices to form the variance estimate. There are three kernels available for `nl`:

`nwest` | `gallant` | `anderson`

specifies the number of lags. If # is not specified, $N - 2$ is assumed.

`vce(hac` *kernel* [#]`)` is not allowed if weights are specified.

`vce(hc2)` and `vce(hc3)` specify alternative bias corrections for the robust variance calculation. `vce(hc2)` and `vce(hc3)` may not be specified with the `svy` prefix. By default, `vce(robust)` uses $\widehat{\sigma}_j^2 = \{n/(n-k)\}u_j^2$ as an estimate of the variance of the jth observation, where u_j is the calculated residual and $n/(n-k)$ is included to improve the overall estimate's small-sample properties.

`vce(hc2)` instead uses $u_j^2/(1 - h_{jj})$ as the observation's variance estimate, where h_{jj} is the jth diagonal element of the hat (projection) matrix. This produces an unbiased estimate of the covariance matrix if the model is homoskedastic. `vce(hc2)` tends to produce slightly more conservative confidence intervals than `vce(robust)`.

`vce(hc3)` uses $u_j^2/(1 - h_{jj})^2$ as suggested by Davidson and MacKinnon (1993 and 2004), who report that this often produces better results when the model is heteroskedastic. `vce(hc3)` produces confidence intervals that tend to be even more conservative.

See, in particular, Davidson and MacKinnon (2004, 239), who advocate the use of `vce(hc2)` or `vce(hc3)` instead of the plain robust estimator for nonlinear least squares.

Reporting

`level(`*#*`)`; see [R] **estimation options**.

`leave` leaves behind after estimation a set of new variables with the same names as the estimated parameters containing the derivatives of $E(y)$ with respect to the parameters. If the dataset contains an existing variable with the same name as a parameter, then using `leave` causes `nl` to issue an error message with return code 110.

`leave` may not be specified with `vce(cluster` *clustvar*`)` or the `svy` prefix.

`title(`*string*`)` specifies an optional title that will be displayed just above the table of parameter estimates.

`title2(`*string*`)` specifies an optional subtitle that will be displayed between the title specified in `title()` and the table of parameter estimates. If `title2()` is specified but `title()` is not, `title2()` has the same effect as `title()`.

Opt options

optimization_options: iterate(#), [no]log, trace. iterate() specifies the maximum number of iterations, log/nolog specifies whether to show the iteration log, and trace specifies that the iteration log should include the current parameter vector. These options are seldom used.

eps(#) specifies the convergence criterion for successive parameter estimates and for the residual sum of squares. The default is eps(1e-5).

delta(#) specifies the relative change in a parameter to be used in computing the numeric derivatives. The derivative for parameter β_i is computed as $\{f(X, \beta_1, \beta_2, \ldots, \beta_i + d, \beta_{i+1}, \ldots) - f(X, \beta_1, \beta_2, \ldots, \beta_i, \beta_{i+1}, \ldots)\}/d$, where d is $\delta(\beta_i + \delta)$. The default is delta(4e-7).

Remarks

Remarks are presented under the following headings:

Substitutable expressions
Substitutable expression programs
Built-in functions
Log-normal errors
Other uses
Weights
Potential errors
General comments on fitting nonlinear models
Function evaluator programs

nl fits an arbitrary nonlinear function by least squares. The interactive version allows you to enter the function directly on the command line or dialog box using *substitutable expressions*. You can write a *substitutable expression program* for functions that you fit frequently to save yourself time. Finally, *function evaluator programs* give you the most flexibility in defining your nonlinear function, though they are more complicated to use.

The next section explains the substitutable expressions that are used to define the regression function, and the section thereafter explains how to write substitutable expression program files so that you do not need to type in commonly used functions over and over. Later sections highlight other features of nl.

The final section discusses function evaluator programs. If you find substitutable expressions adequate to define your nonlinear function, then you can skip that section entirely. Function evaluator programs are generally needed only for complicated problems, such as multistep estimators. The program receives a vector of parameters at which it is to compute the function and a variable into which the results are to be placed.

Substitutable expressions

You define the nonlinear function to be fitted by nl using a substitutable expression. Substitutable expressions are just like any other mathematical expressions involving scalars and variables, such as those you would use with Stata's generate command, except that the parameters to be estimated are bound in braces. See [U] **13.2 Operators** and [U] **13.3 Functions** for more information on expressions.

For example, suppose that you wish to fit the function

$$y_i = \beta_0(1 - e^{-\beta_1 x_i}) + \epsilon_i$$

where β_0 and β_1 are the parameters to be estimated and ϵ_i is an error term. You would simply type

```
. nl (y = {b0}*(1 - exp(-1*{b1}*x)))
```

You must enclose the entire equation in parentheses. Because b0 and b1 are enclosed in braces, nl knows that they are parameters in the model. nl will initialize b0 and b1 to zero by default. To request that nl initialize b0 to 1 and b1 to 0.25, you would type

```
. nl (y = {b0=1}*(1 - exp(-1*{b1=0.25}*x)))
```

That is, inside the braces denoting a parameter, you put the parameter name followed by an equals sign and the initial value. If a parameter appears in your function multiple times, you need only specify an initial value only once (or never, if you wish to set the initial value to zero). If you do specify more than one initial value for the same parameter, nl will use the *last* value given. Parameter names must follow the same conventions as variable names. See [U] **11.3 Naming conventions**.

Frequently, even nonlinear functions contain linear combinations of variables. As an example, suppose that you wish to fit the function

$$y_i = \beta_0 \left\{1 - e^{-(\beta_1 x_{1i} + \beta_2 x_{2i} + \beta_3 x_{3i})}\right\} + \epsilon_i$$

nl allows you to declare a linear combination of variables by using the shorthand notation

```
. nl (y = {b0=1}*(1 - exp(-1*{xb: x1 x2 x3})))
```

In the syntax {xb: x1 x2 x3}, you are telling nl that you are declaring a linear combination named xb that is a function of three variables, x1, x2, and x3. nl will create three parameters, named xb_x1, xb_x2, and xb_x3, and initialize them to zero. Instead of typing the previous command, you could have typed

```
. nl (y = {b0=1}*(1 - exp(-1*({xb_x1}*x1 + {xb_x2}*x2 + {xb_x3}*x3))))
```

and yielded the same result. You can refer to the parameters created by nl in the linear combination later in the function, though you must declare the linear combination first if you intend to do that. When creating linear combinations, nl ensures that the parameter names it chooses are unique and have not yet been used in the function.

In general, there are three rules to follow when defining substitutable expressions:

1. Parameters of the model are bound in braces: {b0}, {param}, etc.
2. Initial values for parameters are given by including an equal sign and the initial value inside the braces: {b0=1}, {param=3.571}, etc.
3. Linear combinations of variables can be included using the notation {*eqname*:*varlist*}: {xb: mpg price weight}, {score: w x z}, etc. Parameters of linear combinations are initialized to zero.

If you specify initial values by using the initial() option, they override whatever initial values are given within the substitutable expression. Substitutable expressions are so named because, once values are assigned to the parameters, the resulting expression can be handled by generate and replace.

▷ Example 1

We wish to fit the CES production function

$$\ln Q_i = \beta_0 - \frac{1}{\rho} \ln\left\{\delta K_i^{-\rho} + (1-\delta) L_i^{-\rho}\right\} + \epsilon_i \tag{1}$$

where $\ln Q_i$ is the log of output for firm i; K_i and L_i are firm i's capital and labor usage, respectively; and ϵ_i is a regression error term. Because ρ appears in the denominator of a fraction, zero is not a feasible initial value; for a CES production function, $\rho = 1$ is a reasonable choice. Setting $\delta = 0.5$ implies that labor and capital have equal impacts on output, which is also a reasonable choice for an initial value. We type

```
. use http://www.stata-press.com/data/r10/production
. nl (lnoutput = {b0} - 1/{rho=3}*ln({delta=0.5}*capital^(-1*{rho}) +
> (1 - {delta})*labor^(-1*{rho})))
(obs = 100)
Iteration 0:  residual SS =   30.4558
Iteration 1:  residual SS =  29.37008
Iteration 2:  residual SS =  29.36595
Iteration 3:  residual SS =  29.36581
Iteration 4:  residual SS =  29.36581
Iteration 5:  residual SS =  29.36581
Iteration 6:  residual SS =  29.36581
Iteration 7:  residual SS =  29.36581

      Source |       SS       df       MS
-------------+------------------------------         Number of obs =       100
       Model |  91.1449924     2  45.5724962         R-squared     =    0.7563
    Residual |  29.3658055    97  .302740263         Adj R-squared =    0.7513
-------------+------------------------------         Root MSE      =  .5502184
       Total |  120.510798    99  1.21728079         Res. dev.     =  161.2538

------------------------------------------------------------------------------
    lnoutput |      Coef.   Std. Err.      t    P>|t|     [95% Conf. Interval]
-------------+----------------------------------------------------------------
         /b0 |   3.792157   .0996818    38.04   0.000     3.594317    3.989998
        /rho |   1.386991   .4725791     2.93   0.004     .4490524     2.32493
      /delta |   .4823616   .0519789     9.28   0.000     .3791978    .5855253
------------------------------------------------------------------------------
  Parameter b0 taken as constant term in model & ANOVA table
```

`nl` will attempt to find a constant term in the model, and if one is found, mention it at the bottom of the output. `nl` found `b0` to be a constant because the partial derivative $\partial \ln Q_i / \partial \text{b0}$ has a coefficient of variation less than `eps()` in the estimation sample.

The elasticity of substitution for the CES production function is $\sigma = 1/(1+\rho)$; and, having fitted the model, we can use `nlcom` to estimate it:

```
. nlcom (1/(1 + _b[/rho]))
       _nl_1:  1/(1 + _b[/rho])

------------------------------------------------------------------------------
    lnoutput |      Coef.   Std. Err.      t    P>|t|     [95% Conf. Interval]
-------------+----------------------------------------------------------------
       _nl_1 |   .4189374   .0829417     5.05   0.000     .2543211    .5835538
------------------------------------------------------------------------------
```

See [R] **nlcom** and [U] **13.5 Accessing coefficients and standard errors** for more information.

◁

`nl`'s output closely mimics that of `regress`; see [R] **regress** for more information. The R^2, sums of squares, and similar statistics are calculated in the same way that `regress` calculates them. If no "constant" term is specified, the usual caveats apply to the interpretation of the R^2 statistic; see the comments and references in Goldstein (1992). Unlike `regress`, `nl` does not report a model F statistic, because a test of the joint significance of all the parameters except the constant term may not be relevant in a nonlinear model.

Substitutable expression programs

If you fit the same model often or if you want to write an estimator that will operate on whatever variables you specify, then you will want to write a substitutable expression program. That program will return a macro containing a substitutable expression that `nl` can then evaluate, and it may optionally calculate initial values as well. The name of the program must begin with the letters `nl`.

To illustrate, suppose that you use the CES production function often in your work. Instead of typing in the formula each time, you can write a program like this:

```
program nlces, rclass
        version 10
        syntax varlist(min=3 max=3) if
        local logout : word 1 of `varlist'
        local capital : word 2 of `varlist'
        local labor : word 3 of `varlist'
        // Initial value for b0 given delta=0.5 and rho=1
        tempvar y
        generate double `y' = `logout' + ln(0.5*`capital'^-1 + 0.5*`labor'^-1)
        summarize `y' `if', meanonly
        local b0val = r(mean)
        // Terms for substitutable expression
        local capterm "{delta=0.5}*`capital'^(-1*{rho})"
        local labterm "(1-{delta})*`labor'^(-1*{rho})"
        local term2   "1/{rho=1}*ln(`capterm' + `labterm')"
        // Return substitutable expression and title
        return local eq "`logout' = {b0=`b0val'} - `term2'"
        return local title "CES ftn., ln Q=`logout', K=`capital', L=`labor'"
end
```

The program accepts three variables for log output, capital, and labor, and it accepts an `if` *exp* qualifier to restrict the estimation sample. All programs that you write to use with `nl` must accept an `if` *exp* qualifier because, when `nl` calls the program, it passes a binary variable that marks the estimation sample (the variable equals one if the observation is in the sample and zero otherwise). When calculating initial values, you will want to restrict your computations to the estimation sample, and you can do so by using `if` with any commands that accept `if` *exp* qualifiers. Even if your program does not calculate initial values or otherwise use the `if` qualifier, the `syntax` statement must still allow it. See [P] **syntax** for more information on the `syntax` command and the use of `if`.

As in the previous example, reasonable initial values for δ and ρ are 0.5 and 1, respectively. Conditional on those values, (1) can be rewritten as

$$\beta_0 = \ln Q_i + \ln(0.5K_i^{-1} + 0.5L_i^{-1}) - \epsilon_i \tag{2}$$

so a good initial value for β_0 is the mean of the right-hand side of (2) ignoring ϵ_i. Lines 7–10 of the function evaluator program calculate that mean and store it in a local macro. Notice the use of `if` in the `summarize` statement so that the mean is calculated only for the estimation sample.

The final part of the program returns two macros. The macro `title` is optional and defines a short description of the model that will be displayed in the output immediately above the table of parameter estimates. The macro `eq` is required and defines the substitutable expression that `nl` will use. If the expression is short, you can define it all at once. However, because the expression used here is somewhat lengthy, defining local macros and then building up the final expression from them is easier.

To verify that there are no errors in your program, you can call it directly and then use `return list`:

```
. use http://www.stata-press.com/data/r10/production, clear
. nlces lnoutput capital labor
. return list
macros:
             r(title) : "CES ftn., ln Q=lnoutput, K=capital, L=labor"
                r(eq) : "lnoutput = {b0=3.711606264663641} - 1/{rho=1}*ln({delt
> a=0.5}*capital^(-1*{rho}) + (1-{delta})*labor^(-1*{rho}))"
```

The macro `r(eq)` contains the same substitutable expression that we specified at the command line in the preceding example, except for the initial value for `b0`. In short, an `nl` substitutable expression program should return in `r(eq)` the same substitutable expression you would type at the command line. The only difference is that when writing a substitutable expression program, you do not bind the entire expression inside parentheses.

Having written the program, you can use it by typing

```
. nl ces: lnoutput capital labor
```

(There is a space between `nl` and `ces`.) The output is identical to that shown in example 1, save for the title defined in the function evaluator program that appears immediately above the table of parameter estimates.

❑ Technical Note

You will want to store `nlces` as an ado-file called `nlces.ado`. The alternative is to type the code into Stata interactively or to place the code in a do-file. While those alternatives are adequate for occasional use, if you save the program as an ado-file, you can use the function anytime you use Stata without having to redefine the program. When `nl` attempts to execute `nlces`, if the program is not in Stata's memory, Stata will search the disk(s) for an ado-file of the same name, and, if found, automatically load it. All you have to do is name the file with the `.ado` suffix and then place it in a directory where Stata will find it. You should put the file in the directory Stata reserves for user-written ado-files, which, depending on your operating system, is `c:\ado\personal` (Windows), `~/ado/personal` (Unix), or `~:ado:personal` (Macintosh). See [U] **17 Ado-files**. ❑

Sometimes you may want to pass additional options to the substitutable expression program. You can modify the `syntax` statement of your program to accept whatever options you wish. Then when you call `nl` with the syntax

. nl *func_prog*: *varlist*, *options*

any *options* that are not recognized by `nl` (see the table of options at the beginning of this entry) are passed on to your function evaluator program. The only other restriction is that your program cannot accept an option named `at` because `nl` uses that option with function evaluator programs.

Built-in functions

Some functions are used so often that `nl` has them built in so that you do not need to write them yourself. `nl` automatically chooses initial values for the parameters, though you can use the `initial(...)` option to override them.

Three alternatives are provided for exponential regression with one asymptote:

exp3	$y_i = \beta_0 + \beta_1\beta_2^{x_i} + \epsilon_i$
exp2	$y_i = \beta_1\beta_2^{x_i} + \epsilon_i$
exp2a	$y_i = \beta_1\big(1 - \beta_2^{x_i}\big) + \epsilon_i$

For instance, typing `nl exp3: ras dvl` fits the three-parameter exponential model (parameters β_0, β_1, and β_2) using $y_i =$ `ras` and $x_i =$ `dvl`.

Two alternatives are provided for the logistic function (symmetric sigmoid shape; not to be confused with logistic regression):

log4	$y_i = \beta_0 + \beta_1 \Big/ \Big[1 + \exp\big\{-\beta_2(x_i - \beta_3)\big\}\Big] + \epsilon_i$
log3	$y_i = \beta_1 \Big/ \Big[1 + \exp\big\{-\beta_2(x_i - \beta_3)\big\}\Big] + \epsilon_i$

Finally, two alternatives are provided for the Gompertz function (asymmetric sigmoid shape):

gom4	$y_i = \beta_0 + \beta_1 \exp\Big[-\exp\big\{-\beta_2(x_i - \beta_3)\big\}\Big] + \epsilon_i$
gom3	$y_i = \beta_1 \exp\Big[-\exp\big\{-\beta_2(x_i - \beta_3)\big\}\Big] + \epsilon_i$

Log-normal errors

A nonlinear model with errors that are independently and identically distributed normal may be written

$$y_i = f(\mathbf{x}_i, \boldsymbol{\beta}) + u_i, \qquad u_i \sim \mathrm{N}(0, \sigma^2) \tag{3}$$

for $i = 1, \dots, n$. If the y_i are thought to have a k-shifted log-normal instead of a normal distribution—that is, $\ln(y_i - k) \sim \mathrm{N}(\zeta_i, \tau^2)$, and the systematic part $f(\mathbf{x}_i, \boldsymbol{\beta})$ of the original model is still thought appropriate for y_i—the model becomes

$$\ln(y_i - k) = \zeta_i + v_i = \ln\big\{f(\mathbf{x}_i, \boldsymbol{\beta}) - k\big\} + v_i, \quad v_i \sim \mathrm{N}(0, \tau^2) \tag{4}$$

This model is fitted if `lnlsq(`k`)` is specified.

If model (4) is correct, the variance of $(y_i - k)$ is proportional to $\big\{f(\mathbf{x}_i, \boldsymbol{\beta}) - k\big\}^2$. Probably the most common case is $k = 0$, sometimes called "proportional errors" since the standard error of y_i is proportional to its expectation, $f(\mathbf{x}_i, \boldsymbol{\beta})$. Assuming that the value of k is known, (4) is just another nonlinear model in $\boldsymbol{\beta}$, and it may be fitted as usual. However, we may wish to compare the fit of (3) with that of (4) using the residual sum of squares or the deviance D, $D = -2\times$ log-likelihood, from each model. To do so, we must allow for the change in scale introduced by the log transformation.

Assuming, then, the y_i to be normally distributed, Atkinson (1985, 85–87, 184), by considering the Jacobian $\prod |\partial \ln(y_i - k)/\partial y_i|$, showed that multiplying both sides of (4) by the geometric mean of $y_i - k$, $\dot{y}$, gives residuals on the same scale as those of y_i. The geometric mean is given by

$$\dot{y} = e^{n^{-1}\sum \ln(y_i - k)}$$

which is a constant for a given dataset. The residual deviance for (3) and for (4) may be expressed as

$$D(\widehat{\boldsymbol{\beta}}) = \Big\{1 + \ln(2\pi\widehat{\sigma}^2)\Big\}n \tag{5}$$

where $\widehat{\beta}$ is the maximum likelihood estimate (MLE) of β for each model and $n\widehat{\sigma}^2$ is the RSS from (3), or that from (4) multiplied by $\dot{y}^2$.

Since (3) and (4) are models with different error structures but the same functional form, the arithmetic difference in their RSS or deviances is not easily tested for statistical significance. However, if the deviance difference is large (>4, say), we would naturally prefer the model with the smaller deviance. Of course, the residuals for each model should be examined for departures from assumptions (nonconstant variance, nonnormality, serial correlations, etc.) in the usual way.

Alternatively, consider modeling

$$E(y_i) = 1/(C + Ae^{Bx_i}) \tag{6}$$

$$E(1/y_i) = E(y_i') = C + Ae^{Bx_i} \tag{7}$$

where C, A, and B are parameters to be estimated. Using the data $(y, x) = (.04, 5)$, $(.06, 12)$, $(.08, 25)$, $(.1, 35)$, $(.15, 42)$, $(.2, 48)$, $(.25, 60)$, $(.3, 75)$, and $(.5, 120)$ (Danuso 1991), fitting the models yields

Model	C	A	B	RSS	Deviance
(6)	1.781	25.74	−.03926	−.001640	−51.95
(6) with `lnlsq(0)`	1.799	25.45	−.04051	−.001431	−53.18
(7)	1.781	25.74	−.03926	8.197	24.70
(7) with `lnlsq(0)`	1.799	27.45	−.04051	3.651	17.42

There is little to choose between the two versions of the logistic model (6), whereas for the exponential model (7), the fit using `lnlsq(0)` is much better (a deviance difference of 7.28). The reciprocal transformation has introduced heteroskedasticity into y_i', which is countered by the proportional errors property of the lognormal distribution implicit in `lnlsq(0)`. The deviances are not comparable between the logistic and exponential models because the change of scale has not been allowed for, although in principle it could be.

Other uses

Even if you are fitting linear regression models, you may find that `nl` can save you some typing. Since you specify the parameters of your model explicitly, you can impose constraints on them directly.

▷ Example 2

In [R] **cnsreg**, we showed how to fit the model

$$\texttt{mpg} = \beta_0 + \beta_1\texttt{price} + \beta_2\texttt{weight} + \beta_3\texttt{displ} + \beta_4\texttt{gear\_ratio} + \beta_5\texttt{foreign} + \beta_6\texttt{length} + u$$

subject to the constraints

$$\beta_1 = \beta_2 = \beta_3 = \beta_6$$
$$\beta_4 = -\beta_5 = \beta_0/20$$

An alternative way is to use nl:

```
. use http://www.stata-press.com/data/r10/auto, clear
(1978 Automobile Data)

. nl (mpg = {b0} + {b1}*price + {b1}*weight + {b1}*displ +
> {b0}/20*gear_ratio - {b0}/20*foreign + {b1}*length)
(obs = 74)

Iteration 0:  residual SS =  1578.522
Iteration 1:  residual SS =  1578.522

      Source |       SS       df       MS
-------------+------------------------------         Number of obs =        74
       Model |  34429.4777     2  17214.7389         R-squared     =    0.9562
    Residual |  1578.52226    72  21.9239203         Adj R-squared =    0.9549
-------------+------------------------------         Root MSE      =  4.682299
       Total |       36008    74  486.594595         Res. dev.     =  436.4562

------------------------------------------------------------------------------
         mpg |      Coef.   Std. Err.      t    P>|t|     [95% Conf. Interval]
-------------+----------------------------------------------------------------
          b0 |   26.52229   1.375178    19.29   0.000     23.78092    29.26365
          b1 |   -.000923   .0001534    -6.02   0.000    -.0012288   -.0006172
------------------------------------------------------------------------------
```

The point estimates and standard errors for β_0 and β_1 are identical to those reported in [R] **cnsreg**. To get the estimate for β_4, we can use nlcom:

```
. nlcom _b[/b0]/20

       _nl_1:  _b[b0]/20

------------------------------------------------------------------------------
         mpg |      Coef.   Std. Err.      t    P>|t|     [95% Conf. Interval]
-------------+----------------------------------------------------------------
       _nl_1 |   1.326114   .0687589    19.29   0.000     1.189046    1.463183
------------------------------------------------------------------------------
```

The advantage to using nl is that we do not need to use the constraint command six times.

◁

nl is also a useful tool when doing exploratory data analysis. For example, you may want to run a regression of y on a function of x, though you have not decided whether to use sqrt(x) or ln(x). You can use nl to run both regressions without having first to generate two new variables:

```
. nl (y = {b0} + {b1}*ln(x))
. nl (y = {b0} + {b1}*sqrt(x))
```

Weights

Weights are specified the usual way—analytic and frequency weights as well as iweights are supported; see [U] **20.17 Weighted estimation**. Use of analytic weights implies that the y_i have different variances. Therefore, model (3) may be rewritten as

$$y_i = f(\mathbf{x}_i, \boldsymbol{\beta}) + u_i, \qquad u_i \sim \mathrm{N}(0, \sigma^2/w_i) \tag{3a}$$

where w_i are (positive) weights, assumed to be known and normalized such that their sum equals the number of observations. The residual deviance for (3a) is

$$D(\widehat{\boldsymbol{\beta}}) = \left\{1 + \ln(2\pi\widehat{\sigma}^2)\right\}n - \sum \ln(w_i) \tag{5a}$$

(compare with (5)), where

$$n\widehat{\sigma}^2 = \text{RSS} = \sum w_i \{y_i - f(\mathbf{x}_i, \widehat{\boldsymbol{\beta}})\}^2$$

Defining and fitting a model equivalent to (4) when weights have been specified as in (3a) is not straightforward and has not been attempted. Thus deviances using and not using the `lnlsq()` option may not be strictly comparable when analytic weights (other than 0 and 1) are used.

You do not need to modify your substitutable expression in any way to use weights. If, however, you write a substitutable expression program, then you should account for weights when obtaining initial values. When `nl` calls your program, it passes whatever weight expression (if any) was specified by the user. Here is an outline of a substitutable expression program that accepts weights:

```
program nlname, rclass
        version 10
        syntax varlist [aw fw iw] if
        ...
        // Obtain initial values allowing weights
        // Use the syntax [`weight'`exp'].  For example,
        summarize varname [`weight'`exp'] `if'
        regress depvar varlist [`weight'`exp'] `if'
        ...
        // Return substitutable expression
        return local eq "substitutable expression"
        return local title "description of estimator"
end
```

For details on how the `syntax` command processes weight expressions, see [P] **syntax**.

Potential errors

`nl` is reasonably robust to the inability of your nonlinear function to be evaluated at some parameter values. `nl` does assume that your function can be evaluated at the initial values of the parameters. If your function cannot be evaluated at the initial values, an error message is issued with return code 480. Recall that if you do not specify an initial value for a parameter, then `nl` initializes it to zero. Many nonlinear functions cannot be evaluated when some parameters are zero, so in those cases specifying alternative initial values is crucial.

Thereafter, as `nl` changes the parameter values, it monitors your function for unexpected missing values. If these are detected, `nl` backs up. That is, `nl` finds a point between the previous, known-to-be-good parameter vector and the new, known-to-be-bad vector at which the function can be evaluated and continues its iterations from that point.

`nl` requires that once a parameter vector is found where the predictions can be calculated, small changes to the parameter vector be made to calculate numeric derivatives. If a boundary is encountered at this point, an error message is issued with return code 481.

When specifying `lnlsq()`, an attempt to take logarithms of $y_i - k$ when $y_i \leq k$ results in an error message with return code 482.

If `iterate()` iterations are performed and estimates still have not converged, results are presented with a warning, and the return code is set to 430.

If you use the programmed substitutable expression version of `nl` with a function evaluator program, or vice versa, Stata issues an error message. Verify that you are using the syntax appropriate for the program you have.

General comments on fitting nonlinear models

Achieving convergence is often problematic. For example, a unique minimum of the sum-of-squares function may not exist. Much literature exists on different algorithms that have been used, on strategies for obtaining good initial parameter values, and on tricks for parameterizing the model to make its behavior as linear-like as possible. Selected references are Kennedy and Gentle (1980, chap. 10) for computational matters and Ross (1990) and Ratkowsky (1983) for all three aspects. Ratkowsky's book is particularly clear and approachable, with useful discussion on the meaning and practical implications of intrinsic and parameter-effects nonlinearity. An excellent text on nonlinear estimation is Gallant (1987). Also see Davidson and MacKinnon (1993 and 2004).

To enhance the success of `nl`, pay attention to the form of the model fitted, along the lines of Ratkowsky and Ross. For example, Ratkowsky (1983, 49–59) analyzes three possible three-parameter yield-density models for plant growth:

$$E(y_i) = \begin{cases} (\alpha + \beta x_i)^{-1/\theta} \\ (\alpha + \beta x_i + \gamma x_i^2)^{-1} \\ (\alpha + \beta x_i^{\phi})^{-1} \end{cases}$$

All three models give similar fits. However, he shows that the second formulation is dramatically more linear-like than the other two and therefore has better convergence properties. In addition, the parameter estimates are virtually unbiased and normally distributed, and the asymptotic approximation to the standard errors, correlations, and confidence intervals is much more accurate than for the other models. Even within a given model, the way the parameters are expressed (e.g., ϕ^{x_i} or $e^{\theta x_i}$) affects the degree of linearity and convergence behavior.

Function evaluator programs

Occasionally, a nonlinear function may be so complex that writing a substitutable expression for it is impractical. For example, there could be many parameters in the model. Alternatively, if you are implementing a two-step estimator, writing a substitutable expression may be altogether impossible. Function evaluator programs can be used in these situations.

`nl` will pass to your function evaluator program a list of variables, a weight expression, a variable marking the estimation sample, and a vector of parameters. Your program is to replace the dependent variable, which is the first variable in the variables list, with the values of the nonlinear function evaluated at those parameters. As with substitutable expression programs, the first two letters of the name must be `nl`.

To focus on the mechanics of the function evaluator program, again let's compare the CES production function to the previous examples. The function evaluator program is

```
program nlces2
        version 10
        syntax varlist(min=3 max=3) if, at(name)
        local logout : word 1 of `varlist'
        local capital : word 2 of `varlist'
        local labor : word 3 of `varlist'
        // Retrieve parameters out of at matrix
        tempname b0 rho delta
        scalar `b0' = `at'[1, 1]
        scalar `rho' = `at'[1, 2]
        scalar `delta' = `at'[1, 3]
        tempvar kterm lterm
        generate double `kterm' = `delta'*`capital'^(-1*`rho') `if'
        generate double `lterm' = (1-`delta')*`labor'^(-1*`rho') `if'
        // Fill in dependent variable
        replace `logout' = `b0' - 1/`rho'*ln(`kterm' + `lterm') `if'
end
```

Unlike the previous `nlces` program, this one is not declared to be r-class. The `syntax` statement again accepts three variables: one for log output, one for capital, and one for labor. An `if` *exp* is again required because `nl` will pass a binary variable marking the estimation sample. All function evaluator programs must accept an option named `at()` that takes a `name` as an argument—that is how `nl` passes the parameter vector to your program.

The next part of the program retrieves the output, labor, and capital variables from the variables list. It then breaks up the temporary matrix `at` and retrieves the parameters `b0`, `rho`, and `delta`. Pay careful attention to the order in which the parameters refer to the columns of the `at` matrix because that will affect the syntax you use with `nl`. The temporary names you use inside this program are immaterial, however.

The rest of the program computes the nonlinear function, using some temporary variables to hold intermediate results. The final line of the program then replaces the dependent variable with the values of the function. Notice the use of `` `if' `` to restrict attention to the estimation sample. `nl` makes a copy of your dependent variable so that when the command is finished your data are left unchanged.

To use the program and fit your model, you type

```
. use http://www.stata-press.com/data/r10/production, clear
. nl ces2 @ lnoutput capital labor, parameters(b0 rho delta)
> initial(b0 0 rho 1 delta 0.5)
```

The output is again identical to that shown in example 1. The order in which the parameters were specified in the `parameters()` option is the same in which they are retrieved from the `at` matrix in the program. To initialize them, you simply list the parameter name, a space, the initial value, and so on.

If you use the `nparameters()` option instead of the `parameters()` option, the parameters are named `b1`, `b2`, ..., `b`k, where k is the number of parameters. Thus you could have typed

```
. nl ces2 @ lnoutput capital labor, nparameters(3) initial(b1 0 b2 1 b3 0.5)
```

With that syntax, the parameters called `b0`, `rho`, and `delta` in the program will be labeled `b1`, `b2`, and `b3`, respectively. In programming situations or if there are many parameters, instead of listing the parameter names and initial values in the `initial()` option, you may find it more convenient to pass a column vector. In those cases, you could type

```
. matrix myvals = (0, 1, 0.5)
. nl ces2 @ lnoutput capital labor, nparameters(3) initial(myvals)
```

In summary, a function evaluator program receives a list of variables, the first of which is the dependent variable that you are to replace with the values of your nonlinear function. Additionally, it must accept an `if` *exp*, as well as an option named `at` that will contain the vector of parameters at which `nl` wants the function evaluated. You are then free to do whatever is necessary to evaluate your function and replace the dependent variable.

If you wish to use weights, your function evaluator program's `syntax` statement must accept them. If your program consists only of, for example, `generate` statements, you need not do anything with the weights passed to your program. However, if in calculating the nonlinear function you use commands such as `summarize` or `regress`, then you will want to use the weights with those commands.

As with substitutable expression programs, `nl` will pass to it any options specified that `nl` does not accept, providing you with a way to pass more information to your function.

❑ Technical Note

Before version 9 of Stata, the `nl` command used a different syntax, which required you to write a *nlfcn* program, and it did not have a syntax for interactive use other than the seven functions that were built-in. The old syntax of `nl` still works, and you can still use those *nlfcn* programs. If `nl` does not see a colon, at sign, or a set of parentheses surrounding the equation in your command, it assumes that the old syntax is being used.

The current version of `nl` uses scalars and matrices to store intermediate calculations instead of local and global macros as the old version did, so the current version produces more accurate results. In practice, however, any discrepancies are likely to be small.

❑

Saved Results

`nl` saves the following in `e()`:

Scalars

`e(N)`	number of observations	`e(cj)`	position of constant in `e(b)` or `0` if no constant
`e(k)`	number of parameters	`e(r2_a)`	adjusted R-squared
`e(k_eq_model)`	number of equations to use for model test; always `0`	`e(rmse)`	root mean squared error
`e(k_aux)`	number of estimated parameters	`e(converge)`	`1` if converged, `0` otherwise
`e(mss)`	model sum of squares	`e(df_t)`	total degrees of freedom
`e(tss)`	total sum of squares	`e(dev)`	residual deviance
`e(df_m)`	model degrees of freedom	`e(N_clust)`	number of clusters
`e(rss)`	residual sum of squares	`e(ic)`	number of iterations
`e(df_r)`	residual degrees of freedom	`e(lnlsq)`	value of `lnlsq` if specified
`e(mms)`	model mean square	`e(log_t)`	`1` if `lnlsq` specified, `0` otherwise
`e(msr)`	residual mean square	`e(gm_2)`	square of geometric mean of $(y-k)$ if `lnlsq`; `1` otherwise
`e(ll)`	log likelihood assuming i.i.d. normal errors	`e(delta)`	relative change used to compute derivatives
`e(r2)`	R-squared		

Macros

`e(cmd)`	`nl`	`e(sexp)`	substitutable expression
`e(cmdline)`	command as typed	`e(params)`	names of parameters
`e(depvar)`	name of dependent variable	`e(clustvar)`	name of cluster variable
`e(title)`	title in estimation output	`e(funcprog)`	function evaluator program
`e(title_2)`	secondary title in estimation output	`e(vce)`	*vcetype* specified in `vce()`
`e(wtype)`	weight type	`e(vcetype)`	title used to label Std. Err.
`e(wexp)`	weight expression	`e(predict)`	program used to implement `predict`
`e(properties)`	`b V`	`e(rhs)`	right-hand-side variables (function evaluator program version only)
`e(type)`	1 = interactively entered expression 2 = substitutable expression program 3 = function evaluator program		

Matrices

`e(b)`	coefficient vector	`e(V)`	variance–covariance matrix of the estimators
`e(init)`	initial values vector		

Functions

`e(sample)`	marks estimation sample

Methods and Formulas

`nl` is implemented as an ado-file.

The derivation here is based on Davidson and MacKinnon (2004, chap. 6). Let $\boldsymbol{\beta}$ denote the $k \times 1$ vector of parameters, and write the regression function using matrix notation as $\mathbf{y} = \mathbf{f}(\mathbf{x}, \boldsymbol{\beta}) + \mathbf{u}$ so that the objective function can be written as

$$\mathrm{SSR}(\boldsymbol{\beta}) = \{\mathbf{y} - \mathbf{f}(\mathbf{x}, \boldsymbol{\beta})\}' \mathbf{D} \{\mathbf{y} - \mathbf{f}(\mathbf{x}, \boldsymbol{\beta})\}$$

The $\mathbf{D}$ matrix contains the weights and is defined in [R] **regress**; if no weights are specified, then $\mathbf{D}$ is the $N \times N$ identity matrix. Taking a second-order Taylor series expansion centered at $\boldsymbol{\beta}_0$ yields

$$\mathrm{SSR}(\boldsymbol{\beta}) \approx \mathrm{SSR}(\boldsymbol{\beta}_0) + \mathbf{g}'(\boldsymbol{\beta}_0)(\boldsymbol{\beta} - \boldsymbol{\beta}_0) + \frac{1}{2}(\boldsymbol{\beta} - \boldsymbol{\beta}_0)'\mathbf{H}(\boldsymbol{\beta}_0)(\boldsymbol{\beta} - \boldsymbol{\beta}_0) \tag{8}$$

where $\mathbf{g}(\boldsymbol{\beta}_0)$ denotes the $k \times 1$ gradient of $\mathrm{SSR}(\boldsymbol{\beta})$ evaluated at $\boldsymbol{\beta}_0$ and $\mathbf{H}(\boldsymbol{\beta}_0)$ denotes the $k \times k$ Hessian of $\mathrm{SSR}(\boldsymbol{\beta})$ evaluated at $\boldsymbol{\beta}_0$. Letting $\mathbf{X}$ denote the $N \times k$ matrix of derivatives of $\mathbf{f}(\mathbf{x}, \boldsymbol{\beta})$ with respect to $\boldsymbol{\beta}$, the gradient $\mathbf{g}(\boldsymbol{\beta})$ is

$$\mathbf{g}(\boldsymbol{\beta}) = -2\mathbf{X}'\mathbf{D}\mathbf{u} \tag{9}$$

$\mathbf{X}$ and $\mathbf{u}$ are obviously functions of $\boldsymbol{\beta}$, though for notational simplicity that dependence is not shown explicitly. The (m, n) element of the Hessian can be written

$$H_{mn}(\boldsymbol{\beta}) = -2 \sum_{i=1}^{i=N} d_{ii} \left[\frac{\partial^2 f_i}{\partial \beta_m \partial \beta_n} u_i - X_{im} X_{in} \right] \tag{10}$$

where d_{ii} is the ith diagonal element of $\mathbf{D}$. As discussed in Davidson and MacKinnon (2004, chap. 6), the first term inside the brackets of (10) has expectation zero, so the Hessian can be approximated as

$$\mathbf{H}(\boldsymbol{\beta}) = 2\mathbf{X}'\mathbf{D}\mathbf{X} \tag{11}$$

Differentiating the Taylor series expansion of $\text{SSR}(\boldsymbol{\beta})$ shown in (8) yields the first-order condition for a minimum

$$\mathbf{g}(\boldsymbol{\beta}_0) + \mathbf{H}(\boldsymbol{\beta}_0)(\boldsymbol{\beta} - \boldsymbol{\beta}_0) = \mathbf{0}$$

which suggests the iterative procedure

$$\boldsymbol{\beta}_{j+1} = \boldsymbol{\beta}_j - \alpha\mathbf{H}^{-1}(\boldsymbol{\beta}_j)\mathbf{g}(\boldsymbol{\beta}_j) \tag{12}$$

where α is a "step size" parameter chosen at each iteration to improve convergence. Using (9) and (11), we can write (12)

$$\boldsymbol{\beta}_{j+1} = \boldsymbol{\beta}_j + \alpha(\mathbf{X}'\mathbf{D}\mathbf{X})^{-1}\mathbf{X}'\mathbf{D}\mathbf{u} \tag{13}$$

where $\mathbf{X}$ and $\mathbf{u}$ are evaluated at $\boldsymbol{\beta}_j$. Apart from the scalar α, the second term on the right-hand side of (13) can be computed via a (weighted) regression of the columns of $\mathbf{X}$ on the errors. `nl` computes the derivatives numerically and then calls `regress`. At each iteration, α is set to one, and a candidate value $\boldsymbol{\beta}^*_{j+1}$ is computed by (13). If $\text{SSR}(\boldsymbol{\beta}^*_{j+1}) < \text{SSR}(\boldsymbol{\beta}_j)$, then $\boldsymbol{\beta}_{j+1} = \boldsymbol{\beta}^*_{j+1}$ and the iteration is complete. Otherwise, α is halved, a new $\boldsymbol{\beta}^*_{j+1}$ is calculated, and the process is repeated. Convergence is declared when $\alpha|\beta_{j+1,m}| \le \epsilon(|\beta_{jm}| + \tau)$ for all $m = 1 \ldots k$. `nl` uses $\tau = 10^{-3}$ and, by default, $\epsilon = 10^{-5}$, though you can specify an alternative value of ϵ with the `eps()` option.

As derived, for example, in Davidson and MacKinnon (2004, chap. 6), an expedient way to obtain the covariance matrix is to compute $\mathbf{u}$ and the columns of $\mathbf{X}$ at the final estimate $\widehat{\boldsymbol{\beta}}$ and then regress that $\mathbf{u}$ on $\mathbf{X}$. The covariance matrix of the estimated parameters of that regression serves as an estimate of $\text{Var}(\widehat{\boldsymbol{\beta}})$. If that regression employs a robust covariance matrix estimator, then the covariance matrix for the parameters of the nonlinear regression will also be robust.

All other statistics are calculated analogously to those in linear regression, except that the nonlinear function $f(\mathbf{x}_i, \boldsymbol{\beta})$ plays the role of the linear function $\mathbf{x}'_i\boldsymbol{\beta}$. See [R] **regress**.

Acknowledgments

The original version of `nl` was written by Patrick Royston of the MRC Clinical Trials Unit, London, and published in Royston (1992). Francesco Danuso's menu-driven nonlinear regression program (1991) provided the inspiration.

References

Atkinson, A. C. 1985. *Plots, Transformations, and Regression: An Introduction to Graphical Methods of Diagnostic Regression Analysis*. Oxford: Oxford University Press.

Danuso, F. 1991. sg1: Nonlinear regression command. *Stata Technical Bulletin* 1: 17–19. Reprinted in *Stata Technical Bulletin Reprints*, vol. 1, pp. 96–98.

Davidson, R., and J. G. MacKinnon. 1993. *Estimation and Inference in Econometrics*. New York: Oxford University Press.

——. 2004. *Econometric Theory and Methods*. New York: Oxford University Press.

Gallant, A. R. 1987. *Nonlinear Statistical Models*. New York: Wiley.

Goldstein, R. 1992. srd7: Adjusted summary statistics for logarithmic regressions. *Stata Technical Bulletin* 5: 17–21. Reprinted in *Stata Technical Bulletin Reprints*, vol. 1, pp. 178–183.

Kennedy, W. J., Jr., and J. E. Gentle. 1980. *Statistical Computing*. New York: Dekker.

Ratkowsky, D. A. 1983. *Nonlinear Regression Modeling: A Unified Practical Approach*. New York: Dekker.

Ross, G. J. S. 1987. *MLP User Manual, release 3.08*. Oxford: Numerical Algorithms Group.

——. 1990. *Nonlinear Estimation*. New York: Springer.

Royston, P. 1992. sg1.2: Nonlinear regression command. *Stata Technical Bulletin* 7: 11–18. Reprinted in *Stata Technical Bulletin Reprints*, vol. 2, pp. 112–120.

——. 1993. sg1.4: Standard nonlinear curve fits. *Stata Technical Bulletin* 11: 17. Reprinted in *Stata Technical Bulletin Reprints*, vol. 2, p. 121.

Also See

[R] **nl postestimation** — Postestimation tools for nl

[R] **ml** — Maximum likelihood estimation

[R] **nlcom** — Nonlinear combinations of estimators

[R] **nlsur** — Estimation of nonlinear systems of equations

[R] **regress** — Linear regression

[SVY] **svy estimation** — Estimation commands for survey data

[U] **20 Estimation and postestimation commands**

Title

nl postestimation — Postestimation tools for nl

Description

The following postestimation commands are available for `nl`:

command	description
`estat`	AIC, BIC, VCE, and estimation sample summary
`estat` (svy)	postestimation statistics for survey data
`estimates`	cataloging estimation results
`lincom`	point estimates, standard errors, testing, and inference for linear combinations of coefficients
`lrtest`[1]	likelihood-ratio test
`mfx`	marginal effects or elasticities
`nlcom`	point estimates, standard errors, testing, and inference for nonlinear combinations of coefficients
`predict`	predictions and residuals
`predictnl`	point estimates, standard errors, testing, and inference for generalized predictions
`test`	Wald tests for simple and composite linear hypotheses
`testnl`	Wald tests of nonlinear hypotheses

[1] `lrtest` is not appropriate with `svy` estimation results.

See the corresponding entries in the *Stata Base Reference Manual* for details, but see [SVY] **estat** for details about `estat` (svy).

Syntax for predict

`predict` [*type*] *newvar* [*if*] [*in*] [, *statistic*]

`predict` [*type*] { *stub** | *newvar*$_1$ ... *newvar*$_k$ } [*if*] [*in*] , `scores`

where k is the number of parameters in the model.

statistic	description
Main	
`yhat`	fitted values; the default
`residuals`	residuals
`pr(`*a*`,`*b*`)`	$\Pr(y_j \mid a < y_j < b)$
`e(`*a*`,`*b*`)`	$E(y_j \mid a < y_j < b)$
`ystar(`*a*`,`*b*`)`	$E(y_j^*)$, $y_j^* = \max\{a, \min(y_j, b)\}$

These statistics are available both in and out of sample; type `predict ... if e(sample) ...` if wanted only for the estimation sample.

Options for predict

Main

`yhat`, the default, calculates the fitted values.

`residuals` calculates the residuals.

`pr(`*a*`,`*b*`)` calculates $\Pr(a < \mathbf{x}_j\mathbf{b} + u_j < b)$, the probability that $y_j|\mathbf{x}_j$ would be observed in the interval (a, b), assuming that the error term in the model is identically and independently normally distributed.

a and *b* may be specified as numbers or variable names; *lb* and *ub* are variable names;
`pr(20,30)` calculates $\Pr(20 < \mathbf{x}_j\mathbf{b} + u_j < 30)$;
`pr(`*lb*`,`*ub*`)` calculates $\Pr(lb < \mathbf{x}_j\mathbf{b} + u_j < ub)$; and
`pr(20,`*ub*`)` calculates $\Pr(20 < \mathbf{x}_j\mathbf{b} + u_j < ub)$.

a missing ($a \geq .$) means $-\infty$; `pr(.,30)` calculates $\Pr(-\infty < \mathbf{x}_j\mathbf{b} + u_j < 30)$;
`pr(`*lb*`,30)` calculates $\Pr(-\infty < \mathbf{x}_j\mathbf{b} + u_j < 30)$ in observations for which $lb \geq .$
and calculates $\Pr(lb < \mathbf{x}_j\mathbf{b} + u_j < 30)$ elsewhere.

b missing ($b \geq .$) means $+\infty$; `pr(20,.)` calculates $\Pr(+\infty > \mathbf{x}_j\mathbf{b} + u_j > 20)$;
`pr(20,`*ub*`)` calculates $\Pr(+\infty > \mathbf{x}_j\mathbf{b} + u_j > 20)$ in observations for which $ub \geq .$
and calculates $\Pr(20 < \mathbf{x}_j\mathbf{b} + u_j < ub)$ elsewhere.

`e(`*a*`,`*b*`)` calculates $E(\mathbf{x}_j\mathbf{b} + u_j \mid a < \mathbf{x}_j\mathbf{b} + u_j < b)$, the expected value of $y_j|\mathbf{x}_j$ conditional on $y_j|\mathbf{x}_j$ being in the interval (a, b), meaning that $y_j|\mathbf{x}_j$ is censored.
a and *b* are specified as they are for `pr()`.

`ystar(`*a*`,`*b*`)` calculates $E(y_j^*)$, where $y_j^* = a$ if $\mathbf{x}_j\mathbf{b} + u_j \leq a$, $y_j^* = b$ if $\mathbf{x}_j\mathbf{b} + u_j \geq b$, and $y_j^* = \mathbf{x}_j\mathbf{b} + u_j$ otherwise, meaning that y_j^* is truncated. *a* and *b* are specified as they are for `pr()`.

`scores` calculates the scores. The jth new variable created will contain the score for the jth parameter in `e(b)`.

Methods and Formulas

All postestimation commands listed above are implemented as ado-files.

Also See

[R] **nl** — Nonlinear least-squares estimation

[U] **20 Estimation and postestimation commands**

Title

nlcom — Nonlinear combinations of estimators

Syntax

Nonlinear combination of estimators—one expression

`nlcom` [*name*:]*exp* [, *options*]

Nonlinear combinations of estimators—more than one expression

`nlcom` ([*name*:]*exp*) [([*name*:]*exp*) ...] [, *options*]

The second syntax means that if more than one expression is specified, each must be surrounded by parentheses. *exp* is any function of the parameter estimates that is valid syntax for `testnl`; see [R] **testnl**. However, *exp* may not contain an equal sign or a comma. The optional *name* is any valid Stata name and labels the transformations.

options	description
`level(#)`	set confidence level; default is `level(95)`
`iterate(#)`	maximum number of iterations
`post`	post estimation results
†`noheader`	suppress output header

†`noheader` does not appear in the dialog box.

Description

`nlcom` computes point estimates, standard errors, test statistics, significance levels, and confidence intervals for (possibly) nonlinear combinations of parameter estimates after any Stata estimation command. Results are displayed in the usual table format used for displaying estimation results. Calculations are based on the "delta method", an approximation appropriate in large samples.

`nlcom` can be used with `svy` estimation results; see [SVY] **svy postestimation**.

Options

`level(#)` specifies the confidence level, as a percentage, for confidence intervals. The default is `level(95)` or as set by `set level`; see [U] **20.7 Specifying the width of confidence intervals**.

`iterate(#)` specifies the maximum number of iterations used to find the optimal step size in calculating numerical derivatives of the transformation(s) with respect to the original parameters. By default, the maximum number of iterations is 100, but convergence is usually achieved after only a few iterations. You should rarely have to use this option.

`post` causes `nlcom` to behave like a Stata estimation (`eclass`) command. When `post` is specified, `nlcom` will post the vector of transformed estimators and its estimated variance–covariance matrix to `e()`. This option, in essence, makes the transformation permanent. Thus you could, after `posting`, treat the transformed estimation results in the same way as you would treat results from other Stata estimation commands. For example, after posting, you could redisplay the results by typing `nlcom` without any arguments, or use `test` to perform simultaneous tests of hypotheses on linear combinations of the transformed estimators.

Specifying `post` clears out the previous estimation results, which can be recovered only by refitting the original model or by storing the estimation results before running `nlcom` and then restoring them; see [R] **estimates store**.

The following option is available with `nlcom` but is not shown in the dialog box:

`noheader` suppresses the output header.

Remarks

Remarks are presented under the following headings:

Introduction
Basics
Using the post option
Reparameterizing ML estimators for univariate data
nlcom versus eform

Introduction

`nlcom` and `predictnl` are Stata's delta method commands—they take nonlinear transformations of the estimated parameter vector from some fitted model and apply the delta method to calculate the variance, standard error, Wald test statistic, etc., of the transformations. `nlcom` is designed for functions of the parameters, and `predictnl` is designed for functions of the parameters and of the data, that is, for predictions.

`nlcom` generalizes `lincom` (see [R] **lincom**) in two ways. First, `nlcom` allows the transformations to be nonlinear. Second, `nlcom` can be used to simultaneously estimate many transformations (whether linear or nonlinear) and to obtain the estimated variance–covariance matrix of these transformations.

Basics

In [R] **lincom**, the following regression was performed:

(Continued on next page)

```
. use http://www.stata-press.com/data/r10/regress
. regress y x1 x2 x3
      Source |       SS       df       MS              Number of obs =     148
-------------+------------------------------           F(  3,   144) =   96.12
       Model |  3259.3561     3  1086.45203           Prob > F      =  0.0000
    Residual | 1627.56282   144  11.3025196           R-squared     =  0.6670
-------------+------------------------------           Adj R-squared =  0.6600
       Total | 4886.91892   147  33.2443464           Root MSE      =  3.3619

------------------------------------------------------------------------------
           y |      Coef.   Std. Err.      t    P>|t|     [95% Conf. Interval]
-------------+----------------------------------------------------------------
          x1 |   1.457113    1.07461     1.36   0.177     -.666934    3.581161
          x2 |   2.221682   .8610358     2.58   0.011     .5197797    3.923583
          x3 |   -.006139   .0005543   -11.08   0.000    -.0072345   -.0050435
       _cons |   36.10135   4.382693     8.24   0.000     27.43863    44.76407
------------------------------------------------------------------------------
```

Then lincom was used to estimate the difference between the coefficients of x1 and x2:

```
. lincom _b[x2] - _b[x1]
 ( 1)  - x1 + x2 = 0

------------------------------------------------------------------------------
           y |      Coef.   Std. Err.      t    P>|t|     [95% Conf. Interval]
-------------+----------------------------------------------------------------
         (1) |   .7645682   .9950282     0.77   0.444     -1.20218    2.731316
------------------------------------------------------------------------------
```

It was noted, however, that nonlinear expressions are not allowed with lincom:

```
. lincom _b[x2]/_b[x1]
not possible with test
r(131);
```

Nonlinear transformations are instead estimated using nlcom:

```
. nlcom _b[x2]/_b[x1]
       _nl_1:  _b[x2]/_b[x1]

------------------------------------------------------------------------------
           y |      Coef.   Std. Err.      t    P>|t|     [95% Conf. Interval]
-------------+----------------------------------------------------------------
       _nl_1 |   1.524714   .9812848     1.55   0.122    -.4148688    3.464297
------------------------------------------------------------------------------
```

❑ Technical Note

The notation _b[*name*] is the standard way in Stata to refer to regression coefficients; see [U] **13.5 Accessing coefficients and standard errors**. Some commands, such as lincom and test, allow you to drop the _b[] and just refer to the coefficients by *name*. nlcom, however, requires the full specification _b[*name*].

❑

Returning to our linear regression example, nlcom also allows simultaneous estimation of more than one combination:

```
. nlcom (_b[x2]/_b[x1]) (_b[x3]/_b[x1]) (_b[x3]/_b[x2])
       _nl_1:  _b[x2]/_b[x1]
       _nl_2:  _b[x3]/_b[x1]
       _nl_3:  _b[x3]/_b[x2]
```

y	Coef.	Std. Err.	t	P>|t|	[95% Conf.	Interval]
_nl_1	1.524714	.9812848	1.55	0.122	-.4148688	3.464297
_nl_2	-.0042131	.0033483	-1.26	0.210	-.0108313	.002405
_nl_3	.0027632	.0010695	-2.58	0.011	-.0048772	-.0006493

We can also label the transformations to produce more informative names in the estimation table:

```
. nlcom (ratio21:_b[x2]/_b[x1]) (ratio31:_b[x3]/_b[x1]) (ratio32:_b[x3]/_b[x2])
     ratio21:  _b[x2]/_b[x1]
     ratio31:  _b[x3]/_b[x1]
     ratio32:  _b[x3]/_b[x2]
```

y	Coef.	Std. Err.	t	P>|t|	[95% Conf.	Interval]
ratio21	1.524714	.9812848	1.55	0.122	-.4148688	3.464297
ratio31	-.0042131	.0033483	-1.26	0.210	-.0108313	.002405
ratio32	-.0027632	.0010695	-2.58	0.011	-.0048772	-.0006493

`nlcom` saves the vector of estimated combinations and its estimated variance–covariance matrix in `r()`.

```
. matrix list r(b)
r(b)[1,3]
       ratio21     ratio31     ratio32
c1   1.5247143  -.00421315  -.00276324

. matrix list r(V)
symmetric r(V)[3,3]
             ratio21     ratio31     ratio32
ratio21   .96291982
ratio31  -.00287781   .00001121
ratio32  -.00014234   2.137e-06   1.144e-06
```

Using the post option

When used with the `post` option, `nlcom` saves the estimation vector and variance–covariance matrix in `e()`, making the transformation permanent:

```
. quietly nlcom (ratio21:_b[x2]/_b[x1]) (ratio31:_b[x3]/_b[x1])
> (ratio32:_b[x3]/_b[x2]), post

. matrix list e(b)
e(b)[1,3]
       ratio21     ratio31     ratio32
y1   1.5247143  -.00421315  -.00276324

. matrix list e(V)
symmetric e(V)[3,3]
             ratio21     ratio31     ratio32
ratio21   .96291982
ratio31  -.00287781   .00001121
ratio32  -.00014234   2.137e-06   1.144e-06
```

After posting, we can proceed as if we had just run a Stata estimation (eclass) command. For instance, we can replay the results,

```
. nlcom
```

y	Coef.	Std. Err.	t	P>|t|	[95% Conf.	Interval]
ratio21	1.524714	.9812848	1.55	0.122	-.4148688	3.464297
ratio31	-.0042131	.0033483	-1.26	0.210	-.0108313	.002405
ratio32	-.0027632	.0010695	-2.58	0.011	-.0048772	-.0006493

or perform other postestimation tasks in the transformed metric, this time making reference to the new "coefficients":

```
. display _b[ratio31]
-.00421315

. correlate, _coef
             | ratio21  ratio31  ratio32
-------------+---------------------------
     ratio21 |  1.0000
     ratio31 | -0.8759   1.0000
     ratio32 | -0.1356   0.5969   1.0000

. test _b[ratio21] = 1
 ( 1)  ratio21 = 1
       F(  1,   144) =    0.29
            Prob > F =    0.5937
```

We see that testing _b[ratio21]=1 in the transformed metric is equivalent to testing using testnl _b[x2]/_b[x1]=1 in the original metric:

```
. quietly reg y x1 x2 x3

. testnl _b[x2]/_b[x1] = 1
  (1)  _b[x2]/_b[x1] = 1
                F(1, 144) =       0.29
                 Prob > F =       0.5937
```

We needed to refit the regression model to recover the original parameter estimates.

❑ Technical Note

In a previous technical note, we mentioned that commands such as lincom and test permit reference to *name* instead of _b[*name*]. This is not the case when lincom and test are used after nlcom, post. In the above, we used

```
. test _b[ratio21] = 1
```

rather than

```
. test ratio21 = 1
```

which would have returned an error. Consider this a limitation of Stata. For the shorthand notation to work, you need a variable named *name* in the data. In nlcom, however, *name* is just a coefficient label that does not necessarily correspond to any variable in the data.

❑

Reparameterizing ML estimators for univariate data

When run using only a response and no covariates, Stata's maximum likelihood (ML) estimation commands will produce ML estimates of the parameters of some assumed univariate distribution for the response. The parameterization, however, is usually not one we are used to dealing with in a nonregression setting. In such cases, `nlcom` can be used to transform the estimation results from a regression model to those from a maximum likelihood estimation of the parameters of a univariate probability distribution in a more familiar metric.

▷ Example 1

Consider the following univariate data on Y = # of traffic accidents at a certain intersection in a given year:

```
. use http://www.stata-press.com/data/r10/trafint, clear
. summarize accidents
    Variable |       Obs        Mean    Std. Dev.       Min        Max
-------------+--------------------------------------------------------
   accidents |        12    13.83333    14.47778          0         41
```

A quick glance of the output from `summarize` leads us to quickly reject the assumption that Y is distributed as Poisson since the estimated variance of Y is much greater than the estimated mean of Y.

Instead, we choose to model the data as univariate negative binomial, of which a common parameterization is

$$\Pr(Y=y)=\frac{\Gamma(r+y)}{\Gamma(r)\Gamma(y+1)}p^r(1-p)^y \qquad 0\le p\le 1,\quad r>0,\quad y=0,1,\ldots$$

with

$$E(Y)=\frac{r(1-p)}{p} \qquad \mathrm{Var}(Y)=\frac{r(1-p)}{p^2}$$

There exist no closed-form solutions for the maximum likelihood estimates of p and r, yet they may be estimated by the iterative method of Newton–Raphson. One way to get these estimates would be to write our own Newton–Raphson program for the negative binomial. Another way would be to write our own ML evaluator; see [R] **ml**.

The easiest solution, however, would be to use Stata's existing negative binomial ML regression command, `nbreg`. The only problem with this solution is that `nbreg` estimates a different parameterization of the negative binomial, but we can worry about that later.

```
. nbreg accidents
Fitting Poisson model:
Iteration 0:   log likelihood = -105.05361
Iteration 1:   log likelihood = -105.05361
Fitting constant-only model:
Iteration 0:   log likelihood = -43.948619
Iteration 1:   log likelihood = -43.891483
Iteration 2:   log likelihood =  -43.89144
Iteration 3:   log likelihood =  -43.89144
Fitting full model:
Iteration 0:   log likelihood =  -43.89144
Iteration 1:   log likelihood =  -43.89144
```

```
Negative binomial regression                      Number of obs   =         12
                                                  LR chi2(0)      =       0.00
Dispersion     = mean                             Prob > chi2     =          .
Log likelihood = -43.89144                        Pseudo R2       =     0.0000

------------------------------------------------------------------------------
   accidents |      Coef.   Std. Err.      z    P>|z|     [95% Conf. Interval]
-------------+----------------------------------------------------------------
       _cons |   2.627081   .3192233     8.23   0.000     2.001415    3.252747
-------------+----------------------------------------------------------------
    /lnalpha |   .1402425   .4187147                     -.6804233    .9609083
-------------+----------------------------------------------------------------
       alpha |   1.150553   .4817534                      .5064026     2.61407
------------------------------------------------------------------------------
Likelihood-ratio test of alpha=0:  chibar2(01) =  122.32 Prob>=chibar2 = 0.000
. matrix list e(b)
e(b)[1,2]
     accidents:    lnalpha:
         _cons       _cons
y1   2.6270811   .14024253
```

From this output, we see that, when used with univariate data, **nbreg** estimates a regression intercept, β_0, and the logarithm of some parameter α. This parameterization is useful in regression models: β_0 is the intercept meant to be augmented with other terms of the linear predictor, and α is an overdispersion parameter used for comparison with the Poisson regression model.

However, we need to transform $(\beta_0, \ln\alpha)$ to (p, r). Examining *Methods and Formulas* of [R] **nbreg** reveals the transformation as

$$p = \{1 + \alpha \exp(\beta_0)\}^{-1} \qquad r = \alpha^{-1}$$

which we apply using `nlcom`:

```
. nlcom (p:1/(1 + exp([lnalpha]_b[_cons] + _b[_cons]))) (r:exp(-[lnalpha]_b[_cons]))
           p:  1/(1 + exp([lnalpha]_b[_cons] + _b[_cons]))
           r:  exp(-[lnalpha]_b[_cons])

------------------------------------------------------------------------------
   accidents |      Coef.   Std. Err.      z    P>|z|     [95% Conf. Interval]
-------------+----------------------------------------------------------------
           p |   .0591157   .0292857     2.02   0.044     .0017168    .1165146
           r |   .8691474   .3639248     2.39   0.017     .1558679    1.582427
------------------------------------------------------------------------------
```

Given the invariance of maximum likelihood estimators and the properties of the delta method, the above parameter estimates, standard errors, etc. are precisely those we would have obtained had we instead performed the Newton–Raphson optimization in the (p, r) metric.

◁

❑ Technical Note

Note how we referred to the estimate of $\ln\alpha$ in above as `[lnalpha]_b[_cons]`. This is not entirely evident from the output of `nbreg`, which is why we listed the elements of `e(b)` so that we could examine the column labels; see [U] **13.5 Accessing coefficients and standard errors**.

❑

nlcom versus eform

Many Stata estimation commands allow you to display exponentiated regression coefficients, some by default, some optionally. Known as "eform" in Stata terminology, this reparameterization serves many uses: it gives odds ratios for logistic models, hazard ratios in survival models, incidence-rate ratios in Poisson models, and relative risk ratios in multinomial logit models, to name a few.

For example, consider the following estimation taken directly from [R] **poisson**:

```
. use http://www.stata-press.com/data/r10/airline
. gen lnN = ln(n)
. poisson injuries XYZowned lnN
Iteration 0:   log likelihood = -22.333875
Iteration 1:   log likelihood = -22.332276
Iteration 2:   log likelihood = -22.332276
Poisson regression                                Number of obs   =          9
                                                  LR chi2(2)      =      19.15
                                                  Prob > chi2     =     0.0001
Log likelihood = -22.332276                       Pseudo R2       =     0.3001
```

injuries	Coef.	Std. Err.	z	P>|z|	[95% Conf.	Interval]
XYZowned	.6840667	.3895877	1.76	0.079	-.0795111	1.447645
lnN	1.424169	.3725155	3.82	0.000	.6940517	2.154285
_cons	4.863891	.7090501	6.86	0.000	3.474178	6.253603

When we replay results and specify the irr (incidence-rate ratios) option,

```
. poisson, irr
Poisson regression                                Number of obs   =          9
                                                  LR chi2(2)      =      19.15
                                                  Prob > chi2     =     0.0001
Log likelihood = -22.332276                       Pseudo R2       =     0.3001
```

injuries	IRR	Std. Err.	z	P>|z|	[95% Conf.	Interval]
XYZowned	1.981921	.7721322	1.76	0.079	.9235678	4.253085
lnN	4.154402	1.547579	3.82	0.000	2.00181	8.621728

we obtain the exponentiated regression coefficients and their estimated standard errors.

Contrast this with what we obtain if we exponentiate the coefficients manually using nlcom:

```
. nlcom (E_XYZowned:exp(_b[XYZowned])) (E_lnN:exp(_b[lnN]))
  E_XYZowned:  exp(_b[XYZowned])
       E_lnN:  exp(_b[lnN])
```

injuries	Coef.	Std. Err.	z	P>|z|	[95% Conf.	Interval]
E_XYZowned	1.981921	.7721322	2.57	0.010	.4685701	3.495273
E_lnN	4.154402	1.547579	2.68	0.007	1.121203	7.187602

There are three things to note when comparing `poisson, irr` (and `eform` in general) with `nlcom`:

1. The exponentiated coefficients and standard errors are identical. This is certainly good news.
2. The Wald test statistic (`z`) and level of significance are different. When using `poisson, irr` and other related `eform` options, the Wald test does not change from what you would have obtained without the `eform` option, and you can see this by comparing both versions of the `poisson` output given previously.

   When you use `eform`, Stata knows that what is usually desired is a test of

$$H_0\colon \exp(\beta) = 1$$

   and not the uninformative-by-comparison

$$H_0\colon \exp(\beta) = 0$$

   The test of $H_0\colon \exp(\beta) = 1$ is asymptotically equivalent to a test of $H_0\colon \beta = 0$, the Wald test in the original metric, but the latter has better small-sample properties. Thus if you specify `eform`, you get a test of $H_0\colon \beta = 0$.

   `nlcom`, however, is general. It does not attempt to infer the test of greatest interest for a given transformation, and so a test of

$$H_0\colon \text{transformed coefficient} = 0$$

   is always given, regardless of the transformation.

3. You may be surprised to see that, even though the coefficients and standard errors are identical, the confidence intervals (both 95%) are different.

   `eform` confidence intervals are standard confidence intervals with the endpoints transformed. For example, the confidence interval for the coefficient on `lnN` is $(0.694, 2.154)$, whereas the confidence interval for the incidence-rate ratio due to `lnN` is $(\exp(0.694), \exp(2.154)) = (2.002, 8.619)$, which, except for some roundoff error, is what we see from the output of `poisson, irr`. For exponentiated coefficients, confidence intervals based on transform-the-endpoints methodology generally have better small-sample properties than their asymptotically equivalent counterparts.

   The transform-the-endpoints method, however, gives valid coverage only when the transformation is monotonic. `nlcom` uses a more general and asymptotically equivalent method for calculating confidence intervals, as described in *Methods and Formulas*.

Saved Results

`nlcom` saves the following in `r()`:

Scalars

`r(N)`	number of observations
`r(df_r)`	residual degrees of freedom

Matrices

`r(b)`	vector of transformed coefficients
`r(V)`	estimated variance–covariance matrix of the transformed coefficients

If `post` is specified, `nlcom` also saves the following in `e()`:

Scalars

`e(N)`	number of observations
`e(df_r)`	residual degrees of freedom
`e(N_strata)`	number of strata L, if used after `svy`
`e(N_psu)`	number of sampled PSUs n, if used after `svy`

Macros

`e(cmd)`	`nlcom`
`e(predict)`	program used to implement `predict`
`e(properties)`	`b V`

Matrices

`e(b)`	vector of transformed coefficients
`e(V)`	estimated variance–covariance matrix of the transformed coefficients
`e(V_srs)`	simple-random-sampling-without-replacement (co)variance $\widehat{V}_{\text{srswor}}$, if `svy`
`e(V_srswr)`	simple-random-sampling-with-replacement (co)variance $\widehat{V}_{\text{srswr}}$, if `svy` and `fpc()`
`e(V_msp)`	misspecification (co)variance $\widehat{V}_{\text{msp}}$, if `svy` and available

Functions

`e(sample)`	marks estimation sample

Methods and Formulas

`nlcom` is implemented as an ado-file.

Given a $1 \times k$ vector of parameter estimates, $\widehat{\boldsymbol{\theta}} = (\widehat{\theta}_1, \dots, \widehat{\theta}_k)$, consider the estimated p-dimensional transformation

$$g(\widehat{\boldsymbol{\theta}}) = [g_1(\widehat{\boldsymbol{\theta}}), g_2(\widehat{\boldsymbol{\theta}}), \dots, g_p(\widehat{\boldsymbol{\theta}})]$$

The estimated variance–covariance of $g(\widehat{\boldsymbol{\theta}})$ is given by

$$\widehat{\text{Var}}\left\{g(\widehat{\boldsymbol{\theta}})\right\} = \mathbf{GVG}'$$

where $\mathbf{G}$ is the $p \times k$ matrix of derivatives for which

$$\mathbf{G}_{ij} = \left.\frac{\partial g_i(\boldsymbol{\theta})}{\partial \theta_j}\right|_{\boldsymbol{\theta}=\widehat{\boldsymbol{\theta}}} \qquad i = 1, \dots, p \qquad j = 1, \dots, k$$

and $\mathbf{V}$ is the estimated variance–covariance matrix of $\widehat{\boldsymbol{\theta}}$. Standard errors are obtained as the square roots of the variances.

The Wald test statistic for testing

$$H_0\colon g_i(\boldsymbol{\theta}) = 0$$

versus the two-sided alternative is given by

$$Z_i = \frac{g_i(\widehat{\boldsymbol{\theta}})}{\left[\widehat{\text{Var}}_{ii}\left\{g(\widehat{\boldsymbol{\theta}})\right\}\right]^{1/2}}$$

When the variance–covariance matrix of $\widehat{\boldsymbol{\theta}}$ is an asymptotic covariance matrix, Z_i is approximately distributed as Gaussian. For linear regression, Z_i is taken to be approximately distributed as $t_{1,r}$ where r is the residual degrees of freedom from the original fitted model.

A $(1-\alpha)\times 100\%$ confidence interval for $g_i(\boldsymbol{\theta})$ is given by

$$g_i(\widehat{\boldsymbol{\theta}}) \pm z_{\alpha/2}\left[\widehat{\mathrm{Var}}_{ii}\left\{g(\widehat{\boldsymbol{\theta}})\right\}\right]^{1/2}$$

for those cases where Z_i is Gaussian and

$$g_i(\widehat{\boldsymbol{\theta}}) \pm t_{\alpha/2,r}\left[\widehat{\mathrm{Var}}_{ii}\left\{g(\widehat{\boldsymbol{\theta}})\right\}\right]^{1/2}$$

for those cases where Z_i is t-distributed. z_p is the $1-p$ quantile of the standard normal distribution, and $t_{p,r}$ is the $1-p$ quantile of the t distribution with r degrees of freedom.

References

Feiveson, A. H. 1999. FAQ: What is the delta method, and how is it used to estimate the standard error of a transformed parameter? http://www.stata.com/support/faqs/stat/.

Gould, W. W. 1996. crc43: Wald test of nonlinear hypotheses after model estimation. *Stata Technical Bulletin* 29: 2–4. Reprinted in *Stata Technical Bulletin Reprints*, vol. 5, pp. 15–18.

Oehlert, G. W. 1992. A note on the delta method. *American Statistician* 46: 27–29.

Phillips, P. C. B., and J. Y. Park. 1988. On the formulation of Wald tests of nonlinear restrictions. *Econometrica* 56: 1065–1083.

Also See

[R] **lincom** — Linear combinations of estimators

[R] **predictnl** — Obtain nonlinear predictions, standard errors, etc., after estimation

[R] **test** — Test linear hypotheses after estimation

[R] **testnl** — Test nonlinear hypotheses after estimation

[U] **20 Estimation and postestimation commands**

Title

nlogit — Nested logit regression

Syntax

Nested logit regression

`nlogit` *depvar* [*indepvars*] [*if*] [*in*] [*weight*] [|| *lev1_equation*
[|| *lev2_equation* ...]] || *altvar*: [*byaltvarlist*] , `case(`*varname*`)` [*options*]

where the syntax of *lev#_equation* is

altvar: [*byaltvarlist*] [, `base(`*#* | *lbl*`)` `estconst`]

Create variable based on specification of branches

`nlogitgen` *newaltvar* = *altvar* (*branchlist*) [, `nolog`]

where *branchlist* is

branch, *branch* [, *branch* ...]

and *branch* is

[*label*:] *alternative* [| *alternative* [| *alternative* ...]]

Display tree structure

`nlogittree` *altvarlist* [*if*] [*in*] [*weight*] [, `choice(`*depvar*`)` `nolabel` `nobranches`]

(*Continued on next page*)

options	description
Model	
* `case(`*varname*`)`	use *varname* to identify cases
`base(`*#* | *lbl*`)`	use the specified level or label of *altvar* as the base alternative for the bottom level
`noconstant`	suppress the constant terms for the bottom-level alternatives
`nonnormalized`	use the nonnormalized parameterization
`altwise`	use alternative-wise deletion instead of casewise deletion
`constraints(`*constraints*`)`	apply specified linear constraints
`collinear`	keep collinear variables
SE/Robust	
`vce(`*vcetype*`)`	*vcetype* may be `oim`, `robust`, `cluster` *clustvar*, `bootstrap`, or `jackknife`
Reporting	
`level(`*#*`)`	set confidence level; default is `level(95)`
`notree`	suppress display of tree-structure output; see also `nlogittree` options `nolabel` and `nobranches`
Max options	
maximize_options	control the maximization process; seldom used

* `case(`*varname*`)` is required.

`bootstrap`, `by`, `jackknife`, `statsby`, and `xi` are allowed; see [U] **11.1.10 Prefix commands**.

Weights are not allowed with the `bootstrap` prefix.

`fweight`s, `iweight`s, and `pweight`s are allowed with `nlogit`, and `fweight`s are allowed with `nlogittree`; see [U] **11.1.6 weight**.

See [U] **20 Estimation and postestimation commands** for more capabilities of estimation commands.

Description

`nlogit` performs full information maximum-likelihood estimation for nested logit models. These models relax the assumption of independently distributed errors and the independence of irrelevant alternatives inherent in conditional and multinomial logit models by clustering similar alternatives into nests.

By default `nlogit` uses a parameterization that is consistent with random utility maximization (RUM). Before version 10 of Stata, a nonnormalized version of the nested logit model was fitted, which you can request by specifying the `nonnormalized` option.

You must use `nlogitgen` to generate a new categorical variable to specify the branches of the decision tree before calling `nlogit`.

Options

Specification and options for lev#_equation

altvar is a variable identifying alternatives at this level of the hierarchy.

byaltvarlist specifies the variables to be used to compute the by-alternative regression coefficients for that level. For each variable specified in the variable list, there will be one regression coefficient for each alternative of that level of the hierarchy. If the variable is constant across each alternative (a case-specific variable), the regression coefficient associated with the base alternative is not identifiable. These regression coefficients are labeled as (base) in the regression table. If the variable varies among the alternatives, a regression coefficient is estimated for each alternative.

`base(`*#* | *lbl*`)` can be specified in each level equation where it identifies the base alternative to be used at that level. The default is the alternative that has the highest frequency.

If `vce(bootstrap)` or `vce(jackknife)` is specified, you must specify the base alternative for each level that has a *byaltvarlist* or if the constants will be estimated. Doing so ensures that the same model is fitted with each call to `nlogit`.

`estconst` applies to all the level equations except the bottom-level equation. Specifying `estconst` requests that constants for each alternative (except the base alternative) be estimated. By default, no constant is estimated at these levels. Constants can be estimated in only one level of the tree hierarchy. If you specify `estconst` for one of the level equations, you must specify `noconstant` for the bottom-level equation.

Options for nlogit

Model

`case(`*varname*`)` specifies the variable that identifies each case. `case()` is required.

`base(`*#* | *lbl*`)` can be specified in each level equation where it identifies the base alternative to be used at that level. The default is the alternative that has the highest frequency.

If `vce(bootstrap)` or `vce(jackknife)` is specified, you must specify the base alternative for each level that has a *byaltvarlist* or if the constants will be estimated. Doing so ensures that the same model is fitted with each call to `nlogit`.

`noconstant` applies only to the equation defining the bottom level of the hierarchy. By default, constants are estimated for each alternative of *altvar*, less the base alternative. To suppress the constant terms for this level, specify `noconstant`. If you do not specify `noconstant`, you cannot specify `estconst` for the higher-level equations.

`nonnormalized` requests a nonnormalized parameterization of the model that does not scale the inclusive values by the degree of dissimilarity of the alternatives within each nest. Use this option to replicate results from older versions of Stata. The default is to use the random utility maximization (RUM)–consistent parameterization.

`altwise` specifies that alternative-wise deletion be used when marking out observations due to missing values in your variables. The default is to use casewise deletion. This option does not apply to observations that are marked out by the `if` or `in` qualifier or the `by` prefix.

constraints(*constraints*); see [R] **estimation options**.

The inclusive-valued/dissimilarity parameters are parameterized as ml ancillary parameters. They are labeled as [*alternative*_tau]_const, where *alternative* is one of the alternatives defining a branch in the tree. To constrain the inclusive-valued/dissimilarity parameter for alternative a1 to be, say, equal to alternative a2, you would use the following syntax:

```
. constraint 1 [a1_tau]_cons = [a2_tau]_cons
. nlogit ..., constraints(1)
```

collinear prevents collinear variables from being dropped. Use this option when you know that you have collinear variables and you are applying constraints() to handle the rank reduction. See [R] **estimation options** for details on using collinear with constraints().

SE/Robust

vce(*vcetype*) specifies the type of standard error reported, which includes types that are derived from asymptotic theory, that are robust to some kinds of misspecification, that allow for intragroup correlation, and that use bootstrap or jackknife methods; see [R] ***vce_option***. If vce(robust) or vce(cluster *clustvar*) is specified, the likelihood-ratio test for the independence of irrelevant alternatives (IIA) is not computed.

Reporting

level(*#*); see [R] **estimation options**.

notree specifies that the tree structure of the nested logit model not be displayed. See also nolabel and nobranches below for when notree is not specified.

Max options

maximize_options: difficult, technique(*algorithm_spec*), iterate(*#*), [no]log, trace, gradient, showstep, hessian, shownrtolerance, tolerance(*#*), ltolerance(*#*), gtolerance(*#*), nrtolerance(*#*), nonrtolerance; see [R] **maximize**. These options are seldom used.

The option technique(bhhh) is not allowed.

Specification and options for nlogitgen

newaltvar and *altvar* are variables identifying alternatives at each level of the hierarchy.

label defines a label to associate with the branch. If no label is given, a numeric value is used.

alternative specifies an alternative, of *altvar* specified in the syntax, to be included in the branch. It is either a numeric value or the label associated with that value. An example of nlogitgen is

```
. nlogitgen type = restaurant(fast: 1 | 2,
> family: CafeEccell | LosNortenos | WingsNmore, fancy: 6 | 7)
```

nolog suppresses the display of the iteration log.

Specification and options for nlogittree

Main

altvarlist is a list of alternative variables that define the tree hierarchy. The first variable must define bottom-level alternatives, and the order continues to the variable defining the top-level alternatives.

`choice(`*depvar*`)` defines the choice indicator variable and forces `nlogittree` to compute and display choice frequencies for each bottom-level alternative.

`nolabel` forces `nlogittree` to suppress value labels in tree-structure output.

`nobranches` forces `nlogittree` to suppress drawing branches in the tree-structure output.

Remarks

Remarks are presented under the following headings:

Introduction
Data setup and the tree structure
Estimation
Testing for the IIA
Nonnormalized model

Introduction

`nlogit` performs full information maximum-likelihood estimation for nested logit models. These models relax the assumption of independently distributed errors and the independence of irrelevant alternatives inherent in conditional and multinomial logit models by clustering similar alternatives into nests. Because the nested logit model is a direct generalization of the alternative-specific conditional logit model (also known as McFadden's choice model), you may want to read [R] **asclogit** before continuing.

By default `nlogit` uses a parameterization that is consistent with random utility maximization (RUM). Before version 10 of Stata, a nonnormalized version of the nested logit model was fitted, which you can request by specifying the `nonnormalized` option. We recommend using the RUM-consistent version of the model for new projects because it is based on a sound model of consumer behavior.

McFadden (1977, 1981) showed how this model can be derived from a rational choice framework. Amemiya (1985, chap. 9) contains a nice discussion of how this model can be derived under the assumption of utility maximization. Hensher, Rose, and Greene (2005) provides a lucid introduction to choice models including nested logit.

Throughout this entry we consider a model of restaurant choice. We begin by introducing the data.

▷ Example 1

We have fictional data on 300 families and their choice of seven local restaurants. Freebirds and Mama's Pizza are fast food restaurants; Café Eccell, Los Norteños, and Wings 'N More are family restaurants; and Christopher's and Mad Cows are fancy restaurants. We want to model the decision of where to eat as a function of household income (`income`, in thousands of dollars), the number of children in the household (`kids`), the rating of the restaurant according to a local restaurant guide (`rating`, coded 0–5), the average meal cost per person (`cost`), and the distance between the household and the restaurant (`distance`, in miles). `income` and `kids` are attributes of the family, `rating` is an attribute of the alternative (the restaurant), and `cost` and `distance` are attributes of the alternative as perceived by the families—that is, each family has its own cost and distance for each restaurant.

We begin by loading the data and listing some of the variables for the first three families:

```
. use http://www.stata-press.com/data/r10/restaurant
. describe
Contains data from http://www.stata-press.com/data/r10/restaurant.dta
  obs:         2,100
 vars:             8                          10 Mar 2007 01:17
 size:        75,600 (99.0% of memory free)
-------------------------------------------------------------------------------
              storage  display     value
variable name   type   format      label      variable label
-------------------------------------------------------------------------------
family_id       float  %9.0g                  family ID
restaurant      float  %12.0g      names      choices of restaurants
income          float  %9.0g                  household income
cost            float  %9.0g                  average meal cost per person
kids            float  %9.0g                  number of kids in the household
rating          float  %9.0g                  ratings in local restaurant
                                                guide
distance        float  %9.0g                  distance between home and
                                                restaurant
chosen          float  %9.0g                  0 no 1 yes
-------------------------------------------------------------------------------
Sorted by:  family_id
. list family_id restaurant chosen kids rating distance in 1/21, sepby(fam) abbrev(10)

     +--------------------------------------------------------------+
     | family_id     restaurant   chosen   kids   rating   distance |
     |--------------------------------------------------------------|
  1. |         1      Freebirds        1      1        0   1.245553 |
  2. |         1     MamasPizza        0      1        1    2.82493 |
  3. |         1     CafeEccell        0      1        2    4.21293 |
  4. |         1    LosNortenos        0      1        3   4.167634 |
  5. |         1     WingsNmore        0      1        2   6.330531 |
  6. |         1   Christophers        0      1        4   10.19829 |
  7. |         1        MadCows        0      1        5   5.601388 |
     |--------------------------------------------------------------|
  8. |         2      Freebirds        0      3        0   4.162657 |
  9. |         2     MamasPizza        0      3        1   2.865081 |
 10. |         2     CafeEccell        0      3        2   5.337799 |
 11. |         2    LosNortenos        1      3        3   4.282864 |
 12. |         2     WingsNmore        0      3        2   8.133914 |
 13. |         2   Christophers        0      3        4   8.664631 |
 14. |         2        MadCows        0      3        5   9.119597 |
     |--------------------------------------------------------------|
 15. |         3      Freebirds        1      3        0   2.112586 |
 16. |         3     MamasPizza        0      3        1   2.215329 |
 17. |         3     CafeEccell        0      3        2   6.978715 |
 18. |         3    LosNortenos        0      3        3   5.117877 |
 19. |         3     WingsNmore        0      3        2   5.312941 |
 20. |         3   Christophers        0      3        4   9.551273 |
 21. |         3        MadCows        0      3        5   5.539806 |
     +--------------------------------------------------------------+
```

Because each family chose among seven restaurants, there are 7 observations in the dataset for each family. The variable `chosen` is coded 0/1, 1 indicating the chosen restaurant and 0 otherwise.

◁

We could fit a conditional logit model to our data. Since `income` and `kids` are constant within each family, we would use the `asclogit` command instead of `clogit`. However, the conditional logit may be inappropriate. That model assumes that the random errors are independent, and as a result it forces the odds ratio of any two alternatives to be independent of the other alternatives, a property known as the independence of irrelevant alternatives (IIA). We will discuss the IIA assumption in more detail later.

Assuming that unobserved shocks influencing a decision maker's attitude toward one alternative have no effect on his attitudes toward the other alternatives may seem innocuous, but often this assumption is too restrictive. Suppose that when a family was deciding which restaurant to visit, they were pressed for time because of plans to attend a movie later. The unobserved shock (being in a hurry) would raise the likelihood that the family goes to either fast food restaurant (Freebirds or Mama's Pizza). Similarly, another family might be choosing a restaurant to celebrate a birthday and therefore be inclined to attend a fancy restaurant (Christopher's or Mad Cows).

Nested logit models relax the independence assumption and allow us to group alternatives for which unobserved shocks may have concomitant effects. Here we suspect that restaurants should be grouped by type (fast, family, or fancy). The tree structure of a family's decision about where to eat might look like this:

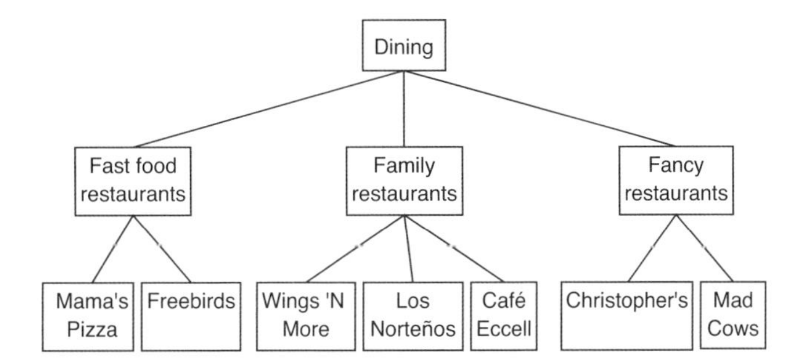

At the bottom of the tree are the individual restaurants, indicating that there are some random shocks that affect a family's decision to eat at each restaurant independently. Above the restaurants are the three types of restaurants, indicating that other random shocks affect the type of restaurant chosen. As is customary when drawing decision trees, at the top level is one box, representing the family making the decision.

We use the following terms to describe nested logit models.

level, or decision level, is the level or stage at which a decision is made. The example above has only two levels. In the first level, a type of restaurant is chosen—fast food, family, or fancy—and in the second level, a specific restaurant is chosen.

bottom level is the level where the final decision is made. In our example, this is when we choose a specific restaurant.

alternative set is the set of all possible alternatives at any given decision level.

bottom alternative set is the set of all possible alternatives at the bottom level. This concept is often referred to as the choice set in the economics-choice literature. In our example, the bottom alternative set is all seven of the specific restaurants.

alternative is a specific alternative within an alternative set. In the first level of our example, "fast food" is an alternative. In the second or bottom level, "Mad Cows" is an alternative. Not all alternatives within an alternative set are available to someone making a choice at a specific stage, only those that are nested within all higher level decisions.

chosen alternative is the alternative from an alternative set that we observe someone having chosen.

❑ Technical Note

Although decision trees in nested logit analysis are often interpreted as implying that the highest-level decisions are made first, followed by decisions at lower levels, and finally the decision among alternatives at the bottom level, no such temporal ordering is implied. See Hensher, Rose, and Greene (2005, chap. 13). In our example, we are not assuming that families first choose whether to attend a fast, family, or fancy restaurant and then choose the particular restaurant; we assume merely that they choose one of the seven restaurants.

❑

Data setup and the tree structure

To fit a nested logit model, you must first create a variable that defines the structure of your decision tree.

▷ Example 2

To run `nlogit`, we need to generate a categorical variable that identifies the first-level set of alternatives: fast food, family restaurants, or fancy restaurants. We can do so easily by using `nlogitgen`.

```
. nlogitgen type = restaurant(fast: Freebirds | MamasPizza,
> family: CafeEccell | LosNortenos| WingsNmore, fancy: Christophers | MadCows)
new variable type is generated with 3 groups
label list lb_type
lb_type:
           1 fast
           2 family
           3 fancy
. nlogittree restaurant type, choice(chosen)
tree structure specified for the nested logit model
 type    N       restaurant     N   k
─────────────────────────────────────────
 fast    600 ─┬─ Freebirds     300  12
              └─ MamasPizza    300  15
 family  900 ─┬─ CafeEccell    300  78
              ├─ LosNortenos   300  75
              └─ WingsNmore    300  69
 fancy   600 ─┬─ Christophers  300  27
              └─ MadCows       300  24
─────────────────────────────────────────
                         total 2100 300
k = number of times alternative is chosen
N = number of observations at each level
```

The new categorical variable is `type`, which takes value 1 (fast) if `restaurant` is Freebirds or Mama's Pizza; value 2 (family) if `restaurant` is Café Eccell, Los Norteños, or Wings 'N More; and value 3 (fancy) otherwise. `nlogittree` displays the tree structure.

◁

❑ Technical Note

We could also use values instead of value labels of `restaurant` in `nlogitgen`. Value labels are optional, and the default value labels for `type` are `type1`, `type2`, and `type3`. The vertical bar is also optional.

```
. use http://www.stata-press.com/data/r10/restaurant, clear
. nlogitgen type = restaurant(1 2, 3 4 5, 6 7)
new variable type is generated with 3 groups
label list lb_type
lb_type:
           1 type1
           2 type2
           3 type3
. nlogittree restaurant type
tree structure specified for the nested logit model
 type    N       restaurant     N
--------------------------------------
 type1 600 ─┬─ Freebirds     300
            └─ MamasPizza    300
 type2 900 ─┬─ CafeEccell    300
            ├─ LosNortenos   300
            └─ WingsNmore    300
 type3 600 ─┬─ Christophers  300
            └─ MadCows       300
--------------------------------------
                      total  2100
N = number of observations at each level
```

❑

In our dataset, every family was able to choose among all seven restaurants. However, in other applications some decision makers may not have been able to choose among all possible alternatives. For example, two cases may have choice hierarchies of

```
     case 1                          case 2
 type       restaurant          type       restaurant
-----------------------        -----------------------
 fast    ─┬─ Freebirds          fast    ─┬─ Freebirds
          └─ MamasPizza                  └─ MamasPizza
 family  ─┬─ CafeEccell         family  ─┬─ LosNortenos
          ├─ LosNortenos                 └─ WingsNmore
          └─ WingsNmore
 fancy   ─┬─ Christophers       fancy   ─── Christophers
          └─ MadCows
```

where the second case does not have the restaurant alternatives Café Eccell or Mad Cows available to them. The only restriction is that the relationships between higher- and lower-level alternative sets be the same for all decision makers. In this two-level example, Freebirds and Mama's Pizza are classified as fast food restaurants for both cases, Café Eccell, Los Norteños, and Wings 'N More are family restaurants; and Christopher's and Mad Cows are fancy restaurants. `nlogit` requires only that hierarchy be maintained for all cases.

Estimation

▷ Example 3

With our `type` variable created that defines the three types of restaurants, we can now examine how the alternative-specific attributes (`cost`, `rating`, and `distance`) apply to the bottom alternative set (the seven restaurants) and how family-specific attributes (`income` and `kid`) apply to the alternative set at the first decision level (the three types of restaurants).

```
. nlogit chosen cost rating distance || type: income kids, base(family) ||
> restaurant:, noconstant case(family_id)
tree structure specified for the nested logit model

 type      N       restaurant     N   k
--------------------------------------------
 fast     600 -+-- Freebirds     300  12
               +-- MamasPizza    300  15
 family   900 -+-- CafeEccell    300  78
               |-- LosNortenos   300  75
               +-- WingsNmore    300  69
 fancy    600 -+-- Christophers  300  27
               +-- MadCows       300  24
--------------------------------------------
                         total  2100 300

k = number of times alternative is chosen
N = number of observations at each level

Iteration 0:   log likelihood = -541.93581
  (output omitted)
Iteration 17:  log likelihood = -485.47331

RUM-consistent nested logit regression          Number of obs      =      2100
Case variable: family_id                        Number of cases    =       300

Alternative variable: restaurant                Alts per case: min =         7
                                                               avg =       7.0
                                                               max =         7

                                                   Wald chi2(7)    =     46.71
Log likelihood = -485.47331                        Prob > chi2     =    0.0000

------------------------------------------------------------------------------
      chosen |      Coef.   Std. Err.      z    P>|z|     [95% Conf. Interval]
-------------+----------------------------------------------------------------
restaurant   |
        cost |  -.1843847   .0933975    -1.97   0.048    -.3674404   -.0013289
      rating |    .463694   .3264935     1.42   0.156    -.1762215     1.10361
    distance |  -.3797474   .1003828    -3.78   0.000    -.5764941   -.1830007
------------------------------------------------------------------------------
type equations
-------------+----------------------------------------------------------------
fast         |
      income |  -.0266038   .0117306    -2.27   0.023    -.0495952   -.0036123
        kids |  -.0872584   .1385026    -0.63   0.529    -.3587184    .1842016
-------------+----------------------------------------------------------------
family       |
      income |     (base)
        kids |     (base)
-------------+----------------------------------------------------------------
fancy        |
      income |   .0461827   .0090936     5.08   0.000     .0283595    .0640059
        kids |  -.3959413   .1220356    -3.24   0.001    -.6351267   -.1567559
------------------------------------------------------------------------------
dissimilarity parameters
```

```
type
   /fast_tau |   1.712878    1.48685                     -1.201295    4.627051
 /family_tau |   2.505113   .9646351                       .614463    4.395763
  /fancy_tau |   4.099844   2.810123                     -1.407896    9.607583

LR test for IIA (tau = 1):             chi2(3) =      6.87    Prob > chi2 = 0.0762
```

First, let's examine how we called `nlogit`. The delimiters (||) separate equations. The first equation specifies the dependent variable, `chosen`, and three alternative-specific variables, `cost`, `rating`, and `distance`. We refer to these variables as alternative-specific because they vary among the bottom-level alternatives, the restaurants. We obtain one parameter estimate for each variable. These estimates are listed in the equation subtable labeled `restaurant`.

For the second equation we specify the variable `type`. It identifies the first-level alternatives, the restaurant types. Following the colon after `type`, we specify two case-specific variables, `income` and `kids`. Here we obtain a parameter estimate for each variable for each alternative at this level. That is why we call these variable lists *by-alternative* variables. Because `income` and `kids` do not vary within each case, to identify the model one alternative's set of parameters must be set to zero. We specified the `base(family)` option with this equation to restrict the parameters for the `family` alternative.

The variable identifying the bottom-level alternatives, `restaurant`, is specified after the second equation delimiter. We do not specify any variables after the colon delimiter at this level. Had we specified variables here, we would have obtained an estimate for each variable in each equation. As we will see below, these variables parameterize the constant term in the utility equation for each bottom-level alternative. The `noconstant` option suppresses bottom-level alternative-specific constant terms.

Near the bottom of the output are the dissimilarity parameters, which measure the degree of correlation of random shocks within each of the three types of restaurants. Dissimilarity parameters greater than one imply that the model is inconsistent with RUM; Hensher, Rose, and Greene (2005, sec. 13.6) discuss this in detail. We will ignore the fact that all our dissimilarity parameters exceed one.

The conditional logit model is a special case of nested logit in which all the dissimilarity parameters are equal to one. At the bottom of the output, we find a likelihood-ratio test of this hypothesis. Here we have mixed evidence for the null hypothesis that all the parameters are one. Equivalently, the property known as the independence of irrelevant alternatives (IIA) imposed by the conditional logit model holds if and only if all dissimilarity parameters are equal to one. We discuss the IIA in more detail now.

◁

Testing for the IIA

The IIA is a property of the multinomial and conditional logit models that forces the odds of choosing one alternative over another to be independent of the other alternatives. For simplicity, suppose that a family was choosing only between Freebirds and Mama's Pizza, and the family was equally likely to choose either of the restaurants. The probability of going to each restaurant is 50%. Now suppose that Bill's Burritos opens up next door to Freebirds, which is also a burrito restaurant. If the IIA holds, then the probability of going to each restaurant must now be 33 1/3% so that the family remains equally likely to go to Mama's Pizza or Freebirds.

The IIA may sometimes be a plausible assumption. However, a more likely scenario would be for the probability of going to Mama's Pizza to remain at 50% and the probabilities of going to Freebirds and Bill's Burritos to be 25% each, since the two restaurants are next door to each other and serve the same food. Nested logit analysis would allow us to relax the IIA assumption of conditional logit. We could group Bill's Burritos and Freebirds into one nest that encompasses all burrito restaurants and create a second nest for pizzerias.

The IIA is a consequence of assuming that the errors are independently and identically distributed (i.i.d.). Since the errors are i.i.d., they cannot contain any alternative-specific unobserved information, and therefore adding a new alternative cannot affect the relationship between a pair of existing alternatives.

In the previous example we saw that a joint test that the dissimilarity parameters were equal to one is one way to test for IIA. However, that test required us to specify a decision tree for the nested logit model, and different specifications could lead to conflicting results of the test. Hausman and McFadden (1984) suggest that if part of the choice set truly is irrelevant with respect to the other alternatives, omitting that subset from the conditional logit model will not lead to inconsistent estimates. Therefore, Hausman's (1978) specification test can be used to test for IIA, and this test will not be sensitive to the tree structure we specify for a nested logit model.

▷ Example 4

We want to test the IIA for the subset of family restaurants against the alternatives of fast food and fancy restaurants. To do so, we need to use Stata's `hausman` command; see [R] **hausman**.

We first run the estimation on the full bottom alternative set, save the results by using `estimates store`, and then run the estimation on the bottom alternative set, excluding the alternatives of family restaurants. We then run the `hausman` test.

```
. gen incFast = (type == 1) * income
. gen incFancy = (type == 3) * income
. gen kidFast = (type == 1) * kids
. gen kidFancy = (type == 3) * kids
. clogit chosen cost rating dist incFast incFancy kidFast kidFancy, group(family_id)
> nolog
Conditional (fixed-effects) logistic regression   Number of obs   =       2100
                                                  LR chi2(7)      =     189.73
                                                  Prob > chi2     =     0.0000
Log likelihood = -488.90834                       Pseudo R2       =     0.1625

------------------------------------------------------------------------------
      chosen |      Coef.   Std. Err.      z    P>|z|     [95% Conf. Interval]
-------------+----------------------------------------------------------------
        cost |  -.1367799   .0358479    -3.82   0.000    -.2070404   -.0665193
      rating |   .3066622   .1418291     2.16   0.031     .0286823     .584642
    distance |  -.1977505   .0471653    -4.19   0.000    -.2901927   -.1053082
     incFast |  -.0390183   .0094018    -4.15   0.000    -.0574455   -.0205911
    incFancy |   .0407053   .0080405     5.06   0.000     .0249462    .0564644
     kidFast |  -.2398757   .1063674    -2.26   0.024     -.448352   -.0313994
    kidFancy |  -.3893862   .1143797    -3.40   0.001    -.6135662   -.1652061
------------------------------------------------------------------------------

. estimates store fullset

. clogit chosen cost rating dist incFast kidFast if type !=2, group(family_id)
> nolog
note: 222 groups (888 obs) dropped due to all positive or
      all negative outcomes.
```

```
Conditional (fixed-effects) logistic regression   Number of obs   =        312
                                                  LR chi2(5)      =      44.35
                                                  Prob > chi2     =     0.0000
Log likelihood = -85.955324                       Pseudo R2       =     0.2051

------------------------------------------------------------------------------
      chosen |      Coef.   Std. Err.      z    P>|z|     [95% Conf. Interval]
-------------+----------------------------------------------------------------
        cost |  -.0616621    .067852    -0.91   0.363    -.1946496    .0713254
      rating |   .1659001   .2832041     0.59   0.558    -.3891698      .72097
    distance |   -.244396   .0995056    -2.46   0.014    -.4394234   -.0493687
     incFast |  -.0737506   .0177444    -4.16   0.000     -.108529   -.0389721
     kidFast |   .4105386   .2137051     1.92   0.055    -.0083157    .8293928
------------------------------------------------------------------------------

. hausman . fullset

                 ---- Coefficients ----
             |      (b)          (B)            (b-B)     sqrt(diag(V_b-V_B))
             |       .         fullset       Difference          S.E.
-------------+----------------------------------------------------------------
        cost |   -.0616621    -.1367799        .0751178        .0576092
      rating |    .1659001     .3066622       -.1407621        .2451308
    distance |    -.244396    -.1977505       -.0466456        .0876173
     incFast |   -.0737506    -.0390183       -.0347323         .015049
     kidFast |    .4105386    -.2398757        .6504143        .1853533
------------------------------------------------------------------------------
                           b = consistent under Ho and Ha; obtained from clogit
            B = inconsistent under Ha, efficient under Ho; obtained from clogit

    Test:  Ho:  difference in coefficients not systematic

                  chi2(5) = (b-B)'[(V_b-V_B)^(-1)](b-B)
                          =       10.70
                Prob>chi2 =      0.0577
                (V_b-V_B is not positive definite)
```

Similar to our findings in example 3, the results of the test of the IIA are mixed. We cannot reject the IIA at the commonly used 5% significance level, but we could at the 10% level. Substantively, a significant test result suggest that the odds of going to one of the fancy restaurants versus going to one of the fast food restaurants changes if we include the family restaurants in the alternative set and that a nested logit specification may be warranted.

◁

Nonnormalized model

Previous versions of Stata fitted a nonnormalized nested logit model that is available via the `nonnormalized` option. The nonnormalized version is presented in, for example, Greene (2003). Here we outline the differences between the RUM-consistent and nonnormalized models. Our discussion follows Heiss (2002) and assumes the decision tree has two levels, with M alternatives at the upper level and a total of J alternatives at the bottom level.

In a RUM framework, by consuming alternative j, decision maker i obtains utility

$$U_{ij} = V_{ij} + \epsilon_{ij} = \alpha_j + \mathbf{x}_{ij}\boldsymbol{\beta}_j + \mathbf{z}_i\boldsymbol{\gamma}_j + \epsilon_{ij}$$

where V_{ij} is the deterministic part of utility and ϵ_{ij} is the random part. $\mathbf{x}_{ij}$ are alternative-specific variables and $\mathbf{z}_i$ are case-specific variables. The set of errors $\epsilon_{i1}, \dots, \epsilon_{iJ}$ are assumed to follow the generalized extreme-value (GEV) distribution, which is a generalization of the type 1 extreme-value distribution that allows for alternatives within nests of the tree structure to be correlated. Let ρ_m denote the correlation in nest m, and define the dissimilarity parameter $\tau_m = \sqrt{1-\rho_m}$. $\tau_m = 0$ implies that the alternatives in nest m are perfectly correlated, whereas $\tau_m = 1$ implies independence.

The *inclusive value* for the mth nest corresponds to the expected value of the utility that decision maker i obtains by consuming an alternative in nest m. Denote this value IV_m:

$$IV_m = \ln \sum_{j \in B_m} \exp\left(V_k/\tau_m\right) \tag{1}$$

where B_m denotes the set of alternatives in nest m. Given the inclusive values, we can show that the probability that random utility–maximizing decision maker i chooses alternative j is

$$\text{Pr}_j = \frac{\exp\{V_j/\tau(j)\}}{\exp\{IV(j)\}} \frac{\exp\{\tau(j)IV(j)\}}{\sum_m \exp\left(\tau_m IV_m\right)}$$

where $\tau(j)$ and $IV(j)$ are the dissimilarity parameter and inclusive value for the nest in which alternative j lies.

In contrast, for the nonnormalized model we have a latent variable

$$\widetilde{V}_{i,j} = \widetilde{\alpha}_j + \mathbf{x}_{i,j}\widetilde{\boldsymbol{\beta}}_j + \mathbf{z}_i\widetilde{\boldsymbol{\gamma}}_j$$

and corresponding inclusive values

$$\widetilde{IV}_m = \ln \sum_{j \in B_m} \exp\left(\widetilde{V}_k\right) \tag{2}$$

The probability of choosing alternative j is

$$\text{Pr}_j = \frac{\exp\left(\widetilde{V}_j\right)}{\exp\left\{\widetilde{IV}(j)\right\}} \frac{\exp\left\{\tau(j)\widetilde{IV}(j)\right\}}{\sum_m \exp\left(\tau_m \widetilde{IV}_m\right)}$$

Equations (1) and (2) represent the key difference between the RUM-consistent and nonnormalized models. By scaling the V_{ij} within each nest, the RUM-consistent model allows utilities to be compared across nests. Without the rescaling, utilities can be compared only for goods within the same nest. Moreover, adding a constant to each V_{ij} for consumer i will not affect the probabilities of the RUM-consistent model, but adding a constant to each $\widetilde{V}_{ij}$ will affect the probabilities from the nonnormalized model. Decisions based on utility maximization can depend only on utility differences and not the scale or zero point of the utility function because utility is an ordinal concept, so the nonnormalized model cannot be consistent with utility maximization.

Heiss (2002) showed that the nonnormalized model can be RUM consistent in the special case where all the variables are specified in the bottom-level equation. Then multiplying the nonnormalized coefficients by the respective dissimilarity parameters results in the RUM-consistent coefficients.

❑ Technical Note

Degenerate nests occur when there is only one alternative in a branch of the tree hierarchy. The associated dissimilarity parameter of the RUM model is not defined. The inclusive-valued parameter of the nonnormalized model will be identifiable if there are alternative-specific variables specified in (1) of the model specification (the *indepvars* in the model syntax). Numerically you can skirt the issue of nonidentifiable/undefined parameters by setting constraints on them. For the RUM model constraint, set the dissimilarity parameter to 1. See the description of `constraints` in *Options* for details on setting constraints on the dissimilarity parameters.

❑

Saved Results

`nlogit` saves the following in `e()`:

Scalars

`e(N)`	number of observations
`e(N_case)`	number of cases
`e(N_clust)`	number of clusters
`e(k_alt)`	number of alternatives for bottom level
`e(k_alt`*j*`)`	number of alternatives for *j*th level
`e(k_indvars)`	number of independent variables
`e(k_ind2vars)`	number of by-alternative variables for bottom level
`e(k_ind2vars`*j*`)`	number of by-alternative variables for *j*th level
`e(k_eq)`	number of equations in `e(b)`
`e(k_eq_model)`	number of equations in model Wald test
`e(const)`	constant indicator for bottom level
`e(const`*j*`)`	constant indicator for *j*th level
`e(i_base)`	base index for bottom level
`e(i_base`*j*`)`	base index for *j*th level
`e(levels)`	number of levels
`e(df_m)`	model degrees of freedom
`e(df_c)`	`clogit` model degrees of freedom
`e(ll)`	log likelihood
`e(ll_c)`	`clogit` model log likelihood
`e(chi2)`	χ^2
`e(chi2_c)`	likelihood-ratio test for IIA
`e(p)`	p-value for model Wald test
`e(p_c)`	p-value for IIA test
`e(alt_min)`	minimum number of alternatives
`e(alt_avg)`	average number of alternatives
`e(alt_max)`	maximum number of alternatives
`e(rank)`	rank of `e(V)`
`e(ic)`	number of iterations
`e(rc)`	return code
`e(converged)`	`1` if converged, `0` otherwise
`e(rum)`	`1` if RUM model, `0` otherwise

(*Continued on next page*)

Macros

`e(cmd)`	`nlogit`
`e(cmdline)`	command as typed
`e(depvar)`	dependent variable
`e(indvars)`	independent variables
`e(ind2vars)`	by-alternative variables for bottom level
`e(ind2vars`*j*`)`	by-alternative variables for *j*th level
`e(case)`	variable defining cases
`e(altvar)`	alternative variable for bottom level
`e(altvar`*j*`)`	alternative variable for *j*th level
`e(alteqs)`	equation names for bottom level
`e(alteqs`*j*`)`	equation names for *j*th level
`e(alt`*i*`)`	*i*th alternative for bottom level
`e(alt`*j_i*`)`	*i*th alternative for *j*th level
`e(title)`	title in estimation output
`e(wtype)`	weight type
`e(wexp)`	weight expression
`e(clustvar)`	name of cluster variable
`e(chi2type)`	`Wald`, type of model χ^2 test
`e(vce)`	*vcetype* specified in `vce()`
`e(vcetype)`	title used to label Std. Err.
`e(opt)`	type of optimization
`e(ml_method)`	type of `ml` method
`e(user)`	name of likelihood-evaluator program
`e(technique)`	maximization technique
`e(crittype)`	optimization criterion
`e(properties)`	`b V`
`e(datasignaturevars)`	variables used in calculation of checksum
`e(datasignature)`	the checksum
`e(predict)`	program used to implement `predict`
`e(estat_cmd)`	program used to implement `estat`

Matrices

`e(b)`	coefficient vector
`e(k_altern)`	number of alternatives at each level
`e(k_branch`*j*`)`	number of branches at each alternative of *j*th level
`e(stats)`	alternative statistics for bottom level
`e(stats`*j*`)`	alternative statistics for *j*th level
`e(altidx`*j*`)`	alternative indices for *j*th level
`e(ilog)`	iteration log (up to 20 iterations)
`e(gradient)`	gradient vector
`e(V)`	variance–covariance matrix of the estimators

Functions

`e(sample)`	marks estimation sample

Methods and Formulas

`nlogit`, `nlogitgen`, and `nlogittree` are implemented as ado-files.

Consider our two-level nested logit model for restaurant choice. We define $T = \{1, 2, 3\}$ to be the set of indices denoting the three restaurant types and $R_1 = \{1, 2\}$, $R_2 = \{3, 4, 5\}$, and $R_3 = \{6, 7\}$ to be the set of indices representing each restaurant within type $t \in T$. Let C_1 and C_2 be the random variables that represent the choices made for the first level, restaurant type, and second level, restaurant, of the hierarchy, where we observe the choices $C_1 = t, t \in T$, and $C_2 = j, j \in R_t$. Let $\mathbf{z}_t$ and $\mathbf{x}_{tj}$, for $t \in T$ and $j \in R_t$ refer to the row vectors of explanatory variables for the first-level alternatives and bottom-level alternatives for one case, respectively. We write the utilities (latent variables) as $U_{tj} = \mathbf{z}_t\boldsymbol{\alpha}_t + \mathbf{x}_{tj}\boldsymbol{\beta}_j + \epsilon_{tj} = \boldsymbol{\eta}_{tj} + \epsilon_{tj}$, where $\boldsymbol{\alpha}_t$ and $\boldsymbol{\beta}_j$ are column vectors and the ϵ_{tj} are random disturbances. When the $\mathbf{x}_{tj}$ are alternative specific, we can drop the indices from $\boldsymbol{\beta}$, where we estimate one coefficient for each alternative in R_t, $t \in T$. These variables are specified in the first equation of the `nlogit` syntax (see *Example 3*).

When the random-utility framework is used to describe the choice behavior, the alternative that is chosen is the alternative that has the highest utility. Assume for our restaurant example that we choose restaurant type $t \in T$. For the RUM parameterization of `nlogit`, the conditional distribution of ϵ_{tj} given choice of restaurant type t is a multivariate version of Gumbel's extreme-value distribution,

$$F_{R|T}(\boldsymbol{\epsilon} \,|\, t) = \exp\left[-\left\{\sum_{m \in R_t} \exp(\epsilon_{tm}/\tau_t)\right\}^{\tau_t}\right] \tag{3}$$

where it has been shown that the ϵ_{tj}, $j \in R_t$ are exchangeable with correlation $1 - \tau_t^2$, for $\tau_t \in (0, 1]$ (Kotz and Nadarajah 2000). For example, the probability of choosing Christopher's, $j = 6$ given type $t = 3$, is

$$\begin{aligned}
\Pr(C_2 = 6 \,|\, C_1 = 3) &= \Pr\left(U_{36} - U_{37} > 0\right) \\
&= \Pr\left(\epsilon_{37} \le \epsilon_{36} + \eta_{36} - \eta_{37}\right) \\
&= \int_{-\infty}^{\infty} \left\{ \int_{-\infty}^{\epsilon_{36}+\eta_{36}-\eta_{37}} f_{R|T}\left(\epsilon_{36}, \epsilon_{37}\right) d\epsilon_{37} \right\} d\epsilon_{36}
\end{aligned}$$

where $f = \dfrac{\partial F}{\partial \epsilon_{36} \partial \epsilon_{37}}$ is the joint density function of $\boldsymbol{\epsilon}$ given t. U_{37} is the utility of eating at Mad Cows, the other fancy ($t = 3$) restaurant. Amemiya (1985) demonstrates that this integral evaluates to the logistic function

$$\begin{aligned}
\Pr(C_2 = 6 \,|\, C_1 = 3) &= \frac{\exp(\boldsymbol{\eta}_{36}/\tau_3)}{\exp(\boldsymbol{\eta}_{36}/\tau_3) + \exp(\boldsymbol{\eta}_{37}/\tau_3)} \\
&= \frac{\exp(\mathbf{x}_{36}\boldsymbol{\beta}_6/\tau_3)}{\exp(\mathbf{x}_{36}\boldsymbol{\beta}_6/\tau_3) + \exp(\mathbf{x}_{37}\boldsymbol{\beta}_7/\tau_3)}
\end{aligned}$$

and in general

$$\Pr(C_2 = j \,|\, C_1 = t) = \frac{\exp(\mathbf{x}_{tj}\boldsymbol{\beta}_j/\tau_t)}{\sum_{m \in R_t} \exp(\mathbf{x}_{tm}\boldsymbol{\beta}_m/\tau_t)} \tag{4}$$

Letting $\tau_t = 1$ in (3) reduces to the product of independent extreme-value distributions, and (4) reduces to the multinomial logistic function.

For the logistic function in (4), we scale the linear predictors by the dissimilarity parameters. Another formulation of the conditional probability of choosing alternative $j \in R_t$ given choice $t \in T$ is the logistic function without this normalization:

$$\Pr(C_2 = j \,|\, C_1 = t) = \frac{\exp(\mathbf{x}_{tj}\boldsymbol{\beta}_j)}{\sum_{m\in R_t} \exp(\mathbf{x}_{tm}\boldsymbol{\beta}_m)}$$

and this is what is used in `nlogit`'s nonnormalized parameterization.

Amemiya (1985) defines the general form for the joint distribution of the ϵ's as

$$F_{T,R}(\boldsymbol{\epsilon}) = \exp\left\{ -\sum_{k\in T} \theta_k \left(\sum_{m\in R_k} \exp(-\epsilon_{km}/\tau_k) \right)^{\tau_k} \right\}$$

from which the probability of choice $t, t \in T$ can be derived as

$$\Pr(C_1 = t) = \frac{\theta_t \left\{\sum_{m\in R_t} \exp(\boldsymbol{\eta}_{tm}/\tau_t)\right\}^{\tau_t}}{\sum_{k\in T} \theta_k \left\{\sum_{m\in R_k} \exp(\boldsymbol{\eta}_{km}/\tau_k)\right\}^{\tau_k}} \tag{5}$$

`nlogit` sets $\theta_t = 1$. Noting that

$$\begin{aligned}
\left\{ \sum_{m\in R_t} \exp(\boldsymbol{\eta}_{tm}/\tau_t) \right\}^{\tau_t} &= \left\{ \sum_{m\in R_t} \exp\left(\frac{\mathbf{z}_t\boldsymbol{\alpha}_t + \mathbf{x}_{tm}\boldsymbol{\beta}_m}{\tau_t} \right) \right\}^{\tau_t} \\
&= \exp(\mathbf{z}_t\boldsymbol{\alpha}_t) \left\{ \sum_{m\in R_t} \exp\left(\mathbf{x}_{tm}\boldsymbol{\beta}_m/\tau_t\right) \right\}^{\tau_t} \\
&= \exp(\mathbf{z}_t\boldsymbol{\alpha}_t + \tau_t I_t)
\end{aligned}$$

we define the inclusive values I_t as

$$I_t = \ln\left\{ \sum_{m\in R_t} \exp(\mathbf{x}_{tm}\boldsymbol{\beta}_m/\tau_t) \right\}$$

and we can view

$$\exp(\tau_t I_t) = \left\{ \sum_{m\in R_t} \exp(x_{tm}\boldsymbol{\beta}_m)^{1/\tau_t} \right\}^{\tau_t}$$

as a weighted average of the $\exp(x_{tm}\boldsymbol{\beta}_m)$, for $m \in R_t$. For the `nlogit` RUM parameterization, we can express (5) as

$$\Pr(C_1 = t) = \frac{\exp(\mathbf{z}_t\boldsymbol{\alpha}_t + \tau_t I_t)}{\sum_{k\in T} \exp(\mathbf{z}_k\boldsymbol{\alpha}_k + \tau_k I_k)}$$

Next we define inclusive values for the nonnormalized model to be

$$\widetilde{I}_t = \ln\left\{ \sum_{m\in R_t} \exp(\mathbf{x}_{tm}\boldsymbol{\beta}_m) \right\}$$

and we express $\Pr(C_1 = t)$ as

$$\Pr(C_1 = t) = \frac{\exp(\mathbf{z}_t\boldsymbol{\alpha}_t + \tau_t\widetilde{I}_t)}{\sum_{k\in T}\exp(\mathbf{z}_k\boldsymbol{\alpha}_k + \tau_k\widetilde{I}_k)} \tag{6}$$

Equation (5) is consistent with (6) only when $\boldsymbol{\eta}_{ij} = \mathbf{x}_{ij}\boldsymbol{\beta}_j$, so in general the `nlogit` nonnormalized model is not consistent with the RUM model.

Now assume that we have N cases where we add a third subscript, i, to denote case i, $i = 1,\dots,N$. Denote y_{itj} to be a binary variable indicating the choice made by case i so that for each i only one y_{itj} is 1 and the rest are 0 for all $t \in T$ and $j \in R_t$. The log likelihood for the two-level RUM-consistent model is

$$\begin{aligned}\log\ell &= \sum_{i=1}^{N}\sum_{k\in T}\sum_{m\in R_k} y_{ikm}\log\left\{\Pr(C_{i1}=i)\Pr(C_{i2}=m|C_{i1}=i)\right\}\\ &= \sum_{i=1}^{N}\sum_{k\in T}\sum_{m\in R_k} y_{ikm}\left[\mathbf{z}_{ik}\boldsymbol{\alpha}_k + \tau_k I_{ik} - \log\left\{\sum_{l\in T}\exp(\mathbf{z}_{il}\boldsymbol{\alpha}_l + \tau_l I_{il})\right\} + \right.\\ &\qquad \left.\mathbf{x}_{ikm}\boldsymbol{\beta}_m/\tau_k - \log\left\{\sum_{l\in R_k}\exp(\mathbf{x}_{ikl}\boldsymbol{\beta}_m/\tau_k)\right\}\right]\end{aligned}$$

The likelihood for the nonnormalized model has a similar form, replacing I with $\widetilde{I}$ and by not scaling $\mathbf{x}_{ikj}\boldsymbol{\beta}_j$ by τ_k.

Three-level nested logit model

Here we define a three-level nested logit model that can be generalized to the four-level and higher models. As before, let the integer set T be the indices for the first level of choices. Let sets S_t, $t \in T$, be mutually exclusive sets of integers representing the choices of the second level of the hierarchy. Finally, let R_j, $j \in S_t$ be the bottom-level choices. Let $U_{tjk} = \eta_{tjk} + \epsilon_{tjk}$, $k \in R_j$, and the distribution of ϵ_{tjk} be Gumbel's multivariate extreme value of the form

$$F(\epsilon) = \exp\left(-\sum_{t\in T}\left[\sum_{j\in S_t}\left\{\sum_{k\in R_j}\exp(-\eta_{tjk}/\tau_j)\right\}^{\tau_j/\upsilon_t}\right]^{\upsilon_j}\right)$$

Let C_1, C_2, and C_3 represent the choice random variables for levels 1, 2, and the bottom, respectively. Then the set of conditional probabilities is

(Continued on next page)

$$\Pr(C_3 = k \mid C_1 = t, C_2 = j) = \frac{\exp(\eta_{tjk}/\tau_j)}{\sum_{l\in R_j}\exp(\eta_{tjl}/\tau_j)}$$

$$\Pr(C_2 = j \mid C_1 = t) = \frac{\left\{\sum_{k\in R_j}\exp(\eta_{tjk}/\tau_j)\right\}^{\tau_j/\upsilon_t}}{\sum_{l\in T_t}\left\{\sum_{k\in R_l}\exp(\eta_{tlk}/\tau_l)\right\}^{\tau_l/\upsilon_t}}$$

$$\Pr(C_1 = t) = \frac{\left[\sum_{j\in S_t}\left\{\sum_{k\in R_j}\exp(\eta_{tjk}/\tau_j)\right\}^{\tau_j/\upsilon_t}\right]^{\upsilon_t}}{\sum_{l\in T}\left[\sum_{j\in S_l}\left\{\sum_{k\in R_j}\exp(\eta_{ljk}/\tau_j)\right\}^{\tau_j/\upsilon_l}\right]^{\upsilon_l}}$$

Assume that we can decompose the linear predictor as $\boldsymbol{\eta}_{tjk} = \mathbf{z}_t\boldsymbol{\alpha}_t + \mathbf{u}_{tj}\boldsymbol{\gamma}_j + \mathbf{x}_{tjk}\boldsymbol{\beta}_k$. Here $\mathbf{z}_t$, $\mathbf{u}_{tj}$, and $\mathbf{x}_{tjk}$ are the row vectors of explanatory variables for the first, second, and bottom levels of the hierarchy, respectively, and $\boldsymbol{\alpha}_t$, $\boldsymbol{\gamma}_j$, and $\boldsymbol{\beta}_k$ are the corresponding column vectors of regression coefficients for $t \in T$, $j \in S_t$, and $k \in R_j$. We then can define the inclusive values for the first and second levels as

$$I_{tj} = \log \sum_{k\in R_j}\exp(\mathbf{x}_{tjk}\boldsymbol{\beta}_j/\tau_j)$$

$$J_t = \log\left(\sum_{j\in S_t}\mathbf{u}_{tj}\boldsymbol{\gamma}_j/\upsilon_t + \frac{\tau_j}{\upsilon_t}I_{tj}\right)$$

and rewrite the probabilities

$$\Pr(C_3 = k \mid C_1 = t, C_2 = j) = \frac{\exp(\mathbf{x}_{tjk}\boldsymbol{\beta}_k/\tau_j)}{\sum_{l\in R_j}\exp(\mathbf{x}_{tjk}\boldsymbol{\beta}_l/\tau_j)}$$

$$\Pr(C_2 = j \mid C_1 = t) = \frac{\exp(\mathbf{u}_{tj}\boldsymbol{\gamma}_j/\upsilon_t + \frac{\tau_j}{\upsilon_t}I_{tj})}{\sum_{l\in T_t}\exp(\mathbf{u}_{tl}\boldsymbol{\gamma}_l/\upsilon_t + \frac{\tau_l}{\upsilon_t}I_{tl})}$$

$$\Pr(C_1 = t) = \frac{\exp(\mathbf{z}_t\boldsymbol{\alpha}_t + \upsilon_t J_t)}{\sum_{l\in T}\exp(\mathbf{z}_l\boldsymbol{\alpha}_l + \upsilon_l J_l)}$$

We add a fourth index i for case and define the indicator variable y_{lijk}, $l = 1,\dots,N$ to indicate the choice made by case i, $t \in T$, $j \in S_t$, and $k \in R_j$. The log likelihood for the `nlogit` RUM-consistent model is

$$\begin{aligned}\ell = \sum_{i=1}^{N}\sum_{t\in T}\sum_{j\in S_t}\sum_{k\in R_j} y_{itjk}\Bigg\{&\mathbf{z}_{it}\boldsymbol{\alpha}_t + \upsilon_t J_{it} - \log\left(\sum_{m\in T}\mathbf{z}_{im}\boldsymbol{\alpha}_m + \upsilon_m J_{im}\right) + \\ &\mathbf{u}_{itj}\boldsymbol{\gamma}_j/\upsilon_t + \frac{\tau_j}{\upsilon_t}I_{itj} - \log\left(\sum_{m\in S_t}\mathbf{u}_{itm}\boldsymbol{\gamma}_m/\upsilon_t + \frac{\tau_m}{\upsilon_t}I_{itm}\right) + \\ &\mathbf{x}_{itjk}\boldsymbol{\beta}_k/\tau_k - \sum_{m\in R_t}\exp(\mathbf{x}_{itjm}\boldsymbol{\beta}_m/\tau_k)\Bigg\}\end{aligned}$$

and for the nonnormalized `nlogit` model the log likelihood is

$$\ell = \sum_{i=1}^{N} \sum_{t\in T} \sum_{j\in S_t} \sum_{k\in R_j} y_{itjk} \Bigg\{ \mathbf{z}_{it}\boldsymbol{\alpha}_t + \upsilon_t J_{it} - \log\Bigg(\sum_{m\in T} \mathbf{z}_{im}\boldsymbol{\alpha}_m + \upsilon_m J_{im} \Bigg) +$$
$$\mathbf{u}_{itj}\boldsymbol{\gamma}_j + \tau_j I_{itj} - \log\Bigg(\sum_{m\in S_t} \mathbf{u}_{itm}\boldsymbol{\gamma}_m + \tau_m I_{itm} \Bigg) +$$
$$\mathbf{x}_{itjk}\boldsymbol{\beta}_k - \sum_{m\in R_t} \exp(\mathbf{x}_{itjm}\boldsymbol{\beta}_m) \Bigg\}$$

Extending the model to more than three levels is straightforward, albeit notationally cumbersome.

References

Amemiya, T. 1985. *Advanced Econometrics.* Cambridge, MA: Harvard University Press.

Greene, W. H. 2003. *Econometric Analysis.* 5th ed. Upper Saddle River, NJ: Prentice Hall.

Hausman, J. A. 1978. Specification tests in econometrics. *Econometrica* 46: 1251–1271.

Hausman, J., and D. McFadden. 1984. Specification tests in econometrics. *Econometrica* 52: 1219–1240.

Heiss, F. 2002. Structural choice analysis with nested logit models. *Stata Journal* 2: 227–252.

Hensher, D. A., Rose, J. M., and Greene, W. H. 2005. *Applied Choice Analysis: A Primer.* New York: Cambridge University Press.

Kotz, S., and S. Nadarajah. 2000. *Extreme Value Distributions: Theory and Applications.* London: Imperial College Press.

Maddala, G. S. 1983. *Limited-Dependent and Qualitative Variables in Econometrics.* Cambridge: Cambridge University Press.

McFadden, D. 1977. Quantitative methods for analyzing behavior of individuals: Some recent developments. Cowles Foundation Discussion Paper no. 474.

——. 1981. Econometric models of probabilistic choice. In *Structural Analysis of Discrete Data with Econometric Applications*, pp. 198–272. Cambridge, MA: MIT Press.

Also See

[R] **nlogit postestimation** — Postestimation tools for nlogit

[R] **asclogit** — Alternative-specific conditional logit (McFadden's choice) model

[R] **mlogit** — Multinomial (polytomous) logistic regression

[R] **ologit** — Ordered logistic regression

[R] **rologit** — Rank-ordered logistic regression

[R] **slogit** — Stereotype logistic regression

[U] **20 Estimation and postestimation commands**

Title

nlogit postestimation — Postestimation tools for nlogit

Description

The following postestimation commands are of special interest after `nlogit`:

command	description
`estat alternatives`	alternative summary statistics

For information about this command, see [R] **asmprobit postestimation**.

The following standard postestimation commands are also available:

command	description
`estat`	AIC, BIC, VCE, and estimation sample summary
`estimates`	cataloging estimation results
`hausman`	Hausman's specification test
`lincom`	point estimates, standard errors, testing, and inference for linear combinations of coefficients
`lrtest`	likelihood-ratio test
`nlcom`	point estimates, standard errors, testing, and inference for nonlinear combinations of coefficients
`predict`	predictions, residuals, influence statistics, and other diagnostic measures
`predictnl`	point estimates, standard errors, testing, and inference for generalized predictions
`test`	Wald tests for simple and composite linear hypotheses
`testnl`	Wald tests of nonlinear hypotheses

See the corresponding entries in the *Stata Base Reference Manual* for details.

Syntax for predict

predict [*type*] *newvar* [*if*] [*in*] [, *statistic* hlevel(#) altwise]

predict [*type*] { *stub** | *newvarlist* } [*if*] [*in*], scores

statistic	description
Main	
pr	predicted probabilities of choosing the alternatives at all levels of the hierarchy or at level #, where # is specified by hlevel(#); the default
xb	linear predictors for all levels of the hierarchy or at level #, where # is specified by hlevel(#)
condp	predicted conditional probabilities at all levels of the hierarchy or at level #, where # is specified by hlevel(#)
iv	inclusive values for levels 2, ..., e(levels) or for hlevel(#)

The inclusive value for the first-level alternatives is not used in estimation; therefore, it is not calculated.

These statistics are available both in and out of sample; type predict ... if e(sample) ... if wanted only for the estimation sample.

Options for predict

Main

pr calculates the probability of choosing each alternative at each level of the hierarchy. Use the hlevel(#) option to compute the alternative probabilities at level #. When hlevel(#) is not specified, j new variables must be given, where j is the number of levels, or use the *stub** option to have predict generate j variables with the prefix *stub* and numbered from 1 to j. Option pr is the default and if one new variable is given, the probability of the bottom-level alternatives are computed. Otherwise, probabilities for all levels are computed and the *stub** option is still valid.

xb calculates the linear prediction for each alternative at each level. Use the hlevel(#) option to compute the linear predictor at level #. When hlevel(#) is not specified, j new variables must be given, where j is the number of levels, or use the *stub** option to have predict generate j variables with the prefix *stub* and numbered from 1 to j.

condp calculates the conditional probabilities for each alternative at each level. Use the hlevel(#) option to compute the conditional probabilities of the alternatives at level #. When hlevel(#) is not specified, j new variables must be given, where j is the number of levels, or use the *stub** option to have predict generate j variables with the prefix *stub* and numbered from 1 to j.

iv calculates the inclusive value for each alternative at each level. Use the hlevel(#) option to compute the inclusive value at level #. There is no inclusive value at level 1. If hlevel(#) is not used, $j - 1$ new variables are required, where j is the number of levels, or use *stub** to have predict generate $j - 1$ variables with the prefix *stub* and numbered from 2 to j. See *Methods and Formulas* in [R] **nlogit** for a definition of the inclusive values.

hlevel(#) calculates the prediction only for hierarchy level #.

altwise specifies that alternative-wise deletion be used when marking out observations due to missing values in your variables. The default is to use casewise deletion. The option xb always uses alternative-wise deletion.

`scores` calculates the scores for each coefficient in `e(b)`. This option requires a new-variable list of length equal to the number of columns in `e(b)`. Otherwise, use the *stub** option to have `predict` generate enumerated variables with prefix *stub*.

Remarks

`predict` may be used after `nlogit` to obtain the predicted values of the probabilities, the conditional probabilities, the linear predictions, and the inclusive values for each level of the nested logit model. Predicted probabilities for `nlogit` must be interpreted carefully. Probabilities are estimated for each case as a whole and not for individual observations.

▷ Example 1

Continuing with our model in example 3 of [R] **nlogit**, we refit the model and then examine a summary of the alternatives and their frequencies in the estimation sample.

```
. use http://www.stata-press.com/data/r10/restaurant
. nlogitgen type = restaurant(fast: Freebirds | MamasPizza,
> family: CafeEccell | LosNortenos | WingsNmore, fancy: Christophers | MadCows)
  (output omitted )
. nlogit chosen cost rating distance || type: income kids, base(family) ||
> restaurant:, noconst case(family_id)
  (output omitted )
. estat alternatives
  Alternatives summary for type
```

index	Alternative value	label	Cases present	Frequency selected	Percent selected
1	1	fast	600	27	9.00
2	2	family	900	222	74.00
3	3	fancy	600	51	17.00

Alternatives summary for restaurant

index	Alternative value	label	Cases present	Frequency selected	Percent selected
1	1	Freebirds	300	12	4.00
2	2	MamasPizza	300	15	5.00
3	3	CafeEccell	300	78	26.00
4	4	LosNortenos	300	75	25.00
5	5	WingsNmore	300	69	23.00
6	6	Christophers	300	27	9.00
7	7	MadCows	300	24	8.00

Next we predict p2 = Pr(`restaurant`); p1 = Pr(`type`); `condp` = Pr(`restaurant` | `type`); `xb2`, the linear prediction for the bottom-level alternatives; `xb1`, the linear prediction for the first-level alternatives; and `iv`, the inclusive values for the bottom-level alternatives.

```
. predict p*
(option pr assumed)
. predict condp, condp hlevel(2)
. sort family_id type restaurant
. list restaurant type chosen p2 p1 condp in 1/14, sepby(family_id) divider
```

	restaurant	type	chosen	p2	p1	condp
1.	Freebirds	fast	1	.0642332	.1189609	.5399519
2.	MamasPizza	fast	0	.0547278	.1189609	.4600481
3.	CafeEccell	family	0	.284409	.7738761	.3675124
4.	LosNortenos	family	0	.3045242	.7738761	.3935051
5.	WingsNmore	family	0	.1849429	.7738761	.2389825
6.	Christophers	fancy	0	.0429508	.107163	.4007991
7.	MadCows	fancy	0	.0642122	.107163	.5992009
8.	Freebirds	fast	0	.0183578	.0488948	.3754559
9.	MamasPizza	fast	0	.030537	.0488948	.6245441
10.	CafeEccell	family	0	.2832149	.756065	.3745907
11.	LosNortenos	family	1	.3038883	.756065	.4019341
12.	WingsNmore	family	0	.1689618	.756065	.2234752
13.	Christophers	fancy	0	.1041277	.1950402	.533878
14.	MadCows	fancy	0	.0909125	.1950402	.466122

```
. predict xb*, xb
. predict iv, iv
. list restaurant type chosen xb* iv in 1/14, sepby(family_id) divider
```

	restaurant	type	chosen	xb1	xb2	iv
1.	Freebirds	fast	1	-1.124805	-1.476914	-.9026417
2.	MamasPizza	fast	0	-1.124805	-1.751229	-.9026417
3.	CafeEccell	family	0	0	-2.181112	.1303341
4.	LosNortenos	family	0	0	-2.00992	.1303341
5.	WingsNmore	family	0	0	-3.259229	.1303341
6.	Christophers	fancy	0	1.405185	-6.804211	-.4025908
7.	MadCows	fancy	0	1.405185	-5.155514	-.4025908
8.	Freebirds	fast	0	-1.804794	-2.552233	-1.564074
9.	MamasPizza	fast	0	-1.804794	-1.680583	-1.564074
10.	CafeEccell	family	0	0	-2.400434	.0237072
11.	LosNortenos	family	1	0	-2.223939	.0237072
12.	WingsNmore	family	0	0	-3.694409	.0237072
13.	Christophers	fancy	0	1.490775	-5.35932	-.3159956
14.	MadCows	fancy	0	1.490775	-5.915751	-.3159956

◁

Methods and Formulas

All postestimation commands listed above are implemented as ado-files.

Also See

[R] **nlogit** — Nested logit regression

[U] **20 Estimation and postestimation commands**

Title

nlsur — Estimation of nonlinear systems of equations

Syntax

Interactive version

nlsur (*depvar_1* = *<sexp_1>*) (*depvar_2* = *<sexp_2>*) ... [*if*] [*in*] [*weight*] [, *options*]

Programmed substitutable expression version

nlsur *sexp_prog* : *depvar_1 depvar_2* ... [*varlist*] [*if*] [*in*] [*weight*] [, *options*]

Function evaluator program version

nlsur *func_prog* @ *depvar_1 depvar_2* ... [*varlist*] [*if*] [*in*] [*weight*] ,

nequations(#) { parameters(*namelist*) | nparameters(#)} [*options*]

where

depvar_j is the dependent variable for equation j;

<sexp>_j is the substitutable expression for equation j;

sexp_prog is a substitutable expression program; and

func_prog is a function evaluator program.

(*Continued on next page*)

options	description
Model	
fgnls	use two-step FGNLS estimator; the default
ifgnls	use iterative FGNLS estimator
nls	use NLS estimator
variables(*varlist*)	variables in model
initial(*initial_values*)	initial values for parameters
nequations(#)	number of equations in model (function evaluator program version only)
* parameters(*namelist*)	parameters in model (function evaluator program version only)
* nparameters(#)	number of parameters in model (function evaluator program version only)
sexp_options	options for substitutable expression program
func_options	options for function evaluator program
SE/Robust	
vce(*vcetype*)	*vcetype* may be gnr, robust, cluster *clustvar*, bootstrap, or jackknife
Reporting	
level(#)	set confidence level; default is level(95)
title(*string*)	display *string* as title above the table of parameter estimates
title2(*string*)	display *string* as subtitle
Opt options	
optimization_options	control the optimization process; seldom used
eps(#)	specify # for convergence criteria; default is eps(1e-5)
ifgnlsiterate(#)	set maximum number of FGNLS iterations
ifgnlseps(#)	specify # for FGNLS convergence criterion; default is ifgnlseps(1e-10)
delta(#)	specify stepsize # for computing derivatives; default is delta(4e-7)
noconstants	no equations have constant terms
hasconstants(*namelist*)	use *namelist* as constant terms

* You must specify parameters(*namelist*), nparameters(#), or both.

bootstrap, by, jackknife, rolling, and statsby are allowed; see [U] **11.1.10 Prefix commands**.

Weights are not allowed with the bootstrap prefix.

aweights are not allowed with the jackknife prefix.

fweights, aweights, pweights, and iweights are allowed; see [U] **11.1.6 weight**.

See [U] **20 Estimation and postestimation commands** for more capabilities of estimation commands.

Description

nlsur fits a system of nonlinear equations by feasible generalized nonlinear least squares (FGNLS). With the interactive version of the command, you enter the system of equations on the command line or in the dialog box, using substitutable expressions. If you have a system that you fit regularly, you can write a substitutable expression program and use the second syntax to avoid having to reenter the system every time. The function evaluator program version gives you the most flexibility in exchange for increased complexity; with this version, your program is given a vector of parameters and a variable list, and your program computes the system of equations.

When you write a substitutable expression program or a function evaluator program, the first five letters of the name must be `nlsur`. *sexp_prog* and *func_prog* refer to the name of the program without the first five letters. For example, if you wrote a function evaluator program named `nlsurregss`, you would type `nlsur regss @ ...` to estimate the parameters.

Options

Model

`fgnls` requests the two-step feasible generalized nonlinear least-squares (FGNLS) estimator. This is the default.

`ifgnls` requests the iterative FGNLS estimator. For the nonlinear systems estimator, this is equivalent to maximum likelihood estimation.

`nls` requests the nonlinear least-squares (NLS) estimator.

`variables(`*varlist*`)` specifies the variables in the system. `nlsur` ignores observations for which any of these variables has missing values. If you do not specify `variables()`, `nlsur` issues an error if the estimation sample contains any missing values.

`initial(`*initial_values*`)` specifies the initial values to begin the estimation. You can specify a $1 \times k$ matrix, where k is the total number of parameters in the system, or you can specify a parameter name, its initial value, another parameter name, its initial value, and so on. For example, to initialize `alpha` to 1.23 and `delta` to 4.57, you would type

```
. nlsur ..., initial(alpha 1.23 delta 4.57) ...
```

Initial values declared using this option override any that are declared within substitutable expressions. If you specify a matrix, the values must be in the same order that the parameters are declared in your model. `nlsur` ignores the row and column names of the matrix.

`nequations(`*#*`)` specifies the number of equations in the system.

`parameters(`*namelist*`)` specifies the names of the parameters in the system. The names of the parameters must adhere to the naming conventions of Stata's variables; see [U] **11.3 Naming conventions**. If you specify both `parameters()` and `nparameters()`, the number of names in the former must match the number specified in the latter.

`nparameters(`*#*`)` specifies the number of parameters in the system. If you do not specify names with the `parameters()` options, `nlsur` names them `b1`, `b2`, ..., `b`*#*. If you specify both `parameters()` and `nparameters()`, the number of names in the former must match the number specified in the latter.

sexp_options refer to any options allowed by your *sexp_prog*.

func_options refer to any options allowed by your *func_prog*.

SE/Robust

`vce(`*vcetype*`)` specifies the type of standard error reported, which includes types that are derived from asymptotic theory, that are robust to some kinds of misspecification, that allow for intragroup correlation, and that use bootstrap or jackknife methods; see [R] ***vce_option***.

`vce(gnr)`, the default, uses the conventionally derived variance estimator for nonlinear models fit using Gauss–Newton regression.

Reporting

`level(`*#*`)`; see [R] **estimation options**.

`title(`*string*`)` specifies an optional title that will be displayed just above the table of parameter estimates.

`title2(`*string*`)` specifies an optional subtitle that will be displayed between the title specified in `title()` and the table of parameter estimates. If `title2()` is specified but `title()` is not, `title2()` has the same effect as `title()`.

Opt options

optimization_options: `iterate(#)`, `[no]log`, `trace`. `iterate()` specifies the maximum number of iterations to use for NLS at each round of FGNLS estimation. This option is different from `ifgnlsiterate()`, which controls the maximum rounds of FGNLS estimation to use when the `ifgnls` option is specified. `log`/`nolog` specifies whether to show the iteration log, and `trace` specifies that the iteration log should include the current parameter vector.

`eps(#)` specifies the convergence criterion for successive parameter estimates and for the residual sum of squares (RSS). The default is `eps(1e-5)` (0.00001). `eps()` also specifies the convergence criterion for successive parameter estimates between rounds of iterative FGNLS estimation when `ifgnls` is specified.

`ifgnlsiterate(#)` specifies the maximum number of FGNLS iterations to perform. The default is the number set using `set maxiter`, which is 16,000 by default. To use this option, you must also specify the `ifgnls` option.

`ifgnlseps(#)` specifies the convergence criterion for successive estimates of the error covariance matrix during iterative FGNLS estimation. The default is `ifgnlseps(1e-10)`. To use this option, you must also specify the `ifgnls` option.

`delta(#)` specifies the relative change in a parameter, δ, to be used in computing the numeric derivatives. The derivative for parameter β_i is computed as

$$\{f_i(\mathbf{x}_i, \beta_1, \beta_2, \ldots, \beta_i + d, \beta_{i+1}, \ldots) - f_i(\mathbf{x}_i, \beta_1, \beta_2, \ldots, \beta_i, \beta_{i+1}, \ldots)\}/d$$

where $d = \delta(|\beta_i| + \delta)$. The default is `delta(4e-7)`.

`noconstants` indicates that none of the equations in the system include constant terms. This option is generally not needed, even if there are no constant terms in the system; though in rare cases without this option, `nlsur` may claim that there is one or more constant terms even if there are none.

`hasconstants(`*namelist*`)` indicates the parameters that are to be treated as constant terms in the system of equations. The number of elements of *namelist* must equal the number of equations in the system. The ith entry of *namelist* specifies the constant term in the ith equation. If an equation does not include a constant term, specify a period (.) instead of a parameter name. This option is seldom needed with the interactive and programmed substitutable expression versions, because in those cases `nlsur` can almost always find the constant terms automatically.

Remarks

Remarks are presented under the following headings:

Introduction

`nlsur` fits a system of nonlinear equations by feasible generalized nonlinear least squares. It can be viewed as a nonlinear variant of Zellner's seemingly unrelated regression model (Zellner 1962, Zellner and Huang 1962, Zellner 1963) and is therefore commonly called nonlinear SUR or nonlinear SURE. The model is also discussed in textbooks such as Davidson and MacKinnon (1993, 2004) and Greene (2003). Formally, the model fitted by `nlsur` is

$$
\begin{aligned}
y_{i1} &= f_1(\mathbf{x}_i, \beta) + u_{i1} \\
y_{i2} &= f_2(\mathbf{x}_i, \beta) + u_{i2} \\
\vdots &= \vdots \\
y_{iM} &= f_M(\mathbf{x}_i, \beta) + u_{iM}
\end{aligned}
$$

for $i = 1, \dots, N$ observations and $m = 1, \dots, M$ equations. The errors for the ith observation, $u_{i1}, u_{i2}, \dots, u_{iM}$ may be correlated, so fitting the m equations jointly may lead to more efficient estimates. Moreover, fitting the equations jointly allows us to impose cross-equation restrictions on the parameters. Not all elements of the parameter vector β and data vector $\mathbf{x}_i$ must appear in all of the equations, though each element of β must appear in at least one equation for β to be identified. For this model, iterative FGNLS estimation is equivalent to maximum likelihood estimation with multivariate normal disturbances.

The syntax you use with `nlsur` closely mirrors that used with `nl`. In particular, you use substitutable expressions with the interactive and programmed substitutable expression versions to define the functions in your system. See [R] **nl** for more information on substitutable expressions. Here we reiterate the three rules that you must follow:

1. Parameters of the model are bound in braces: `{b0}`, `{param}`, etc.
2. Initial values for parameters are given by including an equals sign and the initial value inside the braces: `{b0=1}`, `{param=3.571}`, etc. If you do not specify an initial value, that parameter is initialized to zero. The `initial()` option overrides initial values in substitutable expressions.
3. Linear combinations of variables can be included using the notation {*eqname*:*varlist*}: `{xb: mpg price weight}`, `{score: w x z}`, etc. Parameters of linear combinations are initialized to zero.

▷ Example 1: Interactive version using two-step FGNLS estimator

We have data from an experiment in which two closely-related types of bacteria were placed in a Petri dish, and the number of each type of bacteria were recorded every hour. We suspect a two-parameter exponential growth model can be used to model each type of bacteria, but because they shared the same dish, we want to allow for correlation in the error terms. We want to fit the system of equations

$$
\begin{aligned}
p_1 &= \beta_1 {\beta_2}^t + u_1 \\
p_2 &= \gamma_1 {\gamma_2}^t + u_2
\end{aligned}
$$

where p_1 and p_2 are the two populations and t is time, and we want to allow for nonzero correlation between u_1 and u_2. We type

```
. use http://www.stata-press.com/data/r10/petridish
. nlsur (p1 = {b1}*{b2}^t) (p2 = {g1}*{g2}^t)
(obs = 25)
Calculating NLS estimates...
Iteration 0:  Residual SS =  335.5286
Iteration 1:  Residual SS =  333.8583
Iteration 2:  Residual SS =  219.9233
Iteration 3:  Residual SS =  127.9355
Iteration 4:  Residual SS =  14.86765
Iteration 5:  Residual SS =  8.628459
Iteration 6:  Residual SS =  8.281268
Iteration 7:  Residual SS =   8.28098
Iteration 8:  Residual SS =  8.280979
Iteration 9:  Residual SS =  8.280979
Calculating FGNLS estimates...
Iteration 0:  Scaled RSS =  49.99892
Iteration 1:  Scaled RSS =  49.99892
Iteration 2:  Scaled RSS =  49.99892
FGNLS regression
```

	Equation	Obs	Parms	RMSE	R-sq	Constant
1	p1	25	2	.4337019	0.9734*	(none)
2	p2	25	2	.3783479	0.9776*	(none)

`* Uncentered R-sq`

	Coef.	Std. Err.	z	P>|z|	[95% Conf.	Interval]
/b1	.3926631	.064203	6.12	0.000	.2668275	.5184987
/b2	1.119593	.0088999	125.80	0.000	1.102149	1.137036
/g1	.5090441	.0669495	7.60	0.000	.3778256	.6402626
/g2	1.102315	.0072183	152.71	0.000	1.088167	1.116463

The header of the output contains a summary of each equation, including the number of observations and parameters and the root mean squared error of the residuals. `nlsur` checks to see whether each equation contains a constant term, and if an equation does contain a constant term, an R^2 statistic is presented. If an equation does not have a constant term, an uncentered R^2 is instead reported. The R^2 statistic for each equation measures the percentage of variance explained by the nonlinear function and may be useful for descriptive purposes, though it does not have the same formal interpretation in the context of FGNLS as it does with NLS estimation. As we would expect, β_2 and γ_2 are both greater than one, indicating the two bacterial populations increased in size over time.

◁

The model we fit in the next three examples is in fact linear in the parameters, so it could be fit using the `sureg` command. However, we will fit the model using `nlsur` so that we can focus on the mechanics of using the command. Moreover, using `nlsur` will obviate the need to generate several variables as well as the need to use the `constraint` command to impose parameter restrictions.

▷ Example 2: Interactive version using iterative FGNLS estimator — the translog production function

Greene (1997, sec. 15.6) discusses the transcendental logarithmic (translog) cost function and provides cost and input price data for capital, labor, energy, and materials for the United States economy. One way to fit the translog production function to this data is to fit the system of three equations

$$s_k = \beta_k + \delta_{kk} \ln\left(\frac{p_k}{p_m}\right) + \delta_{kl} \ln\left(\frac{p_l}{p_m}\right) + \delta_{ke} \ln\left(\frac{p_e}{p_m}\right) + u_1$$
$$s_l = \beta_l + \delta_{kl} \ln\left(\frac{p_k}{p_m}\right) + \delta_{ll} \ln\left(\frac{p_l}{p_m}\right) + \delta_{le} \ln\left(\frac{p_e}{p_m}\right) + u_2$$
$$s_e = \beta_e + \delta_{ke} \ln\left(\frac{p_k}{p_m}\right) + \delta_{le} \ln\left(\frac{p_l}{p_m}\right) + \delta_{ee} \ln\left(\frac{p_e}{p_m}\right) + u_3$$

where s_k is capital's cost share, s_l is labor's cost share, and s_e is energy's cost share; p_k, p_l, p_e, and p_m are the prices of capital, labor, energy, and materials, respectively; the u's are regression error terms, and the βs and δs are parameters to be estimated. There are three cross-equation restrictions on the parameters: δ_{kl}, δ_{ke}, and δ_{le} each appear in two equations. To fit this model by using the iterative FGNLS estimator, we type

```
. use http://www.stata-press.com/data/r10/mfgcost
. nlsur (s_k = {bk} + {dkk}*ln(pk/pm) + {dkl}*ln(pl/pm) + {dke}*ln(pe/pm))
>       (s_l = {bl} + {dkl}*ln(pk/pm) + {dll}*ln(pl/pm) + {dle}*ln(pe/pm))
>       (s_e = {be} + {dke}*ln(pk/pm) + {dle}*ln(pl/pm) + {dee}*ln(pe/pm)),
>       ifgnls
(obs = 25)
Calculating NLS estimates...
Iteration 0:  Residual SS =  .0009989
Iteration 1:  Residual SS =  .0009989
Calculating FGNLS estimates...
Iteration 0:  Scaled RSS =  65.45197
Iteration 1:  Scaled RSS =  65.45197
 (output omitted)
FGNLS iteration 10...
Iteration 0:  Scaled RSS =        75
Iteration 1:  Scaled RSS =        75
Iteration 2:  Scaled RSS =        75
Parameter change         =  4.076e-06
Covariance matrix change =  6.264e-10
FGNLS regression
```

	Equation	Obs	Parms	RMSE	R-sq	Constant
1	s_k	25	4	.0031722	0.4776	bk
2	s_l	25	4	.0053963	0.8171	bl
3	s_e	25	4	.00177	0.6615	be

	Coef.	Std. Err.	z	P>|z|	[95% Conf.	Interval]
/bk	.0568925	.0013454	42.29	0.000	.0542556	.0595294
/dkk	.0294833	.0057956	5.09	0.000	.0181241	.0408425
/dkl	-.0000471	.0038478	-0.01	0.990	-.0075887	.0074945
/dke	-.0106749	.0033882	-3.15	0.002	-.0173157	-.0040341
/bl	.253438	.0020945	121.00	0.000	.2493329	.2575432
/dll	.0754327	.0067572	11.16	0.000	.0621889	.0886766
/dle	-.004756	.002344	-2.03	0.042	-.0093502	-.0001619
/be	.0444099	.0008533	52.04	0.000	.0427374	.0460823
/dee	.0183415	.0049858	3.68	0.000	.0085694	.0281135

We draw your attention to the iteration log at the top of the output. When iterative FGNLS estimation is used, the final scaled RSS will equal the product of the number of observations in the estimation sample and the number of equations; see *Methods and Formulas* for details. Because the RSS is scaled by the error covariance matrix during each round of FGNLS estimation, the scaled RSS is not comparable from one FGNLS iteration to the next.

◁

❑ Technical Note

You may have noticed that we mentioned having data for four factors of production, yet we fitted only three share equations. Because the four shares sum to one, we must drop one of the equations to avoid having a singular error covariance matrix. The iterative FGNLS estimator is equivalent to maximum likelihood estimation and thus it is invariant to which one of the four equations we choose to drop. The (linearly restricted) parameters of the fourth equation can be obtained using the `lincom` command. Nonlinear functions of the parameters, such as the elasticities of substitution, can be computed using `nlcom`.

❑

Substitutable expression programs

If you fit the same model repeatedly or you want to share code with colleagues, you can write a *substitutable expression program* to define your system of equations and avoid having to retype the system every time. The first five letters of the program's name must be `nlsur`, and the program must set the r-class macro `r(n_eq)` to the number of equations in your system. The first equation's substitutable expression must be returned in `r(eq_1)`, the second equation's in `r(eq_2)`, and so on. You may optionally set `r(title)` to label your output; that has the same effect as specifying the `title()` option.

▷ Example 3: Programmed substitutable expression version

We return to our translog cost function, for which a substitutable expression program is

```
program nlsurtranslog, rclass
        version 10
        syntax varlist(min=7 max=7) [if]
        tokenize `varlist'
        args sk sl se pk pl pe pm
        local pkpm ln(`pk'/`pm')
        local plpm ln(`pl'/`pm')
        local pepm ln(`pe'/`pm')
        return scalar n_eq = 3
        return local eq_1 "`sk'= {bk} + {dkk}*`pkpm' + {dkl}*`plpm' + {dke}*`pepm'"
        return local eq_2 "`sl'= {bl} + {dkl}*`pkpm' + {dll}*`plpm' + {dle}*`pepm'"
        return local eq_3 "`se'= {be} + {dke}*`pkpm' + {dle}*`plpm' + {dee}*`pepm'"
        return local title "4-factor translog cost function"
end
```

We made our program accept seven variables, for the three dependent variables s_k, s_l, and s_e, and the four factor prices p_k, p_l, p_m, and p_e. The `tokenize` command assigns to macros `` `1' ``, `` `2' ``, ..., `` `7' `` the seven variables stored in `` `varlist' ``; and the `args` command transfers those numbered macros to macros `` `sk' ``, `` `sl' ``, ..., `` `pm' ``. Because we knew our substitutable expressions were going to be somewhat long, we created local macros to hold the log price ratios. These are simply macros that hold strings such as `ln(pk/pm)`, not variables, and they will save us some repetitious typing when we define our substitutable expressions. Our program returns the number of equations in `r(n_eq)`, and we defined our substitutable expressions in `eq_1`, `eq_2`, and `eq_3`. We do not bind the expressions in parentheses as we do with the interactive version of `nlsur`. Finally, we put a title in `r(title)` to label our output.

Our `syntax` command also accepts an `if` clause, and that is how `nlsur` indicates the estimation sample to our program. In this application, we can safely ignore it, because our program does not compute initial values. However, had we used commands such as `summarize` or `regress` to obtain initial values, then we would need to restrict those commands to analyze only the estimation sample. In those cases typically you simply need to include `` `if' `` with the commands you are using. For example, instead of the command

```
summarize `depvar', meanonly
```

you would use

```
summarize `depvar' `if', meanonly
```

We can check our program by typing

```
. nlsurtranslog s_k s_l s_e pk pl pe pm
. return list
scalars:
              r(n_eq) =  3
macros:
             r(title) : "4-factor translog cost function"
              r(eq_3) : "s_e= {be} + {dke}*ln(pk/pm) + {dle}*ln(pl/pm) + {dee.."
              r(eq_2) : "s_l= {bl} + {dkl}*ln(pk/pm) + {dll}*ln(pl/pm) + {dle.."
              r(eq_1) : "s_k= {bk} + {dkk}*ln(pk/pm) + {dkl}*ln(pl/pm) + {dke.."
```

Now that we know that our program works, we fit our model by typing

```
. nlsur translog: s_k s_l s_e pk pl pe pm, ifgnls
(obs = 25)
Calculating NLS estimates...
Iteration 0:  Residual SS =  .0009989
Iteration 1:  Residual SS =  .0009989
Calculating FGNLS estimates...
Iteration 0:  Scaled RSS =  65.45197
Iteration 1:  Scaled RSS =  65.45197
FGNLS iteration 2...
Iteration 0:  Scaled RSS =  73.28311
Iteration 1:  Scaled RSS =  73.28311
Parameter change           =  6.537e-03
Covariance matrix change =  1.002e-06
  (output omitted)
FGNLS iteration 10...
Iteration 0:  Scaled RSS =        75
Iteration 1:  Scaled RSS =        75
Iteration 2:  Scaled RSS =        75
Parameter change           =  4.076e-06
Covariance matrix change =  6.264e-10
```

```
FGNLS regression

------------------------------------------------------------------------
       Equation |       Obs  Parms        RMSE       R-sq      Constant
----------------+-------------------------------------------------------
1            s_k |        25      4    .0031722     0.4776            bk
2            s_l |        25      4    .0053963     0.8171            bl
3            s_e |        25      4      .00177     0.6615            be
------------------------------------------------------------------------

4-factor translog cost function
------------------------------------------------------------------------------
             |      Coef.   Std. Err.      z    P>|z|     [95% Conf. Interval]
-------------+----------------------------------------------------------------
         /bk |   .0568925   .0013454    42.29   0.000     .0542556    .0595294
        /dkk |   .0294833   .0057956     5.09   0.000     .0181241    .0408425
        /dkl |  -.0000471   .0038478    -0.01   0.990    -.0075887    .0074945
        /dke |  -.0106749   .0033882    -3.15   0.002    -.0173157   -.0040341
         /bl |    .253438   .0020945   121.00   0.000     .2493329    .2575432
        /dll |   .0754327   .0067572    11.16   0.000     .0621889    .0886766
        /dle |   -.004756    .002344    -2.03   0.042    -.0093502   -.0001619
         /be |   .0444099   .0008533    52.04   0.000     .0427374    .0460823
        /dee |   .0183415   .0049858     3.68   0.000     .0085694    .0281135
------------------------------------------------------------------------------
```

Because we set `r(title)` in our substitutable expression program, the coefficient table has a title attached to it. The estimates are identical to those we obtained in example 2.

◁

❑ Technical Note

`nlsur` accepts frequency and analytic weights as well as `pweights` (sampling weights) and `iweights` (importance weights). You do not need to modify your substitutable expressions in any way to perform weighted estimation, though you must make two changes to your substitutable expression program. The general outline of a *sexp_prog* program is

```
program nlsur name, rclass
    version 10
    syntax varlist [fw aw pw iw] [if]
    // Obtain initial values incorporating weights.  For example,
    summarize varname [`weight'`exp'] `if'
    ...
    // Return n_eqn and substitutable expressions
    return scalar n_eq = #
    return local eq_1 = ...
    ...
end
```

First, we wrote the `syntax` statement to accept a weight expression. Here we allow all four types of weights, but if you know that your estimator is valid, say, for only frequency weights, then you should modify the `syntax` line to accept only `fweights`. Second, if your program computes starting values, then any commands you use must incorporate the weights passed to the program; you do that by including `` [`weight'`exp'] `` when calling those commands.

❑

Function evaluator programs

Although substitutable expressions are extremely flexible, there are some problems for which the nonlinear system cannot be defined using them. You can use the function evaluator program version of `nlsur` in these cases. We present two examples, a simple one to illustrate the mechanics of function evaluator programs and a more complicated one to illustrate the power of `nlsur`.

▷ Example 4: Function evaluator program version

Here we write a function evaluator program to fit the translog cost function used in examples 2 and 3. The function evaluator program is

```
program nlsurtranslog2
        version 10
        syntax varlist(min=7 max=7) [if], at(name)
        tokenize `varlist'
        args sk sl se pk pl pe pm
        tempname bk dkk dkl dke bl dll dle be dee
        scalar `bk'  = `at'[1,1]
        scalar `dkk' = `at'[1,2]
        scalar `dkl' = `at'[1,3]
        scalar `dke' = `at'[1,4]
        scalar `bl'  = `at'[1,5]
        scalar `dll' = `at'[1,6]
        scalar `dle' = `at'[1,7]
        scalar `be'  = `at'[1,8]
        scalar `dee' = `at'[1,9]
        local pkpm ln(`pk'/`pm')
        local plpm ln(`pl'/`pm')
        local pepm ln(`pe'/`pm')
        quietly {
                replace `sk' = `bk' + `dkk'*`pkpm' + `dkl'*`plpm' +    ///
                                      `dke'*`pepm' `if'
                replace `sl' = `bl' + `dkl'*`pkpm' + `dll'*`plpm' +    ///
                                      `dle'*`pepm' `if'
                replace `se' = `be' + `dke'*`pkpm' + `dle'*`plpm' +    ///
                                      `dee'*`pepm' `if'
        }
end
```

Unlike the substitutable expression program we wrote in example 3, `nlsurtranslog2` is not declared as r-class because we will not be returning any saved results. We are again expecting seven variables: three shares and four factor prices, and `nlsur` will again mark the estimation sample with an `if` expression.

Our function evaluator program also accepts an option named `at()`, which will receive a parameter vector at which we are to evaluate the system of equations. All function evaluator programs must accept this option. Our model has nine parameters to estimate, and we created nine temporary scalars to hold the elements of the `` `at' `` matrix.

Since our model has three equations, the first three variables passed to our program are the dependent variables that we are to fill in with the function values. We replaced only the observations in our estimation sample by including the `` `if' `` qualifier in the `replace` statements. Here we could have ignored the `` `if' `` qualifier since `nlsur` will skip over observations not in the estimation sample and we did not perform any computations requiring knowledge of the estimation sample. However, including the `` `if' `` is good practice and may result in a slight speed improvement if the functions of your model are complicated and the estimation sample is much smaller than the dataset in memory.

We could have avoided creating temporary scalars to hold our individual parameters by writing the `replace` statements as, for example,

```
replace `sk' = `at'[1,1] + `at'[1,2]*`pkpm' + `at'[1,3]*`plpm' +     ///
                           `at'[1,4]*`pepm' `if'
```

You can use whichever method you find more appealing, though giving the parameters descriptive names reduces the chance for mistakes and makes debugging easier.

To fit our model by using the function evaluator program version of `nlsur`, we type

```
. nlsur translog2 @ s_k s_l s_e pk pl pe pm, ifgnls nequations(3)
>         parameters(bk dkk dkl dke bl dll dle be dee)
>         hasconstants(bk bl be)
(obs = 25)

Calculating NLS estimates...
Iteration 0:  Residual SS =  .0009989
Iteration 1:  Residual SS =  .0009989
Calculating FGNLS estimates...
Iteration 0:  Scaled RSS =  65.45197
Iteration 1:  Scaled RSS =  65.45197
FGNLS iteration 2...
Iteration 0:  Scaled RSS =  73.28311
Iteration 1:  Scaled RSS =  73.28311
Parameter change         =  6.537e-03
Covariance matrix change =  1.002e-06
  (output omitted)
FGNLS iteration 10...
Iteration 0:  Scaled RSS =        75
Iteration 1:  Scaled RSS =        75
Iteration 2:  Scaled RSS =        75
Parameter change         =  4.076e-06
Covariance matrix change =  6.264e-10

FGNLS regression
```

	Equation	Obs	Parms	RMSE	R-sq	Constant
1	s_k	25	.	.0031722	0.4776	bk
2	s_l	25	.	.0053963	0.8171	bl
3	s_e	25	.	.00177	0.6615	be

	Coef.	Std. Err.	z	P>|z|	[95% Conf.	Interval]
/bk	.0568925	.0013454	42.29	0.000	.0542556	.0595294
/dkk	.0294833	.0057956	5.09	0.000	.0181241	.0408425
/dkl	-.0000471	.0038478	-0.01	0.990	-.0075887	.0074945
/dke	-.0106749	.0033882	-3.15	0.002	-.0173157	-.0040341
/bl	.253438	.0020945	121.00	0.000	.2493329	.2575432
/dll	.0754327	.0067572	11.16	0.000	.0621889	.0886766
/dle	-.004756	.002344	-2.03	0.042	-.0093502	-.0001619
/be	.0444099	.0008533	52.04	0.000	.0427374	.0460823
/dee	.0183415	.0049858	3.68	0.000	.0085694	.0281135

When we use the function evaluator program version, `nlsur` requires us to specify the number of equations in `nequations()`, and it requires us to either specify names for each of our parameters or the number of parameters in the model. Here we used the `parameters()` option to name our parameters; the order we specified them in this option is the same as the order in which we extracted them from the `` `at' `` matrix in our program. Had we instead specified `nparameters(9)`, our parameters would have been labeled `/b1`, `/b2`, ..., `/b9` in the output.

`nlsur` has no way of telling how many parameters appear in each equation, so the `Parms` column in the header contains missing values. Moreover, the function evaluator program version of `nlsur` does not attempt to identify constant terms, so we used the `hasconstant` option to tell `nlsur` which parameter in each equation is a constant term.

The estimates are identical to those we obtained in examples 2 and 3.

◁

❑ Technical Note

As with substitutable expression programs, if you intend to do weighted estimation with a function evaluator program, you must modify your *func_prog* program's `syntax` statement to accept weights. Moreover, if you use any statistical commands when computing your nonlinear functions, then you must include the weight expression with those commands.

❑

▷ Example 5: Fitting the basic AIDS model using nlsur

Poi (2002) showed how to fit a quadratic almost ideal demand system (AIDS) by using the `ml` command. Here we show how to fit the basic AIDS model by using `nlsur`. The dataset `food.dta` contains household expenditures, expenditure shares, and log prices for four broad food groups. For a four-good demand system, we need to fit the following system of three equations,

$$w_1 = \alpha_1 + \gamma_{11}\,\ln p_1 + \gamma_{12}\,\ln p_2 + \gamma_{13}\,\ln p_3 + \beta_1 \ln\left\{\frac{m}{P(\mathbf{p})}\right\} + u_1$$

$$w_2 = \alpha_2 + \gamma_{12}\,\ln p_1 + \gamma_{22}\,\ln p_2 + \gamma_{23}\,\ln p_3 + \beta_2 \ln\left\{\frac{m}{P(\mathbf{p})}\right\} + u_2$$

$$w_3 = \alpha_3 + \gamma_{13}\,\ln p_1 + \gamma_{23}\,\ln p_2 + \gamma_{33}\,\ln p_3 + \beta_3 \ln\left\{\frac{m}{P(\mathbf{p})}\right\} + u_3$$

where w_k denotes a household's fraction of expenditures on good k, $\ln p_k$ denotes the logarithm of the price paid for good k, m denotes a household's total expenditure on all four goods, the u's are regression error terms, and

$$\ln P(\mathbf{p}) = \alpha_0 + \sum_{i=1}^{4} \alpha_i \ln p_i + \frac{1}{2}\sum_{i=1}^{4}\sum_{j=1}^{4} \gamma_{ij} \ln p_i \ln p_j$$

The parameters for the fourth good's share equation can be recovered from the following constraints that are imposed by economic theory:

$$\sum_{i=1}^{4} \alpha_i = 1 \qquad \sum_{i=1}^{4} \beta_i = 0 \qquad \gamma_{ij} = \gamma_{ji} \qquad \text{and} \qquad \sum_{i=1}^{4} \gamma_{ij} = 0 \text{ for all } j$$

Our model has a total of 12 unrestricted parameters. We will not estimate α_0 directly. Instead, we will set it equal to 5 as was done in Poi (2002); see Deaton and Muellbauer (1980) for a discussion of why treating α_0 as fixed is acceptable.

Our function evaluator program is

```
program nlsuraids
        version 10
        syntax varlist(min=8 max=8) if, at(name)
        tokenize `varlist'
        args w1 w2 w3 lnp1 lnp2 lnp3 lnp4 lnm
        tempname a1 a2 a3 a4
        scalar `a1' = `at'[1,1]
        scalar `a2' = `at'[1,2]
        scalar `a3' = `at'[1,3]
        scalar `a4' = 1 - `a1' - `a2' - `a3'
        tempname b1 b2 b3
        scalar `b1' = `at'[1,4]
        scalar `b2' = `at'[1,5]
        scalar `b3' = `at'[1,6]
        tempname g11 g12 g13 g14
        tempname g21 g22 g23 g24
        tempname g31 g32 g33 g34
        tempname g41 g42 g43 g44
        scalar `g11' = `at'[1,7]
        scalar `g12' = `at'[1,8]
        scalar `g13' = `at'[1,9]
        scalar `g14' = -`g11'-`g12'-`g13'
        scalar `g21' = `g12'
        scalar `g22' = `at'[1,10]
        scalar `g23' = `at'[1,11]
        scalar `g24' = -`g21'-`g22'-`g23'
        scalar `g31' = `g13'
        scalar `g32' = `g23'
        scalar `g33' = `at'[1,12]
        scalar `g34' = -`g31'-`g32'-`g33'
        scalar `g41' = `g14'
        scalar `g42' = `g24'
        scalar `g43' = `g34'
        scalar `g44' = -`g41'-`g42'-`g43'
        quietly {
                tempvar lnpindex
                gen double `lnpindex' = 5 + `a1'*`lnp1' + `a2'*`lnp2' + ///
                                        `a3'*`lnp3' + `a4'*`lnp4'
                forvalues i = 1/4 {
                        forvalues j = 1/4 {
                                replace `lnpindex' = `lnpindex' + ///
                                        0.5*`g`i'`j''*`lnp`i''*`lnp`j''
                        }
                }
                replace `w1' = `a1' + `g11'*`lnp1' + `g12'*`lnp2' + ///
                                      `g13'*`lnp3' + `g14'*`lnp4' + ///
                                      `b1'*(`lnm' - `lnpindex')
                replace `w2' = `a2' + `g21'*`lnp1' + `g22'*`lnp2' + ///
                                      `g23'*`lnp3' + `g24'*`lnp4' + ///
                                      `b2'*(`lnm' - `lnpindex')
                replace `w3' = `a3' + `g31'*`lnp1' + `g32'*`lnp2' + ///
                                      `g33'*`lnp3' + `g34'*`lnp4' + ///
                                      `b3'*(`lnm' - `lnpindex')
        }
end
```

The `syntax` statement accepts eight variables: three expenditure share variables, all four log-price variables, and a variable for log expenditures ($\ln m$). Most of the code simply extracts the parameters from the 'at' matrix. Although we are estimating only 12 parameters, to calculate the price index term and the expenditure share equations, we need the restricted parameters as well. Notice how we impose the constraints on the parameters. We then created a temporary variable to hold $\ln P(\mathbf{p})$, and we filled the three dependent variables with the predicted expenditure shares.

To fit our model, we type

```
. use http://www.stata-press.com/data/r10/food, clear
. nlsur aids @ w1 w2 w3 lnp1 lnp2 lnp3 lnp4 lnexp,
>              parameters(a1 a2 a3 b1 b2 b3
>                         g11 g12 g13 g22 g32 g33)
>              neq(3) ifgnls
(obs = 4048)
Calculating NLS estimates...
Iteration 0:  Residual SS =  126.9713
Iteration 1:  Residual SS =   125.669
Iteration 2:  Residual SS =   125.669
Iteration 3:  Residual SS =   125.669
Iteration 4:  Residual SS =   125.669
Calculating FGNLS estimates...
Iteration 0:  Scaled RSS =  12080.14
Iteration 1:  Scaled RSS =  12080.14
Iteration 2:  Scaled RSS =  12080.14
Iteration 3:  Scaled RSS =  12080.14
FGNLS iteration 2...
Iteration 0:  Scaled RSS =  12143.99
Iteration 1:  Scaled RSS =  12143.99
Iteration 2:  Scaled RSS =  12143.99
Parameter change         =  1.972e-04
Covariance matrix change =  2.936e-06
FGNLS iteration 3...
Iteration 0:  Scaled RSS =     12144
Iteration 1:  Scaled RSS =     12144
Parameter change         =  2.178e-06
Covariance matrix change =  3.467e-08
FGNLS regression
```

	Equation	Obs	Parms	RMSE	R-sq	Constant
1	w1	4048	.	.1333175	0.9017*	(none)
2	w2	4048	.	.1024166	0.8480*	(none)
3	w3	4048	.	.053777	0.7906*	(none)

* Uncentered R-sq

	Coef.	Std. Err.	z	P>|z|	[95% Conf.	Interval]
/a1	.3163958	.0073871	42.83	0.000	.3019175	.3308742
/a2	.2712501	.0056938	47.64	0.000	.2600904	.2824097
/a3	.1039898	.0029004	35.85	0.000	.0983051	.1096746
/b1	.0161044	.0034153	4.72	0.000	.0094105	.0227983
/b2	-.0260771	.002623	-9.94	0.000	-.0312181	-.0209361
/b3	.0014538	.0013776	1.06	0.291	-.0012463	.004154
/g11	.1215838	.0057186	21.26	0.000	.1103756	.1327921
/g12	-.0522943	.0039305	-13.30	0.000	-.0599979	-.0445908
/g13	-.0351292	.0021788	-16.12	0.000	-.0393996	-.0308588
/g22	.0644298	.0044587	14.45	0.000	.0556909	.0731687
/g32	-.0011786	.0019767	-0.60	0.551	-.0050528	.0026957
/g33	.0424381	.0017589	24.13	0.000	.0389909	.0458854

To get the restricted parameters for the fourth share equation, we can use `lincom`. For example, to obtain α_4, we type

```
. lincom 1 - [a1]_cons - [a2]_cons - [a3]_cons
 ( 1) - [a1]_cons - [a2]_cons - [a3]_cons = -1
```

	Coef.	Std. Err.	z	P>|z|	[95% Conf.	Interval]
(1)	.3083643	.0052611	58.61	0.000	.2980528	.3186758

For more information on `lincom`, see [R] **lincom**.

◁

(*Continued on next page*)

Saved Results

nl saves the following in e():

Scalars	
e(N)	number of observations
e(rss_#)	RSS for equation #
e(mss_#)	model sum of squares for equation #
e(rmse_#)	root mean squared error for equation #
e(r2_#)	R^2 for equation #
e(ll)	Gaussian log likelihood (iflgs version only)
e(k)	number of parameters in system
e(k_#)	number of parameters for equation #
e(k_eq)	number of equation names in e(b)
e(k_eq_model)	number of equations in model Wald test
e(k_aux)	number of estimates parameters
e(n_eq)	number of equations
e(N_clust)	number of clusters
e(converge)	1 if converged; 0 otherwise
Macros	
e(cmd)	nlsur
e(cmdline)	command as typed
e(method)	fgnls, ifgnls, or nls
e(sexp_#)	substitutable expression for equation #
e(depvar)	names of dependent variables
e(depvar_#)	dependent variable for equation #
e(rhs)	contents of variables(), if specified
e(params)	names of all parameters
e(params_#)	parameters in equation #
e(wtype)	weight type
e(wexp)	weight expression
e(title)	title in estimation output
e(title_2)	secondary title in estimation output
e(clustvar)	name of cluster variable
e(vce)	*vcetype* specified in vce()
e(vcetype)	title used in label Std. Err.
e(sexpprog)	substitutable expression program
e(funcprog)	function evaluator program
e(constants)	identifies constant terms
e(type)	1 = interactively entered expression 2 = substitutable expression program 3 = function evaluator program
e(properties)	b V
e(predict)	program used to implement predict
Matrices	
e(b)	coefficient vector
e(V)	variance–covariance matrix of the estimators
e(init)	initial values vector
e(Sigma)	error covariance matrix ($\widehat{\Sigma}$)
Functions	
e(sample)	marks estimation sample

Methods and Formulas

`nlsur` is implemented as an ado-file.

Write the system of equations for the ith observation as

$$\mathbf{y}_i = \mathbf{f}(\mathbf{x}_i, \beta) + \mathbf{u}_i \tag{1}$$

where $\mathbf{y}_i$ and $\mathbf{u}_i$ are $1 \times M$ vectors, for $i = 1, \ldots, N$, $\mathbf{f}$ is a function that returns a $1 \times M$ vector, $\mathbf{x}_i$ represents all the exogenous variables in the system, and β is a $1 \times k$ vector of parameters. The generalized nonlinear least-squares system estimator is defined as

$$\widehat{\beta} \equiv \text{argmin}_\beta \sum_{i=1}^{N} \{\mathbf{y}_i - \mathbf{f}(\mathbf{x}_i, \beta)\} \boldsymbol{\Sigma}^{-1} \{\mathbf{y}_i - \mathbf{f}(\mathbf{x}_i, \beta)\}'$$

where $\boldsymbol{\Sigma} = E(\mathbf{u}_i'\mathbf{u}_i)$ is an $M \times M$ positive-definite weight matrix. Let $\mathbf{T}$ be the Cholesky decomposition of $\boldsymbol{\Sigma}^{-1}$; that is, $\mathbf{T}\mathbf{T}' = \boldsymbol{\Sigma}^{-1}$. Postmultiply (1) by $\mathbf{T}$:

$$\mathbf{y}_i\mathbf{T} = \mathbf{f}(\mathbf{x}_i, \beta)\mathbf{T} + \mathbf{u}_i\mathbf{T} \tag{2}$$

Because $E(\mathbf{T}'\mathbf{u}_i'\mathbf{u}_i\mathbf{T}) = \mathbf{I}$, we can "stack" the columns of (2) and write

$$\begin{aligned}
\mathbf{y}_1\mathbf{T}_1 &= \mathbf{f}(\mathbf{x}_1, \beta)\mathbf{T}_1 + \widetilde{u}_{11} \\
\mathbf{y}_1\mathbf{T}_2 &= \mathbf{f}(\mathbf{x}_1, \beta)\mathbf{T}_2 + \widetilde{u}_{12} \\
\vdots &= \vdots \\
\mathbf{y}_1\mathbf{T}_M &= \mathbf{f}(\mathbf{x}_1, \beta)\mathbf{T}_M + \widetilde{u}_{1M} \\
\vdots &= \vdots \\
\mathbf{y}_N\mathbf{T}_1 &= \mathbf{f}(\mathbf{x}_N, \beta)\mathbf{T}_1 + \widetilde{u}_{N1} \\
\mathbf{y}_N\mathbf{T}_2 &= \mathbf{f}(\mathbf{x}_N, \beta)\mathbf{T}_2 + \widetilde{u}_{N2} \\
\vdots &= \vdots \\
\mathbf{y}_N\mathbf{T}_M &= \mathbf{f}(\mathbf{x}_N, \beta)\mathbf{T}_M + \widetilde{u}_{NM}
\end{aligned} \tag{3}$$

where $\mathbf{T}_j$ denotes the jth column of $\mathbf{T}$. By construction, all $\widetilde{u}_{ij}$ are independently distributed with unit variance. As a result, by transforming the model in (1) to that shown in (3), we have reduced the multivariate generalized nonlinear least-squares system estimator to a univariate nonlinear least-squares problem; and the same parameter estimation technique used by `nl` can be used here. See [R] **nl** for the details. Moreover, because the $\widetilde{u}_{ij}$ all have variance 1, the final scaled RSS reported by `nlsur` is equal to NM.

To make the estimator feasible, we require an estimate $\widehat{\boldsymbol{\Sigma}}$ of $\boldsymbol{\Sigma}$. `nlsur` first sets $\widehat{\boldsymbol{\Sigma}} = \mathbf{I}$. Although not efficient, the resulting estimate $\widehat{\beta}_{\text{NLS}}$ is consistent. If the `nls` option is specified, estimation is complete. Otherwise, the residuals

$$\widehat{\mathbf{u}}_i = \mathbf{y}_i - \mathbf{f}(\mathbf{x}_i, \widehat{\beta}_{\text{NLS}})$$

are calculated and used to compute

$$\widehat{\boldsymbol{\Sigma}} = \frac{1}{N}\sum_{i=1}^{N}\widehat{\mathbf{u}}_i'\widehat{\mathbf{u}}_i$$

With $\widehat{\boldsymbol{\Sigma}}$ in hand, a new estimate $\widehat{\beta}$ is then obtained.

If the `ifgnls` option is specified, the new $\widehat{\beta}$ is used to recompute the residuals and obtain a new estimate of $\widehat{\boldsymbol{\Sigma}}$, from which $\widehat{\beta}$ can then be reestimated. Iterations stop when the relative change in $\widehat{\beta}$ is less than `eps()`, the relative change in $\widehat{\boldsymbol{\Sigma}}$ is less than `ifgnlseps()`, or if `ifgnlsiterate()` iterations have been performed.

If the `vce(robust)` and `vce(cluster` *clustvar*`)` options were not specified, then

$$V(\widehat{\beta}) = \left(\sum_{i=1}^{N}\mathbf{X}_i'\widehat{\boldsymbol{\Sigma}}^{-1}\mathbf{X}_i\right)^{-1}$$

where the $M \times k$ matrix $\mathbf{X}_i$ has typical element X_{ist}, the derivative of the sth element of $\mathbf{f}$ with respect to the tth element of β, evaluated at $\mathbf{x}_i$ and $\widehat{\beta}$. As a practical matter, once the model is written in the form of (3), the variance–covariance matrix can be calculated via a Gauss–Newton regression; see Davidson and MacKinnon (1993, chap. 6).

If `robust` is specified, then

$$V_R(\widehat{\beta}) = \left(\sum_{i=1}^{N}\mathbf{X}_i'\widehat{\boldsymbol{\Sigma}}^{-1}\mathbf{X}_i\right)^{-1}\sum_{i=1}^{N}\mathbf{X}_i'\widehat{\boldsymbol{\Sigma}}^{-1}\widehat{\mathbf{u}}_i'\widehat{\mathbf{u}}_i\widehat{\boldsymbol{\Sigma}}^{-1}\mathbf{X}_i\left(\sum_{i=1}^{N}\mathbf{X}_i'\widehat{\boldsymbol{\Sigma}}^{-1}\mathbf{X}_i\right)^{-1}$$

The cluster–robust variance matrix is

$$V_C(\widehat{\beta}) = \left(\sum_{i=1}^{N}\mathbf{X}_i'\widehat{\boldsymbol{\Sigma}}^{-1}\mathbf{X}_i\right)^{-1}\sum_{c=1}^{N_C}\mathbf{w}_c'\mathbf{w}_c\left(\sum_{i=1}^{N}\mathbf{X}_i'\widehat{\boldsymbol{\Sigma}}^{-1}\mathbf{X}_i\right)^{-1}$$

where N_C is the number of clusters and

$$\mathbf{w}_c = \sum_{j \in C_k}\mathbf{X}_j'\widehat{\boldsymbol{\Sigma}}^{-1}\widehat{\mathbf{u}}_j'$$

with C_k denoting the set of observations in the kth cluster. In evaluating these formulas, we use the value of $\widehat{\boldsymbol{\Sigma}}$ used in calculating the final estimate of $\widehat{\beta}$. That is, we do not recalculate $\widehat{\boldsymbol{\Sigma}}$ after we obtain the final value of $\widehat{\beta}$.

The RSS for the jth equation, RSS_j, is

$$\text{RSS}_j = \sum_{i=1}^{N}(\widehat{y}_{ij} - y_{ij})^2$$

where $\widehat{y}_{ij}$ is the predicted value of the ith observation on the jth dependent variable; the total sum of squares for the jth equation, TSS_j, is

$$\text{TSS}_j = \sum_{i=1}^{N} (y_{ij} - \bar{y}_j)^2$$

if there is a constant term in the jth equation, where $\bar{y}_j$ is the sample mean of the jth dependent variable, and

$$\text{TSS}_j = \sum_{i=1}^{N} y_{ij}^2$$

if there is no constant term in the jth equation; and the model sum of squares for the jth equation, MSS_j, is $\text{TSS}_j - \text{RSS}_j$.

The R^2 for the jth equation is $\text{MSS}_j/\text{TSS}_j$. If an equation does not have a constant term, then the reported R^2 for that equation is "uncentered" and based on the latter definition of TSS_j.

Under the assumption that the $\mathbf{u}_i$ are independently and identically distributed $N(\mathbf{0}, \widehat{\boldsymbol{\Sigma}})$, the log likelihood for the model is

$$\ln L = -\frac{MN}{2}\left\{1 + \ln(2\pi)\right\} - \frac{N}{2}\ln\left|\widehat{\boldsymbol{\Sigma}}\right|$$

The log likelihood is reported only when the `ifgnls` option is specified.

References

Davidson, R., and J. G. MacKinnon. 1993. *Estimation and Inference in Econometrics*. New York: Oxford University Press.

——. 2004. *Econometric Theory and Methods*. New York: Oxford University Press.

Deaton, A. S., and J. Muellbauer. 1980. An almost ideal demand system. *American Economic Review* 70: 312–326.

Greene, W. H. 1997. *Econometric Analysis*. 3rd ed. Upper Saddle River, NJ: Prentice Hall.

——. 2003. *Econometric Analysis*. 5th ed. Upper Saddle River, NJ: Prentice Hall.

Poi, B. P. 2002. From the help desk: Demand system estimation. *Stata Journal* 2: 403–410.

Zellner, A. 1962. An efficient method of estimating seemingly unrelated regressions and tests for aggregation bias. *Journal of the American Statistical Association* 57: 348–368.

——. 1963. Estimators for seemingly unrelated regression equations: Some exact finite sample results. *Journal of the American Statistical Association* 58: 977–992.

Zellner, A., and D. S. Huang. 1962. Further properties of efficient estimators for seemingly unrelated regression equations. *International Economic Review* 3: 300–313.

Also See

[R] **nl** — Nonlinear least-squares estimation

[R] **sureg** — Zellner's seemingly unrelated regression

[R] **reg3** — Three-stage estimation for systems of simultaneous equations

[U] **20 Estimation and postestimation commands**

Title

nlsur postestimation — Postestimation tools for nlsur

Description

The following postestimation commands are available for `nlsur`:

command	description
`estat`	AIC[1], BIC[1], VCE, and estimation sample summary
`estimates`	cataloging estimation results
`lincom`	point estimates, standard errors, testing, and inference for linear combinations of coefficients
`lrtest`[1]	likelihood-ratio test
`mfx`	marginal effects or elasticities
`nlcom`	point estimates, standard errors, testing, and inference for nonlinear combinations of coefficients
`predict`	predictions, residuals, influence statistics, and other diagnostic measures
`predictnl`	point estimates, standard errors, testing, and inference for generalized predictions
`test`	Wald tests for simple and composite linear hypotheses
`testnl`	Wald tests of nonlinear hypotheses

[1]The AIC, BIC, and likelihood-ratio tests are available only if `ifgnls` was used with `nlsur`.

See the corresponding entries in the *Stata Base Reference Manual* for details.

Syntax for predict

`predict` [*type*] *newvar* [*if*] [*in*] [, `equation(#`*eqno*`)` `yhat` `residuals`]

These statistics are available both in and out of sample; type `predict ... if e(sample) ...` if wanted only for the estimation sample.

Options for predict

Main

`equation(#`*eqno*`)` specifies to which equation you are referring. `equation(#1)` would mean that the calculation is to be made for the first equation, `equation(#2)` would mean the second, and so on. If you do not specify `equation()`, results are the same as if you had specified `equation(#1)`.

`yhat`, the default, calculates the fitted values for the specified equation.

`residuals` calculates the residuals for the specified equation.

(*Continued on next page*)

Remarks

▷ Example 1

In example 2 of [R] **nlsur**, we fit a four-factor translog cost function to data for the U.S. economy. The own-price elasticity for a factor measures the percentage change in its usage as a result of a 1% increase in the factor's price, assuming that output is held constant. For the translog production function, the own-price factor elasticities are

$$\eta_i = \frac{\delta_{ii} + s_i(s_i - 1)}{s_i}$$

Here we compute the elasticity for capital at the sample mean of capital's factor share. First, we use `summarize` to get the mean of `s_k` and store that value in a scalar:

```
. summarize s_k

    Variable |       Obs        Mean    Std. Dev.       Min        Max
-------------+--------------------------------------------------------
         s_k |        25     .053488    .0044795     .04602     .06185
. scalar kmean = r(mean)
```

Now we can use `nlcom` to calculate the elasticity:

```
. nlcom (([dkk]_cons + kmean*(kmean-1)) / kmean)

       _nl_1:  ([dkk]_cons + kmean*(kmean-1)) / kmean

------------------------------------------------------------------------------
             |      Coef.   Std. Err.      z    P>|z|     [95% Conf. Interval]
-------------+----------------------------------------------------------------
       _nl_1 |  -.3952986   .1083535    -3.65   0.000    -.6076676   -.1829295
------------------------------------------------------------------------------
```

If the price of capital increases by 1%, its usage will decrease by about 0.4%. To maintain its current level of output, a firm would increase its usage of other inputs to compensate for the lower capital usage. The standard error reported by `nlcom` reflects the sampling variance of the estimated parameter $\widehat{\delta_{kk}}$, but `nlcom` treats the sample mean of `s_k` as a fixed parameter that does not contribute to the sampling variance of the estimated elasticity.

◁

Methods and Formulas

All postestimation commands listed above are implemented as ado-files.

Also See

[R] **nlsur** — Estimation of nonlinear systems of equations

[U] **20 Estimation and postestimation commands**

Title

nptrend — Test for trend across ordered groups

Syntax

`nptrend` *varname* [*if*] [*in*] , `by(`*groupvar*`)` [`nodetail score(`*scorevar*`)`]

Description

`nptrend` performs a nonparametric test for trend across ordered groups.

Options

Main

`by(`*groupvar*`)` is required; it specifies the group on which the data are to be ordered.

`nodetail` suppresses the listing of group rank sums.

`score(`*scorevar*`)` defines scores for groups. When it is not specified, the values of *groupvar* are used for the scores.

Remarks

`nptrend` performs the nonparametric test for trend across ordered groups developed by Cuzick (1985), which is an extension of the Wilcoxon rank-sum test (`ranksum`; see [R] **signrank**). A correction for ties is incorporated into the test. `nptrend` is a useful adjunct to the Kruskal–Wallis test; see [R] **kwallis**.

In addition to `nptrend`, the `signtest` and `spearman` commands can be useful for nongrouped data; see [R] **signrank** and [R] **spearman**. The Cox and Stuart test, for instance, applies the sign test to differences between equally spaced observations of *varname*. The Daniels test calculates Spearman's rank correlation of *varname* with a time index. Under appropriate conditions, the Daniels test is more powerful than the Cox and Stuart test. See Conover (1999) for a discussion of these tests and their asymptotic relative efficiency.

▷ Example 1

The following data (Altman 1991, 217) show ocular exposure to ultraviolet radiation for 32 pairs of sunglasses classified into three groups according to the amount of visible light transmitted.

Group	Transmission of visible light	Ocular exposure to ultraviolet radiation
1	$< 25\%$	1.4 1.4 1.4 1.6 2.3 2.3
2	25 to 35%	0.9 1.0 1.1 1.1 1.2 1.2 1.5 1.9 2.2 2.6 2.6
		2.6 2.8 2.8 3.2 3.5 4.3 5.1
3	$> 35\%$	0.8 1.7 1.7 1.7 3.4 7.1 8.9 13.5

Entering these data into Stata, we have

```
. use http://www.stata-press.com/data/r10/sg
. list, sep(6)

     +------------------+
     | group   exposure |
     |------------------|
  1. |     1        1.4 |
  2. |     1        1.4 |
  3. |     1        1.4 |
  4. |     1        1.6 |
  5. |     1        2.3 |
  6. |     1        2.3 |
     |------------------|
  7. |     2         .9 |
  (output omitted )
     |------------------|
 31. |     3        8.9 |
 32. |     3       13.5 |
     +------------------+
```

We use nptrend to test for a trend of (increasing) exposure across the three groups by typing

```
. nptrend exposure, by(group)
     group      score        obs      sum of ranks
         1          1          6            76
         2          2         18           290
         3          3          8           162

           z  =  1.52
  Prob > |z|  = 0.129
```

When the groups are given any equally spaced scores (such as -1, 0, 1), we will obtain the same answer as above. To illustrate the effect of changing scores, an analysis of these data with scores 1, 2, and 5 (admittedly not sensible here) produces

```
. gen mysc = cond(group==3,5,group)
. nptrend exposure, by(group) score(mysc)
     group      score        obs      sum of ranks
         1          1          6            76
         2          2         18           290
         3          5          8           162

           z  =  1.46
  Prob > |z|  = 0.143
```

This example suggests that the analysis is not all that sensitive to the scores chosen.

◁

❑ Technical Note

The grouping variable may be either a string variable or a numeric variable. If it is a string variable and no score variable is specified, the natural numbers 1, 2, 3, ... are assigned to the groups in the sort order of the string variable. This may not always be what you expect. For example, the sort order of the strings “one”, “two”, “three” is “one”, “three”, “two”.

❑

Saved Results

nptrend saves the following in r():

Scalars

r(N)	number of observations	r(z)	z statistic
r(p)	two-sided p-value	r(T)	test statistic

Methods and Formulas

nptrend is implemented as an ado-file.

nptrend is based on a method in Cuzick (1985). The following description of the statistic is from Altman (1991, 215–217). We have k groups of sample sizes n_i ($i = 1, \ldots, k$). The groups are given scores, l_i, which reflect their ordering, such as 1, 2, and 3. The scores do not have to be equally spaced, but they usually are. $N = \sum n_i$ observations are ranked from 1 to N, and the sums of the ranks in each group, R_i, are obtained. L, the weighted sum of all the group scores, is

$$L = \sum_{i=1}^{k} l_i n_i$$

The statistic T is calculated as

$$T = \sum_{i=1}^{k} l_i R_i$$

Under the null hypothesis, the expected value of T is $E(T) = .5(N+1)L$, and its standard error is

$$\text{se}(T) = \sqrt{\frac{n+1}{12}\left(N\sum_{i=1}^{k} l_i^2 n_i - L^2\right)}$$

so that the test statistic, z, is given by $z = \{\,T - E(T)\,\}/\text{se}(T)$, which has an approximately standard normal distribution when the null hypothesis of no trend is true.

The correction for ties affects the standard error of T. Let $\widetilde{N}$ be the number of unique values of the variable being tested ($\widetilde{N} \le N$), and let t_j be the number of times the jth unique value of the variable appears in the data. Define

$$a = \frac{\sum_{j=1}^{\widetilde{N}} t_j(t_j^2 - 1)}{N(N^2-1)}$$

The corrected standard error of T is $\widetilde{\text{se}}(T) = \sqrt{1-a}\ \text{se}(T)$.

Acknowledgments

nptrend was written by K. A. Stepniewska and D. G. Altman (1992) of the Imperial Cancer Research Fund, London.

References

Altman, D. G. 1991. *Practical Statistics for Medical Research.* London: Chapman & Hall/CRC.

Conover, W. J. 1999. *Practical Nonparametric Statistics.* 3rd ed. New York: Wiley.

Cuzick, J. 1985. A Wilcoxon-type test for trend. *Statistics in Medicine* 4: 87–90.

Sasieni, P. 1996. snp12: Stratified test for trend across ordered groups. *Stata Technical Bulletin* 33: 24–27. Reprinted in *Stata Technical Bulletin Reprints*, vol. 6, pp. 196–200.

Sasieni, P., K. A. Stepniewska, and D. G. Altman. 1996. snp11: Test for trend across ordered groups revisited. *Stata Technical Bulletin* 32: 27–29. Reprinted in *Stata Technical Bulletin Reprints*, vol. 6, pp. 193–196.

Stepniewska, K. A., and D. G. Altman. 1992. snp4: Nonparametric test for trend across ordered groups. *Stata Technical Bulletin* 9: 21–22. Reprinted in *Stata Technical Bulletin Reprints*, vol. 2, p. 169.

Also See

[R] **kwallis** — Kruskal–Wallis equality-of-populations rank test

[R] **signrank** — Equality tests on matched data

[R] **spearman** — Spearman's and Kendall's correlations

[R] **symmetry** — Symmetry and marginal homogeneity tests

[ST] **epitab** — Tables for epidemiologists

[ST] **strate** — Tabulate failure rates and rate ratios

Title

ologit — Ordered logistic regression

Syntax

`ologit` *depvar* [*indepvars*] [*if*] [*in*] [*weight*] [, *options*]

options	description
Model	
`offset(`*varname*`)`	include *varname* in model with coefficient constrained to 1
SE/Robust	
`vce(`*vcetype*`)`	*vcetype* may be `oim`, `robust`, `cluster` *clustvar*, `bootstrap`, or `jackknife`
Reporting	
`level(`*#*`)`	set confidence level; default is `level(95)`
`or`	report odds ratios
Max options	
maximize_options	control the maximization process; seldom used

`bootstrap`, `by`, `jackknife`, `nestreg`, `rolling`, `statsby`, `stepwise`, `svy`, and `xi` are allowed; see [U] **11.1.10 Prefix commands**.

Weights are not allowed with the `bootstrap` prefix.

`vce()` and weights are not allowed with the `svy` prefix.

`fweight`s, `iweight`s, and `pweight`s are allowed; see [U] **11.1.6 weight**.

See [U] **20 Estimation and postestimation commands** for more capabilities of estimation commands.

Description

`ologit` fits ordered logit models of ordinal variable *depvar* on the independent variables *indepvars*. The actual values taken on by the dependent variable are irrelevant, except that larger values are assumed to correspond to "higher" outcomes. Up to 50 outcomes are allowed in Stata/MP, Stata/SE, and Stata/IC, and up to 20 outcomes are allowed in Small Stata.

See [R] **logistic** for a list of related estimation commands.

Options

Model

`offset(`*varname*`)`; see [R] **estimation options**.

SE/Robust

`vce(`*vcetype*`)` specifies the type of standard error reported, which includes types that are derived from asymptotic theory, that are robust to some kinds of misspecification, that allow for intragroup correlation, and that use bootstrap or jackknife methods; see [R] ***vce_option***.

Reporting

`level(#)`; see [R] **estimation options**.

`or` reports the estimated coefficients transformed to odds ratios, i.e., e^b rather than b. Standard errors and confidence intervals are similarly transformed. This option affects how results are displayed, not how they are estimated. `or` may be specified at estimation or when replaying previously estimated results.

Max options

maximize_options: `iterate(#)`, `[no]log`, `trace`, `tolerance(#)`, `ltolerance(#)`; see [R] **maximize**. These options are seldom used.

Remarks

Ordered logit models are used to estimate relationships between an ordinal dependent variable and a set of independent variables. An *ordinal* variable is a variable that is categorical and ordered, for instance, "poor", "good", and "excellent", which might indicate a person's current health status or the repair record of a car. If there are only two outcomes, see [R] **logistic**, [R] **logit**, and [R] **probit**. This entry is concerned only with more than two outcomes. If the outcomes cannot be ordered (e.g., residency in the north, east, south, or west), see [R] **mlogit**. This entry is concerned only with models in which the outcomes can be ordered.

In ordered logit, an underlying score is estimated as a linear function of the independent variables and a set of cutpoints. The probability of observing outcome i corresponds to the probability that the estimated linear function, plus random error, is within the range of the cutpoints estimated for the outcome:

$$\Pr(\text{outcome}_j = i) = \Pr(\kappa_{i-1} < \beta_1 x_{1j} + \beta_2 x_{2j} + \cdots + \beta_k x_{kj} + u_j \le \kappa_i)$$

u_j is assumed to be logistically distributed in ordered logit. In either case, we estimate the coefficients β_1, β_2, ..., β_k together with the cutpoints κ_1, κ_2, ..., κ_{k-1}, where k is the number of possible outcomes. κ_0 is taken as $-\infty$, and κ_k is taken as $+\infty$. All of this is a direct generalization of the ordinary two-outcome logit model.

▷ Example 1

We wish to analyze the 1977 repair records of 66 foreign and domestic cars. The data are a variation of the automobile dataset described in [U] **1.2.1 Sample datasets**. The 1977 repair records, like those in 1978, take on values "Poor", "Fair", "Average", "Good", and "Excellent". Here is a cross-tabulation of the data:

```
. use http://www.stata-press.com/data/r10/fullauto
(Automobile Models)

. tabulate rep77 foreign, chi2

    Repair
    Record          Foreign
      1977   Domestic    Foreign       Total

      Poor          2          1           3
      Fair         10          1          11
   Average         20          7          27
      Good         13          7          20
 Excellent          0          5           5

     Total         45         21          66
          Pearson chi2(4) =  13.8619   Pr = 0.008
```

Although it appears that `foreign` takes on the values "`Domestic`" and "`Foreign`", it is actually a numeric variable taking on the values 0 and 1. Similarly, `rep77` takes on the values 1, 2, 3, 4, and 5, corresponding to "`Poor`", "`Fair`", and so on. The more meaningful words appear because we have attached value labels to the data; see [U] **12.6.3 Value labels**.

Since the chi-squared value is significant, we could claim that there is a relationship between `foreign` and `rep77`. Literally, however, we can only claim that the distributions are different; the chi-squared test is not directional. One way to model these data is to model the categorization that took place when the data were created. Cars have a true frequency of repair, which we will assume is given by $S_j = \beta\,\texttt{foreign}_j + u_j$, and a car is categorized as "poor" if $S_j \le \kappa_0$, as "fair" if $\kappa_0 < S_j \le \kappa_1$, and so on:

```
. ologit rep77 foreign

Iteration 0:   log likelihood = -89.895098
Iteration 1:   log likelihood = -85.951765
Iteration 2:   log likelihood = -85.908227
Iteration 3:   log likelihood = -85.908161

Ordered logistic regression                       Number of obs   =         66
                                                  LR chi2(1)      =       7.97
                                                  Prob > chi2     =     0.0047
Log likelihood = -85.908161                       Pseudo R2       =     0.0444

       rep77       Coef.   Std. Err.      z    P>|z|     [95% Conf. Interval]

     foreign    1.455878   .5308946     2.74   0.006     .4153436    2.496412

       /cut1   -2.765562   .5988207                     -3.939229   -1.591895
       /cut2   -.9963603   .3217704                     -1.627019   -.3657019
       /cut3    .9426153   .3136396                      .3278929    1.557338
       /cut4    3.123351   .5423237                      2.060416    4.186286
```

Our model is $S_j = 1.46\,\texttt{foreign}_j + u_j$; the expected value for foreign cars is 1.46 and, for domestic cars, 0; foreign cars have better repair records.

The estimated cutpoints tell us how to interpret the score. For a foreign car, the probability of a poor record is the probability that $1.46 + u_j \le -2.77$, or equivalently, $u_j \le -4.23$. Making this calculation requires familiarity with the logistic distribution: the probability is $1/(1 + e^{4.23}) = .014$. On the other hand, for domestic cars, the probability of a poor record is the probability $u_j \le -2.77$, which is .059.

This, it seems to us, is a far more reasonable prediction than we would have made based on the table alone. The table showed that 2 of 45 domestic cars had poor records, whereas 1 of 21 foreign cars had poor records—corresponding to probabilities $2/45 = .044$ and $1/21 = .048$. The predictions from our model imposed a smoothness assumption—foreign cars should not, overall, have better repair records without the difference revealing itself in each category. In our data, the fractions of foreign and domestic cars in the poor category are virtually identical only because of the randomness associated with small samples.

Thus if we were asked to predict the true fractions of foreign and domestic cars that would be classified in the various categories, we would choose the numbers implied by the ordered logit model:

	tabulate		logit	
	Domestic	Foreign	Domestic	Foreign
Poor	.044	.048	.059	.014
Fair	.222	.048	.210	.065
Average	.444	.333	.450	.295
Good	.289	.333	.238	.467
Excellent	.000	.238	.043	.159

See [R] **ologit postestimation** for a more complete explanation of how to generate predictions from an ordered logit model.

◁

❑ Technical Note

Here ordered logit provides an alternative to ordinary two-outcome logistic models with an arbitrary dichotomization, which might otherwise have been tempting. We could, for instance, have summarized these data by converting the five-outcome `rep77` variable to a two-outcome variable, combining cars in the average, fair, and poor categories to make one outcome and combining cars in the good and excellent categories to make the second.

Another even less appealing alternative would have been to use ordinary regression, arbitrarily labeling "excellent" as 5, "good" as 4, and so on. The problem is that with different but equally valid labellings (say 10 for "excellent"), we would obtain different estimates. We would have no way of choosing one metric over another. That assertion is not, however, true of `ologit`. The actual values used to label the categories make no difference other than through the order they imply.

In fact, our labeling was 5 for "excellent", 4 for "good", and so on. The words "excellent" and "good" appear in our output because we attached a value label to the variables; see [U] **12.6.3 Value labels**. If we were to now go back and type `replace rep77=10 if rep77==5`, changing all the 5s to 10s, we would still obtain the same results when we refitted our model.

❑

▷ Example 2

In the example above, we used ordered logit as a way to model a table. We are not, however, limited to including only one explanatory variable or to including only categorical variables. We can explore the relationship of `rep77` with any of the variables in our data. We might, for instance, model `rep77` not only in terms of the origin of manufacture, but also including `length` (a proxy for size) and `mpg`:

```
. ologit rep77 foreign length mpg
Iteration 0:   log likelihood = -89.895098
Iteration 1:   log likelihood = -78.775147
Iteration 2:   log likelihood = -78.256299
Iteration 3:   log likelihood = -78.250722
Iteration 4:   log likelihood = -78.250719
Ordered logistic regression                       Number of obs   =         66
                                                  LR chi2(3)      =      23.29
                                                  Prob > chi2     =     0.0000
Log likelihood = -78.250719                       Pseudo R2       =     0.1295

------------------------------------------------------------------------------
       rep77 |      Coef.   Std. Err.      z    P>|z|     [95% Conf. Interval]
-------------+----------------------------------------------------------------
     foreign |   2.896807   .7906411     3.66   0.000     1.347179    4.446435
      length |   .0828275     .02272     3.65   0.000     .0382972    .1273579
         mpg |   .2307677   .0704548     3.28   0.001     .0926788    .3688566
-------------+----------------------------------------------------------------
       /cut1 |   17.92748   5.551191                      7.047344    28.80761
       /cut2 |   19.86506    5.59648                      8.896161    30.83396
       /cut3 |   22.10331   5.708935                        10.914    33.29262
       /cut4 |   24.69213   5.890754                      13.14647     36.2378
------------------------------------------------------------------------------
```

foreign still plays a role—and an even larger role than previously. We find that larger cars tend to have better repair records, as do cars with better mileage ratings.

◁

Saved Results

ologit saves the following in e():

Scalars

e(N)	number of observations	e(r2_p)	pseudo-R-squared
e(k_cat)	number of categories	e(ll)	log likelihood
e(N_cd)	number of completely determined observations	e(ll_0)	log likelihood, constant-only model
		e(N_clust)	number of clusters
e(k_eq)	number of equations in e(b)	e(chi2)	χ^2
e(k_aux)	number of auxiliary parameters	e(converged)	1 if converged, 0 otherwise
e(df_m)	model degrees of freedom		

Macros

e(cmd)	ologit	e(offset)	offset
e(cmdline)	command as typed	e(chi2type)	Wald or LR; type of model χ^2 test
e(depvar)	name of dependent variable	e(vce)	*vcetype* specified in vce()
e(wtype)	weight type	e(vcetype)	title used to label Std. Err.
e(wexp)	weight expression	e(crittype)	optimization criterion
e(title)	title in estimation output	e(predict)	program used to implement predict
e(clustvar)	name of cluster variable	e(properties)	b V

Matrices

e(b)	coefficient vector	e(V)	variance–covariance matrix of the estimators
e(cat)	category values		

Functions

e(sample)	marks estimation sample

Methods and Formulas

ologit is implemented as an ado-file.

See Long and Freese (2006, chap. 5) for a discussion of models for ordinal outcomes and examples that use Stata. A straightforward textbook description of the model fit by ologit, as well as the models fit by oprobit, clogit, and mlogit, can be found in Greene (2003, chap. 21). When you have a qualitative dependent variable, several estimation procedures are available. A popular choice is multinomial logistic regression (see [R] **mlogit**), but if you use this procedure when the response variable is ordinal, you are discarding information because multinomial logit ignores the ordered aspect of the outcome. Ordered logit and probit models provide a means to exploit the ordering information.

There is more than one "ordered logit" model. The model fitted by ologit, which we will call the ordered logit model, is also known as the proportional odds model. Another popular choice, not fitted by ologit, is known as the stereotype model; see [R] **slogit**. All ordered logit models have been derived by starting with a binary logit/probit model and generalizing it to allow for more than two outcomes.

The proportional-odds ordered logit model is so called because, if we consider the odds $\mathrm{odds}(k) = P(Y \leq k)/P(Y > k)$, then $\mathrm{odds}(k_1)$ and $\mathrm{odds}(k_2)$ have the same ratio for all independent variable combinations. The model is based on the principle that the only effect of combining adjoining categories in ordered categorical regression problems should be a loss of efficiency in estimating the regression parameters (McCullagh 1980). This model was also described by Zavoina and McKelvey (1975) and, previously, by Aitchison and Silvey (1957) in a different algebraic form. Brant (1990) offers a set of diagnostics for the model.

Peterson and Harrell (1990) suggest a model that allows nonproportional odds for a subset of the explanatory variables. ologit does not allow this, but a model similar to this was implemented by Fu (1998).

The stereotype model rejects the principle on which the ordered logit model is based. Anderson (1984) argues that there are two distinct types of ordered categorical variables: "grouped continuous", such as income, where the "type a" model applies; and "assessed", such as extent of pain relief, where the stereotype model applies. Greenland (1985) independently developed the same model. The stereotype model starts with a multinomial logistic regression model and imposes constraints on this model.

Goodness of fit for ologit can be evaluated by comparing the likelihood value with that obtained by fitting the model with mlogit. Let $\ln L_1$ be the log-likelihood value reported by ologit, and let $\ln L_0$ be the log-likelihood value reported by mlogit. If there are p independent variables (excluding the constant) and k categories, mlogit will estimate $p(k-1)$ additional parameters. We can then perform a "likelihood-ratio test", i.e., calculate $-2(\ln L_1 - \ln L_0)$, and compare it to $\chi^2\{p(k-2)\}$. This test is only suggestive because the ordered logit model is not nested within the multinomial logit model. A large value of $-2(\ln L_1 - \ln L_0)$ should, however, be taken as evidence of poorness of fit. Marginally large values, on the other hand, should not be taken too seriously.

The coefficients and cutpoints are estimated using maximum likelihood as described in [R] **maximize**. In our parameterization, no constant appears, as the effect is absorbed into the cutpoints.

ologit and oprobit begin by tabulating the dependent variable. Category $i = 1$ is defined as the minimum value of the variable, $i = 2$ as the next ordered value, and so on, for the empirically determined k categories.

The probability of a given observation for ordered logit is

$$p_{ij} = \Pr(y_j = i) = \Pr\left(\kappa_{i-1} < \mathbf{x}_j\boldsymbol{\beta} + u \le \kappa_i\right)$$
$$= \frac{1}{1+\exp(-\kappa_i + \mathbf{x}_j\boldsymbol{\beta})} - \frac{1}{1+\exp(-\kappa_{i-1} + \mathbf{x}_j\boldsymbol{\beta})}$$

κ_0 is defined as $-\infty$ and κ_k as $+\infty$.

For ordered probit, the probability of a given observation is

$$p_{ij} = \Pr(y_j = i) = \Pr\left(\kappa_{i-1} < \mathbf{x}_j\boldsymbol{\beta} + u \le \kappa_i\right)$$
$$= \Phi\left(\kappa_i - \mathbf{x}_j\boldsymbol{\beta}\right) - \Phi\left(\kappa_{i-1} - \mathbf{x}_j\boldsymbol{\beta}\right)$$

where $\Phi()$ is the standard normal cumulative distribution function.

The log likelihood is

$$\ln L = \sum_{j=1}^{N} w_j \sum_{i=1}^{k} I_i(y_j)\ln p_{ij}$$

where w_j is an optional weight and

$$I_i(y_j) = \begin{cases} 1, \text{ if } y_j = i \\ 0, \text{ otherwise} \end{cases}$$

References

Aitchison, J., and S. D. Silvey. 1957. The generalization of probit analysis to the case of multiple responses. *Biometrika* 44: 131–140.

Anderson, J. A. 1984. Regression and ordered categorical variables (with discussion). *Journal of the Royal Statistical Society, Series B* 46: 1–30.

Brant, R. 1990. Assessing proportionality in the proportional odds model for ordinal logistic regression. *Biometrics* 46: 1171–1178.

Fu, V. K. 1998. sg88: Estimating generalized ordered logit models. *Stata Technical Bulletin* 44: 27–30. Reprinted in *Stata Technical Bulletin Reprints*, vol. 8, pp. 160–164.

Goldstein, R. 1997. sg59: Index of ordinal variation and Neyman–Barton GOF. *Stata Technical Bulletin* 33: 10–12. Reprinted in *Stata Technical Bulletin Reprints*, vol. 6, pp. 145–147.

Greene, W. H. 2003. *Econometric Analysis*. 5th ed. Upper Saddle River, NJ: Prentice Hall.

Greenland, S. 1985. An application of logistic models to the analysis of ordinal response. *Biometrical Journal* 27: 189–197.

Kleinbaum, D. G., and M. Klein. 2002. *Logistic Regression: A Self-Learning Text*. 2nd ed. New York: Springer.

Long, J. S. 1997. *Regression Models for Categorical and Limited Dependent Variables*. Thousand Oaks, CA: Sage.

Long, J. S., and J. Freese. 2006. *Regression Models for Categorical Dependent Variables Using Stata*. 2nd ed. College Station, TX: Stata Press.

Lunt, M. 2001. sg163: Stereotype ordinal regression. *Stata Technical Bulletin* 61: 12–18. Reprinted in *Stata Technical Bulletin Reprints*, vol. 10, pp. 298–307.

McCullagh, P. 1977. A logistic model for paired comparisons with ordered categorical data. *Biometrika* 64: 449–453.

——. 1980. Regression models for ordinal data (with discussion). *Journal of the Royal Statistical Society, Series B* 42: 109–142.

McCullagh, P., and J. A. Nelder. 1989. *Generalized Linear Models*. 2nd ed. London: Chapman & Hall.

Miranda, A., and S. Rabe-Hesketh. 2006. Maximum likelihood estimation of endogenous switching and sample selection models for binary, ordinal, and count variables. *Stata Journal* 6: 285–308.

Peterson, B., and F. E. Harrell, Jr. 1990. Partial proportional odds models for ordinal response variables. *Applied Statistics* 39: 205–217.

Williams, R. 2006. Generalized ordered logit/partial proportional odds models for ordinal dependent variables. *Stata Journal* 6: 58–82.

Wolfe, R. 1998. sg86: Continuation-ratio models for ordinal response data. *Stata Technical Bulletin* 44: 18–21. Reprinted in *Stata Technical Bulletin Reprints*, vol. 8, pp. 149–153.

Wolfe, R., and W. W. Gould. 1998. sg76: An approximate likelihood-ratio test for ordinal response models. *Stata Technical Bulletin* 42: 24–27. Reprinted in *Stata Technical Bulletin Reprints*, vol. 7, pp. 199–204.

Xu, J., and J. S. Long. 2005. Confidence intervals for predicted outcomes in regression models for categorical outcomes. *Stata Journal* 5: 537–559.

Zavoina, W., and R. D. McKelvey. 1975. A statistical model for the analysis of ordinal level dependent variables. *Journal of Mathematical Sociology* 4: 103–120.

Also See

[R] **ologit postestimation** — Postestimation tools for ologit

[R] **clogit** — Conditional (fixed-effects) logistic regression

[R] **logistic** — Logistic regression, reporting odds ratios

[R] **logit** — Logistic regression, reporting coefficients

[R] **mlogit** — Multinomial (polytomous) logistic regression

[R] **oprobit** — Ordered probit regression

[R] **rologit** — Rank-ordered logistic regression

[R] **slogit** — Stereotype logistic regression

[SVY] **svy estimation** — Estimation commands for survey data

[U] **20 Estimation and postestimation commands**

Title

ologit postestimation — Postestimation tools for ologit

Description

The following postestimation commands are available for `ologit`:

command	description
`adjust`	adjusted predictions of $\mathbf{x}\beta$ or $\exp(\mathbf{x}\beta)$
`estat`	AIC, BIC, VCE, and estimation sample summary
`estat` (svy)	postestimation statistics for survey data
`estimates`	cataloging estimation results
`lincom`	point estimates, standard errors, testing, and inference for linear combinations of coefficients
`linktest`	link test for model specification
`lrtest`[1]	likelihood-ratio test
`mfx`	marginal effects or elasticities
`nlcom`	point estimates, standard errors, testing, and inference for nonlinear combinations of coefficients
`predict`	predictions, residuals, influence statistics, and other diagnostic measures
`predictnl`	point estimates, standard errors, testing, and inference for generalized predictions
`suest`	seemingly unrelated estimation
`test`	Wald tests for simple and composite linear hypotheses
`testnl`	Wald tests of nonlinear hypotheses

[1] `lrtest` is not appropriate with `svy` estimation results.

See the corresponding entries in the *Stata Base Reference Manual* for details, but see [SVY] **estat** for details about `estat` (svy).

Syntax for predict

`predict` [*type*] *newvars* [*if*] [*in*] [, *statistic* `outcome(`*outcome*`)` `nooffset`]

`predict` [*type*] { *stub** | $newvar_{\text{reg}}$ $newvar_{\kappa_1}$ ... $newvar_{\kappa_{k-1}}$ } [*if*] [*in*] , `scores`

where k is the number of outcomes in the model.

statistic	description
Main	
`pr`	predicted probabilities; the default
`xb`	linear prediction
`stdp`	standard error of the linear prediction

With the `pr` option, you specify one or k new variables depending upon whether the `outcome()` option is also specified. With `xb` and `stdp`, one new variable is specified.

These statistics are available both in and out of sample; type `predict ... if e(sample) ...` if wanted only for the estimation sample.

Options for predict

Main

`pr`, the default, calculates the predicted probabilities. If you do not also specify the `outcome()` option, you must specify k new variables, where k is the number of categories of the dependent variable. Say that you fitted a model by typing `ologit result x1 x2`, and `result` takes on three values. Then you could type `predict p1 p2 p3` to obtain all three predicted probabilities. If you specify the `outcome()` option, you must specify one new variable. Say that `result` takes on the values 1, 2, and 3. Typing `predict p1, outcome(1)` would produce the same `p1`.

`xb` calculates the linear prediction. You specify one new variable, for example, `predict linear, xb`. The linear prediction is defined, ignoring the contribution of the estimated cutpoints.

`stdp` calculates the standard error of the linear prediction. You specify one new variable, for example, `predict se, stdp`.

`outcome(`*outcome*`)` specifies for which outcome the predicted probabilities are to be calculated. `outcome()` should contain either one value of the dependent variable or one of `#1`, `#2`, ..., with `#1` meaning the first category of the dependent variable, `#2` the second category, etc.

`nooffset` is relevant only if you specified `offset(`*varname*`)` for `ologit`. It modifies the calculations made by `predict` so that they ignore the offset variable; the linear prediction is treated as $\mathbf{x}_j\mathbf{b}$ rather than as $\mathbf{x}_j\mathbf{b} + \text{offset}_j$.

`scores` calculates equation-level score variables. The number of score variables created will equal the number of outcomes in the model. If the number of outcomes in the model were k, then

the first new variable will contain $\partial \ln L / \partial(\mathbf{x}_j\mathbf{b})$;

the second new variable will contain $\partial \ln L / \partial \kappa_1$;

the third new variable will contain $\partial \ln L / \partial \kappa_2$;

...

the kth new variable will contain $\partial \ln L / \partial \kappa_{k-1}$, where κ_i refers to the ith cutpoint.

Remarks

See [U] **20 Estimation and postestimation commands** for instructions on obtaining the variance–covariance matrix of the estimators, predicted values, and hypothesis tests. Also see [R] **lrtest** for performing likelihood-ratio tests.

▷ Example 1

In example 2 of [R] **ologit**, we fitted the model `ologit rep77 foreign length mpg`. The `predict` command can be used to obtain the predicted probabilities.

We type `predict` followed by the names of the new variables to hold the predicted probabilities, ordering the names from low to high. In our data, the lowest outcome is "poor", and the highest is "excellent". We have five categories, so we must type five names following `predict`; the choice of names is up to us:

```
. predict poor fair avg good exc
(option pr assumed; predicted probabilities)
```

```
. list exc good make model rep78 if rep77>=., sep(4) divider
```

	exc	good	make	model	rep78
3.	.0033341	.0393056	AMC	Spirit	.
10.	.0098392	.1070041	Buick	Opel	.
32.	.0023406	.0279497	Ford	Fiesta	Good
44.	.015697	.1594413	Merc.	Monarch	Average
53.	.065272	.4165188	Peugeot	604	.
56.	.005187	.059727	Plym.	Horizon	Average
57.	.0261461	.2371826	Plym.	Sapporo	.
63.	.0294961	.2585825	Pont.	Phoenix	.

The eight cars listed were introduced after 1977, so they do not have 1977 repair records in our data. We predicted what their 1977 repair records might have been using the fitted model. We see that, based on its characteristics, the Peugeot 604 had about a $41.65 + 6.53 \approx 48.2\%$ chance of a good or excellent repair record. The Ford Fiesta, which had only a 3% chance of a good or excellent repair record, in fact, had a good record when it was introduced in the following year.

◁

❑ Technical Note

For ordered logit, `predict, xb` produces $S_j = x_{1j}\beta_1 + x_{2j}\beta_2 + \cdots + x_{kj}\beta_k$. The ordered-logit predictions are then the probability that $S_j + u_j$ lies between a pair of cutpoints, κ_{i-1} and κ_i. Some handy formulas are

$$\Pr(S_j + u_j < \kappa) = 1/(1 + e^{S_j - \kappa})$$
$$\Pr(S_j + u_j > \kappa) = 1 - 1/(1 + e^{S_j - \kappa})$$
$$\Pr(\kappa_1 < S_j + u_j < \kappa_2) = 1/(1 + e^{S_j - \kappa_2}) - 1/(1 + e^{S_j - \kappa_1})$$

Rather than using `predict` directly, we could calculate the predicted probabilities by hand. If we wished to obtain the predicted probability that the repair record is excellent and the probability that it is good, we look back at `ologit`'s output to obtain the cutpoints. We find that "good" corresponds to the interval `/cut3` $< S_j + u <$ `/cut4` and "excellent" to the interval $S_j + u >$ `/cut4`:

```
. predict score, xb
. generate probgood = 1/(1+exp(score-_b[/cut4])) - 1/(1+exp(score-_b[/cut3]))
. generate probexc = 1 - 1/(1+exp(score-_b[/cut4]))
```

The results of our calculation will be the same as those produced in the previous example. We refer to the estimated cutpoints just as we would any coefficient, so `_b[/cut3]` refers to the value of the `/cut3` coefficient; see [U] **13.5 Accessing coefficients and standard errors**.

❑

Methods and Formulas

All postestimation commands listed above are implemented as ado-files.

Also See

[R] **ologit** — Ordered logistic regression

[U] **20 Estimation and postestimation commands**

Title

oneway — One-way analysis of variance

Syntax

oneway *response_var factor_var* [*if*] [*in*] [*weight*] [, *options*]

options	description
Main	
bonferroni	Bonferroni multiple-comparison test
scheffe	Schéffe multiple-comparison test
sidak	Šidák multiple-comparison test
tabulate	produce summary table
[no]means	include or suppress means; default is means
[no]standard	include or suppress standard deviations; default is standard
[no]freq	include or suppress frequencies; default is freq
[no]obs	include or suppress number of obs; default is obs if data are weighted
noanova	suppress the ANOVA table
nolabel	show numeric codes, not labels
wrap	do not break wide tables
missing	treat missing values as categories

by is allowed; see [D] **by**.

aweights and fweights are allowed; see [U] **11.1.6 weight**.

Description

The oneway command reports one-way analysis-of-variance (ANOVA) models and performs multiple-comparison tests.

If you wish to fit more complicated ANOVA layouts or wish to fit analysis-of-covariance (ANCOVA) models, see [R] **anova**.

See [D] **encode** for examples of fitting ANOVA models on string variables.

See [R] **loneway** for an alternative oneway command with slightly different features.

Options

Main

bonferroni reports the results of a Bonferroni multiple-comparison test.

scheffe reports the results of a Scheffé multiple-comparison test.

sidak reports the results of a Šidák multiple-comparison test.

`tabulate` produces a table of summary statistics of the *response_var* by levels of the *factor_var*. The table includes the mean, standard deviation, frequency, and, if the data are weighted, the number of observations. Individual elements of the table may be included or suppressed by using the [no]`means`, [no]`standard`, [no]`freq`, and [no]`obs` options. For example, typing

```
oneway response factor, tabulate means standard
```

produces a summary table that contains only the means and standard deviations. You could achieve the same result by typing

```
oneway response factor, tabulate nofreq
```

[no]`means` includes or suppresses only the means from the table produced by the `tabulate` option. See `tabulate` above.

[no]`standard` includes or suppresses only the standard deviations from the table produced by the `tabulate` option. See `tabulate` above.

[no]`freq` includes or suppresses only the frequencies from the table produced by the `tabulate` option. See `tabulate` above.

[no]`obs` includes or suppresses only the reported number of observations from the table produced by the `tabulate` option. If the data are not weighted, only the frequency is reported. If the data are weighted, the frequency refers to the sum of the weights. See `tabulate` above.

`noanova` suppresses the display of the ANOVA table.

`nolabel` causes the numeric codes to be displayed rather than the value labels in the ANOVA and multiple-comparison test tables.

`wrap` requests that Stata not break up wide tables to make them more readable.

`missing` requests that missing values of *factor_var* be treated as a category rather than as observations to be omitted from the analysis.

Remarks

Remarks are presented under the following headings:

Introduction

The `oneway` command reports one-way ANOVA models. To perform a one-way layout of a variable called `endog` on `exog`, type `oneway endog exog`.

▷ Example 1

We run an experiment varying the amount of fertilizer used in growing apple trees. We test four concentrations, using each concentration in three groves of 12 trees each. Later in the year, we measure the average weight of the fruit.

If all had gone well, we would have had 3 observations on the average weight for each of the four concentrations. Instead, two of the groves were mistakenly leveled by a confused man on a large bulldozer. We are left with the following dataset:

```
. use http://www.stata-press.com/data/r10/apple
(Apple trees)
. describe
Contains data from http://www.stata-press.com/data/r10/apple.dta
  obs:            10                          Apple trees
 vars:             2                          16 Jan 2007 11:23
 size:           140 (99.9% of memory free)
-------------------------------------------------------------------------------
              storage  display     value
variable name   type   format      label      variable label
-------------------------------------------------------------------------------
treatment       int    %8.0g                  Fertilizer
weight          double %10.0g                 Average weight in grams
-------------------------------------------------------------------------------
Sorted by:

. list, abbreviate(10)

     +----------------------+
     | treatment   weight   |
     |----------------------|
  1. |         1    117.5   |
  2. |         1    113.8   |
  3. |         1    104.4   |
  4. |         2     48.9   |
  5. |         2     50.4   |
     |----------------------|
  6. |         2     58.9   |
  7. |         3     70.4   |
  8. |         3     86.9   |
  9. |         4     87.7   |
 10. |         4     67.3   |
     +----------------------+
```

To obtain the one-way analysis-of-variance results, we type

```
. oneway weight treatment
                        Analysis of Variance
    Source              SS         df      MS            F     Prob > F
------------------------------------------------------------------------
Between groups      5295.54433      3   1765.18144     21.46     0.0013
 Within groups      493.591667      6   82.2652778
------------------------------------------------------------------------
    Total             5789.136      9   643.237333
Bartlett's test for equal variances:  chi2(3) =   1.3900  Prob>chi2 = 0.708
```

We find significant (at better than the 1% level) differences among the four concentrations.

◁

❑ Technical Note

Rather than using the `oneway` command, we could have performed this analysis by using `anova`. The first example in [R] **anova** repeats this same analysis. You may wish to compare the output.

You will find the `oneway` command quicker than the `anova` command, and, as you will learn, `oneway` allows you to perform multiple-comparison tests. On the other hand, `anova` will let you generate predictions, examine the covariance matrix of the estimators, and perform more general hypothesis tests.

❑

❑ Technical Note

Although the output is a usual ANOVA table, let's run through it anyway. The between-group sum of squares for the model is 5295.5 with 3 degrees of freedom, resulting in a mean square of $5295.5/3 \approx 1765.2$. The corresponding F statistic is 21.46 and has a significance level of 0.0013. Thus the model appears to be significant at the 0.13% level.

The second line summarizes the within-group (residual) variation. The within-group sum of squares is 493.59 with 6 degrees of freedom, resulting in a mean squared error of 82.27.

The between- and residual-group variations sum to the total sum of squares, which is reported as 5789.1 in the last line of the table. This is the total sum of squares of `weight` after removal of the mean. Similarly, the between plus residual degrees of freedom sum to the total degrees of freedom, 9. Remember that there are 10 observations. Subtracting 1 for the mean, we are left with 9 total degrees of freedom.

At the bottom of the table, Bartlett's test for equal variances is reported. The value of the statistic is 1.39. The corresponding significance level (χ^2 with 3 degrees of freedom) is 0.708, so we cannot reject the assumption that the variances are homogeneous.

❑

Obtaining observed means

▷ Example 2

We typed `oneway weight treatment` to obtain an ANOVA table of weight of fruit by fertilizer concentration. Although we obtained the table, we obtained no information on which fertilizer seems to work the best. If we add the `tabulate` option, we obtain that additional information:

```
. oneway weight treatment, tabulate

            |  Summary of Average weight in grams
 Fertilizer |        Mean   Std. Dev.       Freq.
------------+------------------------------------
          1 |       111.9   6.7535176           3
          2 |   52.733333   5.3928966           3
          3 |       78.65   11.667262           2
          4 |        77.5   14.424978           2
------------+------------------------------------
      Total |       80.62   25.362124          10

                        Analysis of Variance
    Source              SS         df      MS            F     Prob > F
------------------------------------------------------------------------
Between groups      5295.54433      3   1765.18144     21.46     0.0013
 Within groups      493.591667      6   82.2652778
------------------------------------------------------------------------
    Total             5789.136      9   643.237333

Bartlett's test for equal variances:  chi2(3) =   1.3900  Prob>chi2 = 0.708
```

We find that the average weight was largest when we used fertilizer concentration 1.

◁

Multiple-comparison tests

▷ Example 3

oneway can also perform multiple-comparison tests using either Bonferroni, Scheffé, or Šidák normalizations. For instance, to obtain the Bonferroni multiple-comparison test, we specify the bonferroni option:

```
. oneway weight treatment, bonferroni
                        Analysis of Variance
    Source              SS         df      MS            F     Prob > F
------------------------------------------------------------------------
Between groups      5295.54433      3   1765.18144     21.46     0.0013
 Within groups      493.591667      6   82.2652778
------------------------------------------------------------------------
    Total             5789.136      9   643.237333
Bartlett's test for equal variances:  chi2(3) =   1.3900  Prob>chi2 = 0.708

             Comparison of Average weight in grams by Fertilizer
                                (Bonferroni)
Row Mean-|
Col Mean |          1          2          3
---------+---------------------------------
       2 |   -59.1667
         |      0.001
         |
       3 |     -33.25    25.9167
         |      0.042      0.122
         |
       4 |      -34.4    24.7667      -1.15
         |      0.036      0.146      1.000
```

The results of the Bonferroni test are presented as a matrix. The first entry, −59.17, represents the difference between fertilizer concentrations 2 and 1 (labeled "Row Mean - Col Mean" in the upper stub of the table). Remember that in the previous example we requested the tabulate option. Looking back, we find that the means of concentrations 1 and 2 are 111.90 and 52.73, respectively. Thus $52.73 - 111.90 = -59.17$.

Underneath that number is reported "0.001". This is the Bonferroni-adjusted significance of the difference. The difference is significant at the 0.1% level. Looking down the column, we see that concentration 3 is also worse than concentration 1 (4.2% level), as is concentration 4 (3.6% level).

On the basis of this evidence, we would use concentration 1 if we grew apple trees.

◁

▷ Example 4

We can just as easily obtain the Scheffé-adjusted significance levels. Rather than specifying the bonferroni option, we specify the scheffe option.

We will also add the `noanova` option to prevent Stata from redisplaying the ANOVA table:

```
. oneway weight treatment, noanova scheffe

              Comparison of Average weight in grams by Fertilizer
                                    (Scheffe)
Row Mean-|
Col Mean |          1           2           3
---------+------------------------------------
       2 |   -59.1667
         |      0.001
         |
       3 |     -33.25     25.9167
         |      0.039       0.101
         |
       4 |      -34.4     24.7667       -1.15
         |      0.034       0.118       0.999
```

The differences are the same as those we obtained in the Bonferroni output, but the significance levels are not. According to the Bonferroni-adjusted numbers, the significance of the difference between fertilizer concentrations 1 and 3 is 4.2%. The Scheffé-adjusted significance level is 3.9%.

We will leave it to you to decide which results are more accurate.

◁

▷ Example 5

Let's conclude this example by obtaining the Šidák-adjusted multiple-comparison tests. We do this to illustrate Stata's capabilities to calculate these results, as searching across adjustment methods until you find the results you want is not a valid technique for obtaining significance levels.

```
. oneway weight treatment, noanova sidak

              Comparison of Average weight in grams by Fertilizer
                                     (Sidak)
Row Mean-|
Col Mean |          1           2           3
---------+------------------------------------
       2 |   -59.1667
         |      0.001
         |
       3 |     -33.25     25.9167
         |      0.041       0.116
         |
       4 |      -34.4     24.7667       -1.15
         |      0.035       0.137       1.000
```

We find results that are similar to the Bonferroni-adjusted numbers.

◁

Henry Scheffé (1907–1977) was born in New York. He studied mathematics at the University of Wisconsin, gaining a doctorate with a dissertation on differential equations. He taught mathematics at Wisconsin, Oregon State University, and Reed College, but his interests changed to statistics and he joined Wilks at Princeton. After periods at Syracuse, UCLA, and Columbia, Scheffé settled in Berkeley from 1953. His research increasingly focused on linear models and particularly ANOVA, on which he produced a celebrated monograph. His death was the result of a bicycle accident.

Weighted data

▷ Example 6

oneway can work with both weighted and unweighted data. Let's assume that we wish to perform a one-way layout of the death rate on the four census regions of the United States using state data. Our data contain three variables, drate (the death rate), region (the region), and pop (the population of the state).

To fit the model, we type oneway drate region [weight=pop], although we typically abbreviate weight as w. We will also add the tabulate option to demonstrate how the table of summary statistics differs for weighted data:

```
. use http://www.stata-press.com/data/r10/census8
(1980 Census data by state)

. oneway drate region [w=pop], tabulate
(analytic weights assumed)

     Census  |            Summary of Death Rate
     region  |        Mean   Std. Dev.       Freq.          Obs.
-------------+-------------------------------------------------
         NE  |       97.15        5.82    49135283             9
    N Cntrl  |       88.10        5.58    58865670            12
      South  |       87.05       10.40    74734029            16
       West  |       75.65        8.23    43172490            13
-------------+-------------------------------------------------
      Total  |       87.34       10.43   2.259e+08            50

                        Analysis of Variance
    Source              SS         df      MS            F     Prob > F
------------------------------------------------------------------------
Between groups      2360.92281      3   786.974272     12.17     0.0000
 Within groups      2974.09635     46   64.6542685
------------------------------------------------------------------------
    Total           5335.01916     49   108.877942

Bartlett's test for equal variances:  chi2(3) =   5.4971  Prob>chi2 = 0.139
```

When the data are weighted, the summary table has four columns rather than three. The column labeled "Freq." reports the sum of the weights. The overall frequency is $2.259 \cdot 10^8$, meaning that there are approximately 226 million people in the United States.

The ANOVA table is appropriately weighted. Also see [U] **11.1.6 weight**.

◁

Saved Results

oneway saves the following in r():

Scalars

r(N)	number of observations	r(df_m)	between-group degrees of freedom
r(F)	F statistic	r(rss)	within-group sum of squares
r(df_r)	within-group degrees of freedom	r(chi2bart)	Bartlett's χ^2
r(mss)	between-group sum of squares	r(df_bart)	Bartlett's degrees of freedom

Methods and Formulas

The model of one-way analysis of variance is

$$y_{ij} = \mu + \alpha_i + \epsilon_{ij}$$

for levels $i = 1, \ldots, k$ and observations $j = 1, \ldots, n_i$. Define $\overline{y}_i$ as the (weighted) mean of y_{ij} over j and $\overline{y}$ as the overall (weighted) mean of y_{ij}. Define w_{ij} as the weight associated with y_{ij}, which is 1 if the data are unweighted. w_{ij} is normalized to sum to $n = \sum_i n_i$ if `aweights` are used and is otherwise not normalized. w_i refers to $\sum_j w_{ij}$ and w refers to $\sum_i w_i$.

The between-group sum of squares is then

$$S_1 = \sum_i w_i (\overline{y}_i - \overline{y})^2$$

The total sum of squares is

$$S = \sum_i \sum_j w_{ij} (y_{ij} - \overline{y})^2$$

The within-group sum of squares is given by $S_e = S - S_1$.

The between-group mean square is $s_1^2 = S_1/(k-1)$, and the within-group mean square is $s_e^2 = S_e/(w-k)$. The test statistic is $F = s_1^2/s_e^2$. See, for instance, Snedecor and Cochran (1989).

Bartlett's test

Bartlett's test assumes that you have m independent, normal, random samples and tests the hypothesis $\sigma_1^2 = \sigma_2^2 = \cdots = \sigma_m^2$. The test statistic, M, is defined as

$$M = \frac{(T-m)\ln\widehat{\sigma}^2 - \sum (T_i - 1)\ln\widehat{\sigma}_i^2}{1 + \frac{1}{3(m-1)} \left\{ \left(\sum \frac{1}{T_i - 1} \right) - \frac{1}{T-m} \right\}}$$

where there are T overall observations, T_i observations in the ith group, and

$$(T_i - 1)\widehat{\sigma}_i^2 = \sum_{j=1}^{T_i} (y_{ij} - \overline{y}_i)^2$$

$$(T - m)\widehat{\sigma}^2 = \sum_{i=1}^{m} (T_i - 1)\widehat{\sigma}_i^2$$

An approximate test of the homogeneity of variance is based on the statistic M with critical values obtained from the χ^2 distribution of $m - 1$ degrees of freedom. See Bartlett (1937) or Judge et al. (1985, 447–449).

Multiple-comparison tests

Let's begin by reviewing the logic behind these adjustments. The "standard" t statistic for the comparison of two means is

$$t = \frac{\overline{y}_i - \overline{y}_j}{s\sqrt{\frac{1}{n_i} + \frac{1}{n_j}}}$$

where s is the overall standard deviation, $\overline{y}_i$ is the measured average of y in group i, and n_i is the number of observations in the group. We perform hypothesis tests by calculating this t statistic. We simultaneously choose a critical level α and look up the t statistic corresponding to that level in a table. We reject the hypothesis if our calculated t exceeds the value we looked up. Alternatively, since we have a computer at our disposal, we calculate the significance-level e corresponding to our calculated t statistic, and if $e < \alpha$, we reject the hypothesis.

This logic works well when we are performing a *single* test. Now consider what happens when we perform several separate tests, say, n of them. Let's assume, just for discussion, that we set α equal to 0.05 and that we will perform six tests. For each test, we have a 0.05 probability of falsely rejecting the equality-of-means hypothesis. Overall, then, our chances of falsely rejecting *at least one* of the hypotheses is $1 - (1 - .05)^6 \approx .26$ if the tests are independent.

The idea behind multiple-comparison tests is to control for the fact that we will perform multiple tests and to reduce our overall chances of falsely rejecting each hypothesis to α rather than letting it increase with each additional test. (See Miller 1981 and Hochberg and Tamhane 1987 for rather advanced texts on multiple-comparison procedures.)

The Bonferroni adjustment (see Miller 1981; also see Winer, Brown, and Michels 1991, 158–166) does this by (falsely but approximately) asserting that the critical level we should use, a, is the true critical level α divided by the number of tests n; that is, $a = \alpha/n$. For instance, if we are going to perform six tests, each at the .05 significance level, we want to adopt a critical level of $.05/6 \approx .00833$.

We can just as easily apply this logic to e, the significance level associated with our t statistic, as to our critical level α. If a comparison has a calculated significance of e, then its "real" significance, adjusted for the fact of n comparisons, is $n \cdot e$. If a comparison has a significance level of, say, .012, and we perform six tests, then its "real" significance is .072. If we adopt a critical level of .05, we cannot reject the hypothesis. If we adopt a critical level of .10, we can reject it.

Of course, this calculation can go above 1, but that just means that there is no $\alpha < 1$ for which we could reject the hypothesis. (This situation arises because of the crude nature of the Bonferroni adjustment.) Stata handles this case by simply calling the significance level 1. Thus the formula for the Bonferroni significance level is

$$e_b = \min(1, en)$$

where $n = k(k-1)/2$ is the number of comparisons.

The Šidák adjustment (Šidák 1967; also see Winer, Brown, and Michels 1991, 165–166) is slightly different and provides a tighter bound. It starts with the assertion that

$$a = 1 - (1 - \alpha)^{1/n}$$

Turning this formula around and substituting calculated significance levels, we obtain

$$e_s = \min\Big\{1, 1 - (1 - e)^n\Big\}$$

For example, if the calculated significance is 0.012 and we perform six tests, the "real" significance is approximately 0.07.

The Scheffé test (Scheffé 1953, 1959; also see Winer, Brown, and Michels 1991, 191–195) differs in derivation, but it attacks the same problem. Let there be k means for which we want to make all the pairwise tests. Two means are declared significantly different if

$$t \geq \sqrt{(k-1)F(\alpha; k-1, \nu)}$$

where $F(\alpha; k-1, \nu)$ is the α-critical value of the F distribution with $k-1$ numerator and ν denominator degrees of freedom. Scheffé's test has the nicety that it never declares a contrast significant if the overall F test is not significant.

Turning the test around, Stata calculates a significance level

$$\widehat{e} = F\left(\frac{t^2}{k-1}, k-1, \nu\right)$$

For instance, you have a calculated t statistic of 4.0 with 50 degrees of freedom. The simple t test says that the significance level is .00021. The F test equivalent, 16 with 1 and 50 degrees of freedom, says that the same. If you are comparing three means, however, you calculate an F test of 8.0 with 2 and 50 degrees of freedom, which says the significance level is .0010.

References

Acock, A. C. 2006. *A Gentle Introduction to Stata.* College Station, TX: Stata Press.

Altman, D. G. 1991. *Practical Statistics for Medical Research.* London: Chapman & Hall/CRC.

Bartlett, M. S. 1937. Properties of sufficiency and statistical tests. *Proceedings of the Royal Society, Series A* 160: 268–282.

Daniel, C., and E. L. Lehmann. 1979. Henry Scheffé 1907–1977. *Annals of Statistics* 7: 1149–1161.

Hochberg, Y., and A. C. Tamhane. 1987. *Multiple Comparison Procedures.* New York: Wiley.

Judge, G. G., W. E. Griffiths, R. C. Hill, H. Lütkepohl, and T.-C. Lee. 1985. *The Theory and Practice of Econometrics.* 2nd ed. New York: Wiley.

Marchenko, Y. 2006. Estimating variance components in Stata. *Stata Journal* 6: 1–21.

Miller, R. G., Jr. 1981. *Simultaneous Statistical Inference.* 2nd ed. New York: Springer.

Scheffé, H. 1953. A method for judging all contrasts in the analysis of variance. *Biometrika* 40: 87–104.

——. 1959. *The Analysis of Variance.* New York: Wiley.

Šidák, Z. 1967. Rectangular confidence regions for the means of multivariate normal distributions. *Journal of the American Statistical Association* 62: 626–633.

Snedecor, G. W., and W. G. Cochran. 1989. *Statistical Methods.* 8th ed. Ames, IA: Iowa State University Press.

Winer, B. J., D. R. Brown, and K. M. Michels. 1991. *Statistical Principles in Experimental Design.* 3rd ed. New York: McGraw–Hill.

Also See

[R] **anova** — Analysis of variance and covariance

[R] **loneway** — Large one-way ANOVA, random effects, and reliability

Title

oprobit — Ordered probit regression

Syntax

`oprobit` *depvar* [*indepvars*] [*if*] [*in*] [*weight*] [, *options*]

options	description
Model	
`offset(`*varname*`)`	include *varname* in model with coefficient constrained to 1
SE/Robust	
`vce(`*vcetype*`)`	*vcetype* may be `oim`, `robust`, `cluster` *clustvar*, `bootstrap`, or `jackknife`
Reporting	
`level(`*#*`)`	set confidence level; default is `level(95)`
Max options	
maximize_options	control the maximization process; seldom used

`bootstrap`, `by`, `jackknife`, `nestreg`, `rolling`, `statsby`, `stepwise`, `svy`, and `xi` are allowed; see [U] **11.1.10 Prefix commands**.

Weights are not allowed with the `bootstrap` prefix.

`vce()` and weights are not allowed with the `svy` prefix.

`fweight`s, `iweight`s, and `pweight`s are allowed; see [U] **11.1.6 weight**.

See [U] **20 Estimation and postestimation commands** for more capabilities of estimation commands.

Description

`oprobit` fits ordered probit models of ordinal variable *depvar* on the independent variables *indepvars*. The actual values taken on by the dependent variable are irrelevant, except that larger values are assumed to correspond to "higher" outcomes. Up to 50 outcomes are allowed in Stata/MP, Stata/SE, and Stata/IC, and up to 20 are allowed in Small Stata.

See [R] **logistic** for a list of related estimation commands.

Options

Model

`offset(`*varname*`)`; see [R] **estimation options**.

SE/Robust

`vce(`*vcetype*`)` specifies the type of standard error reported, which includes types that are derived from asymptotic theory, that are robust to some kinds of misspecification, that allow for intragroup correlation, and that use bootstrap or jackknife methods; see [R] ***vce_option***.

Reporting

`level(#)`; see [R] **estimation options**.

Max options

maximize_options: `iterate(#)`, `[no]log`, `trace`, `tolerance(#)`, `ltolerance(#)`; see [R] **maximize**. These options are seldom used.

Remarks

An ordered probit model is used to estimate relationships between an ordinal dependent variable and a set of independent variables. An *ordinal* variable is a variable that is categorical and ordered, for instance, "poor", "good", and "excellent", which might indicate a person's current health status or the repair record of a car. If there are only two outcomes, see [R] **logistic**, [R] **logit**, and [R] **probit**. This entry is concerned only with more than two outcomes. If the outcomes cannot be ordered (e.g., residency in the north, east, south, or west), see [R] **mlogit**. This entry is concerned only with models in which the outcomes can be ordered.

In ordered probit, an underlying score is estimated as a linear function of the independent variables and a set of cutpoints. The probability of observing outcome i corresponds to the probability that the estimated linear function, plus random error, is within the range of the cutpoints estimated for the outcome:

$$\Pr(\text{outcome}_j = i) = \Pr(\kappa_{i-1} < \beta_1 x_{1j} + \beta_2 x_{2j} + \cdots + \beta_k x_{kj} + u_j \le \kappa_i)$$

u_j is assumed to be normally distributed. In either case, we estimate the coefficients β_1, β_2, ..., β_k together with the cutpoints κ_1, κ_2, ..., κ_{I-1}, where I is the number of possible outcomes. κ_0 is taken as $-\infty$, and κ_I is taken as $+\infty$. All of this is a direct generalization of the ordinary two-outcome probit model.

▷ Example 1

In [R] **ologit**, we use a variation of the automobile dataset (see [U] **1.2.1 Sample datasets**) to analyze the 1977 repair records of 66 foreign and domestic cars. We use ordered logit to explore the relationship of `rep77` in terms of `foreign` (origin of manufacture), `length` (a proxy for size), and `mpg`. Here we fit the same model using ordered probit rather than ordered logit:

```
. use http://www.stata-press.com/data/r10/fullauto
(Automobile Models)
. oprobit rep77 foreign length mpg
Iteration 0:   log likelihood = -89.895098
Iteration 1:   log likelihood = -78.141221
Iteration 2:   log likelihood = -78.020314
Iteration 3:   log likelihood = -78.020025
Ordered probit regression                         Number of obs   =         66
                                                  LR chi2(3)      =      23.75
                                                  Prob > chi2     =     0.0000
Log likelihood = -78.020025                       Pseudo R2       =     0.1321
```

rep77	Coef.	Std. Err.	z	P>|z|	[95% Conf.	Interval]
foreign	1.704861	.4246786	4.01	0.000	.8725057	2.537215
length	.0468675	.012648	3.71	0.000	.022078	.0716571
mpg	.1304559	.0378627	3.45	0.001	.0562464	.2046654
/cut1	10.1589	3.076749			4.128586	16.18922
/cut2	11.21003	3.107522			5.119399	17.30066
/cut3	12.54561	3.155228			6.361476	18.72974
/cut4	13.98059	3.218786			7.671888	20.2893

We find that foreign cars have better repair records, as do larger cars and cars with better mileage ratings.

◁

Saved Results

oprobit saves the following in e():

Scalars

e(N)	number of observations	e(r2_p)	pseudo-R-squared
e(k_cat)	number of categories	e(ll)	log likelihood
e(N_cd)	number of completely determined observations	e(ll_0)	log likelihood, constant-only model
e(k_eq)	number of equations in e(b)	e(N_clust)	number of clusters
e(k_aux)	number of auxiliary parameters	e(chi2)	χ^2
e(df_m)	model degrees of freedom	e(converged)	1 if converged, 0 otherwise

Macros

e(cmd)	oprobit	e(offset)	offset
e(cmdline)	command as typed	e(chi2type)	Wald or LR; type of model χ^2 test
e(depvar)	name of dependent variable	e(crittype)	optimization criterion
e(wtype)	weight type	e(vce)	*vcetype* specified in vce()
e(wexp)	weight expression	e(vcetype)	title used to label Std. Err.
e(title)	title in estimation output	e(predict)	program used to implement predict
e(clustvar)	name of cluster variable	e(properties)	b V

Matrices

e(b)	coefficient vector	e(V)	variance–covariance matrix of the estimators
e(cat)	category values		

Functions

e(sample)	marks estimation sample

Methods and Formulas

`oprobit` is implemented as an ado-file.

Please see the *Methods and Formulas* section of [R] **ologit**.

References

Aitchison, J., and S. D. Silvey. 1957. The generalization of probit analysis to the case of multiple responses. *Biometrika* 44: 131–140.

Goldstein, R. 1997. sg59: Index of ordinal variation and Neyman–Barton GOF. *Stata Technical Bulletin* 33: 10–12. Reprinted in *Stata Technical Bulletin Reprints*, vol. 6, pp. 145–147.

Greene, W. H. 2003. *Econometric Analysis*. 5th ed. Upper Saddle River, NJ: Prentice Hall.

Long, J. S. 1997. *Regression Models for Categorical and Limited Dependent Variables*. Thousand Oaks, CA: Sage.

Long, J. S., and J. Freese. 2006. *Regression Models for Categorical Dependent Variables Using Stata*. 2nd ed. College Station, TX: Stata Press.

Lunt, M. 2001. sg163: Stereotype ordinal regression. *Stata Technical Bulletin* 61: 12–18. Reprinted in *Stata Technical Bulletin Reprints*, vol. 10, pp. 298–307.

Miranda, A., and S. Rabe-Hesketh. 2006. Maximum likelihood estimation of endogenous switching and sample selection models for binary, ordinal, and count variables. *Stata Journal* 6: 285–308.

Stewart, M. B. 2004. Semi-nonparametric estimation of extended ordered probit models. *Stata Journal* 4: 27–39.

Williams, R. 2006. Generalized ordered logit/partial proportional odds models for ordinal dependent variables. *Stata Journal* 6: 58–82.

Wolfe, R. 1998. sg86: Continuation-ratio models for ordinal response data. *Stata Technical Bulletin* 44: 18–21. Reprinted in *Stata Technical Bulletin Reprints*, vol. 8, pp. 149–153.

Wolfe, R., and W. W. Gould. 1998. sg76: An approximate likelihood-ratio test for ordinal response models. *Stata Technical Bulletin* 42: 24–27. Reprinted in *Stata Technical Bulletin Reprints*, vol. 7, pp. 199–204.

Xu, J., and J. S. Long. 2005. Confidence intervals for predicted outcomes in regression models for categorical outcomes. *Stata Journal* 5: 537–559.

Also See

[R] **oprobit postestimation** — Postestimation tools for oprobit

[R] **logistic** — Logistic regression, reporting odds ratios

[R] **mlogit** — Multinomial (polytomous) logistic regression

[R] **mprobit** — Multinomial probit regression

[R] **ologit** — Ordered logistic regression

[R] **probit** — Probit regression

[SVY] **svy estimation** — Estimation commands for survey data

[U] **20 Estimation and postestimation commands**

Title

oprobit postestimation — Postestimation tools for oprobit

Description

The following postestimation commands are available for `oprobit`:

command	description
`adjust`	adjusted predictions of $\mathbf{x}\beta$
`estat`	AIC, BIC, VCE, and estimation sample summary
`estat` (svy)	postestimation statistics for survey data
`estimates`	cataloging estimation results
`lincom`	point estimates, standard errors, testing, and inference for linear combinations of coefficients
`linktest`	link test for model specification
`lrtest`[1]	likelihood-ratio test
`mfx`	marginal effects or elasticities
`nlcom`	point estimates, standard errors, testing, and inference for nonlinear combinations of coefficients
`predict`	predictions, residuals, influence statistics, and other diagnostic measures
`predictnl`	point estimates, standard errors, testing, and inference for generalized predictions
`suest`	seemingly unrelated estimation
`test`	Wald tests for simple and composite linear hypotheses
`testnl`	Wald tests of nonlinear hypotheses

[1] `lrtest` is not appropriate with `svy` estimation results.

See the corresponding entries in the *Stata Base Reference Manual* for details, but see [SVY] **estat** for details about `estat` (svy).

Syntax for predict

`predict` [*type*] *newvars* [*if*] [*in*] [, *statistic* `outcome(`*outcome*`)` `nooffset`]

`predict` [*type*] { *stub*∗ | $newvar_{\text{reg}}$ $newvar_{\kappa 1}$ ... $newvar_{\kappa_{k-1}}$ } [*if*] [*in*] , `scores`

where k is the number of outcomes in the model.

statistic	description
Main	
`pr`	predicted probabilities; the default
`xb`	linear prediction
`stdp`	standard error of the linear prediction

With the `pr` option, you specify either one or k new variables depending upon whether the `outcome()` option is also specified. With `xb` and `stdp`, one new variable is specified.

These statistics are available both in and out of sample; type `predict ... if e(sample) ...` if wanted only for the estimation sample.

Options for predict

Main

`pr`, the default, calculates the predicted probabilities. If you do not also specify the `outcome()` option, you must specify k new variables, where k is the number of categories of the dependent variable. Say that you fitted a model by typing `oprobit result x1 x2`, and `result` takes on three values. Then you could type `predict p1 p2 p3` to obtain all three predicted probabilities. If you specify the `outcome()` option, you must specify one new variable. Say that `result` takes on values 1, 2, and 3. Typing `predict p1, outcome(1)` would produce the same `p1`.

`xb` calculates the linear prediction. You specify one new variable, for example, `predict linear, xb`. The linear prediction is defined ignoring the contribution of the estimated cutpoints.

`stdp` calculates the standard error of the linear prediction. You specify one new variable, for example, `predict se, stdp`.

`outcome(`*outcome*`)` specifies for which outcome the predicted probabilities are to be calculated. `outcome()` should contain either one value of the dependent variable or one of `#1`, `#2`, ..., with `#1` meaning the first category of the dependent variable, `#2` the second category, etc.

`nooffset` is relevant only if you specified `offset(`*varname*`)` for `oprobit`. It modifies the calculations made by `predict` so that they ignore the offset variable; the linear prediction is treated as $\mathbf{x}_j\mathbf{b}$ rather than as $\mathbf{x}_j\mathbf{b} + \text{offset}_j$.

`scores` calculates equation-level score variables. The number of score variables created will equal the number of outcomes in the model. If the number of outcomes in the model were k, then

the first new variable will contain $\partial \ln L/\partial(\mathbf{x}_j\mathbf{b})$;

the second new variable will contain $\partial \ln L/\partial\kappa_1$;

the third new variable will contain $\partial \ln L/\partial\kappa_2$;

...

the kth new variable will contain $\partial \ln L/\partial\kappa_{k-1}$, where κ_i refers to the ith cutpoint.

Remarks

See [U] **20 Estimation and postestimation commands** for instructions on obtaining the variance–covariance matrix of the estimators, predicted values, and hypothesis tests. Also see [R] **lrtest** for performing likelihood-ratio tests.

▷ Example 1

In example 1 of [R] **oprobit**, we fitted the model `oprobit rep77 foreign length mpg`. The `predict` command can be used to obtain the predicted probabilities. We type `predict` followed by the names of the new variables to hold the predicted probabilities, ordering the names from low to high. In our data, the lowest outcome is "poor" and the highest is "excellent". We have five categories, so we must type five names following `predict`; the choice of names is up to us:

```
. predict poor fair avg good exc
(option pr assumed; predicted probabilities)
. list make model exc good if rep77>=., sep(4) divider
```

	make	model	exc	good
3.	AMC	Spirit	.0006044	.0351813
10.	Buick	Opel	.0043803	.1133763
32.	Ford	Fiesta	.0002927	.0222789
44.	Merc.	Monarch	.0093209	.1700846
53.	Peugeot	604	.0734199	.4202766
56.	Plym.	Horizon	.001413	.0590294
57.	Plym.	Sapporo	.0197543	.2466034
63.	Pont.	Phoenix	.0234156	.266771

◁

❑ Technical Note

For ordered probit, `predict, xb` produces $S_j = x_{1j}\beta_1 + x_{2j}\beta_2 + \cdots + x_{kj}\beta_k$. Ordered probit is identical to ordered logit, except that we use different distribution functions for calculating probabilities. The ordered-probit predictions are then the probability that $S_j + u_j$ lies between a pair of cutpoints κ_{i-1} and κ_i. The formulas for ordered probit are

$$\Pr(S_j + u < \kappa) = \Phi(\kappa - S_j)$$
$$\Pr(S_j + u > \kappa) = 1 - \Phi(\kappa - S_j) = \Phi(S_j - \kappa)$$
$$\Pr(\kappa_1 < S_j + u < \kappa_2) = \Phi(\kappa_2 - S_j) - \Phi(\kappa_1 - S_j)$$

Rather than using `predict` directly, we could calculate the predicted probabilities by hand.

```
. predict pscore, xb
. generate probexc = normal(pscore-_b[/cut4])
. generate probgood = normal(_b[/cut4]-pscore) - normal(_b[/cut3]-pscore)
```

❑

Methods and Formulas

All postestimation tools listed above are implemented as ado-files.

Also See

[R] **oprobit** — Ordered probit regression

[U] **20 Estimation and postestimation commands**

Title

orthog — Orthogonalize variables and compute orthogonal polynomials

Syntax

Orthogonalize variables

`orthog` [*varlist*] [*if*] [*in*] [*weight*] , `generate(`*newvarlist*`)` [`matrix(`*matname*`)`]

Compute orthogonal polynomial

`orthpoly` *varname* [*if*] [*in*] [*weight*] ,

{ `generate(`*newvarlist*`)` | `poly(`*matname*`)` } [`degree(`*#*`)`]

`orthpoly` requires that `generate(`*newvarlist*`)` or `poly(`*matname*`)`, or both, be specified.

varlist may contain time-series operators; see [U] **11.4.3 Time-series varlists**.

`iweight`s, `fweight`s, `pweight`s, and `aweight`s are allowed, see [U] **11.1.6 weight**.

Description

`orthog` orthogonalizes a set of variables, creating a new set of orthogonal variables (all of type `double`), using a modified Gram–Schmidt procedure (Golub and Van Loan 1996). The order of the variables determines the orthogonalization; hence, the "most important" variables should be listed first.

Execution time is proportional to the square of the number of variables. With many (>10) variables, `orthog` will be fairly slow.

`orthpoly` computes orthogonal polynomials for one variable.

Options for orthog

Main

`generate(`*newvarlist*`)` is required. `generate()` creates new orthogonal variables of type `double`. For `orthog`, *newvarlist* will contain the orthogonalized *varlist*. If *varlist* contains d variables, then so will *newvarlist*. *newvarlist* can be specified by giving a list of exactly d new variable names, or it can be abbreviated using the styles *newvar*`1-` *newvard* or *newvar*`*`. For these two styles of abbreviation, new variables *newvar*1, *newvar*2, ..., *newvard* are generated.

`matrix(`*matname*`)` creates a $(d+1)\times(d+1)$ matrix containing the matrix R defined by $X = QR$, where X is the $N\times(d+1)$ matrix representation of *varlist* plus a column of ones and Q is the $N\times(d+1)$ matrix representation of *newvarlist* plus a column of ones (d = number of variables in *varlist*, and N = number of observations).

Options for orthpoly

Main

`generate(`*newvarlist*`)` or `poly()`, or both, must be specified. `generate()` creates new orthogonal variables of type `double`. *newvarlist* will contain orthogonal polynomials of degree 1, 2, ..., d evaluated at *varname*, where d is as specified by `degree(`d`)`. *newvarlist* can be specified by giving a list of exactly d new variable names, or it can be abbreviated using the styles *newvar*`1`-*newvard* or *newvar*`*`. For these two styles of abbreviation, new variables *newvar*`1`, *newvar*`2`, ..., *newvard* are generated.

`poly(`*matname*`)` creates a $(d+1) \times (d+1)$ matrix called *matname* containing the coefficients of the orthogonal polynomials. The orthogonal polynomial of degree $i \leq d$ is

matname`[`i`, `$d+1$`] + `*matname*`[`i`, 1]*`*varname* `+` *matname*`[`i`, 2]*`*varname*^{2}
`+ ··· +` *matname*`[`i`, `i`]*`*varname*^{i}

The coefficients corresponding to the constant term are placed in the last column of the matrix. The last row of the matrix is all zero, except for the last column, which corresponds to the constant term.

`degree(#)` specifies the highest-degree polynomial to include. Orthogonal polynomials of degree 1, 2, ..., $d = \#$ are computed. The default is $d = 1$.

Remarks

Orthogonal variables are useful for two reasons. The first is numerical accuracy for highly collinear variables. Stata's `regress` and other estimation commands can face much collinearity and still produce accurate results. But, at some point, these commands will drop variables because of collinearity. If you know with certainty that the variables are not perfectly collinear, you may want to retain all their effects in the model. If you use `orthog` or `orthpoly` to produce a set of orthogonal variables, all variables will be present in the estimation results.

Users are more likely to find orthogonal variables useful for the second reason: ease of interpreting results. `orthog` and `orthpoly` create a set of variables such that the "effects" of all the preceding variables have been removed from each variable. For example, if we issue the command

```
. orthog x1 x2 x3, generate(q1 q2 q3)
```

the effect of the constant is removed from `x1` to produce `q1`; the constant and `x1` are removed from `x2` to produce `q2`; and finally the constant, `x1`, and `x2` are removed from `x3` to produce `q3`. Hence,

$$\texttt{q1} = r_{01} + r_{11}\,\texttt{x1}$$
$$\texttt{q2} = r_{02} + r_{12}\,\texttt{x1} + r_{22}\,\texttt{x2}$$
$$\texttt{q3} = r_{03} + r_{13}\,\texttt{x1} + r_{23}\,\texttt{x2} + r_{33}\,\texttt{x3}$$

This effect can be generalized and written in matrix notation as

$$X = QR$$

where X is the $N \times (d+1)$ matrix representation of *varlist* plus a column of ones, and Q is the $N \times (d+1)$ matrix representation of *newvarlist* plus a column of ones (d = number of variables in *varlist* and N = number of observations). The $(d+1) \times (d+1)$ matrix R is a permuted upper-triangular matrix, i.e., R would be upper triangular if the constant were first, but the constant is last, so the first row/column has been permuted with the last row/column. Since Stata's estimation commands list the constant term last, this allows R, obtained via the `matrix()` option, to be used to transform estimation results.

▷ Example 1

Consider Stata's auto.dta dataset. Suppose that we postulate a model in which price depends on the car's length, weight, headroom (headroom), and trunk size (trunk). These predictors are collinear, but not extremely so—the correlations are not that close to 1:

```
. use http://www.stata-press.com/data/r10/auto
(1978 Automobile Data)

. correlate length weight headroom trunk
(obs=74)

             |   length   weight headroom    trunk
-------------+------------------------------------
      length |   1.0000
      weight |   0.9460   1.0000
    headroom |   0.5163   0.4835   1.0000
       trunk |   0.7266   0.6722   0.6620   1.0000
```

regress certainly has no trouble fitting this model:

```
. regress price length weight headroom trunk

      Source |       SS       df       MS              Number of obs =      74
-------------+------------------------------           F(  4,    69) =   10.20
       Model |   236016580     4    59004145           Prob > F      =  0.0000
    Residual |   399048816    69  5783316.17           R-squared     =  0.3716
-------------+------------------------------           Adj R-squared =  0.3352
       Total |   635065396    73  8699525.97           Root MSE      =  2404.9

------------------------------------------------------------------------------
       price |      Coef.   Std. Err.      t    P>|t|     [95% Conf. Interval]
-------------+----------------------------------------------------------------
      length |  -101.7092   42.12534    -2.41   0.018     -185.747   -17.67147
      weight |   4.753066   1.120054     4.24   0.000     2.518619    6.987512
    headroom |  -711.5679   445.0204    -1.60   0.114    -1599.359    176.2236
       trunk |   114.0859   109.9488     1.04   0.303    -105.2559    333.4277
       _cons |   11488.47   4543.902     2.53   0.014     2423.638    20553.31
------------------------------------------------------------------------------
```

However, we may believe a priori that length is the most important predictor, followed by weight, headroom, and trunk. We would like to remove the "effect" of length from all the other predictors, remove weight from headroom and trunk, and remove headroom from trunk. We can do this by running orthog, and then we fit the model again using the orthogonal variables:

```
. orthog length weight headroom trunk, gen(olength oweight oheadroom otrunk) matrix(R)

. regress price olength oweight oheadroom otrunk

      Source |       SS       df       MS              Number of obs =      74
-------------+------------------------------           F(  4,    69) =   10.20
       Model |   236016580     4    59004145           Prob > F      =  0.0000
    Residual |   399048816    69  5783316.17           R-squared     =  0.3716
-------------+------------------------------           Adj R-squared =  0.3352
       Total |   635065396    73  8699525.97           Root MSE      =  2404.9

------------------------------------------------------------------------------
       price |      Coef.   Std. Err.      t    P>|t|     [95% Conf. Interval]
-------------+----------------------------------------------------------------
     olength |   1265.049   279.5584     4.53   0.000     707.3454    1822.753
     oweight |   1175.765   279.5584     4.21   0.000     618.0617    1733.469
   oheadroom |  -349.9916   279.5584    -1.25   0.215    -907.6955    207.7122
      otrunk |   290.0776   279.5584     1.04   0.303    -267.6262    847.7815
       _cons |   6165.257   279.5584    22.05   0.000     5607.553    6722.961
------------------------------------------------------------------------------
```

Using the matrix R, we can transform the results obtained using the orthogonal predictors back to the metric of original predictors:

```
. matrix b = e(b)*inv(R)'
. matrix list b
b[1,5]
        length       weight     headroom        trunk        _cons
y1  -101.70924    4.7530659  -711.56789    114.08591    11488.475
```

◁

❑ Technical Note

The matrix R obtained using the `matrix()` option with `orthog` can also be used to recover X (the original *varlist*) from Q (the orthogonalized *newvarlist*), one variable at a time. Continuing with the previous example, we illustrate how to recover the `trunk` variable:

```
. matrix C = R[1...,"trunk"]'
. matrix score double rtrunk = C
. compare rtrunk trunk
```

	count	difference minimum	difference average	difference maximum
rtrunk>trunk	74	1.42e-14	2.27e-14	3.55e-14
jointly defined	74	1.42e-14	2.27e-14	3.55e-14
total	74			

Here the recovered variable `rtrunk` is almost exactly the same as the original `trunk` variable. When you are orthogonalizing many variables, this procedure can be performed to check the numerical soundness of the orthogonalization. Because of the ordering of the orthogonalization procedure, the last variable and the variables near the end of the *varlist* are the most important ones to check.

❑

The `orthpoly` command effectively does for polynomial terms what the `orthog` command does for an arbitrary set of variables.

▷ Example 2

Again consider the `auto.dta` dataset. Suppose that we wish to fit the model

$$\texttt{mpg} = \beta_0 + \beta_1\,\texttt{weight} + \beta_2\,\texttt{weight}^2 + \beta_3\,\texttt{weight}^3 + \beta_4\,\texttt{weight}^4 + \epsilon$$

We will first compute the regression with natural polynomials:

```
. gen double w1 = weight
. gen double w2 = w1*w1
. gen double w3 = w2*w1
. gen double w4 = w3*w1
```

```
. correlate w1-w4
(obs=74)

             |       w1       w2       w3       w4
-------------+------------------------------------
          w1 |   1.0000
          w2 |   0.9915   1.0000
          w3 |   0.9665   0.9916   1.0000
          w4 |   0.9279   0.9679   0.9922   1.0000

. regress mpg w1-w4

      Source |       SS       df       MS              Number of obs =      74
-------------+------------------------------           F(  4,    69) =   36.06
       Model |  1652.73666     4  413.184164           Prob > F      =  0.0000
    Residual |  790.722803    69  11.4597508           R-squared     =  0.6764
-------------+------------------------------           Adj R-squared =  0.6576
       Total |  2443.45946    73  33.4720474           Root MSE      =  3.3852

------------------------------------------------------------------------------
         mpg |      Coef.   Std. Err.      t    P>|t|     [95% Conf. Interval]
-------------+----------------------------------------------------------------
          w1 |   .0289302   .1161939     0.25   0.804    -.2028704    .2607307
          w2 |  -.0000229   .0000566    -0.40   0.687    -.0001359    .0000901
          w3 |   5.74e-09   1.19e-08     0.48   0.631    -1.80e-08    2.95e-08
          w4 |  -4.86e-13   9.14e-13    -0.53   0.596    -2.31e-12    1.34e-12
       _cons |   23.94421   86.60667     0.28   0.783    -148.8314    196.7198
------------------------------------------------------------------------------
```

Some of the correlations among the powers of `weight` are very large, but this does not create any problems for `regress`. However, we may wish to look at the quadratic trend with the constant removed, the cubic trend with the quadratic and constant removed, etc. `orthpoly` will generate polynomial terms with this property:

```
. orthpoly weight, generate(pw*) deg(4) poly(P)

. regress mpg pw1-pw4

      Source |       SS       df       MS              Number of obs =      74
-------------+------------------------------           F(  4,    69) =   36.06
       Model |  1652.73666     4  413.184164           Prob > F      =  0.0000
    Residual |  790.722803    69  11.4597508           R-squared     =  0.6764
-------------+------------------------------           Adj R-squared =  0.6576
       Total |  2443.45946    73  33.4720474           Root MSE      =  3.3852

------------------------------------------------------------------------------
         mpg |      Coef.   Std. Err.      t    P>|t|     [95% Conf. Interval]
-------------+----------------------------------------------------------------
         pw1 |  -4.638252   .3935245   -11.79   0.000    -5.423312   -3.853192
         pw2 |   .8263545   .3935245     2.10   0.039     .0412947    1.611414
         pw3 |  -.3068616   .3935245    -0.78   0.438    -1.091921    .4781982
         pw4 |   -.209457   .3935245    -0.53   0.596    -.9945168    .5756028
       _cons |    21.2973   .3935245    54.12   0.000     20.51224    22.08236
------------------------------------------------------------------------------
```

Compare the p-values of the terms in the natural-polynomial regression with those in the orthogonal-polynomial regression. With orthogonal polynomials, it is easy to see that the pure cubic and quartic trends are not significant and that the constant, linear, and quadratic terms each have $p < 0.05$.

The matrix P obtained with the `poly()` option can be used to transform coefficients for orthogonal polynomials to coefficients for natural polynomials:

```
. orthpoly weight, poly(P) deg(4)

. matrix b = e(b)*P
```

```
. matrix list b

b[1,5]
          deg1        deg2        deg3        deg4       _cons
y1   .02893016  -.00002291  5.745e-09  -4.862e-13   23.944212
```

◁

Methods and Formulas

orthog and orthpoly are implemented as ado-files.

orthog's orthogonalization can be written in matrix notation as

$$X = QR$$

where X is the $N \times (d+1)$ matrix representation of *varlist* plus a column of ones and Q is the $N \times (d+1)$ matrix representation of *newvarlist* plus a column of ones (d = number of variables in *varlist*, and N = number of observations). The $(d+1) \times (d+1)$ matrix R is a permuted upper-triangular matrix; i.e., R would be upper triangular if the constant were first, but the constant is last, so the first row/column has been permuted with the last row/column.

Q and R are obtained using a modified Gram–Schmidt procedure; see Golub and Van Loan (1996, 218–219) for details. The traditional Gram–Schmidt procedure is notoriously unsound, but the modified procedure is good. orthog performs two passes of this procedure.

orthpoly uses the Christoffel–Darboux recurrence formula (Abramowitz and Stegun 1972).

Both orthog and orthpoly normalize the orthogonal variables such that

$$Q'WQ = MI$$

where $W = \text{diag}(w_1, w_2, \ldots, w_N)$ with weights $w_1, w_2, \ldots, w_N$ (all 1 if weights are not specified), and M is the sum of the weights (the number of observations if weights are not specified).

References

Abramowitz, M., and I. A. Stegun, ed. 1972. *Handbook of Mathematical Functions: with Formulas, Graphs, and Mathematical Tables*, 10th printing. Washington, DC: National Bureau of Standards.

Golub, G. H., and C. F. Van Loan. 1996. *Matrix Computations*. 3rd ed. Baltimore: Johns Hopkins University Press.

Sribney, W. M. 1995. sg37: Orthogonal polynomials. *Stata Technical Bulletin* 25: 17–18. Reprinted in *Stata Technical Bulletin Reprints*, vol. 5, pp. 96–98.

Also See

[R] **regress** — Linear regression

Title

pcorr — Partial correlation coefficients

Syntax

pcorr *varname*$_1$ *varlist* [*if*] [*in*] [*weight*]

varname$_1$ and *varlist* may contain time-series operators; see [U] **11.4.3 Time-series varlists.**

by is allowed; see [D] **by**.

aweights and fweights are allowed; see [U] **11.1.6 weight**.

Description

pcorr displays the partial correlation coefficient of *varname*$_1$ with each variable in *varlist*, holding the other variables in *varlist* constant.

Remarks

Assume that y is determined by x_1, x_2, ..., x_k. The partial correlation between y and x_1 is an attempt to estimate the correlation that would be observed by y and x_1 if the other x's did not vary.

▷ Example 1

Using our automobile dataset (described in [U] **1.2.1 Sample datasets**), we can obtain the simple correlations between price, mpg, weight, and foreign from correlate (see [R] **correlate**):

```
. use http://www.stata-press.com/data/r10/auto
(1978 Automobile Data)

. correlate price mpg weight foreign
(obs=74)

             |    price      mpg   weight  foreign
-------------+------------------------------------
       price |   1.0000
         mpg |  -0.4686   1.0000
      weight |   0.5386  -0.8072   1.0000
     foreign |   0.0487   0.3934  -0.5928   1.0000
```

Although correlate gave us the full correlation matrix, our interest is in just the first column. We find, for instance, that the higher the mpg, the lower the price. We obtain the partial correlation coefficients by using pcorr:

```
. pcorr price mpg weight foreign
(obs=74)

Partial correlation of price with
    Variable |    Corr.     Sig.
-------------+------------------
         mpg |   0.0352    0.769
      weight |   0.5488    0.000
     foreign |   0.5402    0.000
```

We now find that, with `weight` and `foreign` held constant, the partial correlation of `price` with `mpg` is virtually zero. Similarly, in the simple correlations, we found that `price` and `foreign` were virtually uncorrelated. In the partial correlations—holding `mpg` and `weight` constant—we find that `price` and `foreign` are positively correlated.

◁

❑ Technical Note

Some caution is in order when interpreting the above results. As we said at the outset, the partial correlation coefficient is an *attempt* to estimate the correlation that would be observed if the other variables were held constant. `pcorr` makes it too easy to ignore the fact that we are fitting a model. In the example above, the model is

$$\texttt{price} = \beta_0 + \beta_1\texttt{mpg} + \beta_2\texttt{weight} + \beta_3\texttt{foreign} + \epsilon$$

which is, in all honesty, a rather silly model. Even if we accept the implied economic assumptions of the model—that consumers value `mpg`, `weight`, and `foreign`—do we really believe that consumers place equal value on every extra 1,000 pounds of weight? That is, have we correctly parameterized the model? If we have not, then the estimated partial correlation coefficients may not represent what they claim to represent. Partial correlation coefficients are a reasonable way to summarize data if we are convinced that the underlying model is reasonable. We should not, however, pretend that there is no underlying model and that the partial correlation coefficients are unaffected by the assumptions and parameterization.

❑

Methods and Formulas

`pcorr` is implemented as an ado-file.

Results are obtained by fitting a linear regression of *varname*$_1$ on *varlist*; see [R] **regress**. The partial correlation coefficient between *varname*$_1$ and each variable in *varlist* is then defined as

$$\frac{t}{\sqrt{t^2 + n - k}}$$

(Theil 1971, 174), where t is the t statistic, n is the number of observations, and k is the number of independent variables, including the constant but excluding any dropped variables. The significance is given by $2 * \texttt{ttail}(n - k, \texttt{abs}(t))$.

Reference

Theil, H. 1971. *Principles of Econometrics*. New York: Wiley.

Also See

[R] **correlate** — Correlations (covariances) of variables or coefficients

[R] **spearman** — Spearman's and Kendall's correlations

Title

permute — Monte Carlo permutation tests

Syntax

Compute permutation test

`permute` *permvar exp_list* [, *options*] : *command*

Report saved results

`permute` [*varlist*] [`using` *filename*] [, *display_options*]

options	description
Main	
`reps(`*#*`)`	perform # random permutations; default is `reps(100)`
`left` | `right`	compute one-sided *p*-values; default is two-sided
Options	
`strata(`*varlist*`)`	permute within strata
`saving(`*filename*`, ...)`	save results to *filename*; save statistics in double precision; save results to *filename* every # replications
Reporting	
`level(`*#*`)`	set confidence level; default is `level(95)`
`noheader`	suppress table header
`nolegend`	suppress table legend
`verbose`	display full table legend
`nodrop`	do not drop observations
`nodots`	suppress replication dots
`noisily`	display any output from *command*
`trace`	trace *command*
`title(`*text*`)`	use *text* as title for permutation results
Advanced	
`eps(`*#*`)`	numerical tolerance; seldom used
`nowarn`	do not warn when `e(sample)` is not set
`force`	do not check for *weights* or `svy` commands; seldom used
`reject(`*exp*`)`	identify invalid results
`seed(`*#*`)`	set random-number seed to #

weights are not allowed in *command*.

display_options	description
left | right	compute one-sided p-values; default is two-sided
level(#)	set confidence level; default is level(95)
noheader	suppress table header
nolegend	suppress table legend
verbose	display full table legend
title(*text*)	use *text* as title for results
eps(#)	numerical tolerance; seldom used

exp_list contains	(*name*: *elist*)
	elist
	eexp
elist contains	*newvar* = (*exp*)
	(*exp*)
eexp is	*specname*
	[*eqno*]*specname*
specname is	_b
	_b[]
	_se
	_se[]
eqno is	##
	name

exp is a standard Stata expression; see [U] **13 Functions and expressions**.

Distinguish between [], which are to be typed, and [ ], which indicate optional arguments.

Description

permute estimates p-values for permutation tests on the basis of Monte Carlo simulations. Typing

```
. permute permvar exp_list, reps(#): command
```

randomly permutes the values in *permvar* # times, each time executing *command* and collecting the associated values from the expression in *exp_list*.

These p-value estimates can be one-sided: $\Pr(T^* \leq T)$ or $\Pr(T^* \geq T)$. The default is two-sided: $\Pr(|T^*| \geq |T|)$. Here T^* denotes the value of the statistic from a randomly permuted dataset, and T denotes the statistic as computed on the original data.

permvar identifies the variable whose observed values will be randomly permuted.

command defines the statistical command to be executed. Most Stata commands and user-written programs can be used with permute, as long as they follow standard Stata syntax; see [U] **11 Language syntax**. The by prefix may not be part of *command*.

exp_list specifies the statistics to be retrieved after the execution of *command*.

`permute` may be used for replaying results, but this feature is appropriate only when a dataset generated by `permute` is currently in memory or is identified by the `using` option. The variables specified in *varlist* in this context must be present in the respective dataset.

Options

Main

`reps(#)` specifies the number of random permutations to perform. The default is 100.

`left` or `right` requests that one-sided p-values be computed. If `left` is specified, an estimate of $\Pr(T^* \le T)$ is produced, where T^* is the test statistic and T is its observed value. If `right` is specified, an estimate of $\Pr(T^* \ge T)$ is produced. By default, two-sided p-values are computed; that is, $\Pr(|T^*| \ge |T|)$ is estimated.

Options

`strata(`*varlist*`)` specifies that the permutations be performed within each stratum defined by the values of *varlist*.

`saving(`*filename*[`,` *suboptions*]`)` creates a Stata data file (`.dta` file) consisting of, for each statistic in *exp_list*, a variable containing the permutation replicates.

`double` specifies that the results for each replication be stored as `double`s, meaning 8-byte reals. By default, they are stored as `float`s, meaning 4-byte reals.

`every(#)` specifies that results are to be written to disk every #th replication. `every()` should be specified only in conjunction with `saving()` when *command* takes a long time for each replication. This will allow recovery of partial results should some other software crash your computer. See [P] **postfile**.

`replace` indicates that *filename* may exist, and, if it does, it should be overwritten. This option does not appear in the dialog box.

Reporting

`level(#)` specifies the confidence level, as a percentage, for confidence intervals. The default is `level(95)` or as set by `set level`; see [R] **level**.

`noheader` suppresses display of the table header. This option implies `nolegend`.

`nolegend` suppresses display of the table legend. The table legend identifies the rows of the table with the expressions they represent.

`verbose` requests that the full table legend be displayed. By default, coefficients and standard errors are not displayed.

`nodrop` prevents `permute` from dropping observations outside the `if` and `in` qualifiers. `nodrop` will also cause `permute` to ignore the contents of `e(sample)` if it exists as a result of running *command*. By default, `permute` temporarily drops out-of-sample observations.

`nodots` suppresses display of the replication dots. By default, one dot character is displayed for each successful replication. A red 'x' is displayed if *command* returns an error or if one of the values in *exp_list* is missing.

`noisily` requests that any output from *command* be displayed. This option implies `nodots`.

`trace` causes a trace of the execution of *command* to be displayed. This option implies `noisily`.

`title(`*text*`)` specifies a title to be displayed above the table of permutation results; the default title is `Monte Carlo permutation results`.

Advanced

`eps(`*#*`)` specifies the numerical tolerance for testing $|T^*| \geq |T|$, $T^* \leq T$, or $T^* \geq T$. These are considered true if, respectively, $|T^*| \geq |T|-\#$, $T^* \leq T+\#$, or $T^* \geq T-\#$. The default is `1e-7`. You will not have to specify `eps()` under normal circumstances.

`nowarn` suppresses the printing of a warning message when *command* does not set `e(sample)`.

`force` suppresses the restriction that *command* may not specify weights or be a `svy` command. `permute` is not suited for weighted estimation, thus `permute` should not be used with weights or `svy`. `permute` reports an error when it encounters weights or `svy` in *command* if the `force` option is not specified. This is a seldom used option, so use it only if you know what you are doing!

`reject(`*exp*`)` identifies an expression that indicates when results should be rejected. When *exp* is true, the resulting values are reset to missing values.

`seed(`*#*`)` sets the random-number seed. Specifying this option is equivalent to typing the following command prior to calling `permute`:

```
. set seed #
```

Remarks

Permutation tests determine the significance of the observed value of a test statistic in light of rearranging the order (permuting) of the observed values of a variable.

▷ Example 1

Suppose that we conducted an experiment to determine the effect of a treatment on the development of cells. Further suppose that we are restricted to six experimental units because of the extreme cost of the experiment. Thus three units are to be given a placebo, and three units are given the treatment. The measurement is the number of newly developed healthy cells. The following listing gives the hypothetical data, along with some summary statistics.

```
. input y treatment
                y  treatment
  1. 7 0
  2. 9 0
  3. 11 0
  4. 10 1
  5. 12 1
  6. 14 1
  7. end
. sort treatment
. summarize y
```

Variable	Obs	Mean	Std. Dev.	Min	Max
y	6	10.5	2.428992	7	14

```
. by treatment: summarize y

-> treatment = 0
    Variable |       Obs        Mean    Std. Dev.       Min        Max
-------------+--------------------------------------------------------
           y |         3           9           2          7         11

-> treatment = 1
    Variable |       Obs        Mean    Std. Dev.       Min        Max
-------------+--------------------------------------------------------
           y |         3          12           2         10         14
```

Clearly, there are more cells in the treatment group than in the placebo group, but a statistical test is needed to conclude that the treatment does affect the development of cells. If the sum of the treatment measures is our test statistic, we can use **permute** to determine the probability of observing 36 or more cells, given the observed data and assuming that there is no effect due to the treatment.

```
. set seed 1234
. permute y sum=r(sum), saving(permdish) right nodrop nowarn: sum y if treatment
(running summarize on estimation sample)
Permutation replications (100)
----+--- 1 ---+--- 2 ---+--- 3 ---+--- 4 ---+--- 5
..................................................    50
..................................................   100
Monte Carlo permutation results                 Number of obs    =          6
      command:  summarize y if treatment
          sum:  r(sum)
  permute var:  y

T            |     T(obs)       c       n   p=c/n   SE(p) [95% Conf. Interval]
-------------+----------------------------------------------------------------
         sum |         36      10     100  0.1000  0.0300  .0490047   .1762226

Note:  confidence interval is with respect to p=c/n.
Note:  c = #{T >= T(obs)}
```

We see that 10 of the 100 randomly permuted datasets yielded sums from the treatment group larger than or equal to the observed sum of 36. Thus the evidence is not strong enough, at the 5% level, to reject the null hypothesis that there is no effect of the treatment.

Because of the small size of this experiment, we could have calculated the exact permutation p-value from all possible permutations. There are six units, but we want the sum of the treatment units. Thus there are $\binom{6}{3} = 20$ permutation sums from the possible unique permutations.

$$
\begin{array}{rrrr}
7+9+10=26 & 7+10+12=29 & 9+10+11=30 & 9+12+14=35 \\
7+9+11=27 & 7+10+14=31 & 9+10+12=31 & 10+11+12=33 \\
7+9+12=28 & 7+11+12=30 & 9+10+14=33 & 10+11+14=35 \\
7+9+14=30 & 7+11+14=32 & 9+11+12=32 & 10+12+14=36 \\
7+10+11=28 & 7+12+14=33 & 9+11+14=34 & 11+12+14=37
\end{array}
$$

Two of the 20 permutation sums are greater than or equal to 36. Thus the exact p-value for this permutation test is 0.1. Tied values will decrease the number of unique permutations.

When the `saving()` option is supplied, `permute` saves the values of the permutation statistic to the indicated file, in our case, `permdish.dta`. This file can be used to replay the result of `permute`. The `level()` option controls the confidence level of the confidence interval for the permutation p-value. This confidence interval is calculated using `cii` with the reported `n` (number of nonmissing replications) and `c` (the counter for events of significance).

```
. permute using permdish, level(80)
Monte Carlo permutation results                     Number of obs    =          6

      command:  summarize y if treatment
          sum:  r(sum)
  permute var:  y

T            |     T(obs)       c       n   p=c/n   SE(p) [80% Conf. Interval]
-------------+----------------------------------------------------------------
         sum |         36      10     100  0.1000  0.0300  .0631113   .1498826
------------------------------------------------------------------------------
Note:  confidence interval is with respect to p=c/n.
Note:  c = #{|T| >= |T(obs)|}
```

◁

▷ Example 2

Consider some fictional data from a randomized complete-block design in which we wish to determine the significance of five treatments.

```
. use http://www.stata-press.com/data/r10/permute1, clear
. list y treatment in 1/10, abbrev(10)

     +----------------------+
     |        y   treatment |
     |----------------------|
  1. | 4.407557           1 |
  2. | 5.693386           1 |
  3. | 7.099699           1 |
  4. |  3.12132           1 |
  5. | 5.242648           1 |
     |----------------------|
  6. | 4.280349           2 |
  7. | 4.508785           2 |
  8. | 4.079967           2 |
  9. | 5.904368           2 |
 10. | 3.010556           2 |
     +----------------------+
```

These data may be analyzed using `anova`.

```
. anova y treatment subject

                           Number of obs =      50     R-squared     =  0.3544
                           Root MSE      = .914159     Adj R-squared =  0.1213

                  Source |  Partial SS    df       MS           F     Prob > F
              -----------+----------------------------------------------------
                   Model |  16.5182188    13  1.27063221       1.52     0.1574
                         |
               treatment |  13.0226706     9  1.44696341       1.73     0.1174
                 subject |  3.49554813     4  .873887032       1.05     0.3973
                         |
                Residual |  30.0847503    36  .835687509
              -----------+----------------------------------------------------
                   Total |  46.6029691    49  .951081002
```

Suppose that we want to compute the significance of the F statistic for `treatment` by using `permute`. All we need to do is write a short program that will save the result of this statistic for `permute` to use. For example,

```
program panova, rclass
        version 10
        args response fac_intrst fac_other
        anova `response' `fac_intrst' `fac_other'
        return scalar Fmodel = e(F)
        test `fac_intrst'
        return scalar F = r(F)
end
```

Now in `panova`, `test` saves the F statistic for the factor of interest in `r(F)`. This is different from `e(F)`, which is the overall model F statistic for the model fitted by `anova` that `panova` saves in `r(Fmodel)`. In the following example, we use the `strata()` option so that the treatments are randomly rearranged within each `subject`. It should not be too surprising that the estimated p-values are equal for this example, since the two F statistics are equivalent when controlling for differences between subjects. However, we would not expect to always get the same p-values every time we reran `permute`.

```
. set seed 1234
. permute treatment treatmentF=r(F) modelF=e(F), reps(1000) strata(subject)
> saving(permanova) nodots: panova y treatment subject
Monte Carlo permutation results
Number of strata =          5                 Number of obs    =         50
      command:  panova y treatment subject
   treatmentF:  r(F)
       modelF:  e(F)
  permute var:  treatment
```

T	T(obs)	c	n	p=c/n	SE(p)	[95% Conf.	Interval]
treatmentF	1.731465	118	1000	0.1180	0.0102	.0986525	.1396277
modelF	1.520463	118	1000	0.1180	0.0102	.0986525	.1396277

```
Note:  confidence intervals are with respect to p=c/n.
Note:  c = #{|T| >= |T(obs)|}
```

◁

▷ Example 3

As a final example, let's consider estimating the p-value of the Z statistic returned by `ranksum`. Suppose that we collected data from some experiment: `y` is some measure we took on 17 individuals, and `group` identifies the group that an individual belongs to.

```
. use http://www.stata-press.com/data/r10/permute2, clear
. list

     +------------+
     | group    y |
     |------------|
  1. |     1    6 |
  2. |     1   11 |
  3. |     1   20 |
  4. |     1    2 |
  5. |     1    9 |
     |------------|
  6. |     1    5 |
  7. |     0    2 |
  8. |     0    1 |
  9. |     0    6 |
 10. |     0    0 |
     |------------|
 11. |     0    2 |
 12. |     0    3 |
 13. |     0    3 |
 14. |     0   12 |
 15. |     0    4 |
     |------------|
 16. |     0    1 |
 17. |     0    5 |
     +------------+
```

Next we analyze the data using **ranksum** and notice that the observed value of the test statistic (saved as `r(z)`) is -2.02 with an approximate p-value of 0.0434.

```
. ranksum y, by(group)
Two-sample Wilcoxon rank-sum (Mann-Whitney) test

       group |      obs    rank sum    expected
-------------+---------------------------------
           0 |       11          79          99
           1 |        6          74          54
-------------+---------------------------------
    combined |       17         153         153

unadjusted variance        99.00
adjustment for ties        -0.97
                        ----------
adjusted variance          98.03

Ho: y(group==0) = y(group==1)
             z =  -2.020
    Prob > |z| =   0.0434
```

The observed value of the rank-sum statistic is 79, with an expected value (under the null hypothesis of no group effect) of 99. There are 17 observations, so the permutation distribution contains $\binom{17}{6} = 12{,}376$ possible values of the rank-sum statistic if we ignore ties. With ties, we have fewer possible values but still too many to want to count them. Thus we use `permute` with 10,000 replications and see that the Monte Carlo permutation test agrees with the result of the test based on the normal approximation.

(Continued on next page)

```
. set seed 18385766
. permute y z=r(z), reps(10000) nowarn nodots: ranksum y, by(group)
Monte Carlo permutation results                  Number of obs   =         17

      command:  ranksum y, by(group)
            z:  r(z)
  permute var:  y

T            |   T(obs)       c       n   p=c/n   SE(p) [95% Conf. Interval]
-------------+----------------------------------------------------------------
           z | -2.020002     468   10000  0.0468  0.0021  .0427429   .0511236
------------------------------------------------------------------------------
Note:  confidence interval is with respect to p=c/n.
Note:  c = #{|T| >= |T(obs)|}
```

◁

❑ Technical Note

permute reports confidence intervals for p to emphasize that it is based on the binomial estimator for proportions. When the variability implied by the confidence interval makes conclusions difficult, you may increase the number of replications to determine more precisely the significance of the test statistic of interest. In other words, the value of p from permute will converge to the true permutation p-value as the number of replications gets arbitrarily large. ❑

Saved Results

permute saves the following in r():

Scalars

r(N)	sample size	r(k_exp)	number of standard expressions
r(N_reps)	number of requested replications	r(k_eexp)	number of _b/_se expressions
r(level)	confidence level		

Macros

r(cmd)	permute	r(left)	left or empty
r(command)	*command* following colon	r(right)	right or empty
r(permvar)	permutation variable	r(seed)	initial random-number seed
r(title)	title in output	r(event)	T <= T(obs), T >= T(obs), or |T| <= |T(obs)|
r(exp#)	#th expression		

Matrices

r(b)	observed statistics	r(p)	observed proportions
r(c)	count when r(event) is true	r(se)	standard errors of observed proportions
r(reps)	number of nonmissing results	r(ci)	confidence intervals of observed proportions

Methods and Formulas

permute is implemented as an ado-file.

Reference

Good, P. I. 2006. *Resampling Methods: A Practical Guide to Data Analysis*. 3rd ed. Boston: Birkhäuser.

Also See

[R] **bootstrap** — Bootstrap sampling and estimation

[R] **jackknife** — Jackknife estimation

[R] **simulate** — Monte Carlo simulations

Title

pk — Pharmacokinetic (biopharmaceutical) data

Description

The term pk refers to pharmacokinetic data and the Stata commands, all of which begin with the letters pk, designed to do some of the analyses commonly performed in the pharmaceutical industry. The system is intended for the analysis of pharmacokinetic data, although some of the commands are for general use.

The pk commands are

`pkexamine`	[R] **pkexamine**	Calculate pharmacokinetic measures
`pksumm`	[R] **pksumm**	Summarize pharmacokinetic data
`pkshape`	[R] **pkshape**	Reshape (pharmacokinetic) Latin-square data
`pkcross`	[R] **pkcross**	Analyze crossover experiments
`pkequiv`	[R] **pkequiv**	Perform bioequivalence tests
`pkcollapse`	[R] **pkcollapse**	Generate pharmacokinetic measurement dataset

Remarks

Several types of clinical trials are commonly performed in the pharmaceutical industry. Examples include combination trials, multicenter trials, equivalence trials, and active control trials. For each type of trial, there is an optimal study design for estimating the effects of interest. Currently, the pk system can be used to analyze equivalence trials, which are usually conducted using a crossover design; however, it is possible to use a parallel design and still draw conclusions about equivalence.

Equivalence trials assess bioequivalence between two drugs. Although proving that two drugs behave the same is impossible, the United States Food and Drug Administration believes that if the absorption properties of two drugs are similar, the two drugs will produce similar effects and have similar safety profiles. Generally, the goal of an equivalence trial is to assess the equivalence of a generic drug to an existing drug. This goal is commonly accomplished by comparing a confidence interval about the difference between a pharmacokinetic measurement of two drugs with a confidence limit constructed from U.S. federal regulations. If the confidence interval is entirely within the confidence limit, the drugs are declared bioequivalent. Another approach to assessing bioequivalence is to use the method of interval hypotheses testing. `pkequiv` is used to conduct these tests of bioequivalence.

Several pharmacokinetic measures can be used to ascertain how available a drug is for cellular absorption. The most common measure is the area under the time-versus-concentration curve (AUC). Another common measure of drug availability is the maximum concentration ($C_{\max}$) achieved by the drug during the follow-up period. Stata reports these and other less common measures of drug availability, including the time at which the maximum drug concentration was observed and the duration of the period during which the subject was being measured. Stata also reports the elimination rate, that is, the rate at which the drug is metabolized, and the drug's half-life, that is, the time it takes for the drug concentration to fall to one-half of its maximum concentration.

pkexamine computes and reports all the pharmacokinetic measures that Stata produces, including four calculations of the area under the time-versus-concentration curve. The standard area under the curve from 0 to the maximum observed time ($\mathrm{AUC}_{0,t_{\max}}$) is computed using cubic splines or the trapezoidal rule. Additionally, pkexamine also computes the area under the curve from 0 to infinity by extending the standard time-versus-concentration curve from the maximum observed time by using three different methods. The first method simply extends the standard curve by using a least-squares linear fit through the last few data points. The second method extends the standard curve by fitting a decreasing exponential curve through the last few data points. Finally, the third method extends the curve by fitting a least-squares linear regression line on the log concentration. The mathematical details of these extensions are described in *Methods and Formulas* of [R] **pkexamine**.

Data from an equivalence trial may also be analyzed using methods appropriate to the particular study design. When you have a crossover design, pkcross can be used to fit an appropriate ANOVA model. As an aside, a crossover design is simply a restricted Latin square; therefore, pkcross can also be used to analyze any Latin-square design.

There are some practical concerns when dealing with data from equivalence trials. Primarily, the data must be organized in a manner that Stata can use. The pk commands include pkcollapse and pkshape, which are designed to help transform data from a common format to one that is suitable for analysis with Stata.

In the following example, we illustrate several different data formats that are often encountered in pharmaceutical research and describe how these formats can be transformed to formats that can be analyzed with Stata.

▷ Example 1

Assume that we have one subject and are interested in determining the drug profile for that subject. A reasonable experiment would be to give the subject the drug and then measure the concentration of the drug in the subject's blood over a given period. For example, here is a part of a dataset from Chow and Liu (2000, 11):

```
. use http://www.stata-press.com/data/r10/auc
. list, abbrev(14)
```

	id	time	concentration
1.	1	0	0
2.	1	.5	0
3.	1	1	2.8
4.	1	1.5	4.4
5.	1	2	4.4
6.	1	3	4.7
7.	1	4	4.1
8.	1	6	4
9.	1	8	3.6
10.	1	12	3
11.	1	16	2.5
12.	1	24	2
13.	1	32	1.6

Examining these data, we notice that the concentration quickly increases, plateaus for a short period, and then slowly decreases over time. pkexamine is used to calculate the pharmacokinetic measures of interest. pkexamine is explained in detail in [R] **pkexamine**. The output is

```
. pkexamine time conc
                                Maximum concentration =        4.7
                        Time of maximum concentration =          3
                      Time of last observation (Tmax) =         32
                                     Elimination rate =     0.0279
                                            Half life =    24.8503

Area under the curve
```

AUC [0, Tmax]	AUC [0, inf.) Linear of log conc.	AUC [0, inf.) Linear fit	AUC [0, inf.) Exponential fit
85.24	142.603	107.759	142.603

```
Fit based on last 3 points.
```

Clinical trials, however, require that data be collected on more than one subject. There are several ways to enter raw measured data collected on several subjects. It would be reasonable to enter for each subject the drug concentration value at specific points in time. Such data could be

```
id   conc1   conc2   conc3   conc4    conc5   conc6   conc7
 1       0       1       4       7        5       3       1
 2       0       2       6       5        4       3       2
 3       0       1       2       3        5       4       1
```

where `conc1` is the concentration at the first measured time, `conc2` is the concentration at the second measured time, etc. This format requires that each drug concentration measurement be made at the same time on each subject. Another more flexible way to enter the data is to have an observation with three variables for each time measurement on a subject. Each observation would have a subject ID, the time at which the measurement was made, and the corresponding drug concentration at that time. The data would be

```
. use http://www.stata-press.com/data/r10/pkdata
. list id concA time, sepby(id)

     +-----------------------+
     | id      concA    time |
     |-----------------------|
  1. |  1          0       0 |
  2. |  1   3.073403      .5 |
  3. |  1   5.188444       1 |
  4. |  1   5.898577     1.5 |
  5. |  1   5.096378       2 |
  6. |  1   6.094085       3 |
  7. |  1   5.158772       4 |
  8. |  1     5.7065       6 |
  9. |  1   5.272467       8 |
 10. |  1     4.4576      12 |
 11. |  1   5.146423      16 |
 12. |  1   4.947427      24 |
 13. |  1   1.920421      32 |
     |-----------------------|
 14. |  2          0       0 |
 15. |  2    2.48462      .5 |
 16. |  2   4.883569       1 |
 17. |  2   7.253442     1.5 |
 18. |  2   5.849345       2 |
 19. |  2   6.761085       3 |
 20. |  2    4.33839       4 |
 21. |  2    5.04199       6 |
 22. |  2    4.25128       8 |
 23. |  2   6.205004      12 |
 24. |  2   5.566165      16 |
 25. |  2   3.689007      24 |
 26. |  2   3.644063      32 |
     |-----------------------|
 27. |  3          0       0 |
       (output omitted )
207. | 20   4.673281      24 |
208. | 20   3.487347      32 |
     +-----------------------+
```

Stata expects the data to be organized in the second form. If your data are organized as described in the first dataset, you will need to use `reshape` to change the data to the second form; see [D] **reshape**. Because the data in the second (or long) format contain information for one drug on several subjects, `pksumm` can be used to produce summary statistics of the pharmacokinetic measurements. The output is

```
. pksumm id time concA
................

Summary statistics for the pharmacokinetic measures

                                        Number of observations =      16

     Measure |    Mean     Median      Variance   Skewness    Kurtosis   p-value
-------------+-------------------------------------------------------------------
         auc |  151.63     152.18        127.58      -0.34        2.07      0.55
     aucline |  397.09     219.83     178276.59       2.69        9.61      0.00
      aucexp |  668.60     302.96     720356.98       2.67        9.54      0.00
      auclog |  665.95     298.03     752573.34       2.71        9.70      0.00
        half |   90.68      29.12      17750.70       2.36        7.92      0.00
          ke |    0.02       0.02          0.00       0.88        3.87      0.08
        cmax |    7.37       7.42          0.40      -0.64        2.75      0.36
        tomc |    3.38       3.00          7.25       2.27        7.70      0.00
        tmax |   32.00      32.00          0.00          .           .         .
```

Until now, we have been concerned with the profile of only one drug. We have characterized the profile of that drug by individual subjects by using pkexamine and by a group of subjects by using pksumm. The goal of an equivalence trial, however, is to compare two drugs, which we will do in the rest of this example.

For equivalence trials, the study design most often used is the crossover design. For a complete discussion of crossover designs, see Ratkowsky, Evans, and Alldredge (1993).

In brief, crossover designs require that each subject be given both treatments at two different times. The order in which the treatments are applied changes between groups. For example, if we had 20 subjects numbered 1–20, the first 10 would receive treatment A during the first period of the study, and then they would be given treatment B. The second 10 subjects would be given treatment B during the first period of the study, and then they would be given treatment A. Each subject in the study will have four variables that describe the observation: a subject identifier, a sequence identifier that indicates the order of treatment, and two outcome variables, one for each treatment. The outcome variables for each subject are the pharmacokinetic measures. The data must be transformed from a series of measurements on individual subjects to data containing the pharmacokinetic measures for each subject. In Stata parlance, this is referred to as a collapse, which can be done with pkcollapse; see [R] **pkcollapse**.

Here is a part of our data:

```
. list, sepby(id)

     +-------------------------------------------+
     | id   seq   time       concA       concB   |
     |-------------------------------------------|
  1. |  1     1      0           0           0   |
  2. |  1     1     .5    3.073403    3.712592   |
  3. |  1     1      1    5.188444    6.230602   |
  4. |  1     1    1.5    5.898577    7.885944   |
  5. |  1     1      2    5.096378    9.241735   |
  6. |  1     1      3    6.094085    13.10507   |
  7. |  1     1      4    5.158772     .169429   |
  8. |  1     1      6      5.7065    8.759894   |
  9. |  1     1      8    5.272467    7.985409   |
 10. |  1     1     12      4.4576    7.740126   |
 11. |  1     1     16    5.146423    7.607208   |
 12. |  1     1     24    4.947427    7.588428   |
 13. |  1     1     32    1.920421    2.791115   |
     |-------------------------------------------|
 14. |  2     1      0           0           0   |
 15. |  2     1     .5     2.48462    .9209593   |
 16. |  2     1      1    4.883569    5.925818   |
 17. |  2     1    1.5    7.253442    8.710549   |
 18. |  2     1      2    5.849345    10.90552   |
 19. |  2     1      3    6.761085    8.429898   |
 20. |  2     1      4     4.33839    5.573152   |
 21. |  2     1      6     5.04199     6.32341   |
 22. |  2     1      8     4.25128    .5251224   |
 23. |  2     1     12    6.205004    7.415988   |
 24. |  2     1     16    5.566165    6.323938   |
 25. |  2     1     24    3.689007    1.133553   |
 26. |  2     1     32    3.644063    5.759489   |
     |-------------------------------------------|
 27. |  3     1      0           0           0   |
  (output omitted)
207. | 20     2     24    4.673281    6.059818   |
208. | 20     2     32    3.487347    5.213639   |
     +-------------------------------------------+
```

This format is similar to the second format described above, except that now we have measurements for two drugs at each time for each subject. We transform these data with pkcollapse:

```
. pkcollapse time concA concB, id(id) keep(seq) stat(auc)
................................
. list, sep(8) abbrev(10)
```

	id	seq	auc_concA	auc_concB
1.	1	1	150.9643	218.5551
2.	2	1	146.7606	133.3201
3.	3	1	160.6548	126.0635
4.	4	1	157.8622	96.17461
5.	5	1	133.6957	188.9038
6.	7	1	160.639	223.6922
7.	8	1	131.2604	104.0139
8.	9	1	168.5186	237.8962
9.	10	2	137.0627	139.7382
10.	12	2	153.4038	202.3942
11.	13	2	163.4593	136.7848
12.	14	2	146.0462	104.5191
13.	15	2	158.1457	165.8654
14.	18	2	147.1977	139.235
15.	19	2	164.9988	166.2391
16.	20	2	145.3823	158.5146

For this example, we chose to use the AUC for two drugs as our pharmacokinetic measure. We could have used any of the measures computed by pkexamine. In addition to the AUCs, the dataset also contains a sequence variable for each subject indicating when each treatment was administered.

The data produced by pkcollapse are in what Stata calls wide format. That is, there is one observation per subject containing two or more outcomes. To use pkcross and pkequiv, we need to transform these data to long format. This goal can be accomplished using pkshape; see [R] **pkshape**.

Consider the first subject in the dataset. This subject is in sequence one, which means that treatment A was applied during the first period of the study and treatment B was applied in the second period of the study. We need to split the first observation into two observations so that the outcome measure is only in one variable. Also we need two new variables, one indicating the treatment the subject received and another recording the period of the study when the subject received that treatment. We might expect the expansion of the first subject to be

```
id   sequence        auc      treat     period
 1          1   150.9643          A          1
 1          1   218.5551          B          2
```

We see that subject number 1 was in sequence 1, had an AUC of 150.9643 when treatment A was applied in the first period of the study, and had an AUC of 218.5551 when treatment B was applied.

Similarly, the expansion of subject 10 (the first subject in sequence 2) would be

```
id   sequence        auc      treat     period
10          2   137.0627          B          1
10          2   139.7382          A          2
```

Here treatment B was applied to the subject during the first period of the study, and treatment A was applied to the subject during the second period of the study.

An additional complication is common in crossover study designs. The treatment applied in the first period of the study might still have some effect on the outcome in the second period. In this example, each subject was given one treatment followed by another treatment. To get accurate estimates of treatment effects, it is necessary to account for the effect that the first treatment has in the second period of the study. This is called the carryover effect. We must, therefore, have a variable that indicates which treatment was applied in the first treatment period. `pkshape` creates a variable that indicates the carryover effect. For treatments applied during the first treatment period, there will never be a carryover effect. Thus the expanded data created by `pkshape` for subject 1 will be

id	sequence	outcome	treat	period	carry
1	1	150.9643	A	1	0
1	1	218.5551	B	2	A

and the data for subject 10 will be

id	sequence	outcome	treat	period	carry
10	2	137.0627	B	1	0
10	2	139.7382	A	2	B

We `pkshape` the data:

```
. pkshape id seq auc*, order(ab ba)
. sort id sequence period
. list, sep(16)
```

	id	sequence	outcome	treat	carry	period
1.	1	1	150.9643	1	0	1
2.	1	1	218.5551	2	1	2
3.	2	1	146.7606	1	0	1
4.	2	1	133.3201	2	1	2
5.	3	1	160.6548	1	0	1
6.	3	1	126.0635	2	1	2
7.	4	1	157.8622	1	0	1
8.	4	1	96.17461	2	1	2
9.	5	1	133.6957	1	0	1
10.	5	1	188.9038	2	1	2
11.	7	1	160.639	1	0	1
12.	7	1	223.6922	2	1	2
13.	8	1	131.2604	1	0	1
14.	8	1	104.0139	2	1	2
15.	9	1	168.5186	1	0	1
16.	9	1	237.8962	2	1	2
17.	10	2	137.0627	2	0	1
18.	10	2	139.7382	1	2	2
19.	12	2	153.4038	2	0	1
20.	12	2	202.3942	1	2	2
21.	13	2	163.4593	2	0	1
22.	13	2	136.7848	1	2	2
23.	14	2	146.0462	2	0	1
24.	14	2	104.5191	1	2	2
25.	15	2	158.1457	2	0	1
26.	15	2	165.8654	1	2	2
27.	18	2	147.1977	2	0	1
28.	18	2	139.235	1	2	2
29.	19	2	164.9988	2	0	1
30.	19	2	166.2391	1	2	2
31.	20	2	145.3823	2	0	1
32.	20	2	158.5146	1	2	2

As an aside, crossover designs do not require that each subject receive each treatment, but if they do, the crossover design is referred to as a complete crossover design.

The last dataset is organized in a manner that can be analyzed with Stata. To fit an ANOVA model to these data, we can use `anova` or `pkcross`. To conduct equivalence tests, we can use `pkequiv`. This example is further analyzed in [R] **pkcross** and [R] **pkequiv**.

◁

References

Chow, S. C., and J. P. Liu. 2000. *Design and Analysis of Bioavailability and Bioequivalence Studies*. 2nd ed, Revised and Explanded. New York: Dekker.

Ratkowsky, D. A., M. A. Evans, and J. R. Alldredge. 1993. *Cross-over Experiments: Design, Analysis and Application*. New York: Dekker.

Title

pkcollapse — Generate pharmacokinetic measurement dataset

Syntax

`pkcollapse` *time concentration* [*if*] , `id(`*id_var*`)` [*options*]

options	description
Main	
* `id(`*id_var*`)`	subject ID variable
`stat(`*measures*`)`	create specified *measures*; default is all
`trapezoid`	use trapezoidal rule; default is cubic splines
`fit(`#`)`	use # points to estimate $AUC_{0,\infty}$; default is `fit(3)`
`keep(`*varlist*`)`	keep variables in *varlist*
`force`	force collapse
`nodots`	suppress dots during calculation

* `id(`*id_var*`)` is required.

measures	description
`auc`	area under the concentration-time curve ($AUC_{0,\infty}$)
`aucline`	area under the concentration-time curve from 0 to ∞ using a linear extension
`aucexp`	area under the concentration-time curve from 0 to ∞ using an exponential extension
`auclog`	area under the log-concentration-time curve extended with a linear fit
`half`	half-life of the drug
`ke`	elimination rate
`cmax`	maximum concentration
`tmax`	time at last concentration
`tomc`	time of maximum concentration

Description

`pkcollapse` generates new variables with the pharmacokinetic summary measures of interest.

`pkcollapse` is one of the pk commands. Please read [R] **pk** before reading this entry.

Options

Main

`id(`*id_var*`)` is required and specifies the variable that contains the subject ID over which `pkcollapse` is to operate.

`stat(`*measures*`)` specifies the measures to be generated. The default is to generate all the measures.

`trapezoid` tells Stata to use the trapezoidal rule when calculating the AUC. The default is to use cubic splines, which give better results for most functions. When the curve is irregular, `trapezoid` may give better results.

`fit(`*#*`)` specifies the number of points to use in estimating the $\mathrm{AUC}_{0,\infty}$. The default is `fit(3)`, the last three points. This number should be viewed as a minimum; the appropriate number of points will depend on your data.

`keep(`*varlist*`)` specifies the variables to be kept during the collapse. Variables not specified with the `keep()` option will be dropped. When `keep()` is specified, the keep variables are checked to ensure that all values of the variables are the same within *id_var*.

`force` forces the collapse, even when the values of the `keep()` variables are different within the *id_var*.

`nodots` suppresses the display of dots during calculation.

Remarks

`pkcollapse` generates all the summary pharmacokinetic measures.

▷ Example 1

We demonstrate the use of `pkcollapse` with the data described in [R] **pk**. We have drug concentration data on 15 subjects. Each subject is measured at 13 time points over a 32-hour period. Some of the records are

```
. use http://www.stata-press.com/data/r10/pkdata
. list, sep(0)
```

	id	seq	time	concA	concB
1.	1	1	0	0	0
2.	1	1	.5	3.073403	3.712592
3.	1	1	1	5.188444	6.230602
4.	1	1	1.5	5.898577	7.885944
5.	1	1	2	5.096378	9.241735
6.	1	1	3	6.094085	13.10507
			(output omitted)		
14.	2	1	0	0	0
15.	2	1	.5	2.48462	.9209593
16.	2	1	1	4.883569	5.925818
17.	2	1	1.5	7.253442	8.710549
18.	2	1	2	5.849345	10.90552
19.	2	1	3	6.761085	8.429898
			(output omitted)		
207.	20	2	24	4.673281	6.059818
208.	20	2	32	3.487347	5.213639

Although `pksumm` allows us to view all the pharmacokinetic measures, we can create a dataset with the measures by using `pkcollapse`.

```
. pkcollapse time concA concB, id(id) stat(auc) keep(seq)
...............................
```

```
. list, sep(8) abbrev(10)
```

	id	seq	auc_concA	auc_concB
1.	1	1	150.9643	218.5551
2.	2	1	146.7606	133.3201
3.	3	1	160.6548	126.0635
4.	4	1	157.8622	96.17461
5.	5	1	133.6957	188.9038
6.	7	1	160.639	223.6922
7.	8	1	131.2604	104.0139
8.	9	1	168.5186	237.8962
9.	10	2	137.0627	139.7382
10.	12	2	153.4038	202.3942
11.	13	2	163.4593	136.7848
12.	14	2	146.0462	104.5191
13.	15	2	158.1457	165.8654
14.	18	2	147.1977	139.235
15.	19	2	164.9988	166.2391
16.	20	2	145.3823	158.5146

The resulting dataset, which we will call `pkdata2`, contains 1 observation per subject. This dataset is in wide format. If we want to use `pkcross` or `pkequiv`, we must transform these data to long format, which we do in the last example of [R] **pkshape**.

◁

Methods and Formulas

`pkcollapse` is implemented as an ado-file.

The statistics generated by `pkcollapse` are described in [R] **pkexamine**.

Also See

[R] **pk** — Pharmacokinetic (biopharmaceutical) data

Title

pkcross — Analyze crossover experiments

Syntax

`pkcross` *outcome* [*if*] [*in*] [, *options*]

options	description
Model	
`sequence(`*varname*`)`	sequence variable; default is `sequence(sequence)`
`treatment(`*varname*`)`	treatment variable; default is `treatment(treat)`
`period(`*varname*`)`	period variable; default is `period(period)`
`id(`*varname*`)`	ID variable
`carryover(`*varname*`)`	name of carryover variable; default is `carryover(carry)`
`carryover(none)`	omit carryover effects from model; default is `carryover(carry)`
`model(`*string*`)`	specify the model to fit
`sequential`	estimate sequential instead of partial sums of squares
Parameterization	
`param(3)`	estimate mean and the period, treatment, and sequence effects; assume no carryover effects exist; the default
`param(1)`	estimate mean and the period, treatment, and carryover effects; assume no sequence effects exist
`param(2)`	estimate mean, period and treatment effects, and period-by-treatment interaction; assume no sequence or carryover effects exist
`param(4)`	estimate mean, period and treatment effects, and period-by-treatment interaction; assume no period or crossover effects exist

Description

`pkcross` analyzes data from a crossover design experiment. When analyzing pharmaceutical trial data, if the treatment, carryover, and sequence variables are known, the omnibus test for separability of the treatment and carryover effects is calculated.

`pkcross` is one of the pk commands. Please read [R] **pk** before reading this entry.

Options

Model

`sequence(`*varname*`)` specifies the variable that contains the sequence in which the treatment was administered. If this option is not specified, `sequence(sequence)` is assumed.

`treatment(`*varname*`)` specifies the variable that contains the treatment information. If this option is not specified, `treatment(treat)` is assumed.

`period(`*varname*`)` specifies the variable that contains the period information. If this option is not specified, `period(period)` is assumed.

`id(`*varname*`)` specifies the variable that contains the subject identifiers. If this option is not specified, `id(id)` is assumed.

`carryover(`*varname* | `none)` specifies the variable that contains the carryover information. If `carry(none)` is specified, the carryover effects are omitted from the model. If this option is not specified, `carryover(carry)` is assumed.

`model(`*string*`)` specifies the model to be fitted. For higher-order crossover designs, this option can be useful if you want to fit a model other than the default. However, `anova` (see [R] **anova**) can also be used to fit a crossover model. The default model for higher-order crossover designs is outcome predicted by sequence, period, treatment, and carryover effects. By default, the model statement is `model(sequence period treat carry)`.

`sequential` specifies that sequential sums of squares be estimated.

Parameterization

`param(`*#*`)` specifies which of the four parameterizations to use for the analysis of a 2×2 crossover experiment. This option is ignored with higher-order crossover designs. The default is `param(3)`. See the technical note for 2×2 crossover designs for more details.

`param(1)` estimates the overall mean, the period effects, the treatment effects, and the carryover effects, assuming that no sequence effects exist.

`param(2)` estimates the overall mean, the period effects, the treatment effects, and the period-by-treatment interaction, assuming that no sequence or carryover effects exist.

`param(3)` estimates the overall mean, the period effects, the treatment effects, and the sequence effects, assuming that no carryover effects exist. This is the default parameterization.

`param(4)` estimates the overall mean, the sequence effects, the treatment effects, and the sequence-by-treatment interaction, assuming that no period or crossover effects exist. When the sequence by treatment is equivalent to the period effect, this reduces to the third parameterization.

Remarks

`pkcross` is designed to analyze crossover experiments. Use `pkshape` first to reshape your data; see [R] **pkshape**. `pkcross` assumes that the data were reshaped by `pkshape` or are organized in the same manner as produced with `pkshape`. Washout periods are indicated by the number 0. See the technical note in this entry for more information on analyzing 2×2 crossover experiments.

❑ Technical Note

The 2×2 crossover design cannot be used to estimate more than four parameters because there are only four pieces of information (the four cell means) collected. `pkcross` uses ANOVA models to analyze the data, so one of the four parameters must be the overall mean of the model, leaving just 3 degrees of freedom to estimate the remaining effects (period, sequence, treatment, and carryover). Thus the model is overparameterized. Estimation of treatment and carryover effects requires the assumption of either no period effects or no sequence effects. Some researchers maintain that it estimating carryover effects at the expense of other effects is a bad idea. This is a limitation of this design. `pkcross` implements four parameterizations for this model. They are numbered sequentially from one to four and are described in the *Options* section of this entry.

❑

▷ Example 1

Consider the example data published in Chow and Liu (2000, 73) and described in [R] **pkshape**. We have entered and reshaped the data with `pkshape` and have variables that identify the subjects, periods, treatments, sequence, and carryover treatment. To compute the ANOVA table, use `pkcross`:

```
. use http://www.stata-press.com/data/r10/chowliu
. pkshape id seq period1 period2, order(ab ba)
. pkcross outcome
                                       sequence variable = sequence
                                         period variable = period
                                      treatment variable = treat
                                      carryover variable = carry
                                             id variable = id

              Analysis of variance (ANOVA) for a 2x2 crossover study
Source of Variation |      SS        df          MS           F      Prob > F
____________________|_________________________________________________________
                    |
Intersubjects       |
    Sequence effect |    276.00       1        276.00       0.37      0.5468
          Residuals |  16211.49      22        736.89       4.41      0.0005
____________________|_________________________________________________________
                    |
Intrasubjects       |
   Treatment effect |     62.79       1         62.79       0.38      0.5463
      Period effect |     35.97       1         35.97       0.22      0.6474
          Residuals |   3679.43      22        167.25
____________________|_________________________________________________________
                    |
              Total |  20265.68      47

Omnibus measure of separability of treatment and carryover =   29.2893%
```

There is evidence of intersubject variability, but there are no other significant effects. The omnibus test for separability is a measure reflecting the degree to which the study design allows the treatment effects to be estimated independently of the carryover effects. The measure of separability of the treatment and carryover effects indicates approximately 29% separability, which can be interpreted as the degree to which the treatment and carryover effects are orthogonal. This is a characteristic of the design of the study. For a complete discussion, see Ratkowsky, Evans, and Alldredge (1993). Compared to the output in Chow and Liu (2000), the sequence effect is mislabeled as a carryover effect. See Ratkowsky, Evans, and Alldredge (1993, section 3.2) for a complete discussion of the mislabeling.

By specifying `param(1)`, we obtain parameterization 1 for this model.

```
. pkcross outcome, param(1)
                                       sequence variable = sequence
                                         period variable = period
                                      treatment variable = treat
                                      carryover variable = carry
                                             id variable = id

              Analysis of variance (ANOVA) for a 2x2 crossover study
Source of Variation | Partial SS   df          MS           F      Prob > F
____________________|_________________________________________________________
                    |
   Treatment effect |     301.04    1        301.04       0.67      0.4189
      Period effect |     255.62    1        255.62       0.57      0.4561
   Carryover effect |     276.00    1        276.00       0.61      0.4388
          Residuals |   19890.92   44        452.07
____________________|_________________________________________________________
                    |
              Total |   20265.68   47

Omnibus measure of separability of treatment and carryover =   29.2893%
```

◁

▷ Example 2

Consider the case of a two-treatment, four-sequence, two-period crossover design. This design is commonly referred to as Balaam's design. Ratkowsky, Evans, and Alldredge (1993) published the following data from an amantadine trial:

```
. use http://www.stata-press.com/data/r10/balaam, clear
. list, sep(0)

     +-----------------------------------------+
     | id   seq   period1   period2   period3 |
     |-----------------------------------------|
  1. |  1   -ab         9      8.75      8.75 |
  2. |  2   -ab        12      10.5      9.75 |
  3. |  3   -ab        17        15      18.5 |
  4. |  4   -ab        21        21      21.5 |
  5. |  1   -ba        23        22        18 |
  6. |  2   -ba        15        15        13 |
  7. |  3   -ba        13        14     13.75 |
  8. |  4   -ba        24     22.75      21.5 |
  9. |  5   -ba        18     17.75     16.75 |
 10. |  1   -aa        14      12.5        14 |
 11. |  2   -aa        27     24.25      22.5 |
 12. |  3   -aa        19     17.25     16.25 |
 13. |  4   -aa        30     28.25     29.75 |
 14. |  1   -bb        21        20     19.51 |
 15. |  2   -bb        11      10.5        10 |
 16. |  3   -bb        20      19.5     20.75 |
 17. |  4   -bb        25      22.5      23.5 |
     +-----------------------------------------+
```

The sequence identifier must be a string with zeros to indicate washout or baseline periods, or a number. If the sequence identifier is numeric, the order option must be specified with pkshape. If the sequence identifier is a string, pkshape will create sequence, period, and treatment identifiers without the order option. In this example, the dash is used to indicate a baseline period, which is an invalid code for this purpose. As a result, the data must be encoded; see [D] **encode**.

```
. encode seq, gen(num_seq)
. pkshape id num_seq period1 period2 period3, order(0aa 0ab 0ba 0bb)
. pkcross outcome, se
                                     sequence variable = sequence
                                       period variable = period
                                    treatment variable = treat
                                    carryover variable = carry
                                           id variable = id

              Analysis of variance (ANOVA) for a crossover study
    Source of Variation |      SS        df        MS           F     Prob > F
    --------------------+-------------------------------------------------------
    Intersubjects       |
         Sequence effect |   285.82      3       95.27       1.01      0.4180
               Residuals |  1221.49     13       93.96      59.96      0.0000
    --------------------+-------------------------------------------------------
    Intrasubjects       |
           Period effect |    15.13      2        7.56       6.34      0.0048
        Treatment effect |     8.48      1        8.48       8.86      0.0056
        Carryover effect |     0.11      1        0.11       0.12      0.7366
               Residuals |    29.56     30        0.99
    --------------------+-------------------------------------------------------
                   Total |  1560.59     50
    Omnibus measure of separability of treatment and carryover =   64.6447%
```

In this example, the sequence specifier used dashes instead of zeros to indicate a baseline period during which no treatment was given. For `pkcross` to work, we need to encode the string sequence variable and then use the `order` option with `pkshape`. A word of caution: `encode` does not necessarily choose the first sequence to be sequence 1, as in this example. Always double-check the sequence numbering when using `encode`.

◁

▷ Example 3

Continuing with the example from [R] **pkshape**, we fit an ANOVA model.

```
. use http://www.stata-press.com/data/r10/pkdata3, clear
. list, sep(8)
```

	id	sequence	outcome	treat	carry	period
1.	1	1	150.9643	A	0	1
2.	2	1	146.7606	A	0	1
3.	3	1	160.6548	A	0	1
4.	4	1	157.8622	A	0	1
5.	5	1	133.6957	A	0	1
6.	7	1	160.639	A	0	1
7.	8	1	131.2604	A	0	1
8.	9	1	168.5186	A	0	1
9.	10	2	137.0627	B	0	1
10.	12	2	153.4038	B	0	1
11.	13	2	163.4593	B	0	1
12.	14	2	146.0462	B	0	1
13.	15	2	158.1457	B	0	1
14.	18	2	147.1977	B	0	1
15.	19	2	164.9988	B	0	1
16.	20	2	145.3823	B	0	1
17.	1	1	218.5551	B	A	2
18.	2	1	133.3201	B	A	2
19.	3	1	126.0635	B	A	2
20.	4	1	96.17461	B	A	2
21.	5	1	188.9038	B	A	2
22.	7	1	223.6922	B	A	2
23.	8	1	104.0139	B	A	2
24.	9	1	237.8962	B	A	2
25.	10	2	139.7382	A	B	2
26.	12	2	202.3942	A	B	2
27.	13	2	136.7848	A	B	2
28.	14	2	104.5191	A	B	2
29.	15	2	165.8654	A	B	2
30.	18	2	139.235	A	B	2
31.	19	2	166.2391	A	B	2
32.	20	2	158.5146	A	B	2

The ANOVA model is fitted using pkcross:

```
. pkcross outcome
                                            sequence variable = sequence
                                              period variable = period
                                           treatment variable = treat
                                           carryover variable = carry
                                                  id variable = id

               Analysis of variance (ANOVA) for a 2x2 crossover study
 Source of Variation |      SS        df         MS           F     Prob > F
---------------------+-------------------------------------------------------
Intersubjects        |
     Sequence effect |     378.04      1       378.04       0.29     0.5961
           Residuals |   17991.26     14      1285.09       1.40     0.2691
---------------------+-------------------------------------------------------
Intrasubjects        |
    Treatment effect |     455.04      1       455.04       0.50     0.4931
       Period effect |     419.47      1       419.47       0.46     0.5102
           Residuals |   12860.78     14       918.63
---------------------+-------------------------------------------------------
               Total |   32104.59     31
Omnibus measure of separability of treatment and carryover =   29.2893%
```

◁

▷ Example 4

Consider the case of a six-treatment crossover trial in which the squares are not variance balanced. The following dataset is from a partially balanced crossover trial published by Ratkowsky, Evans, and Alldredge (1993):

```
. use http://www.stata-press.com/data/r10/nobalance
. list, sep(4)
```

	cow	seq	period1	period2	period3	period4	block
1.	1	adbe	38.7	37.4	34.3	31.3	1
2.	2	baed	48.9	46.9	42	39.6	1
3.	3	ebda	34.6	32.3	28.5	27.1	1
4.	4	deab	35.2	33.5	28.4	25.1	1
5.	1	dafc	32.9	33.1	27.5	25.1	2
6.	2	fdca	30.4	29.4	26.7	23.1	2
7.	3	cfda	30.8	29.3	26.4	23.2	2
8.	4	acdf	25.7	26.1	23.4	18.7	2
9.	1	efbc	25.4	26	23.9	19.9	3
10.	2	becf	21.8	23.9	21.7	17.6	3
11.	3	fceb	21.4	22	19.4	16.6	3
12.	4	cbfe	22.8	21	18.6	16.1	3

When there is no variance balance in the design, a square or blocking variable is needed to indicate in which treatment cell a sequence was observed, but the mechanical steps are the same.

```
. pkshape cow seq period1 period2 period3 period4
. pkcross outcome, model(block cow|block period|block treat carry) se
                         Number of obs =      48     R-squared     =  0.9966
                         Root MSE      = .730751     Adj R-squared =  0.9906

        Source |   Seq. SS     df        MS              F     Prob > F
  -------------+-----------------------------------------------------
         Model |  2650.0419    30  88.3347302        165.42     0.0000
               |
         block |  1607.17045    2  803.585226       1504.85     0.0000
     cow|block |  628.621899    9  69.8468777        130.80     0.0000
  period|block |  407.531876    9  45.2813195         84.80     0.0000
         treat |  2.48979215    5  .497958429          0.93     0.4846
         carry |  4.22788534    5  .845577068          1.58     0.2179
               |
      Residual |  9.07794631   17  .533996842
  -------------+-----------------------------------------------------
         Total |  2659.11985   47  56.5770181
```

When the model statement is used and the omnibus measure of separability is desired, specify the variables in the `treatment()`, `carryover()`, and `sequence()` options to `pkcross`. ◁

Methods and Formulas

`pkcross` is implemented as an ado-file.

`pkcross` uses ANOVA to fit models for crossover experiments; see [R] **anova**.

The omnibus measure of separability is

$$S = 100(1 - V)\%$$

where V is Cramér's V and is defined as

$$V = \left\{ \frac{\frac{\chi^2}{N}}{\min(r-1, c-1)} \right\}^{\frac{1}{2}}$$

The χ^2 is calculated as

$$\chi^2 = \sum_i \sum_j \left\{ \frac{(O_{ij} - E_{ij})^2}{E_{ij}} \right\}$$

where O and E are the observed and expected counts in a table of the number of times each treatment is followed by the other treatments.

References

Chow, S. C., and J. P. Liu. 2000. *Design and Analysis of Bioavailability and Bioequivalence Studies*. 2nd ed, Revised and Expanded. New York: Dekker.

Kutner, M. H., C. J. Nachtsheim, and J. Neter. 2005. *Applied Linear Statistical Models*. 5th ed. Chicago: McGraw–Hill/Irwin.

Ratkowsky, D. A., M. A. Evans, and J. R. Alldredge. 1993. *Cross-over Experiments: Design, Analysis and Application.* New York: Dekker.

Also See

[R] **pk** — Pharmacokinetic (biopharmaceutical) data

Title

pkequiv — Perform bioequivalence tests

Syntax

`pkequiv` *outcome treatment period sequence id* [*if*] [*in*] [, *options*]

options	description
Options	
`compare(`*string*`)`	compare the two specified values of the treatment variable
`limit(`*#*`)`	equivalence limit (between .10 and .99); default is .2
`level(`*#*`)`	set confidence level; default is `level(90)`
`fieller`	calculate confidence interval by Fieller's theorem
`symmetric`	calculate symmetric equivalence interval
`anderson`	Anderson and Hauck hypothesis test for bioequivalence
`tost`	two one-sided hypothesis tests for bioequivalence
`noboot`	do not estimate probability that CI lies within confidence limits

Description

`pkequiv` performs bioequivalence testing for two treatments. By default, `pkequiv` calculates a standard confidence interval symmetric about the difference between the two treatment means. `pkequiv` also calculates confidence intervals symmetric about zero and intervals based on Fieller's theorem. Also `pkequiv` can perform interval hypothesis tests for bioequivalence.

`pkequiv` is one of the pk commands. Please read [R] **pk** before reading this entry.

Options

Options

`compare(`*string*`)` specifies the two treatments to be tested for equivalence. Sometimes there may be more than two treatments, but the equivalence can be determined only between any two treatments.

`limit(`*#*`)` specifies the equivalence limit. The default is .2. The equivalence limit can be changed only symmetrically; that is, it is not possible to have a .15 lower limit and a .2 upper limit in the same test.

`level(`*#*`)` specifies the confidence level, as a percentage, for confidence intervals. The default is `level(90)`. This setting is not controlled by the `set level` command.

`fieller` specifies that an equivalence interval on the basis of Fieller's theorem be calculated.

`symmetric` specifies that a symmetric equivalence interval be calculated.

`anderson` specifies that the Anderson and Hauck hypothesis test for bioequivalence be computed. This option is ignored when calculating equivalence intervals based on Fieller's theorem or when calculating a confidence interval that is symmetric about zero.

`tost` specifies that the two one-sided hypothesis tests for bioequivalence be computed. This option is ignored when calculating equivalence intervals based on Fieller's theorem or when calculating a confidence interval that is symmetric about zero.

`noboot` prevents the estimation of the probability that the confidence interval lies within the confidence limits. If this option is not specified, this probability is estimated by resampling the data.

Remarks

`pkequiv` is designed to conduct tests for bioequivalence based on data from a crossover experiment. `pkequiv` requires that the user specify the *outcome*, *treatment*, *period*, *sequence*, and *id* variables. The data must be in the same format as that produced by `pkshape`; see [R] **pkshape**.

▷ Example 1

Continuing with the example from [R] **pkshape**, we will conduct equivalence testing.

```
. use http://www.stata-press.com/data/r10/pkdata3
. list, sep(4)
```

	id	sequence	outcome	treat	carry	period
1.	1	1	150.9643	A	0	1
2.	2	1	146.7606	A	0	1
3.	3	1	160.6548	A	0	1
4.	4	1	157.8622	A	0	1
5.	5	1	133.6957	A	0	1
6.	7	1	160.639	A	0	1
7.	8	1	131.2604	A	0	1
8.	9	1	168.5186	A	0	1
9.	10	2	137.0627	B	0	1
10.	12	2	153.4038	B	0	1
11.	13	2	163.4593	B	0	1
12.	14	2	146.0462	B	0	1
13.	15	2	158.1457	B	0	1
14.	18	2	147.1977	B	0	1
15.	19	2	164.9988	B	0	1
16.	20	2	145.3823	B	0	1
17.	1	1	218.5551	B	A	2
18.	2	1	133.3201	B	A	2
19.	3	1	126.0635	B	A	2
20.	4	1	96.17461	B	A	2
21.	5	1	188.9038	B	A	2
22.	7	1	223.6922	B	A	2
23.	8	1	104.0139	B	A	2
24.	9	1	237.8962	B	A	2
25.	10	2	139.7382	A	B	2
26.	12	2	202.3942	A	B	2
27.	13	2	136.7848	A	B	2
28.	14	2	104.5191	A	B	2
29.	15	2	165.8654	A	B	2
30.	18	2	139.235	A	B	2
31.	19	2	166.2391	A	B	2
32.	20	2	158.5146	A	B	2

Now we can conduct a bioequivalence test between `treat` $= A$ and `treat` $= B$.

```
. set seed 1
. pkequiv outcome treat period seq id
      Classic confidence interval for bioequivalence
      -------------------------------------------------------------------------
                        [equivalence limits]          [     test limits    ]
      -------------------------------------------------------------------------
           difference:    -30.296      30.296          -11.332       26.416
                ratio:        80%        120%          92.519%     117.439%
      -------------------------------------------------------------------------
      probability test limits are within equivalence limits =    0.6410
      note: reference treatment = 1
```

The default output for **pkequiv** shows a confidence interval for the difference of the means (test limits), the ratio of the means, and the federal equivalence limits. The classic confidence interval can be constructed around the difference between the average measure of effect for the two drugs or around the ratio of the average measure of effect for the two drugs. **pkequiv** reports both the difference measure and the ratio measure. For these data, U.S. federal government regulations state that the confidence interval for the difference must be entirely contained within the range $[-30.296, 30.296]$ and between 80% and 120% for the ratio. Here the test limits are within the equivalence limits. Although the test limits are inside the equivalence limits, there is only a 64% assurance that the observed confidence interval will be within the equivalence limits in the long run. This is an interesting case because, although this sample shows bioequivalence, the evaluation of the long-run performance indicates possible problems. These fictitious data were generated with high intersubject variability, which causes poor long-run performance.

If we conduct a bioequivalence test with the data published in Chow and Liu (2000, 73), which we introduced in [R] **pk** and fully described in [R] **pkshape**, we observe that the probability that the test limits are within the equivalence limits is high.

```
. use http://www.stata-press.com/data/r10/chowliu2
. set seed 1
. pkequiv outcome treat period seq id
      Classic confidence interval for bioequivalence
      -------------------------------------------------------------------------
                        [equivalence limits]          [     test limits    ]
      -------------------------------------------------------------------------
           difference:    -16.512      16.512           -8.698        4.123
                ratio:        80%        120%          89.464%     104.994%
      -------------------------------------------------------------------------
      probability test limits are within equivalence limits =    0.9980
      note: reference treatment = 1
```

For these data, the test limits are well within the equivalence limits, and the probability that the test limits are within the equivalence limits is 99.8%.

◁

(Continued on next page)

▷ Example 2

We compute a confidence interval that is symmetric about zero:

```
. pkequiv outcome treat period seq id, symmetric

        Westlake's symmetric confidence interval for bioequivalence

                          [Equivalence limits]      [    Test mean    ]

        Test formulation:      75.145    89.974              80.272

        note: reference treatment = 1
```

The reported equivalence limit is constructed symmetrically about the reference mean, which is equivalent to constructing a confidence interval symmetric about zero for the difference in the two drugs. In the output above, we see that the test formulation mean of 80.272 is within the equivalence limits, indicating that the test drug is bioequivalent to the reference drug.

pkequiv displays interval hypothesis tests of bioequivalence if you specify the tost and/or the anderson options. For example,

```
. set seed 1

. pkequiv outcome treat period seq id, tost anderson
        Classic confidence interval for bioequivalence

                          [equivalence limits]        [    test limits    ]

              difference:   -16.512       16.512        -8.698        4.123
                   ratio:       80%         120%       89.464%     104.994%

        probability test limits are within equivalence limits =     0.9980
        Schuirmann's two one-sided tests

        upper test statistic =    -5.036                 p-value =      0.000
        lower test statistic =     3.810                 p-value =      0.001
        Anderson and Hauck's test

        noncentrality parameter =     4.423
                 test statistic =    -0.613      empirical p-value =     0.0005
        note: reference treatment = 1
```

Both of Schuirmann's one-sided tests are highly significant, suggesting that the two drugs are bioequivalent. A similar conclusion is drawn from the Anderson and Hauck test of bioequivalence.
◁

Saved Results

pkequiv saves the following in r():

Scalars

r(stddev)	pooled-sample standard deviation of period differences from both sequences
r(uci)	upper confidence interval for a classic interval
r(lci)	lower confidence interval for a classic interval
r(delta)	delta value used in calculating a symmetric confidence interval
r(u3)	upper confidence interval for Fieller's confidence interval
r(l3)	lower confidence interval for Fieller's confidence interval

Methods and Formulas

pkequiv is implemented as an ado-file.

The lower confidence interval for the difference in the two treatments for the classic shortest confidence interval is

$$L_1 = \left(\overline{Y}_T - \overline{Y}_R\right) - t_{(\alpha,n_1+n_2-2)}\widehat{\sigma}_d\sqrt{\frac{1}{n_1}+\frac{1}{n_2}}$$

The upper limit is

$$U_1 = \left(\overline{Y}_T - \overline{Y}_R\right) + t_{(\alpha,n_1+n_2-2)}\widehat{\sigma}_d\sqrt{\frac{1}{n_1}+\frac{1}{n_2}}$$

The limits for the ratio measure are

$$L_2 = \left(\frac{L_1}{\overline{Y}_R}+1\right)100\%$$

and

$$U_2 = \left(\frac{U_1}{\overline{Y}_R}+1\right)100\%$$

where $\overline{Y}_T$ is the mean of the test formulation of the drug, $\overline{Y}_R$ is the mean of the reference formulation of the drug, and $t_{(\alpha,n_1+n_2-2)}$ is the t distribution with n_1+n_2-2 degrees of freedom. $\widehat{\sigma}_d$ is the pooled sample variance of the period differences from both sequences, defined as

$$\widehat{\sigma}_d = \frac{1}{n_1+n_2-2}\sum_{k=1}^{2}\sum_{i=1}^{n_k}\left(d_{ik}-\overline{d}_{.k}\right)^2$$

The upper and lower limits for the symmetric confidence interval are $\overline{Y}_R+\Delta$ and $\overline{Y}_R-\Delta$, where

$$\Delta = k_1\widehat{\sigma}_d\sqrt{\frac{1}{n_1}+\frac{1}{n_2}} - \left(\overline{Y}_T - \overline{Y}_R\right)$$

and (simultaneously)

$$\Delta = -k_2\widehat{\sigma}_d\sqrt{\frac{1}{n_1}+\frac{1}{n_2}} + 2\left(\overline{Y}_T - \overline{Y}_R\right)$$

and k_1 and k_2 are computed iteratively to satisfy the above equalities and the condition

$$\int_{k_1}^{k_2} f(t)dt = 1-2\alpha$$

where $f(t)$ is the probability density function of the t distribution with n_1+n_2-2 degrees of freedom.

See Chow and Liu (2000, 88) for details about calculating the confidence interval based on Fieller's theorem.

The two test statistics for the two one-sided tests of equivalence are

$$T_L = \frac{(\overline{Y}_T - \overline{Y}_R) - \theta_L}{\widehat{\sigma}_d \sqrt{\frac{1}{n_1} + \frac{1}{n_2}}}$$

and

$$T_U = \frac{(\overline{Y}_T - \overline{Y}_R) - \theta_U}{\widehat{\sigma}_d \sqrt{\frac{1}{n_1} + \frac{1}{n_2}}}$$

where $-\theta_L = \theta_U$ and are the regulated confidence limits.

The logic of the Anderson and Hauck test is tricky; see Chow and Liu (2000) for a complete explanation. However, the test statistic is

$$T_{AH} = \frac{(\overline{Y}_T - \overline{Y}_R) - \left(\frac{\theta_L + \theta_U}{2}\right)}{\widehat{\sigma}_d \sqrt{\frac{1}{n_1} + \frac{1}{n_2}}}$$

and the noncentrality parameter is estimated by

$$\widehat{\delta} = \frac{\theta_U - \theta_L}{2\widehat{\sigma}_d \sqrt{\frac{1}{n_1} + \frac{1}{n_2}}}$$

The empirical p-value is calculated as

$$p = F_t\left(|T_{AH}| - \widehat{\delta}\right) - F_t\left(-|T_{AH}| - \widehat{\delta}\right)$$

where F_t is the cumulative distribution function of the t distribution with $n_1 + n_2 - 2$ degrees of freedom.

References

Chow, S. C., and J. P. Liu. 2000. *Design and Analysis of Bioavailability and Bioequivalence Studies.* 2nd ed, Revised and Expanded. New York: Dekker.

Kutner, M. H., C. J. Nachtsheim, and J. Neter. 2005. *Applied Linear Statistical Models.* 5th ed. Chicago: McGraw–Hill/Irwin.

Ratkowsky, D. A., M. A. Evans, and J. R. Alldredge. 1993. *Cross-over Experiments: Design, Analysis and Application.* New York: Dekker.

Also See

[R] **pk** — Pharmacokinetic (biopharmaceutical) data

Title

pkexamine — Calculate pharmacokinetic measures

Syntax

`pkexamine` *time concentration* [*if*] [*in*] [, *options*]

options	description
Main	
`fit(`*#*`)`	use # points to estimate $AUC_{0,\infty}$; default is `fit(3)`
`trapezoid`	use trapezoidal rule; default is cubic splines
`graph`	graph the AUC
`line`	graph the linear extension
`log`	graph the log extension
`exp(`*#*`)`	plot the exponential fit for the $AUC_{0,\infty}$
AUC plot	
cline_options	affect rendition of plotted points connected by lines
marker_options	change look of markers (color, size, etc.)
marker_label_options	add marker labels; change look or position
Add plots	
`addplot(`*plot*`)`	add other plots to the generated graph
Y axis, X axis, Titles, Legend, Overall	
twoway_options	any options other than `by()` documented in [G] ***twoway_options***

by is allowed; see [D] **by**.

Description

`pkexamine` calculates pharmacokinetic measures from time-and-concentration subject-level data. `pkexamine` computes and displays the maximum measured concentration, the time at the maximum measured concentration, the time of the last measurement, the elimination time, the half-life, and the area under the concentration-time curve (AUC). Three estimates of the area under the concentration-time curve from 0 to infinity ($AUC_{0,\infty}$) are also calculated.

`pkexamine` is one of the pk commands. Please read [R] **pk** before reading this entry.

Options

Main

`fit(`*#*`)` specifies the number of points, counting back from the last measurement, to use in fitting the extension to estimate the $AUC_{0,\infty}$. The default is `fit(3)`, or the last three points. This value should be viewed as a minimum; the appropriate number of points will depend on your data.

`trapezoid` specifies that the trapezoidal rule be used to calculate the AUC. The default is cubic splines, which give better results for most functions. When the curve is irregular, `trapezoid` may give better results.

`graph` tells `pkexamine` to graph the concentration-time curve.

`line` and `log` specify the estimates of the $AUC_{0,\infty}$ to display when graphing the $AUC_{0,\infty}$. These options are ignored, unless they are specified with the `graph` option.

`exp(#)` specifies that the exponential fit for the $AUC_{0,\infty}$ be plotted. You must specify the maximum time value to which you want to plot the curve, and this time value must be greater than the maximum time measurement in the data. If you specify 0, the curve will be plotted to the point at which the linear extension would cross the x axis. This option is not valid with the `line` or `log` option and is ignored, unless the `graph` option is also specified.

AUC plot

cline_options affect the rendition of the plotted points connected by lines; see [G] ***cline_options***.

marker_options specify the look of markers. This look includes the marker symbol, the marker size, and its color and outline; see [G] ***marker_options***.

marker_label_options specify if and how the markers are to be labeled; see [G] ***marker_label_options***.

Add plots

`addplot(`*plot*`)` provides a way to add other plots to the generated graph. See [G] ***addplot_option***.

Y axis, X axis, Titles, Legend, Overall

twoway_options are any of the options documented in [G] ***twoway_options***, excluding `by()`. These include options for titling the graph (see [G] ***title_options***) and for saving the graph to disk (see [G] ***saving_option***).

Remarks

`pkexamine` computes summary statistics for a given patient in a pharmacokinetic trial. If `by` *idvar*`:` is specified, statistics will be displayed for each subject in the data.

▷ Example 1

Chow and Liu (2000, 11) present data on a study examining primidone concentrations versus time for a subject over a 32-hour period after dosing.

```
. use http://www.stata-press.com/data/r10/auc
. list, abbrev(14)
```

	id	time	concentration
1.	1	0	0
2.	1	.5	0
3.	1	1	2.8
4.	1	1.5	4.4
5.	1	2	4.4
6.	1	3	4.7
7.	1	4	4.1
8.	1	6	4
9.	1	8	3.6
10.	1	12	3
11.	1	16	2.5
12.	1	24	2
13.	1	32	1.6

We use pkexamine to produce the summary statistics.

```
. pkexamine time conc, graph
                        Maximum concentration =         4.7
                Time of maximum concentration =           3
              Time of last observation (Tmax) =          32
                             Elimination rate =      0.0279
                                    Half life =     24.8503

Area under the curve
```

AUC [0, Tmax]	AUC [0, inf.) Linear of log conc.	AUC [0, inf.) Linear fit	AUC [0, inf.) Exponential fit
85.24	142.603	107.759	142.603

```
Fit based on last 3 points.
```

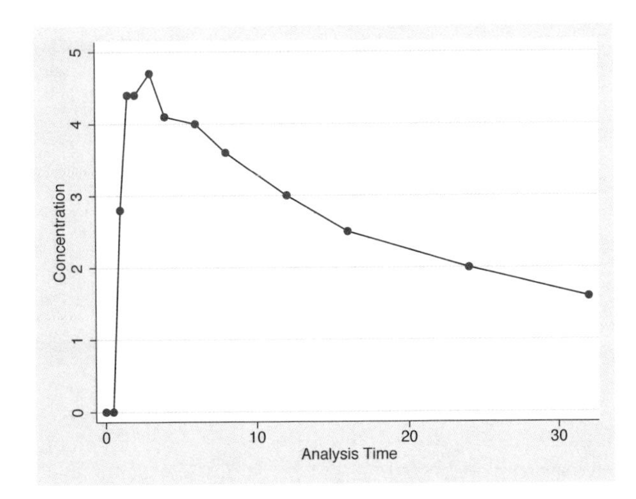

The maximum concentration of 4.7 occurs at time 3, and the time of the last observation (Tmax) is 32. In addition to the AUC, which is calculated from 0 to the maximum value of `time`, `pkexamine` also reports the area under the curve, computed by extending the curve with each of three methods: a linear fit to the log of the concentration, a linear regression line, and a decreasing exponential regression line. See *Methods and Formulas* for details on these three methods.

By default, all extensions to the AUC are based on the last three points. Looking at the graph for these data, it seems more appropriate to use the last seven points to estimate the $\text{AUC}_{0,\infty}$:

```
. pkexamine time conc, fit(7)

                           Maximum concentration =        4.7
                   Time of maximum concentration =          3
                Time of last observation (Tmax) =         32
                                Elimination rate =     0.0349
                                       Half life =    19.8354

     Area under the curve
```

AUC [0, Tmax]	AUC [0, inf.) Linear of log conc.	AUC [0, inf.) Linear fit	AUC [0, inf.) Exponential fit
85.24	131.027	96.805	129.181

```
     Fit based on last 7 points.
```

This approach decreased the estimate of the $\text{AUC}_{0,\infty}$ for all extensions. To see a graph of the $\text{AUC}_{0,\infty}$ using a linear extension, specify the `graph` and `line` options.

```
. pkexamine time conc, fit(7) graph line
                           Maximum concentration =        4.7
                   Time of maximum concentration =          3
                Time of last observation (Tmax) =         32
                                Elimination rate =     0.0349
                                       Half life =    19.8354

     Area under the curve
```

AUC [0, Tmax]	AUC [0, inf.) Linear of log conc.	AUC [0, inf.) Linear fit	AUC [0, inf.) Exponential fit
85.24	131.027	96.805	129.181

```
     Fit based on last 7 points.
```

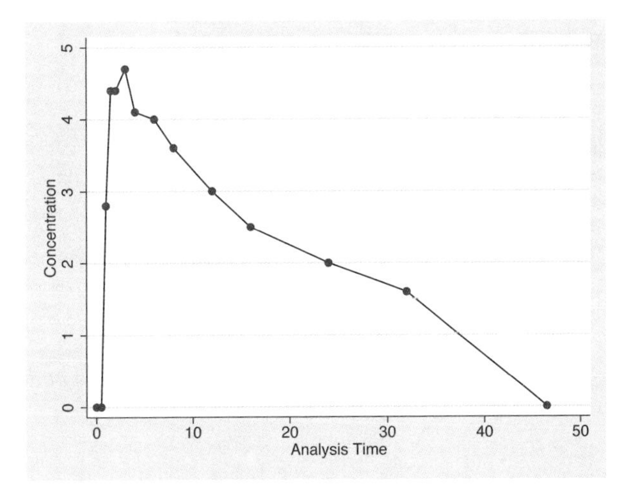

◁

Saved Results

pkexamine saves the following in r():

Scalars

r(auc)	area under the concentration curve
r(half)	half-life of the drug
r(ke)	elimination rate
r(tmax)	time at last concentration measurement
r(cmax)	maximum concentration
r(tomc)	time of maximum concentration
r(auc_line)	$\text{AUC}_{0,\infty}$ estimated with a linear fit
r(auc_exp)	$\text{AUC}_{0,\infty}$ estimated with an exponential fit
r(auc_ln)	$\text{AUC}_{0,\infty}$ estimated with a linear fit of the natural log

Methods and Formulas

pkexamine is implemented as an ado-file.

The $\text{AUC}_{0,t_{\max}}$ is defined as

$$\text{AUC}_{0,t_{\max}} = \int_0^{t_{\max}} C_t dt$$

where C_t is the concentration at time t. By default, the integral is calculated numerically using cubic splines. However, if the trapezoidal rule is used, the $\text{AUC}_{0,t_{\max}}$ is given as

$$\text{AUC}_{0,t_{\max}} = \sum_{i=2}^{k} \frac{C_{i-1} + C_i}{2} (t_i - t_{i-1})$$

The $\text{AUC}_{0,\infty}$ is the $\text{AUC}_{0,t_{\max}}$ + $\text{AUC}_{t_{\max},\infty}$, or

$$\text{AUC}_{0,\infty} = \int_0^{t_{\max}} C_t dt + \int_{t_{\max}}^{\infty} C_t dt$$

When using the linear extension to the $\text{AUC}_{0,t_{\max}}$, the integration is cut off when the line crosses the x axis. The log extension is a linear extension on the log concentration scale. The area for the exponential extension is

$$\text{AUC}_{0,\infty} = \int_{t_{\max}}^{\infty} e^{-(\beta_0+t\beta_1)} dt = -\frac{e^{-(\beta_0+t_{\max}\beta_1)}}{\beta_1}$$

Finally, the elimination rate, K_{eq}, is the negative of the parameter estimate for a linear regression of log time on concentration and is given in the standard manner:

$$K_{\text{eq}} = -\frac{\sum_{i=1}^{k}\left(C_i - \overline{C}\right)\left(\ln t_i - \overline{\ln t}\right)}{\sum_{i=1}^{k}\left(C_i - \overline{C}\right)^2}$$

and

$$t_{1/2} = \frac{\ln 2}{K_{\text{eq}}}$$

Reference

Chow, S. C., and J. P. Liu. 2000. *Design and Analysis of Bioavailability and Bioequivalence Studies*. 2nd ed, Revised and Expanded. New York: Dekker.

Also See

[R] **pk** — Pharmacokinetic (biopharmaceutical) data

Title

pkshape — Reshape (pharmacokinetic) Latin-square data

Syntax

`pkshape` *id sequence period1 period2* [*period list*] [, *options*]

options	description
`order(`*string*`)`	apply treatments in specified order
`outcome(`*newvar*`)`	name for outcome variable; default is `outcome(outcome)`
`treatment(`*newvar*`)`	name for treatment variable; default is `treatment(treat)`
`carryover(`*newvar*`)`	name for carryover variable; default is `carryover(carry)`
`sequence(`*newvar*`)`	name for sequence variable; default is `sequence(sequence)`
`period(`*newvar*`)`	name for period variable; default is `period(period)`

Description

`pkshape` reshapes the data for use with `anova`, `pkcross`, and `pkequiv`. Latin-square and crossover data are often organized in a manner that cannot be analyzed easily with Stata. `pkshape` reorganizes the data in memory for use in Stata.

`pkshape` is one of the pk commands. Please read [R] **pk** before reading this entry.

Options

`order(`*string*`)` specifies the order in which treatments were applied. If the `sequence()` specifier is a string variable that specifies the order, this option is not necessary. Otherwise, `order()` specifies how to generate the treatment and carryover variables. Any string variable can be used to specify the order. For crossover designs, any washout periods can be indicated with the number 0.

`outcome(`*newvar*`)` specifies the name for the outcome variable in the reorganized data. By default, `outcome(outcome)` is used.

`treatment(`*newvar*`)` specifies the name for the treatment variable in the reorganized data. By default, `treatment(treat)` is used.

`carryover(`*newvar*`)` specifies the name for the carryover variable in the reorganized data. By default, `carryover(carry)` is used.

`sequence(`*newvar*`)` specifies the name for the sequence variable in the reorganized data. By default, `sequence(sequence)` is used.

`period(`*newvar*`)` specifies the name for the period variable in the reorganized data. By default, `period(period)` is used.

Remarks

Often data from a Latin-square experiment are naturally organized in a manner that Stata cannot manage easily. pkshape reorganizes Latin-square data so that they can be used with anova (see [R] **anova**) or any pk command. This includes the classic 2×2 crossover design commonly used in pharmaceutical research, as well as many other Latin-square designs.

▷ Example 1

Consider the example data published in Chow and Liu (2000, 73). There are 24 patients, 12 in each sequence. Sequence 1 consists of the reference formulation followed by the test formulation; sequence 2 is the test formulation followed by the reference formulation. The measurements reported are the $\mathrm{AUC}_{0-t_{\max}}$ for each patient and for each period.

```
. use http://www.stata-press.com/data/r10/chowliu
. list, sep(4)
```

	id	seq	period1	period2
1.	1	1	74.675	73.675
2.	4	1	96.4	93.25
3.	5	1	101.95	102.125
4.	6	1	79.05	69.45
5.	11	1	79.05	69.025
6.	12	1	85.95	68.7
7.	15	1	69.725	59.425
8.	16	1	86.275	76.125
9.	19	1	112.675	114.875
10.	20	1	99.525	116.25
11.	23	1	89.425	64.175
12.	24	1	55.175	74.575
13.	2	2	74.825	37.35
14.	3	2	86.875	51.925
15.	7	2	81.675	72.175
16.	8	2	92.7	77.5
17.	9	2	50.45	71.875
18.	10	2	66.125	94.025
19.	13	2	122.45	124.975
20.	14	2	99.075	85.225
21.	17	2	86.35	95.925
22.	18	2	49.925	67.1
23.	21	2	42.7	59.425
24.	22	2	91.725	114.05

Since the outcome for one person is in two different variables, the treatment that was applied to an individual is a function of the period and the sequence. To analyze this treatment using anova, all the outcomes must be in one variable, and each covariate must be in its own variable. To reorganize these data, use pkshape:

```
. pkshape id seq period1 period2, order(ab ba)
. sort seq id treat
```

```
. list, sep(8)
```

	id	sequence	outcome	treat	carry	period
1.	1	1	74.675	1	0	1
2.	1	1	73.675	2	1	2
3.	4	1	96.4	1	0	1
4.	4	1	93.25	2	1	2
5.	5	1	101.95	1	0	1
6.	5	1	102.125	2	1	2
7.	6	1	79.05	1	0	1
8.	6	1	69.45	2	1	2
9.	11	1	79.05	1	0	1
10.	11	1	69.025	2	1	2
11.	12	1	85.95	1	0	1
12.	12	1	68.7	2	1	2
13.	15	1	69.725	1	0	1
14.	15	1	59.425	2	1	2
15.	16	1	86.275	1	0	1
16.	16	1	76.125	2	1	2
17.	19	1	112.675	1	0	1
18.	19	1	114.875	2	1	2
19.	20	1	99.525	1	0	1
20.	20	1	116.25	2	1	2
21.	23	1	89.425	1	0	1
22.	23	1	64.175	2	1	2
23.	24	1	55.175	1	0	1
24.	24	1	74.575	2	1	2
25.	2	2	37.35	1	2	2
26.	2	2	74.825	2	0	1
27.	3	2	51.925	1	2	2
28.	3	2	86.875	2	0	1
29.	7	2	72.175	1	2	2
30.	7	2	81.675	2	0	1
31.	8	2	77.5	1	2	2
32.	8	2	92.7	2	0	1
33.	9	2	71.875	1	2	2
34.	9	2	50.45	2	0	1
35.	10	2	94.025	1	2	2
36.	10	2	66.125	2	0	1
37.	13	2	124.975	1	2	2
38.	13	2	122.45	2	0	1
39.	14	2	85.225	1	2	2
40.	14	2	99.075	2	0	1
41.	17	2	95.925	1	2	2
42.	17	2	86.35	2	0	1
43.	18	2	67.1	1	2	2
44.	18	2	49.925	2	0	1
45.	21	2	59.425	1	2	2
46.	21	2	42.7	2	0	1
47.	22	2	114.05	1	2	2
48.	22	2	91.725	2	0	1

Now the data are organized into separate variables that indicate each factor level for each of the covariates, so the data may be used with `anova` or `pkcross`; see [R] **anova** and [R] **pkcross**.

◁

▷ Example 2

Consider the study of background music on bank teller productivity published in Neter et al. (2005). The data are

Week	Monday	Tuesday	Wednesday	Thursday	Friday
1	18(D)	17(C)	14(A)	21(B)	17(E)
2	13(C)	34(B)	21(E)	16(A)	15(D)
3	7(A)	29(D)	32(B)	27(E)	13(C)
4	17(E)	13(A)	24(C)	31(D)	25(B)
5	21(B)	26(E)	26(D)	31(C)	7(A)

The numbers are the productivity scores, and the letters represent the treatment. We entered the data into Stata:

```
. use http://www.stata-press.com/data/r10/music, clear
. list
```

	id	seq	day1	day2	day3	day4	day5
1.	1	dcabe	18	17	14	21	17
2.	2	cbead	13	34	21	16	15
3.	3	adbec	7	29	32	27	13
4.	4	eacdb	17	13	24	31	25
5.	5	bedca	21	26	26	31	7

We reshape these data with `pkshape`:

```
. pkshape id seq day1 day2 day3 day4 day5
. list, sep(0)
```

	id	sequence	outcome	treat	carry	period
1.	3	1	7	1	0	1
2.	5	2	21	3	0	1
3.	2	3	13	5	0	1
4.	1	4	18	2	0	1
5.	4	5	17	4	0	1
6.	3	1	29	2	1	2
7.	5	2	26	4	3	2
8.	2	3	34	3	5	2
9.	1	4	17	5	2	2
10.	4	5	13	1	4	2
11.	3	1	32	3	2	3
12.	5	2	26	2	4	3
13.	2	3	21	4	3	3
14.	1	4	14	1	5	3
15.	4	5	24	5	1	3
16.	3	1	27	4	3	4
17.	5	2	31	5	2	4
18.	2	3	16	1	4	4
19.	1	4	21	3	1	4
20.	4	5	31	2	5	4
21.	3	1	13	5	4	5
22.	5	2	7	1	5	5
23.	2	3	15	2	1	5
24.	1	4	17	4	3	5
25.	4	5	25	3	2	5

Here the `sequence` variable is a string variable that specifies how the treatments were applied, so the `order` option is not used. When the sequence variable is a string and the `order` is specified, the arguments from the `order` option are used. We could now produce an ANOVA table:

```
. anova outcome seq period treat
                           Number of obs =      25     R-squared     =  0.8666
                           Root MSE      = 3.96232     Adj R-squared =  0.7331

                  Source | Partial SS    df       MS           F     Prob > F
              -----------+----------------------------------------------------
                   Model |     1223.6    12  101.966667       6.49     0.0014
                         |
                sequence |         82     4        20.5       1.31     0.3226
                  period |      477.2     4       119.3       7.60     0.0027
                   treat |      664.4     4       166.1      10.58     0.0007
                         |
                Residual |      188.4    12        15.7
              -----------+----------------------------------------------------
                   Total |       1412    24  58.8333333
```

◁

▷ Example 3

Consider the Latin-square crossover example published in Neter et al. (2005). The example is about apple sales given different methods for displaying apples.

Pattern	Store	Week 1	Week 2	Week 3
1	1	9(B)	12(C)	15(A)
	2	4(B)	12(C)	9(A)
2	1	12(A)	14(B)	3(C)
	2	13(A)	14(B)	3(C)
3	1	7(C)	18(A)	6(B)
	2	5(C)	20(A)	4(B)

We entered the data into Stata:

```
. use http://www.stata-press.com/data/r10/applesales, clear
. list, sep(2)

     +--------------------------------------+
     | id   seq   p1   p2   p3   square |
     |--------------------------------------|
  1. |  1     1    9   12   15        1 |
  2. |  2     1    4   12    9        2 |
     |--------------------------------------|
  3. |  3     2   12   14    3        1 |
  4. |  4     2   13   14    3        2 |
     |--------------------------------------|
  5. |  5     3    7   18    6        1 |
  6. |  6     3    5   20    4        2 |
     +--------------------------------------+
```

Now the data can be reorganized using descriptive names for the outcome variables.

```
. pkshape id seq p1 p2 p3, order(bca abc cab) seq(pattern) period(order)
> treat(displays)
```

```
. anova outcome pattern order display id|pattern

                           Number of obs =      18     R-squared     =  0.9562
                           Root MSE      = 1.59426     Adj R-squared =  0.9069

                  Source | Partial SS    df       MS           F     Prob > F
              -----------+----------------------------------------------------
                   Model | 443.666667     9  49.2962963      19.40     0.0002
                         |
                 pattern | .333333333     2  .166666667       0.07     0.9370
                   order | 233.333333     2  116.666667      45.90     0.0000
                displays |        189     2        94.5      37.18     0.0001
              id|pattern |         21     3           7       2.75     0.1120
                         |
                Residual | 20.3333333     8  2.54166667
              -----------+----------------------------------------------------
                   Total |        464    17  27.2941176
```

These are the same results reported by Neter et al. (2005).

◁

▷ Example 4

We continue with the example from [R] **pkcollapse**; the data are

```
. use http://www.stata-press.com/data/r10/pkdata2, clear
. list, sep(4) abbrev(10)
```

	id	seq	auc_concA	auc_concB
1.	1	1	150.9643	218.5551
2.	2	1	146.7606	133.3201
3.	3	1	160.6548	126.0635
4.	4	1	157.8622	96.17461
5.	5	1	133.6957	188.9038
6.	7	1	160.639	223.6922
7.	8	1	131.2604	104.0139
8.	9	1	168.5186	237.8962
9.	10	2	137.0627	139.7382
10.	12	2	153.4038	202.3942
11.	13	2	163.4593	136.7848
12.	14	2	146.0462	104.5191
13.	15	2	158.1457	165.8654
14.	18	2	147.1977	139.235
15.	19	2	164.9988	166.2391
16.	20	2	145.3823	158.5146

```
. pkshape id seq auc_concA auc_concB, order(ab ba)
. sort period id
```

```
. list, sep(4)
```

	id	sequence	outcome	treat	carry	period
1.	1	1	150.9643	1	0	1
2.	2	1	146.7606	1	0	1
3.	3	1	160.6548	1	0	1
4.	4	1	157.8622	1	0	1
5.	5	1	133.6957	1	0	1
6.	7	1	160.639	1	0	1
7.	8	1	131.2604	1	0	1
8.	9	1	168.5186	1	0	1
9.	10	2	137.0627	2	0	1
10.	12	2	153.4038	2	0	1
11.	13	2	163.4593	2	0	1
12.	14	2	146.0462	2	0	1
13.	15	2	158.1457	2	0	1
14.	18	2	147.1977	2	0	1
15.	19	2	164.9988	2	0	1
16.	20	2	145.3823	2	0	1
17.	1	1	218.5551	2	1	2
18.	2	1	133.3201	2	1	2
19.	3	1	126.0635	2	1	2
20.	4	1	96.17461	2	1	2
21.	5	1	188.9038	2	1	2
22.	7	1	223.6922	2	1	2
23.	8	1	104.0139	2	1	2
24.	9	1	237.8962	2	1	2
25.	10	2	139.7382	1	2	2
26.	12	2	202.3942	1	2	2
27.	13	2	136.7848	1	2	2
28.	14	2	104.5191	1	2	2
29.	15	2	165.8654	1	2	2
30.	18	2	139.235	1	2	2
31.	19	2	166.2391	1	2	2
32.	20	2	158.5146	1	2	2

◁

We call the resulting dataset pkdata3. We conduct equivalence testing on the data in [R] **pkequiv**, and we fit an ANOVA model to these data in the third example of [R] **pkcross**.

Methods and Formulas

pkshape is implemented as an ado-file.

References

Chow, S. C., and J. P. Liu. 2000. *Design and Analysis of Bioavailability and Bioequivalence Studies*. 2nd ed, Revised and Expanded. New York: Dekker.

Kutner, M. H., C. J. Nachtsheim, and J. Neter. 2005. *Applied Linear Statistical Models*. 5th ed. Chicago: McGraw–Hill/Irwin.

Also See

[R] **pk** — Pharmacokinetic (biopharmaceutical) data

Title

pksumm — Summarize pharmacokinetic data

Syntax

`pksumm` *id time concentration* [*if*] [*in*] [, *options*]

options	description
Main	
`trapezoid`	use trapezoidal rule to calculate AUC; default is cubic splines
`fit(#)`	use # points to estimate AUC; default is `fit(3)`
`notimechk`	do not check whether follow-up time for all subjects is the same
`nodots`	suppress the dots during calculation
`graph`	graph the distribution of *statistic*
`stat(`*statistic*`)`	graph the specified statistic; default is `stat(auc)`
Histogram, Density plots, Y axis, X axis, Titles, Legend, Overall	
histogram_options	any option other than `by()` documented in [R] **histogram**

statistic	description
`auc`	area under the concentration-time curve ($\text{AUC}_{0,\infty}$); the default
`aucline`	area under the concentration-time curve from 0 to ∞ using a linear extension
`aucexp`	area under the concentration-time curve from 0 to ∞ using an exponential extension
`auclog`	area under the log-concentration-time curve extended with a linear fit
`half`	half-life of the drug
`ke`	elimination rate
`cmax`	maximum concentration
`tmax`	time at last concentration
`tomc`	time of maximum concentration

Description

`pksumm` obtains summary measures based on the first four moments from the empirical distribution of each pharmacokinetic measurement and tests the null hypothesis that the distribution of that measurement is normally distributed.

`pksumm` is one of the pk commands. Please read [R] **pk** before reading this entry.

(*Continued on next page*)

Options

Main

`trapezoid` specifies that the trapezoidal rule be used to calculate the AUC. The default is cubic splines, which give better results for most situations. When the curve is irregular, the trapezoidal rule may give better results.

`fit(#)` specifies the number of points, counting back from the last time measurement, to use in fitting the extension to estimate the $\text{AUC}_{0,\infty}$. The default is `fit(3)`, the last three points. This default should be viewed as a minimum; the appropriate number of points will depend on the data.

`notimechk` suppresses the check that the follow-up time for all subjects is the same. By default, `pksumm` expects the maximum follow-up time to be equal for all subjects.

`nodots` suppresses the progress dots during calculation. By default, a period is displayed for every call to calculate the pharmacokinetic measures.

`graph` requests a graph of the distribution of the statistic specified with `stat()`.

`stat(`*statistic*`)` specifies the statistic that `pksumm` should graph. The default is `stat(auc)`. If the `graph` option is not specified, this option is ignored.

Histogram, Density plots, Y axis, X axis, Titles, Legend, Overall

histogram_options are any of the options documented in [R] **histogram**, excluding `by()`. For `pksumm`, `fraction` is the default, not `density`.

Remarks

`pksumm` produces summary statistics for the distribution of nine common pharmacokinetic measurements. If there are more than eight subjects, `pksumm` also computes a test for normality on each measurement. The nine measurements summarized by `pksumm` are listed above and are described in *Methods and Formulas* of [R] **pkexamine** and [R] **pk**.

▷ Example 1

We demonstrate the use of `pksumm` on a variation of the data described in [R] **pk**. We have drug concentration data on 15 subjects, each measured at 13 time points over a 32-hour period. A few of the records are

```
. use http://www.stata-press.com/data/r10/pksumm
. list, sep(0)
```

	id	time	conc
1.	1	0	0
2.	1	.5	3.073403
3.	1	1	5.188444
4.	1	1.5	5.898577
5.	1	2	5.096378
6.	1	3	6.094085
	(output omitted)		
183.	15	0	0
184.	15	.5	3.86493
185.	15	1	6.432444
186.	15	1.5	6.969195
187.	15	2	6.307024
188.	15	3	6.509584
189.	15	4	6.555091
190.	15	6	7.318319
191.	15	8	5.329813
192.	15	12	5.411624
193.	15	16	3.891397
194.	15	24	5.167516
195.	15	32	2.649686

We can use **pksumm** to view the summary statistics for all the pharmacokinetic parameters.

```
. pksumm id time conc
...............
Summary statistics for the pharmacokinetic measures
                                          Number of observations =    15
```

Measure	Mean	Median	Variance	Skewness	Kurtosis	p-value
auc	150.74	150.96	123.07	-0.26	2.10	0.69
aucline	408.30	214.17	188856.87	2.57	8.93	0.00
aucexp	691.68	297.08	762679.94	2.56	8.87	0.00
auclog	688.98	297.67	797237.24	2.59	9.02	0.00
half	94.84	29.39	18722.13	2.26	7.37	0.00
ke	0.02	0.02	0.00	0.89	3.70	0.09
cmax	7.36	7.42	0.42	-0.60	2.56	0.44
tomc	3.47	3.00	7.62	2.17	7.18	0.00
tmax	32.00	32.00	0.00	.	.	.

For the 15 subjects, the mean $\text{AUC}_{0,t_{\max}}$ is 150.74, and $\sigma^2 = 123.07$. The skewness of -0.26 indicates that the distribution is slightly skewed left. The p-value of 0.69 for the χ^2 test of normality indicates that we cannot reject the null hypothesis that the distribution is normal.

If we were to consider any of the three variants of the $\text{AUC}_{0,\infty}$, we would see that there is huge variability and that the distribution is heavily skewed. A skewness different from 0 and a kurtosis different from 3 are expected because the distribution of the $\text{AUC}_{0,\infty}$ is not normal.

We now graph the distribution of $\text{AUC}_{0,t_{\max}}$ by specifying the **graph** option.

(Continued on next page)

```
. pksumm id time conc, graph bin(20)
...............

 Summary statistics for the pharmacokinetic measures

                                        Number of observations =      15
     Measure      Mean     Median      Variance  Skewness   Kurtosis   p-value

         auc    150.74     150.96        123.07     -0.26       2.10      0.69
     aucline    408.30     214.17     188856.87      2.57       8.93      0.00
      aucexp    691.68     297.08     762679.94      2.56       8.87      0.00
      auclog    688.98     297.67     797237.24      2.59       9.02      0.00
        half     94.84      29.39      18722.13      2.26       7.37      0.00
          ke      0.02       0.02          0.00      0.89       3.70      0.09
        cmax      7.36       7.42          0.42     -0.60       2.56      0.44
        tomc      3.47       3.00          7.62      2.17       7.18      0.00
        tmax     32.00      32.00          0.00         .          .         .
```

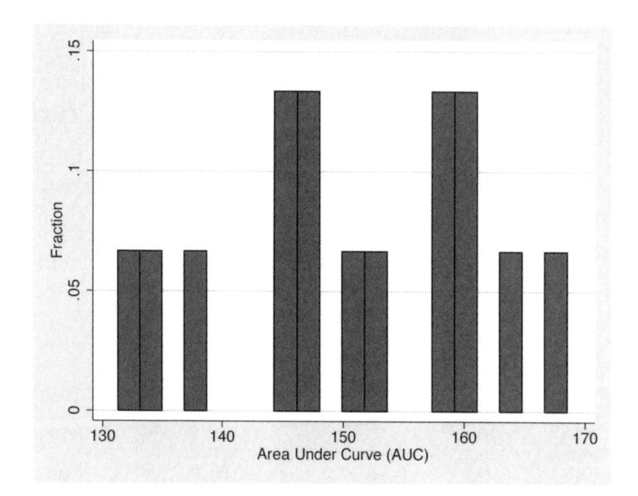

graph, by default, plots $\text{AUC}_{0,t_{\max}}$. To plot a graph of one of the other pharmacokinetic measurements, we need to specify the stat() option. For example, we can ask Stata to produce a plot of the $\text{AUC}_{0,\infty}$ using the log extension:

```
. pksumm id time conc, stat(auclog) graph bin(20)
...............

 Summary statistics for the pharmacokinetic measures

                                        Number of observations =      15
     Measure      Mean     Median      Variance  Skewness   Kurtosis   p-value

         auc    150.74     150.96        123.07     -0.26       2.10      0.69
     aucline    408.30     214.17     188856.87      2.57       8.93      0.00
      aucexp    691.68     297.08     762679.94      2.56       8.87      0.00
      auclog    688.98     297.67     797237.24      2.59       9.02      0.00
        half     94.84      29.39      18722.13      2.26       7.37      0.00
          ke      0.02       0.02          0.00      0.89       3.70      0.09
        cmax      7.36       7.42          0.42     -0.60       2.56      0.44
        tomc      3.47       3.00          7.62      2.17       7.18      0.00
        tmax     32.00      32.00          0.00         .          .         .
```

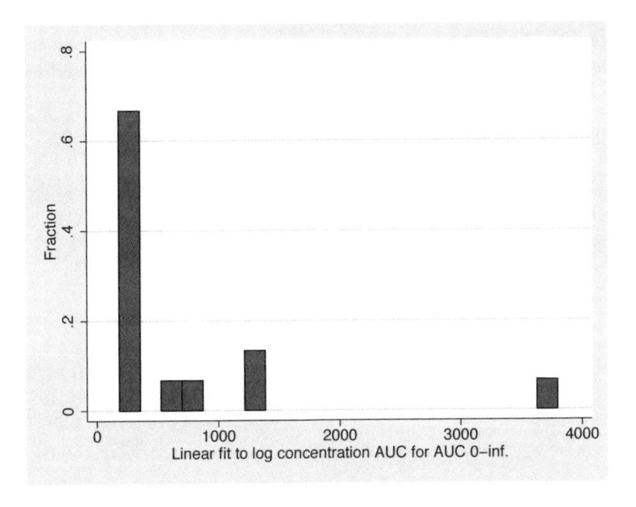

◁

Methods and Formulas

pksumm is implemented as an ado-file.

The χ^2 test for normality is conducted with sktest; see [R] **sktest** for more information on the test of normality.

The statistics reported by pksumm are identical to those reported by summarize and sktest; see [R] **summarize** and [R] **sktest**.

Also See

[R] **pk** — Pharmacokinetic (biopharmaceutical) data

Title

poisson — Poisson regression

Syntax

`poisson` *depvar* [*indepvars*] [*if*] [*in*] [*weight*] [, *options*]

options	description
Model	
`noconstant`	suppress constant term
`exposure(`*varname*$_e$`)`	include ln(*varname*$_e$) in model with coefficient constrained to 1
`offset(`*varname*$_o$`)`	include *varname*$_o$ in model with coefficient constrained to 1
`constraints(`*constraints*`)`	apply specified linear constraints
`collinear`	keep collinear variables
SE/Robust	
`vce(`*vcetype*`)`	*vcetype* may be `oim`, `robust`, `cluster` *clustvar*, `opg`, `bootstrap`, or `jackknife`
Reporting	
`level(`*#*`)`	set confidence level; default is `level(95)`
`irr`	report incidence-rate ratios
Max options	
maximize_options	control the maximization process; seldom used

depvar, *indepvars*, *varname*$_e$, and *varname*$_o$ may contain time-series operators; see [U] **11.4.3 Time-series varlists**.

`bootstrap`, `by`, `jackknife`, `nestreg`, `rolling`, `statsby`, `stepwise`, `svy`, and `xi` are allowed; see [U] **11.1.10 Prefix commands**.

Weights are not allowed with the `bootstrap` prefix.

`vce()` and weights are not allowed with the `svy` prefix.

`fweights`, `iweights`, and `pweights` are allowed; see [U] **11.1.6 weight**.

See [U] **20 Estimation and postestimation commands** for more capabilities of estimation commands.

Description

`poisson` fits a Poisson regression of *depvar* on *indepvars*, where *depvar* is a nonnegative count variable.

If you have panel data, see [XT] **xtpoisson**.

Options

Model

`noconstant`, `exposure(`*varname*$_e$`)`, `offset(`*varname*$_o$`)`, `constraints(`*constraints*`)`, `collinear`; see [R] **estimation options**.

SE/Robust

vce(*vcetype*) specifies the type of standard error reported, which includes types that are derived from asymptotic theory, that are robust to some kinds of misspecification, that allow for intragroup correlation, and that use bootstrap or jackknife methods; see [R] ***vce_option***.

Reporting

level(*#*); see [R] **estimation options**.

irr reports estimated coefficients transformed to incidence-rate ratios, that is, e^{β_i} rather than β_i. Standard errors and confidence intervals are similarly transformed. This option affects how results are displayed, not how they are estimated or stored. irr may be specified at estimation or when replaying previously estimated results.

Max options

maximize_options: difficult, technique(*algorithm_spec*), iterate(*#*), [no]log, trace, gradient, showstep, hessian, shownrtolerance, tolerance(*#*), ltolerance(*#*), gtolerance(*#*), nrtolerance(*#*), nonrtolerance, from(*init_specs*); see [R] **maximize**. These options are seldom used.

Setting the optimization type to technique(bhhh) resets the default *vcetype* to vce(opg).

Remarks

The basic idea of Poisson regression was outlined by Coleman (1964, 378–379). See Cameron and Trivedi (1998) and Feller (1968, 156–164) for information about the Poisson distribution. See Cameron and Trivedi (1998), Long (1997, chap. 8), Long and Freese (2006, chap. 8), McNeil (1996, chap. 6), and Selvin (2004, chap. 9) for an introduction to Poisson regression. Also see Selvin (2004, chap. 5) for a discussion of the analysis of spatial distributions, which includes a discussion of the Poisson distribution. An early example of Poisson regression was Cochran (1940).

Poisson regression fits models of the number of occurrences (counts) of an event. The Poisson distribution has been applied to diverse events, such as the number of soldiers kicked to death by horses in the Prussian army (von Bortkewitsch 1898); the pattern of hits by buzz bombs launched against London during World War II (Clarke 1946); telephone connections to a wrong number (Thorndike 1926); and disease incidence, typically with respect to time, but occasionally with respect to space. The basic assumptions are as follows:

1. There is a quantity called the *incidence rate* that is the rate at which events occur. Examples are 5 per second, 20 per 1,000 person-years, 17 per square meter, and 38 per cubic centimeter.
2. The incidence rate can be multiplied by *exposure* to obtain the expected number of observed events. For example, a rate of 5 per second multiplied by 30 seconds means that 150 events are expected; a rate of 20 per 1,000 person-years multiplied by 2,000 person-years means that 40 events are expected; and so on.
3. Over very small exposures ϵ, the probability of finding more than one event is small compared with ϵ.
4. Nonoverlapping exposures are mutually independent.

With these assumptions, to find the probability of k events in an exposure of size E, you divide E into n subintervals E_1, E_2, ..., E_n, and approximate the answer as the binomial probability of observing k successes in n trials. If you let $n \rightarrow \infty$, you obtain the Poisson distribution.

In the Poisson regression model, the incidence rate for the jth observation is assumed to be given by

$$r_j = e^{\beta_0+\beta_1 x_{1,j}+\cdots+\beta_k x_{k,j}}$$

If E_j is the exposure, the expected number of events, C_j, will be

$$\begin{aligned} C_j &= E_j e^{\beta_0+\beta_1 x_{1,j}+\cdots+\beta_k x_{k,j}} \\ &= e^{\ln(E_j)+\beta_0+\beta_1 x_{1,j}+\cdots+\beta_k x_{k,j}} \end{aligned}$$

This model is fitted by `poisson`. Without the `exposure()` or `offset()` options, E_j is assumed to be 1 (equivalent to assuming that exposure is unknown), and controlling for exposure, if necessary, is your responsibility.

Comparing rates is most easily done by calculating *incidence-rate ratios* (IRRs). For instance, what is the relative incidence rate of chromosome interchanges in cells as the intensity of radiation increases; the relative incidence rate of telephone connections to a wrong number as load increases; or the relative incidence rate of deaths due to cancer for females relative to males? That is, you want to hold all the x's in the model constant except one, say, the ith. The IRR for a one-unit change in x_i is

$$\frac{e^{\ln(E)+\beta_1 x_1+\cdots+\beta_i(x_i+1)+\cdots+\beta_k x_k}}{e^{\ln(E)+\beta_1 x_1+\cdots+\beta_i x_i+\cdots+\beta_k x_k}} = e^{\beta_i}$$

More generally, the IRR for a Δx_i change in x_i is $e^{\beta_i \Delta x_i}$. The `lincom` command can be used after `poisson` to display incidence-rate ratios for any group relative to another; see [R] **lincom**.

▷ Example 1

Chatterjee and Hadi (2006, 162) give the number of injury incidents and the proportion of flights for each airline out of the total number of flights from New York for nine major U.S. airlines in one year:

```
. use http://www.stata-press.com/data/r10/airline
. list
```

	airline	injuries	n	XYZowned
1.	1	11	0.0950	1
2.	2	7	0.1920	0
3.	3	7	0.0750	0
4.	4	19	0.2078	0
5.	5	9	0.1382	0
6.	6	4	0.0540	1
7.	7	3	0.1292	0
8.	8	1	0.0503	0
9.	9	3	0.0629	1

To their data, we have added a fictional variable, `XYZowned`. We will imagine that an accusation is made that the airlines owned by XYZ Company have a higher injury rate.

```
. poisson injuries XYZowned, exposure(n) irr
Iteration 0:   log likelihood = -23.027197
Iteration 1:   log likelihood = -23.027177
Iteration 2:   log likelihood = -23.027177
Poisson regression                                Number of obs   =          9
                                                  LR chi2(1)      =       1.77
                                                  Prob > chi2     =     0.1836
Log likelihood = -23.027177                       Pseudo R2       =     0.0370

------------------------------------------------------------------------------
    injuries |        IRR   Std. Err.      z    P>|z|     [95% Conf. Interval]
-------------+----------------------------------------------------------------
    XYZowned |   1.463467    .406872     1.37   0.171     .8486578    2.523675
           n | (exposure)
------------------------------------------------------------------------------
```

We specified `irr` to see the IRRs rather than the underlying coefficients. We estimate that XYZ Airlines' injury rate is 1.46 times larger than that for other airlines, but the 95% confidence interval is .85 to 2.52; we cannot even reject the hypothesis that XYZ Airlines has a lower injury rate.

◁

❑ Technical Note

In example 1, we assumed that each airline's exposure was proportional to its fraction of flights out of New York. What if "large" airlines, however, also used larger planes, and so had even more passengers than would be expected, given this measure of exposure? A better measure would be each airline's fraction of passengers on flights out of New York, a number that we do not have. Even so, we suppose that `n` represents this number to some extent, so a better estimate of the effect might be

```
. gen lnN=ln(n)
. poisson injuries XYZowned lnN
Iteration 0:   log likelihood = -22.333875
Iteration 1:   log likelihood = -22.332276
Iteration 2:   log likelihood = -22.332276
Poisson regression                                Number of obs   =          9
                                                  LR chi2(2)      =      19.15
                                                  Prob > chi2     =     0.0001
Log likelihood = -22.332276                       Pseudo R2       =     0.3001

------------------------------------------------------------------------------
    injuries |      Coef.   Std. Err.      z    P>|z|     [95% Conf. Interval]
-------------+----------------------------------------------------------------
    XYZowned |   .6840667   .3895877     1.76   0.079    -.0795111    1.447645
         lnN |   1.424169   .3725155     3.82   0.000     .6940517    2.154285
       _cons |   4.863891   .7090501     6.86   0.000     3.474178    6.253603
------------------------------------------------------------------------------
```

Here rather than specifying the `exposure()` option, we explicitly included the variable that would normalize for exposure in the model. We did not specify the `irr` option, so we see coefficients rather than IRRs. We started with the model

$$\text{rate} = e^{\beta_0+\beta_1\text{XYZowned}}$$

The observed counts are therefore

$$\text{count} = ne^{\beta_0+\beta_1\text{XYZowned}} = e^{\ln(n)+\beta_0+\beta_1\text{XYZowned}}$$

which amounts to constraining the coefficient on ln(n) to 1. This is what was estimated when we specified the `exposure(n)` option. In the above model, we included the normalizing exposure ourselves and, rather than constraining the coefficient to be 1, estimated the coefficient.

The estimated coefficient is 1.42, a respectable distance away from 1, and is consistent with our speculation that larger airlines also use larger airplanes. With this small amount of data, however, we also have a wide confidence interval that includes 1.

Our estimated *coefficient* on `XYZowned` is now .684, and the implied IRR is $e^{.684} \approx 1.98$ (which we could also see by typing `poisson, irr`). The 95% confidence interval for the coefficient still includes 0 (the interval for the IRR includes 1), so although the point estimate is now larger, we still cannot be certain of our results.

Our expert opinion would be that, although there is not enough evidence to support the charge, there is enough evidence to justify collecting more data.

❑

▷ Example 2

In a famous age-specific study of coronary disease deaths among male British doctors, Doll and Hill (1966) reported the following data (reprinted in Rothman and Greenland 1998, 259):

	Smokers		Nonsmokers	
Age	Deaths	Person-years	Deaths	Person-years
35–44	32	52,407	2	18,790
45–54	104	43,248	12	10,673
55–64	206	28,612	28	5,710
65–74	186	12,663	28	2,585
75–84	102	5,317	31	1,462

The first step is to enter these data into Stata, which we have done:

```
. use http://www.stata-press.com/data/r10/dollhill3, clear
. list

     +-----------------------------------+
     | agecat   smokes   deaths   pyears |
     |-----------------------------------|
  1. |      1        1       32   52,407 |
  2. |      2        1      104   43,248 |
  3. |      3        1      206   28,612 |
  4. |      4        1      186   12,663 |
  5. |      5        1      102    5,317 |
     |-----------------------------------|
  6. |      1        0        2   18,790 |
  7. |      2        0       12   10,673 |
  8. |      3        0       28    5,710 |
  9. |      4        0       28    2,585 |
 10. |      5        0       31    1,462 |
     +-----------------------------------+
```

`agecat` 1 corresponds to 35–44, `agecat` 2 to 45–54, and so on. The most "natural" analysis of these data would begin by introducing indicator variables for each age category and one indicator for smoking:

```
. tab agecat, gen(a)

     agecat |      Freq.     Percent        Cum.
------------+-----------------------------------
          1 |          2       20.00       20.00
          2 |          2       20.00       40.00
          3 |          2       20.00       60.00
          4 |          2       20.00       80.00
          5 |          2       20.00      100.00
------------+-----------------------------------
      Total |         10      100.00

. poisson deaths smokes a2-a5, exposure(pyears) irr
Iteration 0:   log likelihood = -33.823284
Iteration 1:   log likelihood = -33.600471
Iteration 2:   log likelihood = -33.600153
Iteration 3:   log likelihood = -33.600153
Poisson regression                                Number of obs   =         10
                                                  LR chi2(5)      =     922.93
                                                  Prob > chi2     =     0.0000
Log likelihood = -33.600153                       Pseudo R2       =     0.9321

------------------------------------------------------------------------------
      deaths |        IRR   Std. Err.      z    P>|z|     [95% Conf. Interval]
-------------+----------------------------------------------------------------
      smokes |   1.425519   .1530638     3.30   0.001     1.154984    1.759421
          a2 |   4.410584   .8605197     7.61   0.000     3.009011    6.464997
          a3 |    13.8392   2.542638    14.30   0.000     9.654328    19.83809
          a4 |   28.51678   5.269878    18.13   0.000     19.85177    40.96395
          a5 |   40.45121   7.775511    19.25   0.000     27.75326    58.95885
      pyears | (exposure)
------------------------------------------------------------------------------
```

In the above, we began by using `tabulate` to create the indicator variables. `tabulate` created `a1` equal to 1 when `agecat` = 1 and 0, otherwise; `a2` equal to 1 when `agecat` = 2 and 0, otherwise; and so on. See [U] **25 Dealing with categorical variables**.

We then fitted our model, specifying `irr` to obtain IRRs. We estimate that smokers have 1.43 times the mortality rate of nonsmokers. See, however, example 1 in [R] **poisson postestimation**. ◁

Siméon-Denis Poisson (1781–1840) was a French mathematician and physicist who contributed to several fields: his name is perpetuated in Poisson brackets, Poisson's constant, Poisson's differential equation, Poisson's integral, and Poisson's ratio. Among many other results, he produced a version of the law of large numbers. His rather misleadingly titled *Recherches sur la probabilité des jugements* embraces a complete treatise on probability, as the subtitle indicates, including what is now known as the Poisson distribution. That, however, was discovered earlier by the Huguenot–British mathematician Abraham de Moivre (1667–1754).

(Continued on next page)

Saved Results

poisson saves the following in e():

Scalars

e(N)	number of observations	e(ll_0)	log likelihood, constant-only model
e(k)	number of parameters	e(N_clust)	number of clusters
e(k_eq)	number of equations	e(chi2)	χ^2
e(k_eq_model)	number of equations in model Wald test	e(p)	significance
		e(rank)	rank of e(V)
e(k_dv)	number of dependent variables	e(ic)	number of iterations
e(df_m)	model degrees of freedom	e(rc)	return code
e(r2_p)	pseudo-R-squared	e(converged)	1 if converged, 0 otherwise
e(ll)	log likelihood		

Macros

e(cmd)	poisson	e(vce)	*vcetype* specified in vce()
e(cmdline)	command as typed	e(vcetype)	title used to label Std. Err.
e(depvar)	name of dependent variable	e(opt)	type of optimization
e(wtype)	weight type	e(ml_method)	type of ml method
e(wexp)	weight expression	e(user)	name of likelihood-evaluator program
e(title)	title in estimation output	e(technique)	maximization technique
e(clustvar)	name of cluster variable	e(crittype)	optimization criterion
e(offset)	offset	e(properties)	b V
e(chi2type)	Wald or LR; type of model χ^2 test	e(estat_cmd)	program used to implement estat
		e(predict)	program used to implement predict

Matrices

e(b)	coefficient vector	e(V)	variance–covariance matrix of the estimators
e(ilog)	iteration log (up to 20 iterations)		
e(gradient)	gradient vector		

Functions

e(sample)	marks estimation sample

Methods and Formulas

poisson is implemented as an ado-file.

The log likelihood (with weights w_j and offsets) is given by

$$\Pr(Y = y) = \frac{e^{-\lambda}\lambda^y}{y!}$$

$$\xi_j = \mathbf{x}_j\boldsymbol{\beta} + \text{offset}_j$$

$$f(y_j) = \frac{e^{-\exp(\xi_j)}e^{\xi_j y_j}}{y_j!}$$

$$\ln L = \sum_{j=1}^{n} w_j \left\{ -e^{\xi_j} + \xi_j y_j - \ln(y_j!) \right\}$$

References

Bru, B. 2001. Siméon-Denis Poisson. In *Statisticians of the Centuries*, ed. C. C. Heyde and E. Seneta, 123–126. New York: Springer.

Cameron, A. C., and P. K. Trivedi. 1998. *Regression Analysis of Count Data*. Cambridge: Cambridge University Press.

Chatterjee, S., and A. S. Hadi. 2006. *Regression Analysis by Example*. 4th ed. New York: Wiley.

Clarke, R. D. 1946. An application of the Poisson distribution. *Journal of the Institute of Actuaries* 22: 48.

Cochran, W. G. 1940. The analysis of variance when experimental errors follow the Poisson or binomial laws. *Annals of Mathematical Statistics* 11: 335–347. Reprinted as paper 22 in Cochran (1982).

——.1982 *Contributions to Statistics*. New York: Wiley.

Coleman, J. S. 1964. *Introduction to Mathematical Sociology*. New York: Free Press.

Doll, R., and A. B. Hill. 1966. Mortality of British doctors in relation to smoking; observations on coronary thrombosis. In *Epidemiological Approaches to the Study of Cancer and Other Chronic Diseases*, ed. W. Haenszel. *National Cancer Institute Monograph* 19: 204–268.

Feller, W. 1968. *An Introduction to Probability Theory and Its Applications*, vol. 1. 3rd ed. New York: Wiley.

Hilbe, J. 1998. sg91: Robust variance estimators for MLE Poisson and negative binomial regression. *Stata Technical Bulletin* 45: 26–28. Reprinted in *Stata Technical Bulletin Reprints*, vol. 8, pp. 177–180.

——.1999. sg102: Zero-truncated Poisson and negative binomial regression. *Stata Technical Bulletin* 47: 37–40. Reprinted in *Stata Technical Bulletin Reprints*, vol. 8, pp. 233–236.

Hilbe, J., and D. H. Judson. 1998. sg94: Right, left, and uncensored Poisson regression. *Stata Technical Bulletin* 46: 18–20. Reprinted in *Stata Technical Bulletin Reprints*, vol. 8, pp. 186–189.

Long, J. S. 1997. *Regression Models for Categorical and Limited Dependent Variables*. Thousand Oaks, CA: Sage.

Long, J. S., and J. Freese. 2001. Predicted probabilities for count models. *Stata Journal* 1: 51–57.

——. 2006. *Regression Models for Categorical Dependent Variables Using Stata*. 2nd ed. College Station, TX: Stata Press.

McNeil, D. 1996. *Epidemiological Research Methods*. Chichester, UK: Wiley.

Miranda, A., and S. Rabe-Hesketh. 2006. Maximum likelihood estimation of endogenous switching and sample selection models for binary, ordinal, and count variables. *Stata Journal* 6: 285–308.

Newman, S. C. 2001. *Biostatistical Methods in Epidemiology*. New York: Wiley.

Poisson, S. D. 1837. *Recherches sur la probabilité des jugements en matière criminelle et en matière civile, précédés des règles générales du calcul des probabilités*. Paris: Bachelier.

Rodríguez, G. 1993. sbe10: An improvement to poisson. *Stata Technical Bulletin* 11: 11–14. Reprinted in *Stata Technical Bulletin Reprints*, vol. 2, pp. 94–98.

Rogers, W. H. 1991. sbe1: Poisson regression with rates. *Stata Technical Bulletin* 1: 11–12. Reprinted in *Stata Technical Bulletin Reprints*, vol. 1, pp. 62–64.

Rothman, K. J., and S. Greenland. 1998. *Modern Epidemiology*. 2nd ed. Philadelphia: Lippincott–Raven.

Rutherford, E., J. Chadwick, and C. D. Ellis. 1930. *Radiations from Radioactive Substances*. Cambridge: Cambridge University Press.

Schonlau, M. 2005. Boosted regression (boosting): An introductory tutorial and a Stata plugin. *Stata Journal* 5: 330–354.

Selvin, S. 2004. *Statistical Analysis of Epidemiologic Data*. 3rd ed. New York: Oxford University Press.

Thorndike, F. 1926. Applications of Poisson's probability summation. *Bell System Technical Journal* 5: 604–624.

Tobias, A., and M. J. Campbell. 1998. sts13: Time-series regression for counts allowing for autocorrelation. *Stata Technical Bulletin* 46: 33–37. Reprinted in *Stata Technical Bulletin Reprints*, vol. 8, pp. 291–296.

von Bortkewitsch, L. 1898. *Das Gesetz der Kleinen Zahlen*. Leipzig: Teubner.

Also See

[R] **poisson postestimation** — Postestimation tools for poisson

[R] **glm** — Generalized linear models

[R] **nbreg** — Negative binomial regression

[R] **zip** — Zero-inflated Poisson regression

[R] **ztp** — Zero-truncated Poisson regression

[SVY] **svy estimation** — Estimation commands for survey data

[XT] **xtpoisson** — Fixed-effects, random-effects, and population-averaged Poisson models

[U] **20 Estimation and postestimation commands**

Title

poisson postestimation — Postestimation tools for poisson

Description

The following postestimation command is of special interest after poisson:

command	description
estat gof	goodness-of-fit test

estat gof is not appropriate after the svy prefix. For information about estat gof, see below.

The following standard postestimation commands are also available:

command	description
adjust[1]	adjusted predictions of $\mathbf{x}\beta$ or $\exp(\mathbf{x}\beta)$
estat	AIC, BIC, VCE, and estimation sample summary
estat (svy)	postestimation statistics for survey data
estimates	cataloging estimation results
lincom	point estimates, standard errors, testing, and inference for linear combinations of coefficients
linktest	link test for model specification
lrtest[2]	likelihood-ratio test
mfx	marginal effects or elasticities
nlcom	point estimates, standard errors, testing, and inference for nonlinear combinations of coefficients
predict	predictions, residuals, influence statistics, and other diagnostic measures
predictnl	point estimates, standard errors, testing, and inference for generalized predictions
suest	seemingly unrelated estimation
test	Wald tests for simple and composite linear hypotheses
testnl	Wald tests of nonlinear hypotheses

[1] adjust is not appropriate with time-series operators.
[2] lrtest is not appropriate with svy estimation results.

See the corresponding entries in the *Stata Base Reference Manual* for details, but see [SVY] **estat** for details about estat (svy).

Special-interest postestimation command

estat gof performs a goodness-of-fit test of the model. The default is the deviance statistic; specifying option pearson will give the Pearson statistic. If the test is significant, the Poisson regression model is inappropriate. Then you could try a negative binomial model; see [R] **nbreg**.

Syntax for predict

```
predict [type] newvar [if] [in] [, statistic nooffset ]
```

statistic	description
Main	
n	number of events; the default
ir	incidence rate
xb	linear prediction
stdp	standard error of the linear prediction
score	first derivative of the log likelihood with respect to $\mathbf{x}_j\boldsymbol{\beta}$

These statistics are available both in and out of sample; type `predict ... if e(sample) ...` if wanted only for the estimation sample.

Options for predict

Main

`n`, the default, calculates the predicted number of events, which is $\exp(\mathbf{x}_j\boldsymbol{\beta})$ if neither `offset()` nor `exposure()` was specified when the model was fitted; $\exp(\mathbf{x}_j\boldsymbol{\beta} + \text{offset}_j)$ if `offset()` was specified; or $\exp(\mathbf{x}_j\boldsymbol{\beta}) \times \text{exposure}_j$ if `exposure()` was specified.

`ir` calculates the incidence rate, $\exp(\mathbf{x}_j\boldsymbol{\beta})$, which is the predicted number of events when exposure is 1. Specifying `ir` is equivalent to specifying `n` when neither `offset()` nor `exposure()` was specified when the model was fitted.

`xb` calculates the linear prediction, which is $\mathbf{x}_j\boldsymbol{\beta}$ if neither `offset()` nor `exposure()` was specified; $\mathbf{x}_j\boldsymbol{\beta} + \text{offset}_j$ if `offset()` was specified; or $\mathbf{x}_j\boldsymbol{\beta} + \ln(\text{exposure}_j)$ if `exposure()` was specified; see `nooffset` below.

`stdp` calculates the standard error of the linear prediction.

`score` calculates the equation-level score, $\partial \ln L/\partial(\mathbf{x}_j\boldsymbol{\beta})$.

`nooffset` is relevant only if you specified `offset()` or `exposure()` when you fitted the model. It modifies the calculations made by `predict` so that they ignore the offset or exposure variable; the linear prediction is treated as $\mathbf{x}_j\boldsymbol{\beta}$ rather than as $\mathbf{x}_j\boldsymbol{\beta} + \text{offset}_j$ or $\mathbf{x}_j\boldsymbol{\beta} + \ln(\text{exposure}_j)$. Specifying `predict ..., nooffset` is equivalent to specifying `predict ..., ir`.

Syntax for estat gof

```
estat gof [, pearson ]
```

Option for estat gof

`pearson` requests that `estat gof` calculate the Pearson statistic rather than the deviance statistic.

Remarks

▷ Example 1

Continuing with example 2 of [R] **poisson**, we use `estat gof` to determine whether the model fits the data well.

```
. estat gof

         Goodness-of-fit chi2  =  12.13244
         Prob > chi2(4)        =    0.0164
```

The goodness-of-fit χ^2 tells us that, given the model, we can reject the hypothesis that these data are Poisson distributed at the 1.64% significance level.

So let us now back up and be more careful. We can most easily obtain the incidence-rate ratios within age categories by using `ir`; see [ST] **epitab**:

```
. ir deaths smokes pyears, by(agecat) nocrude nohet

          agecat |       IRR       [95% Conf. Interval]   M-H Weight
-----------------+-------------------------------------------------
               1 |   5.736638       1.463557   49.40468     1.472169 (exact)
               2 |   2.138812       1.173714   4.272545     9.624747 (exact)
               3 |    1.46824       .9863624   2.264107     23.34176 (exact)
               4 |    1.35606       .9081925   2.096412     23.25315 (exact)
               5 |   .9047304       .6000757   1.399687     24.31435 (exact)
-----------------+-------------------------------------------------
     M-H combined |   1.424682       1.154703   1.757784
```

We find that the mortality incidence ratios are greatly different within age category, being highest for the youngest categories and actually dropping below 1 for the oldest. (In the last case, we might argue that those who smoke and who have not died by age 75 are self-selected to be particularly robust.)

Seeing this, we will now parameterize the smoking effects separately for each age category, although we will begin by combining age categories 3 and 4:

```
. gen sa1 = smokes*(agecat==1)
. gen sa2 = smokes*(agecat==2)
. gen sa34 = smokes*(agecat==3 | agecat==4)
. gen sa5 = smokes*(agecat==5)
```

(Continued on next page)

```
. poisson deaths sa1 sa2 sa34 sa5 a2-a5, exposure(pyears) irr
Iteration 0:   log likelihood = -31.635422
Iteration 1:   log likelihood = -27.788819
Iteration 2:   log likelihood = -27.573604
Iteration 3:   log likelihood = -27.572645
Iteration 4:   log likelihood = -27.572645
Poisson regression                                Number of obs   =         10
                                                  LR chi2(8)      =     934.99
                                                  Prob > chi2     =     0.0000
Log likelihood = -27.572645                       Pseudo R2       =     0.9443

------------------------------------------------------------------------------
      deaths |        IRR   Std. Err.      z    P>|z|     [95% Conf. Interval]
-------------+----------------------------------------------------------------
         sa1 |   5.736638   4.181257     2.40   0.017     1.374811    23.93711
         sa2 |   2.138812   .6520701     2.49   0.013     1.176691    3.887609
        sa34 |   1.412229   .2017485     2.42   0.016     1.067343    1.868557
         sa5 |   .9047304   .1855513    -0.49   0.625     .6052658     1.35236
          a2 |    10.5631   8.067702     3.09   0.002     2.364153    47.19624
          a3 |     47.671    34.3741     5.36   0.000     11.60056    195.8978
          a4 |   98.22766   70.85013     6.36   0.000     23.89324    403.8245
          a5 |     199.21   145.3357     7.26   0.000     47.67694     832.365
      pyears | (exposure)
------------------------------------------------------------------------------

. estat gof
         Goodness-of-fit chi2  =  .0774185
         Prob > chi2(1)        =    0.7808
```

The goodness-of-fit χ^2 is now small; we are no longer running roughshod over the data. Let us now consider simplifying the model. The point estimate of the incidence-rate ratio for smoking in age category 1 is much larger than that for smoking in age category 2, but the confidence interval for `sa1` is similarly wide. Is the difference real?

```
. test sa1=sa2
 ( 1)  [deaths]sa1 - [deaths]sa2 = 0
           chi2(  1) =    1.56
         Prob > chi2 =    0.2117
```

The point estimates may be far apart, but there is insufficient data, and we may be observing random differences. With that success, might we also combine the smokers in age categories 3 and 4 with those in 1 and 2?

```
. test sa34=sa2, accum
 ( 1)  [deaths]sa1 - [deaths]sa2 = 0
 ( 2) - [deaths]sa2 + [deaths]sa34 = 0
           chi2(  2) =    4.73
         Prob > chi2 =    0.0938
```

Combining age categories 1–4 may be overdoing it—the 9.38% significance level is enough to stop us, although others may disagree.

Thus we now fit our final model:

```
. generate sa12 = (sa1|sa2)
. poisson deaths sa12 sa34 sa5 a2-a5, exposure(pyears) irr
Iteration 0:   log likelihood = -31.967194
Iteration 1:   log likelihood = -28.524666
Iteration 2:   log likelihood = -28.514535
Iteration 3:   log likelihood = -28.514535
Poisson regression                                Number of obs   =         10
                                                  LR chi2(7)      =     933.11
                                                  Prob > chi2     =     0.0000
Log likelihood = -28.514535                       Pseudo R2       =     0.9424

------------------------------------------------------------------------------
      deaths |        IRR   Std. Err.      z    P>|z|     [95% Conf. Interval]
-------------+----------------------------------------------------------------
        sa12 |   2.636259   .7408403     3.45   0.001     1.519791    4.572907
        sa34 |   1.412229   .2017485     2.42   0.016     1.067343    1.868557
         sa5 |   .9047304   .1855513    -0.49   0.625     .6052658     1.35236
          a2 |   4.294559   .8385329     7.46   0.000     2.928987    6.296797
          a3 |   23.42263   7.787716     9.49   0.000     12.20738    44.94164
          a4 |   48.26309   16.06939    11.64   0.000     25.13068    92.68856
          a5 |   97.87965   34.30881    13.08   0.000     49.24123     194.561
      pyears | (exposure)
------------------------------------------------------------------------------
```

The above strikes us as a fair representation of the data.

◁

Methods and Formulas

All postestimation commands listed above are implemented as ado-files.

In the following, we use the same notation as in [R] **poisson**.

The equation-level scores are given by

$$\text{score}(\mathbf{x}\boldsymbol{\beta})_j = y_j - e^{\xi_j}$$

The deviance (D) and Pearson (P) goodness-of-fit statistics are given by

$$\ln L_{\max} = \sum_{j=1}^{n} w_j \left[-y_j\{\ln(y_j) - 1\} - \ln(y_j!)\right]$$

$$\chi^2_D = -2\{\ln L - \ln L_{\max}\}$$

$$\chi^2_P = \sum_{j=1}^{n} \frac{w_j(y_j - e^{\xi_j})^2}{e^{\xi_j}}$$

Also See

[R] **poisson** — Poisson regression

[U] **20 Estimation and postestimation commands**

Title

predict — Obtain predictions, residuals, etc., after estimation

Syntax

After single-equation (SE) *models*

predict [*type*] *newvar* [*if*] [*in*] [, *single_options*]

After multiple-equation (ME) *models*

predict [*type*] *newvar* [*if*] [*in*] [, *multiple_options*]

predict [*type*] { *stub** | *newvar*$_1$ ... *newvar*$_q$ } [*if*] [*in*] , scores

single_options	description
Main	
xb	calculate linear prediction
stdp	calculate standard error of the prediction
score	calculate first derivative of the log likelihood with respect to $\mathbf{x}_j\boldsymbol{\beta}$
Options	
nooffset	ignore any offset() or exposure() variable
other_options	command-specific options

multiple_options	description
Main	
equation(*eqno*[,*eqno*])	specify equations
xb	calculate linear prediction
stdp	calculate standard error of the prediction
stddp	calculate the difference in linear predictions
Options	
nooffset	ignore any offset() or exposure() variable
other_options	command-specific options

Description

predict calculates predictions, residuals, influence statistics, and the like after estimation. Exactly what predict can do is determined by the previous estimation command; command-specific options are documented with each estimation command. Regardless of command-specific options, the actions of predict share certain similarities across estimation commands:

1) predict *newvar* creates *newvar* containing "predicted values"—numbers related to the $E(y_j|\mathbf{x}_j)$. For instance, after linear regression, predict *newvar* creates $\mathbf{x}_j\mathbf{b}$ and, after probit, creates the probability $\Phi(\mathbf{x}_j\mathbf{b})$.

2) `predict` *newvar*, `xb` creates *newvar* containing $\mathbf{x}_j\mathbf{b}$. This may be the same result as (1) (e.g., linear regression) or different (e.g., probit), but regardless, option `xb` is allowed.

3) `predict` *newvar*, `stdp` creates *newvar* containing the standard error of the linear prediction $\mathbf{x}_j\mathbf{b}$.

4) `predict` *newvar*, *other_options* may create *newvar* containing other useful quantities; see `help` or the *Reference* manual entry for the particular estimation command to find out about other available options.

5) `nooffset` added to any of the above commands requests that the calculation ignore any offset or exposure variable specified by including the `offset(`*varname*$_o$`)` or `exposure(`*varname*$_e$`)` option when you fitted the model.

`predict` can be used to make in-sample or out-of-sample predictions:

6) `predict` calculates the requested statistic for all possible observations, whether they were used in fitting the model or not. `predict` does this for standard options (1)–(3) and generally does this for estimator-specific options (4).

7) `predict` *newvar* `if e(sample),` ... restricts the prediction to the estimation subsample.

8) Some statistics make sense only with respect to the estimation subsample. In such cases, the calculation is automatically restricted to the estimation subsample, and the documentation for the specific option states this. Even so, you can still specify `if e(sample)` if you are uncertain.

9) `predict` can make out-of-sample predictions even using other datasets. In particular, you can

```
. use ds1
. (fit a model)
. use two                           /* another dataset         */
. predict yhat, ...                 /* fill in the predictions */
```

Options

Main

`xb` calculates the linear prediction from the fitted model. All models can be thought of as estimating a set of parameters b_1, b_2, ..., b_k, and the linear prediction is $\widehat{y}_j = b_1x_{1j} + b_2x_{2j} + \cdots + b_kx_{kj}$, often written in matrix notation as $\widehat{\mathbf{y}}_j = \mathbf{x}_j\mathbf{b}$. For linear regression, the values $\widehat{y}_j$ are called the predicted values or, for out-of-sample predictions, the forecast. For logit and probit, for example, $\widehat{y}_j$ is called the logit or probit index.

x_{1j}, x_{2j}, ..., x_{kj} are obtained from the data currently in memory and do not necessarily correspond to the data on the independent variables used to fit the model (obtaining b_1, b_2, ..., b_k).

`stdp` calculates the standard error of the linear prediction. Here the prediction means the same thing as the "index", namely $\mathbf{x}_j\mathbf{b}$. The statistic produced by `stdp` can be thought of as the standard error of the predicted expected value, or mean index, for the observation's covariate pattern. The standard error of the prediction is also commonly referred to as the standard error of the fitted value. The calculation can be made in or out of sample.

`stddp` is allowed only after you have previously fitted a multiple-equation model. The standard error of the difference in linear predictions $(\mathbf{x}_{1j}\mathbf{b} - \mathbf{x}_{2j}\mathbf{b})$ between equations 1 and 2 is calculated. This option requires that `equation(`*eqno*$_1$`,`*eqno*$_2$`)` be specified.

`score` calculates the equation-level score, $\partial \ln L/\partial(\mathbf{x}_j\boldsymbol{\beta})$. Here $\ln L$ refers to the log-likelihood function.

`scores` is the ME model equivalent of the `score` option, resulting in multiple equation-level score variables. An equation-level score variable is created for each equation in the model; ancillary parameters—such as $\ln\sigma$ and $\operatorname{atanh}\rho$—make up separate equations.

`equation(`*eqno*[`,`*eqno*]`)`—synonym `outcome()`—is relevant only when you have previously fitted a multiple-equation model. It specifies the equation to which you are referring.

`equation()` is typically filled in with one *eqno*—it would be filled in that way with options `xb` and `stdp`, for instance. `equation(#1)` would mean the calculation is to be made for the first equation, `equation(#2)` would mean the second, and so on. You could also refer to the equations by their names. `equation(income)` would refer to the equation named income and `equation(hours)` to the equation named hours.

If you do not specify `equation()`, results are the same as if you specified `equation(#1)`.

Other statistics, such as `stddp`, refer to between-equation concepts. In those cases, you might specify `equation(#1,#2)` or `equation(income,hours)`. When two equations must be specified, `equation()` is required.

Options

`nooffset` may be combined with most statistics and specifies that the calculation should be made, ignoring any offset or exposure variable specified when the model was fitted.

This option is available, even if it is not documented for `predict` after a specific command. If neither the `offset(`*varname*$_o$`)` option nor the `exposure(`*varname*$_e$`)` option was specified when the model was fitted, specifying `nooffset` does nothing.

other_options refers to command-specific options that are documented with each command.

Remarks

Remarks are presented under the following headings:

Estimation-sample predictions
Out-of-sample predictions
Residuals
Single-equation (SE) models
SE model scores
Multiple-equation (ME) models
ME model scores

Most of the examples are presented using linear regression, but the general syntax is applicable to all estimators.

You can think of any estimation command as estimating a set of coefficients $b_1, b_2, \ldots, b_k$ corresponding to the variables $x_1, x_2, \ldots, x_k$, along with a (possibly empty) set of ancillary statistics $\gamma_1, \gamma_2, \ldots, \gamma_m$. All estimation commands save the b_is and γ_is. `predict` accesses that saved information and combines it with the data currently in memory to make various calculations. For instance, `predict` can calculate the linear prediction, $\widehat{y}_j = b_1x_{1j} + b_2x_{2j} + \cdots + b_kx_{kj}$. The data on which `predict` makes the calculation can be the same data used to fit the model or a different dataset—it does not matter. `predict` uses the saved parameter estimates from the model, obtains the corresponding values of x for each observation in the data, and then combines them to produce the desired result.

Estimation-sample predictions

▷ Example 1

We have a 74-observation dataset on automobiles, including the mileage rating (mpg), the car's weight (weight), and whether the car is foreign (foreign). We fit the model

```
. use http://www.stata-press.com/data/r10/auto
(1978 Automobile Data)

. regress mpg weight if foreign

      Source |       SS       df       MS              Number of obs =      22
-------------+------------------------------           F(  1,    20) =   17.47
       Model |  427.990298     1  427.990298           Prob > F      =  0.0005
    Residual |  489.873338    20  24.4936669           R-squared     =  0.4663
-------------+------------------------------           Adj R-squared =  0.4396
       Total |  917.863636    21  43.7077922           Root MSE      =  4.9491

------------------------------------------------------------------------------
         mpg |      Coef.   Std. Err.      t    P>|t|     [95% Conf. Interval]
-------------+----------------------------------------------------------------
      weight |   -.010426   .0024942    -4.18   0.000    -.0156287   -.0052232
       _cons |    48.9183   5.871851     8.33   0.000     36.66983    61.16676
------------------------------------------------------------------------------
```

If we were to type predict pmpg now, we would obtain the linear predictions for all 74 observations. To obtain the predictions just for the sample on which we fitted the model, we could type

```
. predict pmpg if e(sample)
(option xb assumed; fitted values)
(52 missing values generated)
```

Here e(sample) is true only for foreign cars because we typed if foreign when we fitted the model and because there are no missing values among the relevant variables. If there had been missing values, e(sample) would also account for those.

By the way, the if e(sample) restriction can be used with any Stata command, so we could obtain summary statistics on the estimation sample by typing

```
. summarize if e(sample)
  (output omitted )
```

◁

Out-of-sample predictions

By out-of-sample predictions, we mean predictions extending beyond the estimation sample. In the example above, typing predict pmpg would generate linear predictions using all 74 observations.

predict will work on other datasets, too. You can use a new dataset and type predict to obtain results for that sample.

▷ Example 2

Using the same auto dataset, assume that we wish to fit the model

$$\texttt{mpg} = \beta_1\texttt{weight} + \beta_2\texttt{weight}^2 + \beta_3\texttt{foreign} + \beta_4$$

We first create the weight^2 variable, and then type the `regress` command:

```
. use http://www.stata-press.com/data/r10/auto, clear
(1978 Automobile Data)

. generate weight2=weight^2

. regress mpg weight weight2 foreign

      Source |       SS       df       MS              Number of obs =      74
-------------+------------------------------           F(  3,    70) =   52.25
       Model |  1689.15372     3  563.05124           Prob > F      =  0.0000
    Residual |   754.30574    70  10.7757963           R-squared     =  0.6913
-------------+------------------------------           Adj R-squared =  0.6781
       Total |  2443.45946    73  33.4720474           Root MSE      =  3.2827

------------------------------------------------------------------------------
         mpg |      Coef.   Std. Err.      t    P>|t|     [95% Conf. Interval]
-------------+----------------------------------------------------------------
      weight |  -.0165729   .0039692    -4.18   0.000    -.0244892   -.0086567
     weight2 |   1.59e-06   6.25e-07     2.55   0.013     3.45e-07    2.84e-06
     foreign |    -2.2035   1.059246    -2.08   0.041      -4.3161   -.0909002
       _cons |   56.53884   6.197383     9.12   0.000     44.17855    68.89913
------------------------------------------------------------------------------
```

If we typed `predict pmpg` now, we would obtain predictions for all 74 cars in the current data. Instead, we are going to use a new dataset.

The dataset `newautos.dta` contains the make, weight, and place of manufacture of two cars, the Pontiac Sunbird and the Volvo 260. Let's use the dataset and create the predictions:

```
. use http://www.stata-press.com/data/r10/newautos, clear
(New Automobile Models)

. list

     +------------------------------------+
     |          make   weight    foreign  |
     |------------------------------------|
  1. | Pont. Sunbird     2690   Domestic  |
  2. |     Volvo 260     3170    Foreign  |
     +------------------------------------+

. predict mpg
(option xb assumed; fitted values)
variable weight2 not found
r(111);
```

Things did not work. We typed `predict mpg`, and Stata responded with the message "variable weight2 not found". `predict` can calculate predicted values on a different dataset only if that dataset contains the variables that went into the model. Here our dataset does not contain a variable called `weight2`. `weight2` is just the square of weight, so we can create it and try again:

```
. generate weight2=weight^2

. predict mpg
(option xb assumed; fitted values)

. list

     +---------------------------------------------------------+
     |          make   weight    foreign    weight2        mpg |
     |---------------------------------------------------------|
  1. | Pont. Sunbird     2690   Domestic    7236100   23.47137 |
  2. |     Volvo 260     3170    Foreign   1.00e+07   17.78846 |
     +---------------------------------------------------------+
```

We obtained our predicted values. The Pontiac Sunbird has a predicted mileage rating of 23.5 mpg, whereas the Volvo 260 has a predicted rating of 17.8 mpg.

◁

Residuals

▷ Example 3

With many estimators, `predict` can calculate more than predicted values. With most regression-type estimators, we can, for instance, obtain residuals. Using our regression example, we return to our original data and obtain residuals by typing

```
. use http://www.stata-press.com/data/r10/auto, clear
(1978 Automobile Data)
. generate weight2=weight^2
. regress mpg weight weight2 foreign
  (output omitted )
. predict double resid, residuals
. summarize resid
```

Variable	Obs	Mean	Std. Dev.	Min	Max
resid	74	-1.78e-15	3.214491	-5.636126	13.85172

We could do this without refitting the model. Stata always remembers the last set of estimates, even as we use new datasets.

It was not necessary to type the `double` in `predict double resid, residuals`, but we wanted to remind you that you can specify the type of a variable in front of the variable's name; see [U] **11.4.2 Lists of new variables**. We made the new variable `resid` a `double` rather than the default `float`.

If you want your residuals to have a mean as close to zero as possible, remember to request the extra precision of `double`. If we had not specified `double`, the mean of `resid` would have been roughly 10^{-8} rather than 10^{-14}. Although 10^{-14} sounds more precise than 10^{-8}, the difference really does not matter.

◁

For linear regression, `predict` can also calculate standardized residuals and studentized residuals with the options `rstandard` and `rstudent`; for examples, see [R] **regress postestimation**.

Single-equation (SE) models

If you have not read the discussion above on using `predict` after linear regression, please do so. And `predict`'s default calculation almost always produces a statistic in the same metric as the dependent variable of the fitted model—e.g., predicted counts for Poisson regression. In any case, `xb` can always be specified to obtain the linear prediction.

`predict` can calculate the standard error of the prediction, which is obtained by using the covariance matrix of the estimators.

▷ Example 4

After most binary outcome models (e.g., logistic, logit, probit, cloglog, scobit), predict calculates the probability of a positive outcome if we do not tell it otherwise. We can specify the xb option if we want the linear prediction (also known as the logit or probit index). The odd abbreviation xb is meant to suggest $\mathbf{x}\beta$. In logit and probit models, for example, the predicted probability is $p = F(\mathbf{x}\beta)$, where $F()$ is the logistic or normal cumulative distribution function, respectively.

```
. logistic foreign mpg weight
  (output omitted)
. predict phat
(option pr assumed; Pr(foreign))
. predict idxhat, xb
. summarize foreign phat idxhat
```

Variable	Obs	Mean	Std. Dev.	Min	Max
foreign	74	.2972973	.4601885	0	1
phat	74	.2972973	.3052979	.000729	.8980594
idxhat	74	-1.678202	2.321509	-7.223107	2.175845

Since this is a logit model, we could obtain the predicted probabilities ourselves from the predicted index

```
. generate phat2 = exp(idxhat)/(1+exp(idxhat))
```

but using predict without options is easier.

◁

▷ Example 5

For all models, predict attempts to produce a predicted value in the same metric as the dependent variable of the model. We have seen that for dichotomous outcome models, the default statistic produced by predict is the probability of a success. Similarly, for Poisson regression, the default statistic produced by predict is the predicted count for the dependent variable. You can always specify the xb option to obtain the linear combination of the coefficients with an observation's x values (the inner product of the coefficients and x values). For poisson (without an explicit exposure), this is the natural log of the count.

```
. use http://www.stata-press.com/data/r10/airline, clear
. poisson injuries XYZowned
  (output omitted)
. predict injhat
(option n assumed; predicted number of events)
. predict idx, xb
. generate exp_idx = exp(idx)
. summarize injuries injhat exp_idx idx
```

Variable	Obs	Mean	Std. Dev.	Min	Max
injuries	9	7.111111	5.487359	1	19
injhat	9	7.111111	.8333333	6	7.666667
exp_idx	9	7.111111	.8333333	6	7.666667
idx	9	1.955174	.1225612	1.791759	2.036882

We note that our "hand-computed" prediction of the count (exp_idx) matches what was produced by the default operation of predict.

If our model has an exposure-time variable, we can use predict to obtain the linear prediction with or without the exposure. Let's verify what we are getting by obtaining the linear prediction with and without exposure, transforming these predictions to count predictions and comparing them with the default count prediction from predict. We must remember to multiply by the exposure time when using predict ..., nooffset.

```
. use http://www.stata-press.com/data/r10/airline, clear
. poisson injuries XYZowned, exposure(n)
  (output omitted)
. predict double injhat
(option n assumed; predicted number of events)
. predict double idx, xb
. generate double exp_idx = exp(idx)
. predict double idxn, xb nooffset
. generate double exp_idxn = exp(idxn)*n
. summarize injuries injhat exp_idx exp_idxn idx idxn
```

Variable	Obs	Mean	Std. Dev.	Min	Max
injuries	9	7.111111	5.487359	1	19
injhat	9	7.111111	3.10936	2.919621	12.06158
exp_idx	9	7.111111	3.10936	2.919621	12.06158
exp_idxn	9	7.111111	3.10936	2.919621	12.06158
idx	9	1.869722	.4671044	1.071454	2.490025
idxn	9	4.18814	.1904042	4.061204	4.442013

Looking at the identical means and standard deviations for injhat, exp_idx, and exp_idxn, we see that we can reproduce the default computations of predict for poisson estimations. We have also demonstrated the relationship between the count predictions and the linear predictions with and without exposure.

◁

SE model scores

▷ Example 6

With most maximum likelihood estimators, predict can calculate equation-level scores. The first derivative of the log likelihood with respect to $\mathbf{x}_j\beta$ is the equation-level score.

```
. use http://www.stata-press.com/data/r10/auto, clear
(1978 Automobile Data)
. logistic foreign mpg weight
  (output omitted)
. predict double sc, score
. summarize sc
```

Variable	Obs	Mean	Std. Dev.	Min	Max
sc	74	-1.05e-13	.3533133	-.8760856	.8821309

See [P] **_robust** and [SVY] **variance estimation** for details regarding the role equation-level scores play in linearization-based variance estimators.

◁

❑ Technical Note

`predict` after some estimation commands, such as `regress` and `ivregress`, allows the `score` option as a synonym for the `residuals` option.

❑

Multiple-equation (ME) models

If you have not read the above discussion on using `predict` after SE models, please do so. With the exception of the ability to select specific equations to predict from, the use of `predict` after ME models follows almost the same form that it does for SE models.

▷ Example 7

The details of prediction statistics that are specific to particular ME models are documented with the estimation command. If you are using ME commands that do not have separate discussions on obtaining predictions, read the `predict` section in [R] **mlogit**, even if your interest is not in multinomial logistic regression. As a general introduction to the ME models, we will demonstrate `predict` after `sureg`:

```
. use http://www.stata-press.com/data/r10/auto, clear
(1978 Automobile Data)

. sureg (price foreign displ) (weight foreign length)

Seemingly unrelated regression
----------------------------------------------------------------------
Equation          Obs  Parms        RMSE    "R-sq"       chi2        P
----------------------------------------------------------------------
price              74      2    2202.447    0.4348      45.21   0.0000
weight             74      2    245.5238    0.8988     658.85   0.0000
----------------------------------------------------------------------

------------------------------------------------------------------------------
             |      Coef.   Std. Err.      z    P>|z|     [95% Conf. Interval]
-------------+----------------------------------------------------------------
price        |
     foreign |   3137.894   697.3805     4.50   0.000     1771.054    4504.735
displacement |   23.06938   3.443212     6.70   0.000     16.32081    29.81795
       _cons |   680.8438   859.8142     0.79   0.428    -1004.361    2366.049
-------------+----------------------------------------------------------------
weight       |
     foreign |   -154.883    75.3204    -2.06   0.040    -302.5082   -7.257674
      length |   30.67594   1.531981    20.02   0.000     27.67331    33.67856
       _cons |  -2699.498   302.3912    -8.93   0.000    -3292.173   -2106.822
------------------------------------------------------------------------------
```

`sureg` estimated two equations, one called `price` and the other `weight`; see [R] **sureg**.

```
. predict pred_p, equation(price)
(option xb assumed; fitted values)

. predict pred_w, equation(weight)
(option xb assumed; fitted values)
```

```
. summarize price pred_p weight pred_w

    Variable |       Obs        Mean    Std. Dev.       Min        Max
-------------+--------------------------------------------------------
       price |        74    6165.257    2949.496       3291      15906
      pred_p |        74    6165.257    1678.805    2664.81   10485.33
      weight |        74    3019.459    777.1936       1760       4840
      pred_w |        74    3019.459    726.0468   1501.602   4447.996
```

You may specify the equation by name, as we did above, or by number: `equation(#1)` means the same thing as `equation(price)` in this case.

◁

ME model scores

▷ Example 8

For ME models, `predict` allows you to specify a stub when generating equation-level score variables. `predict` generates new variables using this stub by appending an equation index. Depending upon the command, the index will start with 0 or 1. Here is an example where `predict` starts indexing the score variables with 0.

```
. ologit rep78 mpg weight
  (output omitted)
. predict double sc*, scores
. summarize sc*

    Variable |       Obs        Mean    Std. Dev.       Min        Max
-------------+--------------------------------------------------------
         sc0 |        69    5.81e-18    .5337363  -.9854088    .921433
         sc1 |        69   -1.69e-17     .186919  -.2738537   .9854088
         sc2 |        69    3.30e-17    .4061637  -.5188487   1.130178
         sc3 |        69   -1.41e-17    .5315368  -1.067351   .8194842
         sc4 |        69   -4.83e-18     .360525   -.921433   .6140182
```

Although it involves much more typing, we could also specify the new variable names individually.

```
. predict double (sc_xb sc_1 sc_2 sc_3 sc_4), scores
. summarize sc_*

    Variable |       Obs        Mean    Std. Dev.       Min        Max
-------------+--------------------------------------------------------
       sc_xb |        69    5.81e-18    .5337363  -.9854088    .921433
        sc_1 |        69   -1.69e-17     .186919  -.2738537   .9854088
        sc_2 |        69    3.30e-17    .4061637  -.5188487   1.130178
        sc_3 |        69   -1.41e-17    .5315368  -1.067351   .8194842
        sc_4 |        69   -4.83e-18     .360525   -.921433   .6140182
```

◁

Methods and Formulas

`predict` is implemented as an ado-file.

Denote the previously estimated coefficient vector as $\mathbf{b}$ and its estimated variance matrix as $\mathbf{V}$. `predict` works by recalling various aspects of the model, such as $\mathbf{b}$, and combining that information with the data currently in memory. Let us write $\mathbf{x}_j$ for the jth observation currently in memory.

The *predicted value* (xb option) is defined as $\widehat{y}_j = \mathbf{x}_j\mathbf{b} + \text{offset}_j$

The *standard error of the prediction* (stdp) is defined as $s_{p_j} = \sqrt{\mathbf{x}_j\mathbf{V}\mathbf{x}_j'}$

The *standard error of the difference in linear predictions* between equations 1 and 2 is defined as

$$s_{dp_j} = \{(\mathbf{x}_{1j}, -\mathbf{x}_{2j}, \mathbf{0}, \ldots, \mathbf{0})\, \mathbf{V}\, (\mathbf{x}_{1j}, -\mathbf{x}_{2j}, \mathbf{0}, \ldots, \mathbf{0})'\}^{\frac{1}{2}}$$

See the individual estimation commands for information about calculating command-specific predict statistics.

Also See

[R] **predictnl** — Obtain nonlinear predictions, standard errors, etc., after estimation

[P] **_predict** — Obtain predictions, residuals, etc., after estimation programming command

[U] **20 Estimation and postestimation commands**

Title

predictnl — Obtain nonlinear predictions, standard errors, etc., after estimation

Syntax

`predictnl` [*type*] *newvar* = *pnl_exp* [*if*] [*in*] [, *options*]

options	description
Main	
`se(`*newvar*`)`	create *newvar* containing standard errors
`variance(`*newvar*`)`	create *newvar* containing variances
`wald(`*newvar*`)`	create *newvar* containing the Wald test statistic
`p(`*newvar*`)`	create *newvar* containing the significance level (p-value) of the Wald test
`ci(`*newvars*`)`	create *newvars* containing lower and upper confidence intervals
`level(`*#*`)`	set confidence level; default is `level(95)`
`g(`*stub*`)`	create *stub*`1`, *stub*`2`, ..., *stub*`k` variables containing observation-specific derivatives
Advanced	
`iterate(`*#*`)`	maximum iterations for finding optimal step size; default is 100
`force`	calculate standard errors, etc., even when possibly inappropriate

Description

`predictnl` calculates (possibly) nonlinear predictions after any Stata estimation command and optionally calculates the variances, standard errors, Wald test statistics, significance levels, and confidence limits for these predictions. Unlike its companion nonlinear postestimation commands `testnl` and `nlcom`, `predictnl` generates functions of the data (i.e., predictions), not scalars. The quantities generated by `predictnl` are thus vectorized over the observations in the data.

Consider some general prediction, $g(\boldsymbol{\theta}, \mathbf{x}_i)$, for $i = 1, \ldots, n$, where $\boldsymbol{\theta}$ are the model parameters and $\mathbf{x}_i$ are some data for the ith observation; $\mathbf{x}_i$ is assumed fixed. Typically, $g(\boldsymbol{\theta}, \mathbf{x}_i)$ is estimated by $g(\widehat{\boldsymbol{\theta}}, \mathbf{x}_i)$, where $\widehat{\boldsymbol{\theta}}$ are the estimated model parameters, which are stored in `e(b)` following any Stata estimation command.

In its most common use, `predictnl` generates two variables: one containing the estimated prediction, $g(\widehat{\boldsymbol{\theta}}, \mathbf{x}_i)$, the other containing the estimated standard error of $g(\widehat{\boldsymbol{\theta}}, \mathbf{x}_i)$. The calculation of standard errors (and other obtainable quantities that are based on the standard errors, such as test statistics) is based on the delta method, an approximation appropriate in large samples; see *Methods and Formulas*.

`predictnl` can be used with `svy` estimation results (assuming that `predict` is also allowed), see [SVY] **svy postestimation**.

The specification of $g(\widehat{\boldsymbol{\theta}}, \mathbf{x}_i)$ is handled by specifying *pnl_exp*, and the values of $g(\widehat{\boldsymbol{\theta}}, \mathbf{x}_i)$ are stored in the new variable *newvar* of storage type *type*. *pnl_exp* is any valid Stata expression and may also contain calls to two special functions unique to `predictnl`:

1. `predict(`[*predict_options*]`)`: When you are evaluating *pnl_exp*, `predict()` is a convenience function that replicates the calculation performed by the command

    `predict ...,` *predict_options*

    As such, the `predict()` function may be used either as a shorthand for the formula used to make this prediction or when the formula is not readily available. When used without arguments, `predict()` replicates the default prediction for that particular estimation command.

2. `xb(`[*eqno*]`)`: The `xb()` function replicates the calculation of the linear predictor $\mathbf{x}_i\mathbf{b}$ for equation *eqno*. If `xb()` is specified without *eqno*, the linear predictor for the first equation (or the only equation in single-equation estimation) is obtained.

    For example, `xb(#1)` (or equivalently, `xb()` with no arguments) translates to the linear predictor for the first equation, `xb(#2)` for the second, and so on. You could also refer to the equations by their names, such as `xb(income)`.

    When specifying *pnl_exp*, both of these functions may be used repeatedly, in combination, and in combination with other Stata functions and expressions. See *Remarks* for examples that use both of these functions.

Options

Main

`se(`*newvar*`)` adds *newvar* of storage type *type*, where for each `i` in the prediction sample, *newvar*`[i]` contains the estimated standard error of $g(\widehat{\boldsymbol{\theta}}, \mathbf{x}_i)$.

`variance(`*newvar*`)` adds *newvar* of storage type *type*, where for each `i` in the prediction sample, *newvar*`[i]` contains the estimated variance of $g(\widehat{\boldsymbol{\theta}}, \mathbf{x}_i)$.

`wald(`*newvar*`)` adds *newvar* of storage type *type*, where for each `i` in the prediction sample, *newvar*`[i]` contains the Wald test statistic for the test of the hypothesis $H_o\colon g(\boldsymbol{\theta}, \mathbf{x}_i) = 0$.

`p(`*newvar*`)` adds *newvar* of storage type *type*, where *newvar*`[i]` contains the significance level (p-value) of the Wald test of $H_o\colon g(\boldsymbol{\theta}, \mathbf{x}_i) = 0$ versus the two-sided alternative.

`ci(`*newvars*`)` requires the specification of two *newvars*, such that the ith observation of each will contain the left and right endpoints (respectively) of a confidence interval for $g(\boldsymbol{\theta}, \mathbf{x}_i)$. The level of the confidence intervals is determined by `level(`*#*`)`.

`level(`*#*`)` specifies the confidence level, as a percentage, for confidence intervals. The default is `level(95)` or as set by `set level`; see [U] **20.7 Specifying the width of confidence intervals**.

`g(`*stub*`)` specifies that new variables, *stub*1, *stub*2, ..., *stub*k be created, where `k` is the dimension of $\boldsymbol{\theta}$. *stub*1 will contain the observation-specific derivatives of $g(\boldsymbol{\theta}, \mathbf{x}_i)$ with respect to the first element, θ_1, of $\boldsymbol{\theta}$; *stub*2 will contain the derivatives of $g(\boldsymbol{\theta}, \mathbf{x}_i)$ with respect to θ_2, etc. If the derivative of $g(\boldsymbol{\theta}, \mathbf{x}_i)$ with respect to a particular coefficient in $\boldsymbol{\theta}$ equals zero for all observations in the prediction sample, the *stub* variable for that coefficient is not created. The ordering of the parameters in $\boldsymbol{\theta}$ is precisely that of the stored vector of parameter estimates `e(b)`.

Advanced

`iterate(`*#*`)` specifies the maximum number of iterations used to find the optimal step size in the calculation of numerical derivatives of $g(\boldsymbol{\theta}, \mathbf{x}_i)$ with respect to $\boldsymbol{\theta}$. By default, the maximum number of iterations is 100, but convergence is usually achieved after only a few iterations. You should rarely have to use this option.

`force` forces the calculation of standard errors and other inference-related quantities in situations where `predictnl` would otherwise refuse to do so. The calculation of standard errors takes place by evaluating (at $\widehat{\boldsymbol{\theta}}$) the numerical derivative of $g(\boldsymbol{\theta}, \mathbf{x}_i)$ with respect to $\boldsymbol{\theta}$. If `predictnl` detects that $g()$ is possibly a function of random quantities other than $\widehat{\boldsymbol{\theta}}$, it will refuse to calculate standard errors or any other quantity derived from them. The `force` option forces the calculation to take place anyway. If you use the `force` option, there is no guarantee that any inference quantities (e.g., standard errors) will be correct or that the values obtained can be interpreted.

Remarks

Remarks are presented under the following headings:

Introduction
Nonlinear transformations and standard errors
Using xb() and predict()
Multiple-equation (ME) estimators
Test statistics and significance levels
Manipulability
Confidence intervals

Introduction

`predictnl` and `nlcom` are Stata's delta method commands—they take a nonlinear transformation of the estimated parameter vector from some fitted model and apply the delta method to calculate the variance, standard error, Wald test statistic, etc., of this transformation. `nlcom` is designed for scalar functions of the parameters, and `predictnl` is designed for functions of the parameters and of the data, that is, for predictions.

Nonlinear transformations and standard errors

We begin by fitting a probit model to the low-birthweight data of Hosmer and Lemeshow (2000, 25). The data are described in detail in [R] **logistic**.

```
. use http://www.stata-press.com/data/r10/lbw
(Hosmer & Lemeshow data)

. probit low lwt smoke ptl ht

Iteration 0:   log likelihood =   -117.336
Iteration 1:   log likelihood = -106.76258
Iteration 2:   log likelihood = -106.67855
Iteration 3:   log likelihood = -106.67851

Probit regression                                 Number of obs   =        189
                                                  LR chi2(4)      =      21.31
                                                  Prob > chi2     =     0.0003
Log likelihood = -106.67851                       Pseudo R2       =     0.0908

------------------------------------------------------------------------------
         low |      Coef.   Std. Err.      z    P>|z|     [95% Conf. Interval]
-------------+----------------------------------------------------------------
         lwt |  -.0095164   .0036875    -2.58   0.010    -.0167438   -.0022891
       smoke |   .3487004   .2041771     1.71   0.088    -.0514794    .7488803
         ptl |    .365667   .1921201     1.90   0.057    -.0108814    .7422154
          ht |   1.082355    .410673     2.64   0.008     .2774504    1.887259
       _cons |   .4238985   .4823224     0.88   0.379    -.5214359    1.369233
------------------------------------------------------------------------------
```

After we fit such a model, we first would want to generate the predicted probabilities of a low birthweight, given the covariate values in the estimation sample. This is easily done using `predict` after `probit`, but it doesn't answer the question, "What are the standard errors of those predictions?"

For the time being, we will consider ourselves ignorant of any automated way to obtain the predicted probabilities after `probit`. The formula for the prediction is

$$P(y \neq 0|\mathbf{x}_i) = \Phi(\mathbf{x}_i\boldsymbol{\beta})$$

where Φ is the standard cumulative normal. Thus for this example, $g(\boldsymbol{\theta}, \mathbf{x}_i) = \Phi(\mathbf{x}_i\boldsymbol{\beta})$. Armed with the formula, we can use `predictnl` to generate the predictions and their standard errors:

```
. predictnl phat = normal(_b[_cons] + _b[ht]*ht + _b[ptl]*ptl +
> _b[smoke]*smoke + _b[lwt]*lwt), se(phat_se)
. list phat phat_se lwt smoke ptl ht in -10/l
```

	phat	phat_se	lwt	smoke	ptl	ht
180.	.2363556	.042707	120	0	0	0
181.	.6577712	.1580714	154	0	1	1
182.	.2793261	.0519958	106	0	0	0
183.	.1502118	.0676338	190	1	0	0
184.	.5702871	.0819911	101	1	1	0
185.	.4477045	.079889	95	1	0	0
186.	.2988379	.0576306	100	0	0	0
187.	.4514706	.080815	94	1	0	0
188.	.5615571	.1551051	142	0	0	1
189.	.7316517	.1361469	130	1	0	1

Thus subject 180 in our data has an estimated probability of low birthweight of 23.6% with standard error 4.3%.

Used without options, `predictnl` is not much different from `generate`. By specifying the option `se(phat_se)`, we were able to obtain a variable containing the standard errors of the predictions; therein lies the utility of `predictnl`.

Using xb() and predict()

As was the case above, a prediction is often not a function of a few isolated parameters and their corresponding variables but instead is some (possibly elaborate) function of the entire linear predictor. For models with many predictors, the brute-force expression for the linear predictor can be cumbersome to type. An alternative is to use the inline function `xb()`. `xb()` is a shortcut for having to type `_b[_cons] + _b[ht]*ht + _b[ptl]*ptl + ...`,

```
. drop phat phat_se
. predictnl phat = normal(xb()), se(phat_se)
```

```
. list phat phat_se lwt smoke ptl ht in -10/l

     +-----------------------------------------------------+
     |     phat    phat_se   lwt   smoke   ptl   ht |
     |-----------------------------------------------------|
180. | .2363556    .042707   120       0     0    0 |
181. | .6577712   .1580714   154       0     1    1 |
182. | .2793261   .0519958   106       0     0    0 |
183. | .1502118   .0676338   190       1     0    0 |
184. | .5702871   .0819911   101       1     1    0 |
     |-----------------------------------------------------|
185. | .4477045    .079889    95       1     0    0 |
186. | .2988379   .0576306   100       0     0    0 |
187. | .4514706    .080815    94       1     0    0 |
188. | .5615571   .1551051   142       0     0    1 |
189. | .7316517   .1361469   130       1     0    1 |
     +-----------------------------------------------------+
```

which yields the same results. This approach is easier, produces more readable code, and is less prone to errors, such as forgetting to include a term in the sum.

Here we used `xb()` without arguments since we have only one equation in our model. In multiple-equation (ME) settings, `xb()` (or equivalently `xb(#1)`) yields the linear predictor from the first equation, `xb(#2)` from the second, etc. You can also refer to equations by their names, e.g., `xb(income)`.

❑ Technical Note

Most estimation commands in Stata allow the postestimation calculation of linear predictors and their standard errors via `predict`. For example, to obtain these for the first (or only) equation in the model, you could type

```
predict xbvar, xb
predict stdpvar, stdp
```

Equivalently, you could type

```
predictnl xbvar = xb(), se(stdpvar)
```

but we recommend the first method, as it is faster. As we demonstrated above, however, `predictnl` is more general.

❑

Returning to our probit example, we can further simplify the calculation by using the inline function `predict()`. `predict(`*pred_options*`)` works by substituting, within our `predictnl` expression, the calculation performed by

```
predict ..., pred_options
```

In our example, we are interested in the predicted probabilities after a probit regression, normally obtained via

```
predict ..., p
```

We can obtain these predictions (and standard errors) by using

```
. drop phat phat_se
. predictnl phat = predict(p), se(phat_se)
```

```
. list phat phat_se lwt smoke ptl ht in -10/l
```

	phat	phat_se	lwt	smoke	ptl	ht
180.	.2363556	.042707	120	0	0	0
181.	.6577712	.1580714	154	0	1	1
182.	.2793261	.0519958	106	0	0	0
183.	.1502118	.0676338	190	1	0	0
184.	.5702871	.0819911	101	1	1	0
185.	.4477045	.079889	95	1	0	0
186.	.2988379	.0576306	100	0	0	0
187.	.4514706	.080815	94	1	0	0
188.	.5615571	.1551051	142	0	0	1
189.	.7316517	.1361469	130	1	0	1

which again replicates what we have already done by other means. However, this version did not require knowledge of the formula for the predicted probabilities after a probit regression—`predict(p)` took care of that for us.

Since the predicted probability is the default prediction after `probit`, we could have just used `predict()` without arguments, namely,

```
. predictnl phat = predict(), se(phat_se)
```

Also the expression *pnl_exp* can be inordinately complicated, with multiple calls to `predict()` and `xb()`. For example,

```
. predictnl phat = normal(invnormal(predict()) + predict(xb)/xb() - 1), se(phat_se)
```

is perfectly valid and will give the same result as before, albeit a bit inefficiently.

❑ Technical Note

When using `predict()` and `xb()`, the *formula* for the calculation is substituted within *pnl_exp*, not the values that result from the application of that formula. To see this, note the subtle difference between

```
. predict xbeta, xb
. predictnl phat = normal(xbeta), se(phat_se)
```

and

```
. predictnl phat = normal(xb()), se(phat_se)
```

Both sequences will yield the same `phat`, yet for the first sequence, `phat_se` will equal zero for all observations. The reason is that, once evaluated, `xbeta` will contain the values of the linear predictor, yet these values are treated as fixed and nonstochastic as far as `predictnl` is concerned. By contrast, since `xb()` is shorthand for the formula used to calculate the linear predictor, it contains not values, but references to the estimated regression coefficients and corresponding variables. Thus the second method produces the desired result.

❑

Multiple-equation (ME) estimators

In [R] **mlogit**, data on insurance choice (Tarlov et al. 1989; Wells et al. 1989) were examined, and a multinomial logit was used to assess the effects of age, gender, race, and site of study (one of three sites) on the type of insurance:

```
. use http://www.stata-press.com/data/r10/sysdsn2, clear
(Health insurance data)
. mlogit insure age male nonwhite site2 site3, nolog
Multinomial logistic regression                   Number of obs   =        615
                                                  LR chi2(10)     =      42.99
                                                  Prob > chi2     =     0.0000
Log likelihood = -534.36165                       Pseudo R2       =     0.0387

------------------------------------------------------------------------------
      insure |      Coef.   Std. Err.      z    P>|z|     [95% Conf. Interval]
-------------+----------------------------------------------------------------
Prepaid      |
         age |   -.011745   .0061946    -1.90   0.058    -.0238862    .0003962
        male |   .5616934   .2027465     2.77   0.006     .1643175    .9590693
    nonwhite |   .9747768   .2363213     4.12   0.000     .5115955    1.437958
       site2 |   .1130359   .2101903     0.54   0.591    -.2989296    .5250013
       site3 |  -.5879879   .2279351    -2.58   0.010    -1.034733   -.1412433
       _cons |   .2697127   .3284422     0.82   0.412    -.3740222    .9134476
-------------+----------------------------------------------------------------
Uninsure     |
         age |  -.0077961   .0114418    -0.68   0.496    -.0302217    .0146294
        male |   .4518496   .3674867     1.23   0.219     -.268411     1.17211
    nonwhite |   .2170589   .4256361     0.51   0.610    -.6171725     1.05129
       site2 |  -1.211563   .4705127    -2.57   0.010    -2.133751   -.2893747
       site3 |  -.2078123   .3662926    -0.57   0.570    -.9257327     .510108
       _cons |  -1.286943   .5923219    -2.17   0.030    -2.447872   -.1260135
------------------------------------------------------------------------------
(insure==Indemnity is the base outcome)
```

Of particular interest is the estimation of the relative risk, which, for a given selection, is the ratio of the probability of making that selection to the probability of selecting the base category (`insure==Indemnity` here), given a set of covariate values. In a multinomial logit model, the relative risk (when comparing to the base category) simplifies to the exponentiated linear predictor for that selection.

Using this example, we can estimate the observation-specific relative risks of selecting a prepaid plan over the base category (with standard errors) by either referring to the `Prepaid` equation by name or number,

```
. predictnl RRppaid = exp(xb(Prepaid)), se(SERRppaid)
```

or

```
. predictnl RRppaid = exp(xb(#1)), se(SERRppaid)
```

since `Prepaid` is the first equation in the model.

Those of us for whom the simplified formula for the relative risk doesn't immediately come to mind may prefer to calculate the relative risk directly from its definition, that is, as a ratio of two predicted probabilities. After `mlogit`, the predicted probability for a category may be obtained using `predict`, but we must specify the category as the outcome:

```
. predictnl RRppaid = predict(outcome(Prepaid))/predict(outcome(Indemnity)),
> se(SERRppaid)
(1 missing value generated)
. list RRppaid SERRppaid age male nonwhite site2 site3 in 1/10
```

	RRppaid	SERRpp~d	age	male	nonwhite	site2	site3
1.	.6168578	.1503759	73.722107	0	0	1	0
2.	1.056658	.1790703	27.89595	0	0	1	0
3.	.8426442	.1511281	37.541397	0	0	0	0
4.	1.460581	.3671465	23.641327	0	1	0	1
5.	.9115747	.1324168	40.470901	0	0	1	0
6.	1.034701	.1696923	29.683777	0	0	1	0
7.	.9223664	.1344981	39.468857	0	0	1	0
8.	1.678312	.4216626	26.702255	1	0	0	0
9.	.9188519	.2256017	63.101974	0	1	0	1
10.	.5766296	.1334877	69.839828	0	0	0	0

The "`(1 missing value generated)`" message is not an error; further examination of the data would reveal that `age` is missing in one observation and that the offending observation (among others) is not in the estimation sample. Just as with `predict`, `predictnl` can generate predictions in or out of the estimation sample.

Thus we estimate (among other things) that a white, female, 73-year-old from site 2 is less likely to choose a prepaid plan over an indemnity plan—her relative risk is about 62% with standard error 15%.

Test statistics and significance levels

Often a standard error calculation is just a means to an end, and what is really desired is a test of the hypothesis,

$$H_o : g(\boldsymbol{\theta}, \mathbf{x}_i) = 0$$

versus the two-sided alternative.

We can use `predictnl` to obtain the Wald test statistics and/or significance levels for the above tests, whether or not we want standard errors. To obtain the Wald test statistics, we use the `wald()` option; for significance levels, we use `p()`.

Returning to our `mlogit` example, suppose that we wanted for each observation a test of whether the relative risk of choosing a prepaid plan over an indemnity plan is different from one. One way to do this would be to define $g()$ to be the relative risk minus one and then test whether $g()$ is different from zero.

```
. predictnl RRm1 = exp(xb(Prepaid)) - 1, wald(W_RRm1) p(sig_RRm1)
(1 missing value generated)
note: significance levels are with respect to the chi-squared(1) distribution.
. list RRm1 W_RRm1 sig_RRm1 age male nonwhite in 1/10
```

	RRm1	W_RRm1	sig_RRm1	age	male	nonwhite
1.	-.3831422	6.491778	.0108375	73.722107	0	0
2.	.0566578	.100109	.7516989	27.89595	0	0
3.	-.1573559	1.084116	.2977787	37.541397	0	0
4.	.4605812	1.573743	.2096643	23.641327	0	1
5.	-.0884253	.4459299	.5042742	40.470901	0	0
6.	.0347015	.0418188	.8379655	29.683777	0	0
7.	-.0776336	.3331707	.563798	39.468857	0	0
8.	.6783119	2.587788	.1076906	26.702255	1	0
9.	-.0811482	.1293816	.719074	63.101974	0	1
10.	-.4233705	10.05909	.001516	69.839828	0	0

The newly created variable `W_RRm1` contains the Wald test statistic for each observation, and `sig_RRm1` contains the level of significance. Thus our 73-year-old white female represented by the first observation would have a relative risk of choosing prepaid over indemnity that is significantly different from 1, at least at the 5% level. For this test, it was not necessary to generate a variable containing the standard error of the relative risk minus 1, but we could have done so had we wanted. We could have also omitted specifying `wald(W_RRm1)` if all we cared about were, say, the significance levels of the tests.

In this regard, `predictnl` acts as an observation-specific version of `testnl`, with the test results vectorized over the observations in the data. The significance levels are pointwise—they are not adjusted to reflect any simultaneous testing over the observations in the data.

Manipulability

There are many ways to specify $g(\boldsymbol{\theta}, \mathbf{x}_i)$ to yield tests such that, for multiple specifications of $g()$, the theoretical conditions for which

$$H_o\colon g(\boldsymbol{\theta}, \mathbf{x}_i) = 0$$

is true will be equivalent. However, this does not mean that the tests themselves will be equivalent. This is known as the manipulability of the Wald test for nonlinear hypotheses; also see [R] **boxcox**.

As an example, consider the previous section where we defined $g()$ to be the relative risk between choosing a prepaid plan over an indemnity plan, minus 1. We could also have defined $g()$ to be the risk difference—the probability of choosing a prepaid plan minus the probability of choosing an indemnity plan. Either specification of $g()$ yields a mathematically equivalent specification of $H_o\colon g() = 0$; that is, the risk difference will equal zero when the relative risk equals one. However, the tests themselves do not give the same results:

(Continued on next page)

```
. predictnl RD = predict(outcome(Prepaid)) - predict(outcome(Indemnity)),
> wald(W_RD) p(sig_RD)
(1 missing value generated)
note: significance levels are with respect to the chi-squared(1) distribution.
```

```
. list RD W_RD sig_RD RRm1 W_RRm1 sig_RRm1 in 1/10
```

	RD	W_RD	sig_RD	RRm1	W_RRm1	sig_RRm1
1.	-.2303744	4.230243	.0397097	-.3831422	6.491778	.0108375
2.	.0266902	.1058542	.7449144	.0566578	.100109	.7516989
3.	-.0768078	.9187646	.3377995	-.1573559	1.084116	.2977787
4.	.1710702	2.366535	.1239619	.4605812	1.573743	.2096643
5.	-.0448509	.4072922	.5233471	-.0884253	.4459299	.5042742
6.	.0165251	.0432816	.835196	.0347015	.0418188	.8379655
7.	-.0391535	.3077611	.5790573	-.0776336	.3331707	.563798
8.	.22382	4.539085	.0331293	.6783119	2.587788	.1076906
9.	-.0388409	.1190183	.7301016	-.0811482	.1293816	.719074
10.	-.2437626	6.151558	.0131296	-.4233705	10.05909	.001516

In certain cases (such as subject 8), the difference can be severe enough to potentially change the conclusion. The reason for this inconsistency is that the nonlinear Wald test is actually a standard Wald test of a first-order Taylor approximation of $g()$, and this approximation can differ according to how $g()$ is specified.

As such, keep in mind the manipulability of nonlinear Wald tests when drawing scientific conclusions.

Confidence intervals

We can also use `predictnl` to obtain confidence intervals for the observation-specific $g(\boldsymbol{\theta}, \mathbf{x}_i)$ by using the `ci()` option to specify two new variables to contain the left and right endpoints of the confidence interval, respectively. For example, we could generate confidence intervals for the risk differences calculated previously:

```
. drop RD
. predictnl RD = predict(outcome(Prepaid)) - predict(outcome(Indemnity)),
> ci(RD_lcl RD_rcl)
(1 missing value generated)
note: Confidence intervals calculated using Z critical values.
```

```
. list RD RD_lcl RD_rcl age male nonwhite in 1/10
```

	RD	RD_lcl	RD_rcl	age	male	nonwhite
1.	-.2303744	-.4499073	-.0108415	73.722107	0	0
2.	.0266902	-.1340948	.1874752	27.89595	0	0
3.	-.0768078	-.2338625	.080247	37.541397	0	0
4.	.1710702	-.0468844	.3890248	23.641327	0	1
5.	-.0448509	-.1825929	.092891	40.470901	0	0
6.	.0165251	-.1391577	.1722078	29.683777	0	0
7.	-.0391535	-.177482	.099175	39.468857	0	0
8.	.22382	.0179169	.4297231	26.702255	1	0
9.	-.0388409	-.2595044	.1818226	63.101974	0	1
10.	-.2437626	-.4363919	-.0511332	69.839828	0	0

The confidence level, here, 95%, is either set using the `level()` option or obtained from the current default level, `c(level)`; see [U] **20.7 Specifying the width of confidence intervals**.

From the above output, we can see that, for subjects 1, 8, and 10, a 95% confidence interval for the risk difference does not contain zero, meaning that, for these subjects, there is some evidence of a significant difference in risks.

The confidence intervals calculated by `predictnl` are pointwise; there is no adjustment (such as a Bonferroni correction) made so that these confidence intervals may be considered jointly at the specified level.

Methods and Formulas

`predictnl` is implemented as an ado-file.

For the ith observation, consider the transformation $g(\boldsymbol{\theta}, \mathbf{x}_i)$, estimated by $g(\widehat{\boldsymbol{\theta}}, \mathbf{x}_i)$, for the $1 \times k$ parameter vector $\boldsymbol{\theta}$ and data $\mathbf{x}_i$ ($\mathbf{x}_i$ is assumed fixed). The variance of $g(\widehat{\boldsymbol{\theta}}, \mathbf{x}_i)$ is estimated by

$$\widehat{\text{Var}}\left\{g(\widehat{\boldsymbol{\theta}}, \mathbf{x}_i)\right\} = \mathbf{GVG}'$$

where $\mathbf{G}$ is the vector of derivatives

$$\mathbf{G} = \left\{ \left. \frac{\partial g(\boldsymbol{\theta}, \mathbf{x}_i)}{\partial \boldsymbol{\theta}} \right|_{\boldsymbol{\theta}=\widehat{\boldsymbol{\theta}}} \right\}_{(1\times k)}$$

and $\mathbf{V}$ is the estimated variance–covariance matrix of $\widehat{\boldsymbol{\theta}}$. Standard errors, $\widehat{\text{s.e.}}\{g(\widehat{\boldsymbol{\theta}}, \mathbf{x}_i)\}$, are obtained as the square roots of the variances.

The Wald test statistic for testing

$$H_o\colon g(\boldsymbol{\theta}, \mathbf{x}_i) = 0$$

versus the two-sided alternative is given by

$$W_i = \frac{\left\{g(\widehat{\boldsymbol{\theta}}, \mathbf{x}_i)\right\}^2}{\widehat{\text{Var}}\left\{g(\widehat{\boldsymbol{\theta}}, \mathbf{x}_i)\right\}}$$

When the variance–covariance matrix of $\widehat{\boldsymbol{\theta}}$ is an asymptotic covariance matrix, W_i is approximately distributed as χ^2 with 1 degree of freedom. For linear regression, W_i is taken to be approximately distributed as $F_{1,r}$, where r is the residual degrees of freedom from the original model fit. The levels of significance of the observation-by-observation tests of H_o versus the two-sided alternative are given by

$$p_i = \Pr(T > W_i)$$

where T is either a χ^2- or F-distributed random variable, as described above.

A $(1-\alpha) \times 100\%$ confidence interval for $g(\boldsymbol{\theta}, \mathbf{x}_i)$ is given by

$$g(\widehat{\boldsymbol{\theta}}, \mathbf{x}_i) \pm z_{\alpha/2} \left[\widehat{\text{s.e.}}\left\{g(\widehat{\boldsymbol{\theta}}, \mathbf{x}_i)\right\}\right]$$

when W_i is χ^2-distributed, and

$$g(\widehat{\boldsymbol{\theta}}, \mathbf{x}_i) \pm t_{\alpha/2,r} \left[\widehat{\text{s.e.}}\left\{g(\widehat{\boldsymbol{\theta}}, \mathbf{x}_i)\right\}\right]$$

when W_i is F-distributed. z_p is the $1 - p$ quantile of the standard normal distribution, and $t_{p,r}$ is the $1 - p$ quantile of the t distribution with r degrees of freedom.

References

Gould, W. W. 1996. crc43: Wald test of nonlinear hypotheses after model estimation. *Stata Technical Bulletin* 29: 2–4. Reprinted in *Stata Technical Bulletin Reprints*, vol. 5, pp. 15–18.

Hosmer, D. W., Jr., and S. Lemeshow. 2000. *Applied Logistic Regression*. 2nd ed. New York: Wiley.

Phillips, P. C. B., and J. Y. Park. 1988. On the formulation of Wald tests of nonlinear restrictions. *Econometrica* 56: 1065–1083.

Tarlov, A. R., J. E. Ware, Jr., S. Greenfield, E. C. Nelson, E. Perrin, and M. Zubkoff. 1989. The medical outcomes study. *Journal of the American Medical Association* 262: 925–930.

Wells, K. E., R. D. Hays, M. A. Burnam, W. H. Rogers, S. Greenfield, and J. E. Ware, Jr. 1989. Detection of depressive disorder for patients receiving prepaid or fee-for-service care. *Journal of the American Medical Association* 262: 3298–3302.

Also See

[R] **lincom** — Linear combinations of estimators

[R] **nlcom** — Nonlinear combinations of estimators

[R] **predict** — Obtain predictions, residuals, etc., after estimation

[R] **test** — Test linear hypotheses after estimation

[R] **testnl** — Test nonlinear hypotheses after estimation

[U] **20 Estimation and postestimation commands**

Title

probit — Probit regression

Syntax

Probit regression

probit *depvar* [*indepvars*] [*if*] [*in*] [*weight*] [, *probit_options*]

Probit regression, reporting marginal effects

dprobit [*depvar indepvars* [*if*] [*in*] [*weight*]] [, *dprobit_options*]

probit_options	description
Model	
noconstant	suppress constant term
offset(*varname*)	include *varname* in model with coefficient constrained to 1
asis	retain perfect predictor variables
SE/Robust	
vce(*vcetype*)	*vcetype* may be oim, robust, cluster *clustvar*, bootstrap, or jackknife
Reporting	
level(#)	set confidence level; default is level(95)
Max options	
maximize_options	control the maximization process; seldom used
†nocoef	do not display the coefficient table; seldom used

(*Continued on next page*)

dprobit_options	description
Model	
offset(*varname*)	include *varname* in model with coefficient constrained to 1
at(*matname*)	point at which marginal effects are evaluated
asis	retain perfect predictor variables
classic	calculate mean effects for dummies like those for continuous variables
SE/Robust	
vce(*vcetype*)	*vcetype* may be oim, robust, or cluster *clustvar*
Reporting	
level(#)	set confidence level; default is level(95)
Max options	
maximize_options	control the maximization process; seldom used
†nocoef	do not display the coefficient table; seldom used

depvar and *indepvars* for probit may contain time-series operators; see [U] **11.4.3 Time-series varlists**.

bootstrap, by, jackknife, nestreg, rolling, statsby, stepwise, svy, and xi are allowed with probit, and by, rolling, statsby, and xi are allowed with dprobit; see [U] **11.1.10 Prefix commands**.

Weights are not allowed with the bootstrap prefix.

aweights are not allowed with the jackknife prefix.

vce(), nocoef, and weights are not allowed with the svy prefix.

†nocoef does not appear in the dialog box.

fweights, aweights, iweights, and pweights are allowed with probit, and fweights, aweights, and pweights are allowed with dprobit; see [U] **11.1.6 weight**.

See [U] **20 Estimation and postestimation commands** for more capabilities of estimation commands.

Description

probit fits a maximum-likelihood probit model.

dprobit fits maximum-likelihood probit models and is an alternative to probit. Rather than reporting the coefficients, dprobit reports the marginal effect, that is, the change in the probability for an infinitesimal change in each independent, continuous variable and, by default, reports the discrete change in the probability for dummy variables. probit may be typed without arguments after dprobit estimation to see the model in coefficient form.

If estimating on grouped data, see the bprobit command described in [R] **glogit**.

Several auxiliary commands may be run after probit, logit, or logistic; see [R] **logistic postestimation** for a description of these commands.

See [R] **logistic** for a list of related estimation commands.

Options for probit

Model

noconstant, offset(*varname*); see [R] **estimation options**.

asis specifies that all specified variables and observations be retained in the maximization process. This option is typically not specified and may introduce numerical instability. Normally probit drops variables that perfectly predict success or failure in the dependent variable along with their associated observations. In those cases, the effective coefficient on the dropped variables is infinity (negative infinity) for variables that completely determine a success (failure). Dropping the variable and perfectly predicted observations has no effect on the likelihood or estimates of the remaining coefficients and increases the numerical stability of the optimization process. Specifying this option forces retention of perfect predictor variables and their associated observations.

SE/Robust

vce(*vcetype*) specifies the type of standard error reported, which includes types that are derived from asymptotic theory, that are robust to some kinds of misspecification, that allow for intragroup correlation, and that use bootstrap or jackknife methods; see [R] ***vce_option***.

Reporting

level(*#*); see [R] **estimation options**.

Max options

maximize_options: iterate(*#*), [no]log, trace, tolerance(*#*), ltolerance(*#*); see [R] **maximize**. These options are seldom used.

The following option is available with probit but is not shown in the dialog box:

nocoef specifies that the coefficient table not be displayed. This option is sometimes used by programmers but is of no use interactively.

Options for dprobit

Model

offset(*varname*); see [R] **estimation options**.

at(*matname*) specifies the point at which marginal effects are evaluated. The default is to evaluate at $\overline{\mathbf{x}}$, the mean of the independent variables. If there are k independent variables, *matname* may be $1 \times k$ or $1 \times (k+1)$; that is, it may optionally include final element 1 reflecting the constant. at() may be specified when the model is fitted or when results are redisplayed.

asis; see *probit_options* above.

classic requests that the mean effects always be calculated using the formula $f(\overline{\mathbf{x}}\mathbf{b})b_i$. If classic is not specified, $f(\overline{\mathbf{x}}\mathbf{b})b_i$ is used for continuous variables, but the mean effects for dummy variables are calculated as $\Phi(\overline{\mathbf{x}}_1\mathbf{b}) - \Phi(\overline{\mathbf{x}}_0\mathbf{b})$. Here $\overline{\mathbf{x}}_1 = \overline{\mathbf{x}}$ but with element i set to 1, $\overline{\mathbf{x}}_0 = \overline{\mathbf{x}}$ but with element i set to 0, and $\overline{\mathbf{x}}$ is the mean of the independent variables or the vector specified by at(). classic may be specified at estimation time or when the results are redisplayed. Results calculated without classic may be redisplayed with classic and vice versa.

SE/Robust

`vce(vcetype)` specifies the type of standard error reported, which includes types that are derived from asymptotic theory, that are robust to some kinds of misspecification, and that allow for intragroup correlation; see [R] ***vce_option***.

Reporting

`level(#)`; see [R] **estimation options**.

Max options

maximize_options: `iterate(#)`, `[no]log`, `trace`, `tolerance(#)`, `ltolerance(#)`; see [R] **maximize**. These options are seldom used.

The following option is available with `dprobit` but is not shown in the dialog box:

`nocoef` specifies that the coefficient table not be displayed. This option is sometimes used by programmers but is of no use interactively.

Remarks

Remarks are presented under the following headings:

Robust standard errors
dprobit
Model identification

`probit` fits maximum likelihood models with dichotomous dependent (left-hand-side) variables coded as 0/1 (more precisely, coded as 0 and not 0).

▷ Example 1

Wu have data on the make, weight, and mileage rating of 22 foreign and 52 domestic automobiles. Wu wish to fit a probit model explaining whether a car is foreign based on its weight and mileage. Here is an overview of our data:

```
. use http://www.stata-press.com/data/r10/auto
(1978 Automobile Data)

. keep make mpg weight foreign

. describe

Contains data from http://www.stata-press.com/data/r10/auto.dta
  obs:            74                          1978 Automobile Data
 vars:             4                          13 Apr 2007 17:45
 size:         1,998 (99.7% of memory free)   (_dta has notes)
-------------------------------------------------------------------------------
              storage  display     value
variable name   type   format      label      variable label
-------------------------------------------------------------------------------
make            str18  %-18s                  Make and Model
mpg             int    %8.0g                  Mileage (mpg)
weight          int    %8.0gc                 Weight (lbs.)
foreign         byte   %8.0g       origin     Car type
-------------------------------------------------------------------------------
Sorted by:  foreign
     Note:  dataset has changed since last saved
```

```
. inspect foreign
foreign:  Car type                                Number of Observations
                                             Total   Integers   Nonintegers
|  #                      Negative               -          -             -
|  #                      Zero                  52         52             -
|  #                      Positive              22         22             -
|  #                                         -----      -----         -----
|  #   #                  Total                 74         74             -
|  #   #                  Missing                -
+----------------------                      -----
0                    1                          74
  (2 unique values)
     foreign is labeled and all values are documented in the label.
```

The variable **foreign** takes on two unique values, 0 and 1. The value 0 denotes a domestic car, and 1 denotes a foreign car.

The model that we wish to fit is

$$\Pr(\texttt{foreign} = 1) = \Phi(\beta_0 + \beta_1 \texttt{weight} + \beta_2 \texttt{mpg})$$

where Φ is the cumulative normal distribution.

To fit this model, we type

```
. probit foreign weight mpg
Iteration 0:  log likelihood =  -45.03321
Iteration 1:  log likelihood = -29.244141
  (output omitted )
Iteration 5:  log likelihood = -26.844189
Probit regression                                 Number of obs   =         74
                                                  LR chi2(2)      =      36.38
                                                  Prob > chi2     =     0.0000
Log likelihood = -26.844189                       Pseudo R2       =     0.4039

------------------------------------------------------------------------------
     foreign |      Coef.   Std. Err.      z    P>|z|     [95% Conf. Interval]
-------------+----------------------------------------------------------------
      weight |  -.0023355   .0005661    -4.13   0.000     -.003445   -.0012261
         mpg |  -.1039503   .0515689    -2.02   0.044    -.2050235   -.0028772
       _cons |   8.275464   2.554142     3.24   0.001     3.269438    13.28149
------------------------------------------------------------------------------
```

We find that heavier cars are less likely to be foreign and that cars yielding better gas mileage are also less likely to be foreign, at least holding the weight of the car constant.

See [R] **maximize** for an explanation of the output.

◁

❑ Technical Note

Stata interprets a value of 0 as a negative outcome (failure) and treats all other values (except missing) as positive outcomes (successes). Thus if your dependent variable takes on the values 0 and 1, 0 is interpreted as failure and 1 as success. If your dependent variable takes on the values 0, 1, and 2, 0 is still interpreted as failure, but both 1 and 2 are treated as successes.

If you prefer a more formal mathematical statement, when you type `probit` y x, Stata fits the model

$$\Pr(y_j \neq 0 \mid \mathbf{x}_j) = \Phi(\mathbf{x}_j\boldsymbol{\beta})$$

where Φ is the standard cumulative normal.

❑

Robust standard errors

If you specify the `vce(robust)` option, `probit` reports robust standard errors; see [U] **20.15 Obtaining robust variance estimates**.

▷ Example 2

For the model from example 1, the robust calculation increases the standard error of the coefficient on `mpg` by almost 15%:

```
. probit foreign weight mpg, vce(robust) nolog
Probit regression                                 Number of obs   =         74
                                                  Wald chi2(2)    =      30.26
                                                  Prob > chi2     =     0.0000
Log pseudolikelihood = -26.844189                 Pseudo R2       =     0.4039

------------------------------------------------------------------------------
             |               Robust
     foreign |      Coef.   Std. Err.      z    P>|z|     [95% Conf. Interval]
-------------+----------------------------------------------------------------
      weight |  -.0023355   .0004934    -4.73   0.000    -.0033025   -.0013686
         mpg |  -.1039503   .0593548    -1.75   0.080    -.2202836    .0123829
       _cons |   8.275464   2.539176     3.26   0.001      3.29877    13.25216
------------------------------------------------------------------------------
```

Without `vce(robust)`, the standard error for the coefficient on `mpg` was reported to be .052 with a resulting confidence interval of $[-.21, -.00]$.

◁

▷ Example 3

The `vce(cluster` *clustvar*`)` option can relax the independence assumption required by the probit estimator to independence between clusters. To demonstrate, we will switch to a different dataset.

We are studying unionization of women in the United States and have a dataset with 26,200 observations on 4,434 women between 1970 and 1988. We will use the variables `age` (the women were 14–26 in 1968, and our data span the age range of 16–46), `grade` (years of schooling completed, ranging from 0 to 18), `not_smsa` (28% of the person-time was spent living outside an SMSA—standard metropolitan statistical area), `south` (41% of the person-time was in the South), and `southXt` (`south` interacted with year, treating 1970 as year 0). We also have variable `union`, indicating union membership. Overall, 22% of the person-time is marked as time under union membership, and 44% of these women have belonged to a union.

We fit the following model, ignoring that the women are observed an average of 5.9 times each in these data:

```
. use http://www.stata-press.com/data/r10/union, clear
(NLS Women 14-24 in 1968)
. probit union age grade not_smsa south southXt
Iteration 0:  log likelihood =  -13864.23
Iteration 1:  log likelihood = -13548.436
Iteration 2:  log likelihood = -13547.308
Iteration 3:  log likelihood = -13547.308
Probit regression                                 Number of obs   =      26200
                                                  LR chi2(5)      =     633.84
                                                  Prob > chi2     =     0.0000
Log likelihood = -13547.308                       Pseudo R2       =     0.0229
```

union	Coef.	Std. Err.	z	P>|z|	[95% Conf.	Interval]
age	.0059461	.0015798	3.76	0.000	.0028496	.0090425
grade	.02639	.0036651	7.20	0.000	.0192066	.0335735
not_smsa	-.1303911	.0202523	-6.44	0.000	-.1700848	-.0906975
south	-.4027254	.033989	-11.85	0.000	-.4693426	-.3361081
southXt	.0033088	.0029253	1.13	0.258	-.0024247	.0090423
_cons	-1.113091	.0657808	-16.92	0.000	-1.242019	-.9841628

The reported standard errors in this model are probably meaningless. Women are observed repeatedly, and so the observations are not independent. Looking at the coefficients, we find a large southern effect against unionization and little time trend. The `vce(cluster` *clustvar*`)` option provides a way to fit this model and obtains correct standard errors:

```
. probit union age grade not_smsa south southXt, vce(cluster id)
Iteration 0:   log pseudolikelihood =  -13864.23
Iteration 1:   log pseudolikelihood = -13548.436
Iteration 2:   log pseudolikelihood = -13547.308
Iteration 3:   log pseudolikelihood = -13547.308
Probit regression                                 Number of obs   =      26200
                                                  Wald chi2(5)    =     165.75
                                                  Prob > chi2     =     0.0000
Log pseudolikelihood = -13547.308                 Pseudo R2       =     0.0229
                          (Std. Err. adjusted for 4434 clusters in idcode)
```

union	Coef.	Robust Std. Err.	z	P>|z|	[95% Conf.	Interval]
age	.0059461	.0023567	2.52	0.012	.001327	.0105651
grade	.02639	.0078378	3.37	0.001	.0110282	.0417518
not_smsa	-.1303911	.0404109	-3.23	0.001	-.209595	-.0511873
south	-.4027254	.0514458	-7.83	0.000	-.5035573	-.3018935
southXt	.0033088	.0039793	0.83	0.406	-.0044904	.0111081
_cons	-1.113091	.1188478	-9.37	0.000	-1.346028	-.8801534

These standard errors are roughly 50% larger than those reported by the inappropriate conventional calculation. By comparison, another model we could fit is an equal-correlation population-averaged probit model:

```
. xtprobit union age grade not_smsa south southXt, pa
Iteration 1: tolerance = .04796083
Iteration 2: tolerance = .00352657
Iteration 3: tolerance = .00017886
Iteration 4: tolerance = 8.654e-06
Iteration 5: tolerance = 4.150e-07
```

```
GEE population-averaged model                   Number of obs      =     26200
Group variable:                    idcode       Number of groups   =      4434
Link:                              probit       Obs per group: min =         1
Family:                          binomial                      avg =       5.9
Correlation:                 exchangeable                      max =        12
                                                Wald chi2(5)       =    241.66
Scale parameter:                        1       Prob > chi2        =    0.0000

------------------------------------------------------------------------------
       union |      Coef.   Std. Err.      z    P>|z|     [95% Conf. Interval]
-------------+----------------------------------------------------------------
         age |   .0031597   .0014678     2.15   0.031     .0002829    .0060366
       grade |   .0329992   .0062334     5.29   0.000      .020782    .0452163
    not_smsa |  -.0721799   .0275189    -2.62   0.009    -.1261159   -.0182439
       south |   -.409029   .0372213   -10.99   0.000    -.4819815   -.3360765
     southXt |   .0081828    .002545     3.22   0.001     .0031946    .0131709
       _cons |  -1.184799   .0890117   -13.31   0.000    -1.359259    -1.01034
------------------------------------------------------------------------------
```

The coefficient estimates are similar, but these standard errors are smaller than those produced by `probit, vce(cluster` *clustvar*`)`, as we would expect. If the equal-correlation assumption is valid, the population-averaged probit estimator above should be more efficient.

Is the assumption valid? That is a difficult question to answer. The population-averaged estimates correspond to an assumption of exchangeable correlation within person. It would not be unreasonable to assume an AR(1) correlation within person or to assume that the observations are correlated but that we do not wish to impose any structure. See [XT] `xtgee` for full details.

◁

`probit, vce(cluster` *clustvar*`)` is robust to assumptions about within-cluster correlation. That is, it inefficiently sums within cluster for the standard error calculation rather than attempting to exploit what might be assumed about the within-cluster correlation.

dprobit

A probit model is defined as

$$\Pr(y_j \neq 0 \mid \mathbf{x}_j) = \Phi(\mathbf{x}_j\mathbf{b})$$

where Φ is the standard cumulative normal distribution, and $\mathbf{x}_j\mathbf{b}$ is called the probit score or index.

Since $\mathbf{x}_j\mathbf{b}$ has a normal distribution, interpreting probit coefficients requires thinking in the Z (normal quantile) metric. For instance, say that we estimated the probit equation

$$\Pr(y_j \neq 0) = \Phi(.08233\,x_1 + 1.529\,x_2 - 3.139)$$

Interpreting the x_1 coefficient, we see that each one-unit increase in x_1 increases the probit index by .08233 standard deviations. Learning to think in the Z metric takes practice, and, even if you do, communicating results to others who have not learned to think this way is difficult.

A transformation of the results helps some people better understand them. The change in the probability somehow feels more natural, but how big that change is depends on where we start. Why not choose as a starting point the mean of the data? If $\overline{x}_1 = 21.29$ and $\overline{x}_2 = .42$, then we would report something like .0257, meaning the change in the probability calculated at the mean. We could make the calculation as follows.

The mean normal index is $.08233 \times 21.29 + 1.529 \times .42 - 3.139 = -.7440$, and the corresponding probability is $\Phi(-.7440) = .2284$. Adding our coefficient of .08233 to the index and recalculating the probability, we obtain $\Phi(-.7440 + .08233) = .2541$. The change in the probability is thus $.2541 - .2284 = .0257$.

In practice, people make this calculation somewhat differently and produce a slightly different number. Rather than making the calculation for a one-unit change in x, they calculate the slope of the probability function. Doing a little calculus, they derive that the change in the probability for a change in x_1 ($\partial\Phi/\partial x_1$) is the height of the normal density multiplied by the x_1 coefficient; that is,

$$\frac{\partial\Phi}{\partial x_1} = \phi(\overline{\mathbf{x}}\mathbf{b})b_1$$

Going through this calculation, they obtain .0249.

The difference between .0257 and .0249 is small; they differ because the .0257 is the exact answer for a one-unit increase in x_1, whereas .0249 is the answer for an infinitesimal change extrapolated out.

▷ Example 4

`dprobit` with the `classic` option transforms results as an infinitesimal change extrapolated out. Consider the automobile data again:

```
. use http://www.stata-press.com/data/r10/auto
(1978 Automobile Data)

. generate goodplus = rep78>=4 if rep78 < .
(5 missing values generated)

. dprobit foreign mpg goodplus, classic

Iteration 0:   log likelihood = -42.400729
Iteration 1:   log likelihood = -27.643138
Iteration 2:   log likelihood = -26.953126
Iteration 3:   log likelihood = -26.942119
Iteration 4:   log likelihood = -26.942114

Probit regression, reporting marginal effects           Number of obs =      69
                                                        LR chi2(2)    =   30.92
                                                        Prob > chi2   = 0.0000
Log likelihood = -26.942114                             Pseudo R2     = 0.3646

------------------------------------------------------------------------------
 foreign |      dF/dx   Std. Err.      z    P>|z|     x-bar  [    95% C.I.   ]
---------+--------------------------------------------------------------------
     mpg |   .0249187   .0110853     2.30   0.022   21.2899   .003192  .046646
goodplus |     .46276   .1187437     3.81   0.000    .42029   .230027  .695493
   _cons |  -.9499603   .2281006    -3.82   0.000         1  -1.39703 -.502891
---------+--------------------------------------------------------------------
  obs. P |   .3043478
 pred. P |   .2286624  (at x-bar)
------------------------------------------------------------------------------
    z and P>|z| are the test of the underlying coefficient being 0
```

After estimation with `dprobit`, we can see the untransformed coefficient results by typing `probit` without options:

```
. probit

Probit regression                                 Number of obs   =         69
                                                  LR chi2(2)      =      30.92
                                                  Prob > chi2     =     0.0000
Log likelihood = -26.942114                       Pseudo R2       =     0.3646

------------------------------------------------------------------------------
     foreign |      Coef.   Std. Err.      z    P>|z|     [95% Conf. Interval]
-------------+----------------------------------------------------------------
         mpg |    .082333    .0358292     2.30   0.022     .0121091     .152557
    goodplus |   1.528992    .4010866     3.81   0.000     .7428771    2.315108
       _cons |  -3.138737    .8209689    -3.82   0.000    -4.747807   -1.529668
------------------------------------------------------------------------------
```

In one case, one can argue that the classic, infinitesimal-change–based adjustment could be improved on, and that is in the case of a dummy variable. A dummy variable takes on the values 0 and 1 only—1 indicates that something is true, and 0 indicates that it is not. **goodplus** is such a variable. To understand the effect of **goodplus**, we want to know how much its being true or false affects the outcome probability.

That is, "at the means", the predicted probability of **foreign** for a car with **goodplus** = 0 is $\Phi(.08233\,\overline{x}_1 - 3.139) = .0829$. For the same car with **goodplus** = 1, the probability is $\Phi(.08233\,\overline{x}_1 + 1.529 - 3.139) = .5569$. The difference is thus $.5569 - .0829 = .4740$.

When we do not specify the **classic** option, **dprobit** makes the calculation for dummy variables in this way. Even though we fitted the model with the **classic** option, we can redisplay results with **classic** omitted:

```
. dprobit

Probit regression, reporting marginal effects           Number of obs =     69
                                                        LR chi2(2)    =  30.92
                                                        Prob > chi2   = 0.0000
Log likelihood = -26.942114                             Pseudo R2     = 0.3646

------------------------------------------------------------------------------
 foreign |      dF/dx   Std. Err.      z    P>|z|     x-bar  [    95% C.I.   ]
---------+--------------------------------------------------------------------
     mpg |   .0249187   .0110853     2.30   0.022   21.2899   .003192  .046646
goodplus*|   .4740077   .1114816     3.81   0.000    .42029   .255508  .692508
---------+--------------------------------------------------------------------
  obs. P |   .3043478
 pred. P |   .2286624  (at x-bar)
------------------------------------------------------------------------------
(*) dF/dx is for discrete change of dummy variable from 0 to 1
    z and P>|z| are the test of the underlying coefficient being 0
```

◁

❑ Technical Note

at(*matname***)** allows you to evaluate effects at points other than the means. Let's obtain the effects for the above model at **mpg** = 20 and **goodplus** = 1:

```
. matrix myx = (20,1)
. dprobit, at(myx)
Probit regression, reporting marginal effects           Number of obs =     69
                                                        LR chi2(2)    =  30.92
                                                        Prob > chi2   = 0.0000
Log likelihood = -26.942114                             Pseudo R2     = 0.3646

------------------------------------------------------------------------------
 foreign |      dF/dx   Std. Err.      z    P>|z|       x  [    95% C.I.   ]
---------+--------------------------------------------------------------------
     mpg |   .0328237   .0144157     2.30   0.022      20   .004569  .061078
goodplus*|   .4468843   .1130835     3.81   0.000       1   .225245  .668524
---------+--------------------------------------------------------------------
  obs. P |   .3043478
 pred. P |   .2286624  (at x-bar)
 pred. P |   .5147238  (at x)
------------------------------------------------------------------------------
(*) dF/dx is for discrete change of dummy variable from 0 to 1
    z and P>|z| are the test of the underlying coefficient being 0
```

❑

Model identification

The `probit` command has one more feature, which is probably the most useful. It will automatically check the model for identification and, if the model is underidentified, drop whatever variables and observations are necessary for estimation to proceed.

▷ Example 5

Have you ever fitted a probit model where one or more of your independent variables perfectly predicted one or the other outcome?

For instance, consider the following data:

Outcome y	Independent Variable x
0	1
0	1
0	0
1	0

Say that we wish to predict the outcome on the basis of the independent variable. The outcome is always zero when the independent variable is one. In our data, $\Pr(y = 0 \mid x = 1) = 1$, which means that the probit coefficient on x must be minus infinity with a corresponding infinite standard error. At this point, you may suspect that we have a problem.

Unfortunately, not all such problems are so easily detected, especially if you have many independent variables in your model. If you have ever had such difficulties, then you have experienced one of the more unpleasant aspects of computer optimization. The computer has no idea that it is trying to solve for an infinite coefficient as it begins its iterative process. All it knows is that, at each step, making the coefficient a little bigger, or a little smaller, works wonders. It continues on its merry way until either (1) the whole thing comes crashing to the ground when a numerical overflow error occurs or (2) it reaches some predetermined cutoff that stops the process. Meanwhile, you have been waiting. And the estimates that you finally receive, if any, may be nothing more than numerical roundoff.

Stata watches for these sorts of problems, alerts you, fixes them, and then properly fits the model.

Let's return to our automobile data. Among the variables we have in the data is one called `repair` that takes on three values. A value of 1 indicates that the car has a poor repair record, 2 indicates an average record, and 3 indicates a better-than-average record. Here is a tabulation of our data:

```
. use http://www.stata-press.com/data/r10/repair, clear
(1978 Automobile Data)

. tabulate foreign repair

           |            repair
  Car type |         1          2          3 |     Total
-----------+---------------------------------+----------
  Domestic |        10         27          9 |        46
   Foreign |         0          3          9 |        12
-----------+---------------------------------+----------
     Total |        10         30         18 |        58
```

All the cars with poor repair records (`repair==1`) are domestic. If we were to attempt to predict `foreign` on the basis of the repair records, the predicted probability for the `repair==1` category would have to be zero. This in turn means that the probit coefficient must be minus infinity, and that would set most computer programs buzzing.

Let's try using Stata on this problem. First, we make up two new variables, `rep_is_1` and `rep_is_2`, that indicate the `repair` category.

```
. generate rep_is_1 = repair==1

. generate rep_is_2 = repair==2
```

The statement `generate rep_is_1=repair==1` creates a new variable, `rep_is_1`, that takes on the value 1 when `repair` is 1 and zero otherwise. Similarly, the next `generate` statement creates `rep_is_2` that takes on the value 1 when `repair` is 2 and zero otherwise. We are now ready to fit our model:

```
. probit foreign rep_is_1 rep_is_2
note: rep_is_1 != 0 predicts failure perfectly
      rep_is_1 dropped and 10 obs not used

Iteration 0:   log likelihood = -26.992087
Iteration 1:   log likelihood = -22.276479
Iteration 2:   log likelihood = -22.229184
Iteration 3:   log likelihood = -22.229138

Probit regression                                 Number of obs   =         48
                                                  LR chi2(1)      =       9.53
                                                  Prob > chi2     =     0.0020
Log likelihood = -22.229138                       Pseudo R2       =     0.1765

------------------------------------------------------------------------------
     foreign |      Coef.   Std. Err.      z    P>|z|     [95% Conf. Interval]
-------------+----------------------------------------------------------------
    rep_is_2 |  -1.281552   .4297324    -2.98   0.003    -2.123812   -.4392916
       _cons |   1.21e-16    .295409     0.00   1.000     -.578991     .578991
------------------------------------------------------------------------------
```

Remember that all the cars with poor repair records (`rep_is_1`) are domestic, so the model cannot be fitted. At least it cannot be fitted if we restrict ourselves to finite coefficients. Stata noted that fact reporting, "note: rep_is_1 != 0 predicts failure perfectly". This is Stata's mathematically precise way of saying what we said in English. When `rep_is_1` is not equal to 0, the car is domestic.

Stata then went on to say, "rep_is_1 dropped and 10 obs not used" and eliminated the problem. First, the variable `rep_is_1` had to be removed from the model because it would have an infinite coefficient. Then the 10 observations that led to the problem had to be eliminated as well, so as not to bias the remaining coefficients in the model. The 10 observations that are not used are the 10 domestic cars that have poor repair records.

Finally, Stata fitted what was left of the model, using the remaining observations.

◁

❑ Technical Note

Stata is pretty smart about catching these problems. It will catch "one-way causation by a dummy variable", as we demonstrated above.

Stata also watches for "two-way causation", that is, a variable that perfectly determines the outcome, both successes and failures. Here Stata says that the variable "predicts outcome perfectly" and stops. Statistics dictate that no model can be fitted.

Stata also checks your data for collinear variables; it will say "so-and-so dropped due to collinearity". No observations need to be eliminated here and model fitting will proceed without the offending variable.

It will also catch a subtle problem that can arise with continuous data. For instance, if we were estimating the chances of surviving the first year after an operation, and if we included in our model `age`, and if all the persons over 65 died within the year, Stata will say, "age $>$ 65 predicts failure perfectly". It will then inform us about how it resolves the issue and fit what can be fitted of our model.

`probit` (and `logit`, `logistic`, and `ivprobit`) will also occasionally display messages such as

```
Note: 4 failures and 0 successes completely determined.
```

The cause of this message and what to do if you see it are described in [R] **logit**.

❑

(*Continued on next page*)

Saved Results

probit saves the following in e():

Scalars

e(N)	number of observations	e(r2_p)	pseudo-R-squared
e(N_cds)	number of completely determined successes	e(ll)	log likelihood
e(N_cdf)	number of completely determined failures	e(ll_0)	log likelihood, constant-only model
e(df_m)	model degrees of freedom	e(N_clust)	number of clusters
		e(chi2)	χ^2

Macros

e(cmd)	probit	e(chi2type)	Wald or LR; type of model χ^2 test
e(cmdline)	command as typed	e(vce)	*vcetype* specified in vce()
e(depvar)	name of dependent variable	e(vcetype)	title used to label Std. Err.
e(wtype)	weight type	e(crittype)	optimization criterion
e(wexp)	weight expression	e(properties)	b V
e(title)	title in estimation output	e(estat_cmd)	program used to implement estat
e(clustvar)	name of cluster variable	e(predict)	program used to implement predict

Matrices

e(b)	coefficient vector	e(V)	variance–covariance matrix of the estimators

Functions

e(sample)	marks estimation sample

dprobit saves the following in e():

Scalars

e(N)	number of observations	e(ll)	log likelihood
e(N_cds)	number of completely determined successes	e(ll_0)	log likelihood, constant-only model
e(N_cdf)	number of completely determined failures	e(N_clust)	number of clusters
e(df_m)	model degrees of freedom	e(chi2)	χ^2
e(r2_p)	pseudo-R-squared	e(pbar)	fraction of successes observed in data
		e(xbar)	average probit score
		e(offbar)	average offset

Macros

e(cmd)	dprobit	e(vcetype)	title used to label Std. Err.
e(cmdline)	command as typed	e(dummy)	string of blank-separated 0s and 1s; 0 means that the corresponding independent variable is not a dummy; 1 means that it is
e(depvar)	name of dependent variable		
e(wtype)	weight type		
e(wexp)	weight expression		
e(title)	title in estimation output	e(crittype)	optimization criterion
e(clustvar)	name of cluster variable	e(properties)	b V
e(at)	predicted probability (at x)	e(estat_cmd)	program used to implement estat
e(chi2type)	Wald or LR; type of model χ^2 test	e(predict)	program used to implement predict
e(vce)	*vcetype* specified in vce()		

Matrices

e(b)	coefficient vector	e(dfdx)	marginal effects
e(V)	variance–covariance matrix of the estimators	e(se_dfdx)	standard errors of the marginal effects

Functions

e(sample)	marks estimation sample

Methods and Formulas

`probit` and `dprobit` are implemented as ado-files.

Probit analysis originated in connection with bioassay, and the word probit, a contraction of "probability unit", was suggested by Bliss (1934). For an introduction to probit and logit, see, for example, Aldrich and Nelson (1984), Johnston and DiNardo (1997), Long (1997), Pampel (2000), or Powers and Xie (2000). Long and Freese (2006, chap. 4) provide an introduction to probit and logit, along with Stata examples.

The log-likelihood function for probit is

$$\ln L = \sum_{j\in S} w_j \ln\Phi(\mathbf{x}_j\boldsymbol{\beta}) + \sum_{j\notin S} w_j \ln\Big\{1 - \Phi(\mathbf{x}_j\boldsymbol{\beta})\Big\}$$

where Φ is the cumulative normal and w_j denotes the optional weights. $\ln L$ is maximized, as described in [R] **maximize**.

If robust standard errors are requested, the calculation described in *Methods and Formulas* of [R] **regress** is carried forward with $\mathbf{u}_j = \{\phi(\mathbf{x}_j\mathbf{b})/\Phi(\mathbf{x}_j\mathbf{b})\}\mathbf{x}_j$ for the positive outcomes and $-[\phi(\mathbf{x}_j\mathbf{b})/\{1-\Phi(\mathbf{x}_j\mathbf{b})\}]\mathbf{x}_j$ for the negative outcomes, where ϕ is the normal density. q_c is given by its asymptotic-like formula.

Turning to `dprobit`, which is implemented as an ado-file, let $\mathbf{b}$ and $\mathbf{V}$ denote the coefficients and variance matrix calculated by `probit`. Let b_i refer to the ith element of $\mathbf{b}$. For continuous variables, or for all variables if `classic` is specified, `dprobit` reports

$$b_i^* = \frac{\partial\Phi(\mathbf{x}\mathbf{b})}{\partial x_i}\bigg|_{\mathbf{x}=\overline{\mathbf{x}}} = \phi(\overline{\mathbf{x}}\mathbf{b})b_i$$

The corresponding variance matrix is $\mathbf{DVD}'$, where $\mathbf{D} = \phi(\overline{\mathbf{x}}\mathbf{b})\{\mathbf{I} - (\overline{\mathbf{x}}\mathbf{b})\mathbf{b}\overline{\mathbf{x}}\}$.

For dummy variables taking on values 0 and 1 when `classic` is not specified, `dprobit` makes the discrete calculation associated with the dummy changing from 0 to 1, $b_i^* = \Phi(\overline{\mathbf{x}}_1\mathbf{b}) - \Phi(\overline{\mathbf{x}}_0\mathbf{b})$, where $\overline{\mathbf{x}}_0 = \overline{\mathbf{x}}_1 = \overline{\mathbf{x}}$ except that the ith elements of $\overline{\mathbf{x}}_0$ and $\overline{\mathbf{x}}_1$ are set to 0 and 1, respectively. The variance of b_i is given by $\mathbf{dVd}'$, where $\mathbf{d} = \phi(\overline{\mathbf{x}}_1\mathbf{b})\overline{\mathbf{x}}_1 - \phi(\overline{\mathbf{x}}_0\mathbf{b})\overline{\mathbf{x}}_0$.

`dprobit` always reports test statistics z_i based on the underlying coefficients b_i.

Chester Ittner Bliss (1899–1979) was born in Ohio. He was educated as an entomologist, earning degrees from Ohio State and Columbia, and was employed by the United States Department of Agriculture until 1933. When he lost his job because of the Depression, Bliss then worked with R. A. Fisher in London and at the Institute of Plant Protection in Leningrad before returning to a post at the Connecticut Agricultural Experiment Station in 1938. He was also a lecturer at Yale for 25 years. Among many contributions to biostatistics, his development and application of probit methods to biological problems are outstanding.

References

Aldrich, J. H., and F. D. Nelson. 1984. *Linear Probability, Logit, and Probit Models*. Newbury Park, CA: Sage.

Berkson, J. 1944. Application of the logistic function to bio-assay. *Journal of the American Statistical Association* 39: 357–365.

Bliss, C. I. 1934. The method of probits. *Science* 79: 38–39, 409–410.

Cochran, W. G. 1979. Chester Ittner Bliss 1899–1979. *Biometrics* 35: 715–717.

Finney, D. J. 1979. Chester Ittner Bliss 1899–1979. *Biometrics* 35: 717.

Hilbe, J. 1996. sg54: Extended probit regression. *Stata Technical Bulletin* 32: 20–21. Reprinted in *Stata Technical Bulletin Reprints*, vol. 6, pp. 131–132.

Johnston, J., and J. DiNardo. 1997. *Econometric Methods*. 4th ed. New York: McGraw–Hill.

Judge, G. G., W. E. Griffiths, R. C. Hill, H. Lütkepohl, and T.-C. Lee. 1985. *The Theory and Practice of Econometrics*. 2nd ed. New York: Wiley.

Long, J. S. 1997. *Regression Models for Categorical and Limited Dependent Variables*. Thousand Oaks, CA: Sage.

Long, J. S., and J. Freese. 2006. *Regression Models for Categorical Dependent Variables Using Stata*. 2nd ed. College Station, TX: Stata Press.

Miranda, A., and S. Rabe-Hesketh. 2006. Maximum likelihood estimation of endogenous switching and sample selection models for binary, ordinal, and count variables. *Stata Journal* 6: 285–308.

Pampel, F. C. 2000. *Logistic Regression: A Primer*. Thousand Oaks, CA: Sage.

Powers, D. A., and Y. Xie. 2000. *Statistical Methods for Categorical Data Analysis*. San Diego, CA: Academic Press.

Xu, J., and J. S. Long. 2005. Confidence intervals for predicted outcomes in regression models for categorical outcomes. *Stata Journal* 5: 537–559.

Also See

[R] **probit postestimation** — Postestimation tools for probit and dprobit

[R] **roc** — Receiver operating characteristic (ROC) analysis

[R] **asmprobit** — Alternative-specific multinomial probit regression

[R] **brier** — Brier score decomposition

[R] **biprobit** — Bivariate probit regression

[R] **glm** — Generalized linear models

[R] **hetprob** — Heteroskedastic probit model

[R] **ivprobit** — Probit model with endogenous regressors

[R] **logistic** — Logistic regression, reporting odds ratios

[R] **logit** — Logistic regression, reporting coefficients

[R] **mprobit** — Multinomial probit regression

[R] **scobit** — Skewed logistic regression

[SVY] **svy estimation** — Estimation commands for survey data

[XT] **xtprobit** — Random-effects and population-averaged probit models

[U] **20 Estimation and postestimation commands**

Title

probit postestimation — Postestimation tools for probit and dprobit

Description

The following postestimation commands are of special interest after `probit` and `dprobit`:

command	description
`estat clas`	`estat classification` reports various summary statistics, including the classification table
`estat gof`	Pearson or Hosmer–Lemeshow goodness-of-fit test
`lroc`	graphs the ROC curve and calculates the area under the curve
`lsens`	graphs sensitivity and specificity versus probability cutoff

These commands are not appropriate after the `svy` prefix.

For information about these commands, see [R] **logistic postestimation**.

The following standard postestimation commands are also available:

command	description
`adjust`[1]	adjusted predictions of $\mathbf{x}\beta$, probabilities, or $\exp(\mathbf{x}\beta)$
`estat`	AIC, BIC, VCE, and estimation sample summary
`estat` (svy)	postestimation statistics for survey data
`estimates`	cataloging estimation results
`hausman`	Hausman's specification test
`lincom`	point estimates, standard errors, testing, and inference for linear combinations of coefficients
`linktest`	link test for model specification
`lrtest`[2]	likelihood-ratio test
`mfx`	marginal effects or elasticities
`nlcom`	point estimates, standard errors, testing, and inference for nonlinear combinations of coefficients
`predict`	predictions, residuals, influence statistics, and other diagnostic measures
`predictnl`	point estimates, standard errors, testing, and inference for generalized predictions
`suest`	seemingly unrelated estimation
`test`	Wald tests for simple and composite linear hypotheses
`testnl`	Wald tests of nonlinear hypotheses

[1] `adjust` is not appropriate with time-series operators.

[2] `lrtest` is not appropriate with `svy` estimation results.

See the corresponding entries in the *Stata Base Reference Manual* for details, but see [SVY] **estat** for details about `estat` (svy).

Syntax for predict

predict [*type*] *newvar* [*if*] [*in*] [, *statistic* nooffset rules asif]

statistic	description
Main	
pr	probability of a positive outcome; the default
xb	linear prediction
stdp	standard error of the linear prediction
* deviance	deviance residual
score	first derivative of the log likelihood with respect to $\mathbf{x}_j\beta$

Unstarred statistics are available both in and out of sample; type `predict ... if e(sample) ...` if wanted only for the estimation sample. Starred statistics are calculated only for the estimation sample, even when `if e(sample)` is not specified.

Options for predict

Main

`pr`, the default, calculates the probability of a positive outcome.

`xb` calculates the linear prediction.

`stdp` calculates the standard error of the linear prediction.

`deviance` calculates the deviance residual.

`score` calculates the equation-level score, $\partial \ln L/\partial(\mathbf{x}_j\beta)$.

`nooffset` is relevant only if you specified `offset(`*varname*`)` for `probit`. It modifies the calculations made by `predict` so that they ignore the offset variable; the linear prediction is treated as $\mathbf{x}_j\mathbf{b}$ rather than as $\mathbf{x}_j\mathbf{b} + \text{offset}_j$.

`rules` requests that Stata use any rules that were used to identify the model when making the prediction. By default, Stata calculates missing for excluded observations.

`asif` requests that Stata ignore the rules and exclusion criteria and calculate predictions for all observations possible using the estimated parameter from the model.

Remarks

Remarks are presented under the following headings:

Obtaining predicted values
Performing hypothesis tests

Obtaining predicted values

Once you have fitted a probit model, you can obtain the predicted probabilities by using the `predict` command for both the estimation sample and other samples; see [U] **20 Estimation and postestimation commands** and [R] **predict**. Here we will make only a few additional comments.

`predict` without arguments calculates the predicted probability of a positive outcome. With the `xb` option, `predict` calculates the linear combination $\mathbf{x}_j\mathbf{b}$, where $\mathbf{x}_j$ are the independent variables in the jth observation and $\mathbf{b}$ is the estimated parameter vector. This is known as the index function since the cumulative density indexed at this value is the probability of a positive outcome.

In both cases, Stata remembers any rules used to identify the model and calculates missing for excluded observations unless `rules` or `asif` is specified. This is covered in the following example.

With the `stdp` option, `predict` calculates the standard error of the prediction, which is *not* adjusted for replicated covariate patterns in the data.

You can calculate the unadjusted-for-replicated-covariate-patterns diagonal elements of the hat matrix, or leverage, by typing

```
. predict pred
. predict stdp, stdp
. generate hat = stdp^2*pred*(1-pred)
```

▷ Example 1

In example 5 of [R] **probit**, we fitted the probit model `probit foreign rep_is_1 rep_is_2`. To obtain predicted probabilities, we type

```
. predict p
(option pr assumed; Pr(foreign))
(10 missing values generated)
. summarize foreign p
```

Variable	Obs	Mean	Std. Dev.	Min	Max
foreign	58	.2068966	.4086186	0	1
p	48	.25	.1956984	.1	.5

Stata remembers any rules used to identify the model and sets predictions to missing for any excluded observations. In the previous example, `probit` dropped the variable `rep_is_1` from our model and excluded 10 observations. When we typed `predict p`, those same 10 observations were again excluded and their predictions set to missing.

`predict`'s `rules` option uses the rules in the prediction. During estimation, we were told, "rep_is_1 != 0 predicts failure perfectly", so the rule is that when `rep_is_1` is not zero, we should predict 0 probability of success or a positive outcome:

```
. predict p2, rules
. summarize foreign p p2
```

Variable	Obs	Mean	Std. Dev.	Min	Max
foreign	58	.2068966	.4086186	0	1
p	48	.25	.1956984	.1	.5
p2	58	.2068966	.2016268	0	.5

`predict`'s `asif` option ignores the rules and the exclusion criteria and calculates predictions for all observations possible using the estimated parameters from the model:

```
. predict p3, asif
. summarize for p p2 p3
```

Variable	Obs	Mean	Std. Dev.	Min	Max
foreign	58	.2068966	.4086186	0	1
p	48	.25	.1956984	.1	.5
p2	58	.2068966	.2016268	0	.5
p3	58	.2931034	.2016268	.1	.5

Which is right? By default, `predict` uses the most conservative approach. If many observations had been excluded due to a simple rule, we could be reasonably certain that the `rules` prediction is correct. The `asif` prediction is correct only if the exclusion is a fluke and we would be willing to exclude the variable from the analysis, anyway. Then, however, we should refit the model to include the excluded observations.

◁

Performing hypothesis tests

After estimation with `probit`, you can perform hypothesis tests by using the `test` or `testnl` commands; see [U] **20 Estimation and postestimation commands**.

Methods and Formulas

All postestimation commands listed above are implemented as ado-files.

predict after probit

Let index j be used to index observations, not covariate patterns. Define M_j for each observation as the total number of observations sharing j's covariate pattern. Define Y_j as the total number of positive responses among observations sharing j's covariate pattern. Define p_j as the predicted probability of a positive outcome for observation j.

For $M_j > 1$, the deviance residual d_j is defined as

$$d_j = \pm\left(2\left[Y_j \ln\left(\frac{Y_j}{M_j p_j}\right) + (M_j - Y_j)\ln\left\{\frac{M_j - Y_j}{M_j(1-p_j)}\right\}\right]\right)^{1/2}$$

where the sign is the same as the sign of $(Y_j - M_j p_j)$. In the limiting cases, the deviance residual is given by

$$d_j = \begin{cases} -\sqrt{2M_j|\ln(1-p_j)|} & \text{if } Y_j = 0 \\ \sqrt{2M_j|\ln p_j|} & \text{if } Y_j = M_j \end{cases}$$

Also See

[R] **probit** — Probit regression

[R] **logistic postestimation** — Postestimation tools for logistic

[U] **20 Estimation and postestimation commands**

Title

proportion — Estimate proportions

Syntax

`proportion` *varlist* [*if*] [*in*] [*weight*] [, *options*]

options	description
Model	
`stdize(`*varname*`)`	variable identifying strata for standardization
`stdweight(`*varname*`)`	weight variable for standardization
`nostdrescale`	do not rescale the standard weight variable
`nolabel`	suppress value labels from *varlist*
`missing`	treat missing values like other values
if/in/over	
`over(`*varlist*[, `nolabel`]`)`	group over subpopulations defined by *varlist*; optionally, suppress group labels
SE/Cluster	
`vce(`*vcetype*`)`	*vcetype* may be `analytic`, `cluster` *clustvar*, `bootstrap`, or `jackknife`
Reporting	
`level(`*#*`)`	set confidence level; default is `level(95)`
`noheader`	suppress table header
`nolegend`	suppress table legend

`bootstrap`, `jackknife`, `rolling`, `statsby`, and `svy` are allowed; see [U] **11.1.10 Prefix commands**.

Weights are not allowed with the `bootstrap` prefix.

`vce()` and weights are not allowed with the `svy` prefix.

`fweights`, `iweights`, and `pweights` are allowed; see [U] **11.1.6 weight**.

`proportion` shares the features of all estimation commands; see [U] **20 Estimation and postestimation commands**.

Description

`proportion` produces estimates of proportions, along with standard errors, for the categories identified by the values in each variable of *varlist*.

Options

Model

`stdize(`*varname*`)` specifies that the point estimates be adjusted by direct standardization across the strata identified by *varname*. This option requires the `stdweight()` option.

`stdweight(`*varname*`)` specifies the weight variable associated with the standard strata identified in the `stdize()` option. The standardization weights must be constant within the standard strata.

`nostdrescale` prevents the standardization weights from being rescaled within the `over()` groups. This option requires `stdize()` but is ignored if the `over()` option is not specified.

`nolabel` specifies that value labels attached to the variables in *varlist* be ignored.

`missing` specifies that missing values in *varlist* be treated as valid categories, rather than omitted from the analysis (the default).

if/in/over

`over(`*varlist* [`, nolabel`]`)` specifies that estimates be computed for multiple subpopulations, which are identified by the different values of the variables in *varlist*.

When this option is supplied with one variable name, such as `over(`*varname*`)`, the value labels of *varname* are used to identify the subpopulations. If *varname* does not have labeled values (or there are unlabeled values), the values themselves are used, provided that they are nonnegative integers. Noninteger values, negative values, and labels that are not valid Stata names are substituted with a default identifier.

When `over()` is supplied with multiple variable names, each subpopulation is assigned a unique default identifier.

`nolabel` requests that value labels attached to the variables identifying the subpopulations be ignored.

SE/Cluster

`vce(`*vcetype*`)` specifies the type of standard error reported, which includes types that are derived from asymptotic theory, that allow for intragroup correlation, and that use bootstrap or jackknife methods; see [R] ***vce_option***.

`vce(analytic)`, the default, uses the analytically derived variance estimator associated with the sample proportion.

Reporting

`level(`*#*`)`; see [R] **estimation options**.

`noheader` prevents the table header from being displayed. This option implies `nolegend`.

`nolegend` prevents the table legend identifying the subpopulations from being displayed.

Remarks

▷ Example 1

We can estimate the proportion of each repair rating in the auto data.

```
. use http://www.stata-press.com/data/r10/auto
(1978 Automobile Data)

. proportion rep78

Proportion estimation              Number of obs    =      69

--------------------------------------------------------------
             |  Proportion   Std. Err.     [95% Conf. Interval]
-------------+------------------------------------------------
rep78        |
           1 |   .0289855    .0203446     -.0116115    .0695825
           2 |    .115942    .0388245      .0384689    .1934152
           3 |   .4347826    .0601159      .3148232     .554742
           4 |   .2608696    .0532498      .1546113    .3671278
           5 |   .1594203    .0443922       .070837    .2480036
--------------------------------------------------------------
```

Here we use the missing option to include missing values as a category of rep78.

```
. proportion rep78, missing
Proportion estimation                  Number of obs    =      74

      _prop_6: rep78 = .
```

	Proportion	Std. Err.	[95% Conf.	Interval]
rep78				
1	.027027	.0189796	-.0107994	.0648534
2	.1081081	.0363433	.0356761	.1805401
3	.4054054	.0574637	.2908804	.5199305
4	.2432432	.0502154	.1431641	.3433224
5	.1486486	.0416364	.0656674	.2316299
_prop_6	.0675676	.0293776	.0090181	.1261171

◁

▷ Example 2

We can also estimate proportions over groups.

```
. proportion rep78, over(foreign)
Proportion estimation                  Number of obs    =      69

      _prop_1: rep78 = 1
      _prop_2: rep78 = 2
      _prop_3: rep78 = 3
      _prop_4: rep78 = 4
      _prop_5: rep78 = 5

     Domestic: foreign = Domestic
      Foreign: foreign = Foreign
```

Over	Proportion	Std. Err.	[95% Conf.	Interval]
_prop_1				
Domestic	.0416667	.0291477	-.0164966	.0998299
Foreign	(no observations)			
_prop_2				
Domestic	.1666667	.0543607	.0581916	.2751417
Foreign	(no observations)			
_prop_3				
Domestic	.5625	.0723605	.4181069	.7068931
Foreign	.1428571	.0782461	-.0132805	.2989948
_prop_4				
Domestic	.1875	.0569329	.0738921	.3011079
Foreign	.4285714	.1106567	.2077595	.6493834
_prop_5				
Domestic	.0416667	.0291477	-.0164966	.0998299
Foreign	.4285714	.1106567	.2077595	.6493834

◁

Saved Results

proportion saves the following in e():

Scalars

e(N)	number of observations	e(N_clust)	number of clusters
e(N_over)	number of subpopulations	e(k_eq)	number of equations in e(b)
e(N_stdize)	number of standard strata	e(df_r)	sample degrees of freedom

Macros

e(cmd)	proportion	e(over)	*varlist* from over()
e(cmdline)	command as typed	e(over_labels)	labels from over() variables
e(varlist)	*varlist*	e(over_namelist)	names from e(over_labels)
e(stdize)	*varname* from stdize()	e(namelist)	proportion identifiers
e(stdweight)	*varname* from stdweight()	e(label#)	labels from #th variable in *varlist*
e(wtype)	weight type	e(vce)	*vcetype* specified in vce()
e(wexp)	weight expression	e(vcetype)	title used to label Std. Err.
e(title)	title in estimation output	e(estat_cmd)	program used to implement estat
e(cluster)	name of cluster variable	e(properties)	b V

Matrices

e(b)	vector of proportion estimates
e(V)	(co)variance estimates
e(_N)	vector of numbers of nonmissing observations
e(_N_stdsum)	number of nonmissing observations within the standard strata
e(_p_stdize)	standardizing proportions
e(error)	error code corresponding to e(b)

Functions

e(sample)	marks estimation sample

Methods and Formulas

proportion is implemented as an ado-file.

Proportions are means of indicator variables; see [R] **mean**.

References

Cochran, W. G. 1977. *Sampling Techniques*. 3rd ed. New York: Wiley.

Stuart, A., and J. K. Ord. 1994. *Kendall's Advanced Theory of Statistics: Distribution Theory, Vol. I.* 6th ed. London: Arnold.

Also See

[R] **proportion postestimation** — Postestimation tools for proportion

[R] **mean** — Estimate means

[R] **ratio** — Estimate ratios

[R] **total** — Estimate totals

[SVY] **direct standardization** — Direct standardization of means, proportions, and ratios

[SVY] **poststratification** — Poststratification for survey data

[SVY] **subpopulation estimation** — Subpopulation estimation for survey data

[SVY] **svy estimation** — Estimation commands for survey data

[SVY] **variance estimation** — Variance estimation for survey data

[U] **20 Estimation and postestimation commands**

Title

proportion postestimation — Postestimation tools for proportion

Description

The following postestimation commands are available for `proportion`:

command	description
`estat`	VCE
`estat` (svy)	postestimation statistics for survey data
`estimates`	cataloging estimation results
`lincom`	point estimates, standard errors, testing, and inference for linear combinations of coefficients
`nlcom`	point estimates, standard errors, testing, and inference for nonlinear combinations of coefficients
`test`	Wald tests for simple and composite linear hypotheses
`testnl`	Wald tests of nonlinear hypotheses

See the corresponding entries in the *Stata Base Reference Manual* for details, but see [SVY] **estat** for details about `estat` (svy).

Methods and Formulas

All postestimation commands listed above are implemented as ado-files.

Also See

[R] **proportion** — Estimate proportions

[SVY] **svy postestimation** — Postestimation tools for svy

[U] **20 Estimation and postestimation commands**

Title

prtest — One- and two-sample tests of proportions

Syntax

One-sample test of proportion

prtest *varname* == $\#_p$ [*if*] [*in*] [, level(#)]

Two-sample test of proportion

prtest $varname_1$ == $varname_2$ [*if*] [*in*] [, level(#)]

Two-group test of proportion

prtest *varname* [*if*] [*in*] , by(*groupvar*) [level(#)]

Immediate form of one-sample test of proportion

prtesti $\#_{obs1}$ $\#_{p1}$ $\#_{p2}$ [, level(#) count]

Immediate form of two-sample test of proportion

prtesti $\#_{obs1}$ $\#_{p1}$ $\#_{obs2}$ $\#_{p2}$ [, level(#) count]

by is allowed with prtest; see [D] **by**.

Description

prtest performs tests on the equality of proportions using large-sample statistics.

In the first form, prtest tests that *varname* has a proportion of $\#_p$. In the second form, prtest tests that $varname_1$ and $varname_2$ have the same proportion. In the third form, prtest tests that *varname* has the same proportion within the two groups defined by *groupvar*.

prtesti is the immediate form of prtest; see [U] **19 Immediate commands**.

The bitest command is a better version of the first form of prtest in that it gives exact *p*-values. Researchers should use bitest when possible, especially for small samples; see [R] **bitest**.

Options

Main

by(*groupvar*) specifies a numeric variable that contains the group information for a given observation. This variable must have only two values. Do not confuse the by() option with the by prefix; both may be specified.

level(#) specifies the confidence level, as a percentage, for confidence intervals. The default is level(95) or as set by set level; see [U] **20.7 Specifying the width of confidence intervals**.

`count` specifies that integer counts instead of proportions be used in the immediate forms of `prtest`. In the first syntax, `prtesti` expects that $\#_{\rm obs1}$ and $\#_{p1}$ are counts—$\#_{p1} \leq \#_{\rm obs1}$—and $\#_{p2}$ is a proportion. In the second syntax, `prtesti` expects that all four numbers are integer counts, that $\#_{\rm obs1} \geq \#_{p1}$, and that $\#_{\rm obs2} \geq \#_{p2}$.

Remarks

The `prtest` output follows the output of `ttest` in providing a lot of information. Each proportion is presented along with a confidence interval. The appropriate one- or two-sample test is performed, and the two-sided and both one-sided results are included at the bottom of the output. For a two-sample test, the calculated difference is also presented with its confidence interval. This command may be used for both large-sample testing and large-sample interval estimation.

▷ Example 1: One-sample test of proportion

In the first form, `prtest` tests whether the mean of the sample is equal to a known constant. Assume that we have a sample of 74 automobiles. We wish to test whether the proportion of automobiles that are foreign is different from 40%.

```
. use http://www.stata-press.com/data/r10/auto
(1978 Automobile Data)

. prtest foreign == .4

One-sample test of proportion                    foreign: Number of obs =       74

    Variable |       Mean   Std. Err.                     [95% Conf. Interval]
-------------+----------------------------------------------------------------
     foreign |   .2972973   .0531331                       .1931583    .4014363

    p = proportion(foreign)                                       z =  -1.8034
Ho: p = 0.4

     Ha: p < 0.4                 Ha: p != 0.4                   Ha: p > 0.4
 Pr(Z < z) = 0.0357         Pr(|Z| > |z|) = 0.0713          Pr(Z > z) = 0.9643
```

The test indicates that we cannot reject the hypothesis that the proportion of foreign automobiles is .40 at the 5% significance level.

◁

▷ Example 2: Two-sample test of proportion

We have two headache remedies that we give to patients. Each remedy's effect is recorded as 0 for failing to relieve the headache and 1 for relieving the headache. We wish to test the equality of the proportion of people relieved by the two treatments.

```
. use http://www.stata-press.com/data/r10/cure
. prtest cure1 == cure2
Two-sample test of proportion                     cure1: Number of obs =       50
                                                  cure2: Number of obs =       59
------------------------------------------------------------------------------
    Variable |       Mean   Std. Err.      z    P>|z|     [95% Conf. Interval]
-------------+----------------------------------------------------------------
       cure1 |        .52   .0706541                      .3815205    .6584795
       cure2 |   .7118644   .0589618                      .5963013    .8274275
-------------+----------------------------------------------------------------
        diff |  -.1918644   .0920245                      -.372229   -.0114998
             |  under Ho:   .0931155   -2.06   0.039
------------------------------------------------------------------------------
        diff = prop(cure1) - prop(cure2)                          z =  -2.0605
    Ho: diff = 0

    Ha: diff < 0                 Ha: diff != 0                 Ha: diff > 0
 Pr(Z < z) = 0.0197         Pr(|Z| < |z|) = 0.0394          Pr(Z > z) = 0.9803
```

We find that the proportions are statistically different from each other at any level greater than 3.9%. ◁

▷ Example 3: Immediate form of one-sample test of proportion

`prtesti` is like `prtest`, except that you specify summary statistics rather than variables as arguments. For instance, we are reading an article that reports the proportion of registered voters among 50 randomly selected eligible voters as .52. We wish to test whether the proportion is .7:

```
. prtesti 50 .52 .70
One-sample test of proportion                         x: Number of obs =       50
------------------------------------------------------------------------------
    Variable |       Mean   Std. Err.                     [95% Conf. Interval]
-------------+----------------------------------------------------------------
           x |        .52   .0706541                      .3815205    .6584795
------------------------------------------------------------------------------
    p = proportion(x)                                             z =  -2.7775
Ho: p = 0.7

     Ha: p < 0.7                 Ha: p != 0.7                 Ha: p > 0.7
 Pr(Z < z) = 0.0027         Pr(|Z| > |z|) = 0.0055          Pr(Z > z) = 0.9973
```

◁

▷ Example 4: Immediate form of two-sample test of proportion

To judge teacher effectiveness, we wish to test whether the same proportion of people from two classes will answer an advanced question correctly. In the first classroom of 30 students, 40% answered the question correctly, whereas in the second classroom of 45 students, 67% answered the question correctly.

(Continued on next page)

```
. prtesti 30 .4 45 .67
Two-sample test of proportion                  x: Number of obs =        30
                                               y: Number of obs =        45
------------------------------------------------------------------------------
    Variable |       Mean   Std. Err.       z    P>|z|     [95% Conf. Interval]
-------------+----------------------------------------------------------------
           x |         .4   .0894427                         .2246955    .5753045
           y |        .67   .0700952                          .532616     .807384
-------------+----------------------------------------------------------------
        diff |       -.27   .1136368                        -.4927241   -.0472759
             |  under Ho:   .1169416   -2.31   0.021
------------------------------------------------------------------------------
        diff = prop(x) - prop(y)                                  z =  -2.3088
    Ho: diff = 0

    Ha: diff < 0                 Ha: diff != 0                 Ha: diff > 0
 Pr(Z < z) = 0.0105         Pr(|Z| < |z|) = 0.0210          Pr(Z > z) = 0.9895
```

◁

Saved Results

prtest and prtesti save the following in r():

Scalars

r(z)	z statistic	r(N_#)	number of observations for variable #
r(P_#)	proportion for variable #		

Methods and Formulas

prtest and prtesti are implemented as ado-files.

A large-sample $(1-\alpha)100\%$ confidence interval for a proportion p is

$$\widehat{p} \pm z_{1-\alpha/2}\sqrt{\frac{\widehat{p}\,\widehat{q}}{n}}$$

and a $(1-\alpha)100\%$ confidence for the difference of two proportions is given by

$$(\widehat{p}_1 - \widehat{p}_2) \pm z_{1-\alpha/2}\sqrt{\frac{\widehat{p}_1\widehat{q}_1}{n_1} + \frac{\widehat{p}_2\widehat{q}_2}{n_2}}$$

where $\widehat{q} = 1 - \widehat{p}$ and z is calculated from the inverse normal distribution.

The one-tailed and two-tailed tests of a population proportion use a normally distributed test statistic calculated as

$$z = \frac{\widehat{p} - p_0}{\sqrt{p_0 q_0/n}}$$

where p_0 is the hypothesized proportion. A test of the difference of two proportions also uses a normally distributed test statistic calculated as